Marine and Freshwater Botany

Marine Fungi and Fungal-like Organisms

Edited by E.B. Gareth Jones and Ka-Lai Pang

DE GRUYTER

Editors
E.B. Gareth Jones
Institute of Ocean and Earth Sciences
University of Malaya
Kuala Lumpur 50603
Malaysia
E-mail: torperadgj@gmail.com

Ka-Lai Pang
Institute of Marine Biology and Center of Excellence for Marine Bioenvironment
and Biotechnology
National Taiwan Ocean University
Keelung 20224
Taiwan (R.O.C.)
E-mail: klpang@ntou.edu.tw

This book has 77 figures and 44 tables.

Front cover image
The cover shows three spores of common Ascomycota on wood in a marine environment, *Aniptodera* sp., *Haiyanga salina* and *Torpedospora radiata*. These fungi have evolved different types of ascospore appendages for entanglement and attachment to substrata in a marine environment (National Taiwan Ocean University, Taiwan).

ISBN 978-3-11-026398-5
e-ISBN 978-3-11-026406-7

Library of Congress Cataloging-in-Publication Data
A CIP catalog record for this book has been applied for at the Library of Congress.

Bibliographic information published by the Deutsche Nationalbibliothek
The Deutsche Nationalbibliothek lists this publication in the Deutsche Nationalbibliografie; detailed bibliographic data are available in the Internet at http://dnb.dnb.de.

© 2012 Walter de Gruyter GmbH & Co. KG, Berlin/Boston

Typesetting: Compuscript Ltd., Shannon, Ireland
Printing: Hubert & Co. GmbH & Co. KG, Göttingen

♾ Printed on acid-free paper
Printed in Germany
www.degruyter.com

The citation of registered names, trade names, trade marks, etc. in this work does not imply, even in the absence of a specific statement, that such names are exempt from laws and regulations protecting trade marks etc. and therefore free for general use.

Dedication

This volume on marine fungi and fungal-like organisms is dedicated to the distinguished mycologists who have contributed to the pioneering studies of marine fungi: F.K. Sparrow, I.M. Wilson, S.P. Meyers, D.H. Linder, W. Höhnk, K. Tubaki, T.W. Johnson and J. Kohlmeyer. Each has spent a significant part of their careers collecting and describing new marine fungi and fungal-like organisms. Fred Sparrow discovered the fungal-like organism *Thraustochytrium* in 1936, and remarked "one of the major advances in marine mycology in the decade beginning in 1960, was the discovery of the widespread occurrence of thraustochytridiaceous fungi in ocean waters, in sediments and coastal sands, and on surfaces of algae and organic detritus" (Sparrow 1968). He also co-authored the first book on marine fungi with Terry Johnson in 1961. Another pioneer was David Linder, who described lignicolous marine fungi on driftwood collected in the Massachusetts area. This led to numerous studies on wood-inhabiting marine fungi by Willy Höhnk in Bremerhaven, Germany, Sam Meyers in the Biscayne bay, USA, Irene Wilson at Aberystwyth, Wales, Keisuke Tubaki in Japan and Jan Kohlmeyer in Berlin-Dahlem, Germany.

A second generation of mycologists went on to consolidate the new discipline of marine mycology including: Jack Fell with a major contribution to the study of marine yeasts; Paul Kirk with the study of ascosporogenesis in marine pyrenomycetes; Gill Hughes with observations on the geographical distribution of marine fungi; Alwin Gaertner and Anamarie Ulken furthering studies on the distribution of thraustochytrids and their taxonomy; and Steve Newell on fungi and their role in the decomposition of *Spartina* in salt marsh meadows. Most of these contributions were confined to temperate water marine fungi and it was not until the 1990s that tropical marine fungi received attention by Jan Kohlmeyer, Gareth Jones and Kevin Hyde.

Preface

This book *Marine Fungi and Fungal-like Organisms* is the outcome of recent research by leading world mycologists on selected topics on fungi in oceans, mangroves and estuarine habitats. The main aim of the book is to amalgamate the present state of knowledge about marine fungi, to consider their importance to scientific knowledge, and to highlight the challenges facing marine mycology and future research in the subject. The book covers many subjects not covered in other monographic volumes, such as marine yeasts, fungal-like organisms and their importance as potential sources of omega-3 fatty acids, enzymes from marine fungi, marine-derived fungi, and their roles in the decomposition of recalcitrant compounds and in the productivity of coastal and oceanic waters.

Texts in marine biology rarely include fungi and fungal-like organisms in their volumes and thus ignore their important contributions and roles in marine ecosystems. Their roles in bioremediation, breakdown and sequestration of pollutants in the oceans are often neglected. Fungi are also the predominant organisms in the decomposition of complex organic compounds yielding nutrients for other marine organisms in the web of life of the sea. Mangroves, salt marshes and seagrass meadows are major habitats in the sequestration of carbon dioxide and thus contribute in the alleviation of global warming.

The opening section of the book documents the current knowledge on the phylogeny of obligate marine fungi: Ascomycota (Dothideomycetes, Sordariomycetes and other classes); Basidiomycota; anamorphic fungi; yeasts; Chytridiomycota; and Blastocladiomycota. The second section is devoted to the phylogeny of fungal-like organisms that have been generally studied by mycologists: Cryptomycota and Mesomycetozoea; Hyphochytriomycota, Oomycota and Perkinsozoa; Labyrinthulomycota; and Phytomyxea. The biodiversity of fungi in selected habitats is reviewed in the third section: mangrove, endophytes, algicolous fungi, marine-derived fungi and salt marsh fungi. The concluding section considers more applied subjects: bioactive compounds, fungal enzymes, decomposition of material in the sea and culture collections and their maintenance. The epilogue considers the importance of marine fungi and fungal-like organisms, and the challenges and direction of future research.

It is hoped that this book will be essential reading for mycologists, microbiologists, marine biologists and limnologists interested in marine fungi and will serve as a useful reference work on their occurrence and role in marine ecosystems. With increased interest in the blue carbon initiative, this volume will provide valuable information for potential researchers and stimulate further studies on marine fungi and fungal-like organisms.

We thank all the authors for agreeing to write for this volume and for delivering on time. We are much indebted to Frank Gleason for the suggestions given during its preparation, in particular the section on fungal-like organisms. Our thanks also go to all the staff at De Gruyter for conceiving this volume and for their support in its publication.

List of contributing authors

Mohamed A. Abdel-Wahab
Department of Botany and Microbiology
College of Science
King Saud University
Riyadh 11451
Kingdom of Saudi Arabia
E-mail: maessa@ksu.edu.sa

S. Aisyah Alias
Institute of Biological Sciences
University of Malaya
50603 Kuala Lumpur
Malaysia
E-mail: siti.alias@gmail.com

Ali H.A. Bahkali
Department of Botany and Microbiology
College of Science
King Saud University
Riyadh 11451
Kingdom of Saudi Arabia
E-mail: abahkali@ksu.edu.sa

Margarida Barata
Departamento de Biologia Vegetal
Faculdade de Ciências da Universidade de Lisboa
Campo Grande
Edifício C2, 4° Piso
1749–016 Lisbon
Portugal
E-mail: mstb@fc.ul.pt

Jirayu Buatong
Department of Poultry Diagnostic Laboratory
SAHA FARM CO., LTD99
Saraburi-Lomsak Road
Huayhin, Chaibadarn
Lopburi 15130
Thailand
E-mail: Jirayu.mook@gmail.com

Maria da Luz Calado
Departamento de Biologia Vegetal
Faculdade de Ciências da Universidade de Lisboa
Campo Grande
Edifício C2, 4° Piso
1749–016 Lisbon
Portugal
E-mail: mdcalado@fc.ul.pt

Prapaipit Chaowalit
Medical Molecular Biology Unit
Faculty of Medicine
Mahidol University
Bangkok 10700
Thailand
E-mail: prapaipit_cw@hotmail.com

Rattaket Choeyklin
National Center for Genetic Engineering and Biotechnology (BIOTEC)
113 Thailand Science Park, Phaholyothin Road
Klong 1, Klong Luang
Pathumthani 12120
Thailand
E-mail: rattaket@biotec.or.th

Varada Damare
Institut für Marine Biotechnologie
Biotechnikum
Walter Rathenau Strasse 49A
17489 Greifswald
Germany
E-mail: chimulkarvarada@rediffmail.com

Rainer Ebel
Department of Chemistry
University of Aberdeen
Meston Walk, AB24 3UE
Aberdeen
United Kingdom
E-mail: r.ebel@abdn.ac.uk

List of contributing authors

Jack W. Fell
Rosenstiel School of Marine &
Atmospheric Sciences
University of Miami
4600 Rickenbacker Causeway
Miami FL 33149
USA
E-mail: jfell@rsmas.miami.edu

Claire M.M. Gachon
Scottish Association for Marine Science
Scottish Marine Institute
Oban PA37 1QA
United Kingdom
E-mail: claire.gachon@sams.ac.uk

Frank H. Gleason
School of Biological Sciences Level 5
Carslaw (F07)
The University of Sydney
NSW 2006
Australia
E-mail: frankjanet@ozE-mail.com.au

Sally L. Glockling
135 Brodrick Road
Hampden Park
Eastbourne BN22 9RA
United Kingdom
E-mail: sally@glockling.com

Sung-Yuan Hsieh
Food Industrial Research and
Development Institute
331 Shih-Pin Road
Hsinchu 300
Taiwan (R.O.C.)
E-mail: sungyuan@gmail.com

Kevin D. Hyde
School of Science
Mae Fah Luang University
333 Moo 1, Tambon Tasud
Muang District, Chiang Rai 57100
Thailand
E-mail: kdhyde1@gmail.com

E.B. Gareth Jones
Institute of Ocean and Earth Sciences
University of Malaya
Kuala Lumpur 50603
Malaysia
E-mail: torperadgj@gmail.com

Meredith D.M. Jones
Department of Zoology
Natural History Museum
London SW7 5BD
United Kingdom
E-mail: m.jones@nhm.ac.uk

Martin Kirchmair
Institute of Microbiology
Leopold–Franzens–University Innsbruck
Technikerstr. 25
6020 Innsbruck
Austria
E-mail: martin.kirchmair@uibk.ac.at

Frithjof C. Küpper
Oceanlab
University of Aberdeen
Main St, Newburgh
Aberdeenshire AB41 6AA
United Kingdom
E-mail: fkuepper@abdn.ac.uk

Eduardo M. Leaño
Network of Aquaculture Centres
in Asia-Pacific
Suraswadi Bldg
Kasetsart University Campus
Ladyao, Jatujak
Bangkok 10900
Thailand
E-mail: edleano2004@yahoo.com

Yang Soo Lee
Department of Forest Science and
Technology
Institute of Agricultural Science and
Technology
Chonbuk National University
664–14 Deokjin-dong 1ga
Deokjin-gu, Jeonju-si, Jeonbuk
South Korea
E-mail: ysoolee@chonbuk.ac.kr

Keming Leng
Shenzhen Ocean and Fishery
Environmental Observatory Station
Futian, Shenzhen 518031
P.R. China

Osu Lilje
School of Biological Sciences Level 5
Carslaw (F07)
The University of Sydney
NSW 2006
Australia
E-mail: osu.lilje@sydney.edu.au

Apilux Loilong
National Center for Genetic Engineering
and Biotechnology (BIOTEC)
113 Thailand Science Park, Phaholyothin
Road
Klong 1, Klong Luang
Pathumthani 12120
Thailand
E-mail: apilux25@yahoo.com

Agostina V. Marano
Instituto de Botánica Spegazzini
Universidad Nacional de La Plata
calle 53 N 477
La Plata, 1900
Buenos Aires
Argentina
E-mail: agosvm@hotmail.com

Akira Nakagiri
Fungus/Mushroom Resource and
Research Center (FMRC)
Faculty of Agriculture, Tottori University
4–101 Koyama-Minami
Tottori 680–8553
Japan
E-mail: nakagiri@muses.tottori-u.ac.jp

Sigrid Neuhauser
Institute of Microbiology
Leopold Franzens-University Innsbruck
Technikerstr. 25
6020 Innsbruck
Austria
E-mail: sigrid.neuhauser@uibk.ac.at

Ka-Lai Pang
Institute of Marine Biology and Center of
Excellence for Marine Bioenvironment
and Biotechnology
National Taiwan Ocean University
2 Pei-Ning Road, Keelung 20224
Taiwan (R.O.C.)
E-mail: klpang@ntou.edu.tw

Souwalak Phongpaichit
Natural Products Research Center and
Department of Microbiology
Faculty of Science
Prince of Songkhla University
Hat Yai, Songkhla 90112
Thailand
E-mail: souwalak.p@psu.ac.th

Carmen L.A. Pires-Zottarelli
Instituto de Botânica
CP 3005
01061–970 São Paulo, SP
Brazil
E-mail: zottarelli@uol.com.br

Sita Preedanon
National Center for Genetic Engineering
and Biotechnology (BIOTEC)
113 Thailand Science Park, Phaholyothin
Road
Klong 1, Klong Luang
Pathumthani 12120
Thailand
E-mail: sita.pre@biotec.or.th

Vatcharin Rukachaisirikul
Department of Chemistry
Faculty of Science
Prince of Songkhla University
Hat Yai
Songkhla 90112
Thailand
E-mail: vatcharin.r@psu.ac.th

Nattawut Rungjindamai
National Center for Genetic Engineering
and Biotechnology (BIOTEC)
113 Thailand Science Park, Phaholyothin
Road
Klong 1, Klong Luang
Pathumthani 12120
Thailand
E-mail: nattawut.run@biotec.or.th

Jariya Sakayaroj
National Center for Genetic Engineering
and Biotechnology (BIOTEC)
113 Thailand Science Park, Phaholyothin
Road
Klong 1, Klong Luang
Pathumthani 12120
Thailand
E-mail: jariyask@biotec.or.th

Purnima Singh
Shenzhen Engineering Laboratory for
Algal Biofuel Technology Development
and Application
School of Environment and Energy
Peking University Shenzhen Graduate
School
Shenzhen 518055
P.R. China

José I. de Souza
Instituto de Botânica
Nucleo Pesquisa Micol
Av Miguel Stefano 3687
BR-04301902 Sao Paulo
Brazil
E-mail: jisouza@yahoo.com.br

Kandikere R. Sridhar
Department of Biosciences
Mangalore University
Mangalagangotri 574 199
Mangalore, Karnataka
India
E-mail: sirikr@yahoo.com

Susan J. Stanley
Department of Biology
Hong Kong University
Pokfulam
Hong Kong SAR

Martina Strittmatter
Scottish Association for Marine Science
Scottish Marine Institute
Oban PA37 1QA
United Kingdom
E-mail: martina.strittmatter@sams.ac.uk

Satinee Suetrong
National Center for Genetic Engineering
and Biotechnology (BIOTEC)
113 Thailand Science Park, Phaholyothin
Road
Klong 1, Klong Luang
Pathumthani 12120
Thailand
E-mail: satinee.na@gmail.com

Natarajan Velmurugan
Department of Chemical and
Biomolecular Engineering
The Korea Advanced Institute of
Science and Technology
Yuseong-gu
Taejon 305701
South Korea
E-mail: velmmk@gmail.com

Guangyi Wang
Department of Microbiology
University of Hawaii at Manoa
Honolulu
Hawaii 96822
USA
E-mail: guangyi@hawaii.edu

Xin Wang
Department of Microbiology
University of Hawaii at Manoa
Honolulu, HI 96822
USA
E-mail: wx@hawaii.edu

B3304005

Contents

1. Introduction Marine fungi
   E.B. Gareth Jones and Ka-Lai Pang 1

Phylogeny of marine fungi

2. Phylogeny of the Dothideomycetes and other classes of marine Ascomycota
   *E.B. Gareth Jones, Kevin D. Hyde, Ka-Lai Pang
   and Satinee Suetrong* 17

3. Phylogeny of the marine Sordariomycetes
   Ka-Lai Pang 35

4. Basidiomycota
   E.B. Gareth Jones and Jack W. Fell 49

5. Taxonomy of filamentous anamorphic marine fungi: morphology
   and molecular evidence
   Mohamed A. Abdel-Wahab and Ali H.A. Bahkali 65

6. Yeasts in marine environments
   Jack W. Fell 91

7. Zoosporic true fungi
   Frank H. Gleason, Frithjof C. Küpper and Sally L. Glockling 103

8. Morphology and ultrastructure of marine fungi with special reference
   to the origin of the membrane complex in the marine ascomycete
   Corollospora gracilis
   Sung-Yuan Hsieh and E.B. Gareth Jones 117

Phylogeny of fungal-like organisms

9. An introduction to fungus-like microorganisms
   *Sigrid Neuhauser, Sally L. Glockling, Eduardo M. Leaño, Osu Lilje,
   Agostina V. Marano and Frank H. Gleason* 137

10. Cryptomycota (Rozellida) and Mesomycetozoea (Ichthyosporea)
    the Super-group Opisthokonta
    Sally L. Glockling, Meredith D.M. Jones and Frank H. Gleason 153

11. Hyphochytriomycota, Oomycota and Perkinsozoa (Super-group
    Chromalveolata)
    *Agostina V. Marano, Carmen L.A. Pires-Zottarelli, José I. de Souza,
    Sally L. Glockling, Eduardo Leaño, Claire M.M. Gachon,
    Martina Strittmatter and Frank H. Gleason* 167

12. Labyrinthulomycota
    Eduardo M. Leaño and Varada Damare 215

13. Phytomyxea (Super-group Rhizaria)
    Sigrid Neuhauser, Frank H. Gleason and Martin Kirchmair 245

Biodiversity of marine fungi

14. Mangrove fungi
    Kandikere R. Sridhar, S. Aisyah Alias and Ka-Lai Pang 253

15. Biodiversity of fungi on the palm *Nypa fruticans*
    Apilux Loilong, Jariya Sakayaroj, Nattawut Rungjindamai, Rattaket Choeyklin and E.B. Gareth Jones 273

16. Diversity of endophytic and marine-derived fungi associated with marine plants and animals
    Jariya Sakayaroj, Sita Preedanon, Souwalak Phongpaichit, Jirayu Buatong, Prapaipit Chaowalit and Vatcharin Rukachaisirikul 291

17. Fungi from marine algae
    E.B. Gareth Jones, Ka-Lai Pang and Susan J. Stanley 329

18. Salt marsh fungi
    Maria da Luz Calado and Margarida Barata 345

19. Diversity and ecology of marine-derived fungi
    Purnima Singh, Xin Wang, Keming Leng and Guangyi Wang 383

Application of marine fungi

20. Natural products from marine-derived fungi
    Rainer Ebel 411

21. Enzymes from marine fungi: current research and future prospects
    Natarajan Velmurugan and Yang Soo Lee 441

22. Decomposition of materials in the sea
    Kandikere R. Sridhar 475

23. Culture collections and maintenance of marine fungi
    Akira Nakagiri 501

24. Epilogue: importance and impact of marine mycology and fungal-like organisms: challenges for the future
    Ka-Lai Pang and E.B. Gareth Jones 509

Index 519

1 Introduction
Marine fungi

E.B. Gareth Jones and Ka-Lai Pang

Considerable progress has been achieved during the past two decades in documenting the occurrence, distribution and potential biotechnological application of marine fungi. Although marine fungi were reported as early as 1840–1880, they were few in number, and largely from seaweeds (Hariot 1889; Cotton 1909; Sutherland 1916). The period from 1980 to 1999 saw the description of no fewer than 300 marine fungi from a wide range of substrata and geographical locations (Jones et al. 2009). This period coincided with the discovery of tropical marine fungi, especially from mangrove habitats (Kohlmeyer 1984). Early studies involved the description of species based on morphological characteristics, later aided by ultrastructural observations (Jones et al. 1983; Jones 1995). However, the last decade has seen reclassification of marine fungi based on sequence data (Spatafora and Blackwell 1994; Jones et al. 2009; Suetrong et al. 2009; Abdel-Wahab et al. 2010; Suetrong et al. 2011), and the discovery of many new lineages of marine fungi (Spatafora et al. 1998; Suetrong et al. 2009; Pang et al. 2010a; Suetrong et al. 2011).

In this introductory chapter, we first consider some of the general concepts that apply to all groups of marine fungi and fungal-like organisms. These include the composition of major groups and their origin, numbers of species and distribution. Finally we discuss the general organization of this volume and the contents of each major section. Chapters are written by specialists in the field of marine mycology and incorporate the latest published data.

Origin of fungal-like organisms and the fungal kingdom

Traditionally, marine fungi have been separated into two groups: "higher" marine fungi, which include filamentous fungi in the Basidiomycota and Ascomycota, and "lower" zoosporic fungi which include the Chytridiomycota, Oomycetes, Labyrinthulomycetes and other zoosporic organisms. Recent phylogenetic studies have revealed that many of these zoosporic groups are unrelated to the fungal kingdom, but form disparate deep branches in the early evolution of fungi (Baldauf 2003; James et al. 2006a, 2006b), see Chapters 9, 10, 11, 12 and 13 of this volume.

Fungal-like organisms

Since Baldauf (2003) proposed the organization of the tree of life (Figure 1.1), there have been changes to the higher-level classification of eukaryotes and phylogenetic studies have revealed several new orders containing marine organisms (Adl et al. 2005). The principle change has been the addition of the Rhizaria supergroup, which contains organisms with pseudopodia, including the order Cercozoa to which the Phytomyxea belong

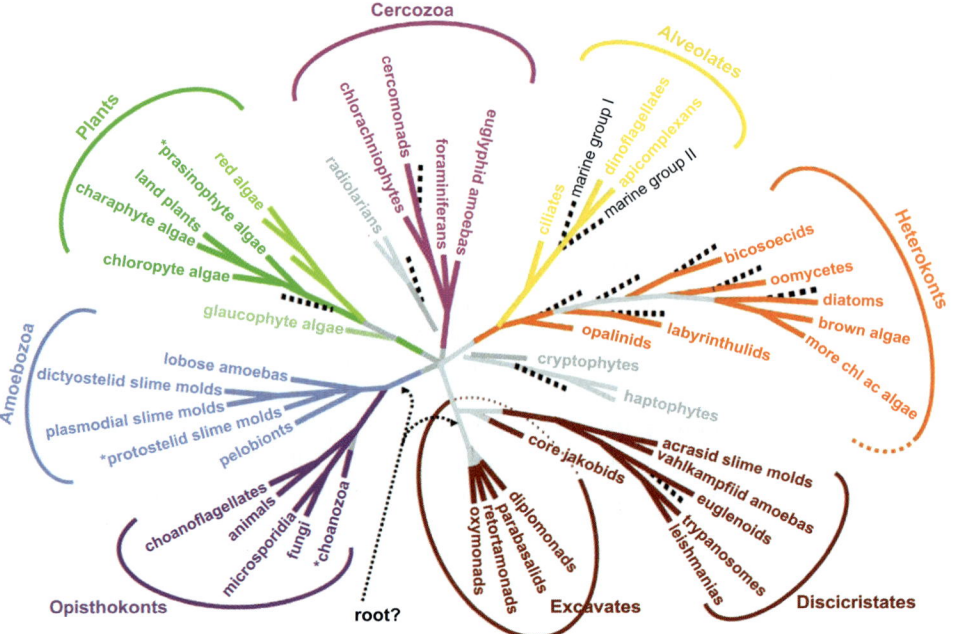

Figure 1.1 A consensus phylogeny of eukaryotes with an arrow pointing to the origin of the fungi (after Baldauf 2003).

(see Chapter 13). Phylogenetic studies have also revealed new orders within the Chromalveolata supergroup. The Hyphochytridiomycota (see Chapter 11) have been shown to be a group within the Straminipila, and basal to the Oomycetes. The Perkinsozoa (see Chapter 11) is an order that belongs to the Alveolata, also within the Chromalveolata supergroup. Two orders can now also be added to the Opisthokonta supergroup. The Mesomycetozoea (see Chapter 10) are thought to have evolved before or around the time that the Fungi and Animalia diverged, while the Cryptomycota have been shown to be a group that is basal to the Fungi.

Kingdom fungi

Based on fossil records, fungi are thought to have been present in late Proterozoic some 900–570 million years ago (MYA), and were certainly present in the Silurian (438–408) and Devonian (408–360) periods (Taylor et al. 1992, 1995). However, estimates based on 18S rDNA data indicate their presence some 800 MYA, while ascomycetes and basidiomycetes are thought to have diverged about 600 MYA (Berbee and Taylor 1993). There are no estimates of when marine fungi migrated into the world's oceans, but Spatafora et al. (1998) provided molecular evidence that the Halosphaeriales are secondary marine ascomycetes. Subsequent research has shown many lineages of marine fungi are derived from terrestrial counterparts. However, the pathway in the adaptation of terrestrial to marine habitats can only be speculated upon (Shearer 1993; Jones 1995).

Introduction | 3

In a phylogenetic analysis of the 18S rDNA sequences of freshwater and marine fungi, Vijaykrishna et al. (2006) postulated that freshwater ascomycetes had evolved from terrestrial fungi, probably circa 390 MYA (in the Paleozoic era). Their data also included the sequences of a number of marine halosphaeriaceous ascomycetes (e.g., *Aniptodera juncicola*, *Halosphaera appendiculata*, and *Lignincola laevis*), which suggested they evolved in the early Mesozoic period (200 MYA). Marine Dothideomycetes may have evolved earlier (Vijaykrishna et al. 2006).

Marine fungi are well adapted to the marine environment and made the transition to this environment along many different pathways. Eleven lineages of marine fungi occur in six classes: Basidiomycota: Ustilaginomycetes (two orders), Agaricomycetes (three orders/clades); Ascomycota: Dothideomycetes (two orders), Eurotiomycetes (one order), Lecanoromycetes (one order), and Sordariomycetes (seven orders) (Figure 1.2). Undoubtedly further marine lineages will be established as further phylogenetic studies are undertaken. For example, the marine origin of a group of cleistothecial ascomycetes recovered from woody substrata often associated with sand: *Biflua*, *Crinigera*, *Dryosphaera* and *Marisolaris* (Koch and Jones 1989). These fungi are rarely collected and difficult

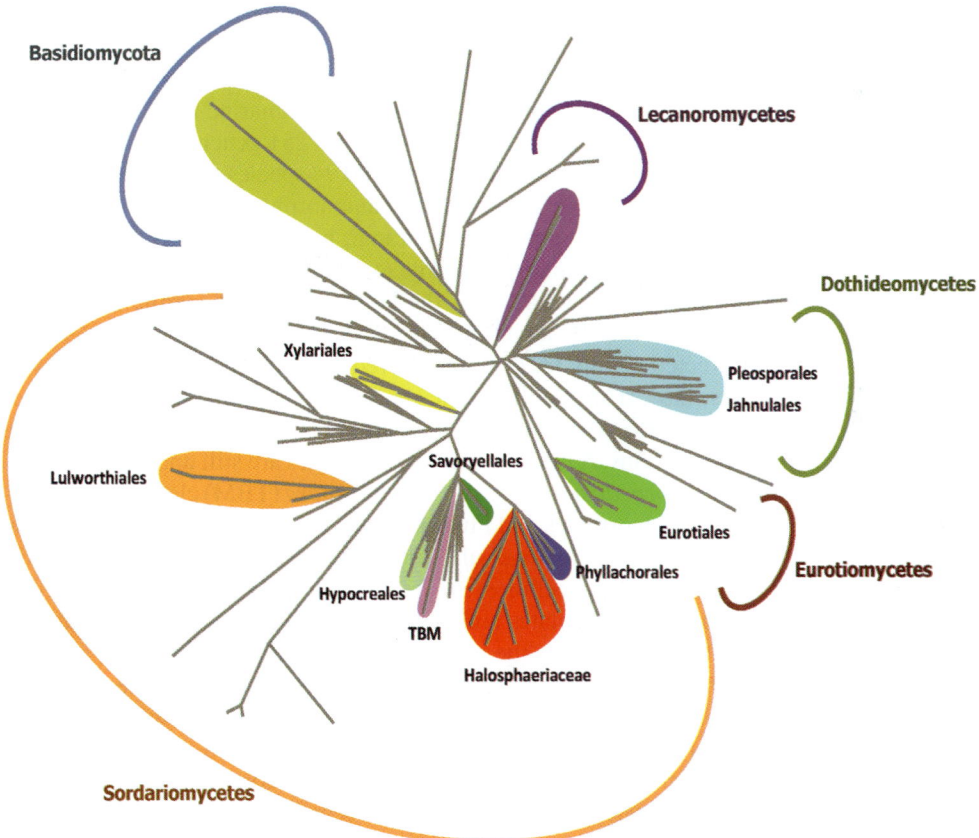

Figure 1.2 Lineages of marine fungi in the Ascomycota and Basidiomycota.

to get into culture, but preliminary sequence data places them in a basal position to the Ascomycota. Jones and Pang (2012) suggest that marine members of the Sordariomycetes evolved from freshwater species, although halosphaeriaceous fungi may have adapted from the Microascales with evanescent asci and are known to possess or assumed to have insect-dispersed ascospores (Malloch and Blackwell 1992; Spatafora et al. 1998). However, few members of the Microascales have been reported from marine habitats (J. Jin and E.B.G. Jones, unpublished data). The dothideomycetous ascomycetes may well have evolved from terrestrial taxa that adapted to a mangrove habitat. Whether marine genera evolved directly from terrestrial taxa remains unknown (Vijaykrishna et al. 2006).

The marine origin of the chytrids is not resolved, as most phylogenetic studies have been of terrestrial and freshwater species (James et al. 2006a). However it seems likely that they evolved from freshwater habitats. Bass et al. (2007) have recently recovered novel lineages of chytrids from environmental DNA from marine ecosystems.

Marine fungi have been found in all the oceans of the world, so how and when did they spread? It is possible that they evolved prior to the split of the continents and then underwent further adaptation. Hyde and Goh (2003) and Vijaykrishna et al. (2006) hypothesize as to the adaptations and dispersal of freshwater fungi around the globe, and similar conditions may equally apply to marine fungi.

Classification of true marine fungi

Jones et al. (2009) have provided an estimate of the number of marine fungi: 530 species in 321 genera of obligate taxa. Among these, 424 species (in 251 genera) belong to Ascomycota, 12 species to Basidiomycota (in nine genera) and 94 species to anamorphic fungi (in 61 genera) (Table 1.1, Figure 1.3). The number of anamorphic fungi has further risen to 143 species (see Chapter 5). No zoosporic true fungi, however, were included in the review by Jones et al. (2009). Most marine ascomycetes belong to the Dothideomycetes and Sordariomycetes. The Halosphaeriaceae is the largest family of marine fungi, with 121 species in 52 genera, of which 32 are monotypic (Sakayaroj et al. 2011), see Chapter 3. The Lulworthiales, erected mainly for genera removed from the Halosphaeriaceae with filamentous ascospores, has 30 species in eight genera (Kohlmeyer et al. 2000; Jones et al. 2009). Other orders in the Ascomycota with marine taxa are only represented by a few genera or even a single genus. A number of marine lineages are still unclassified in the Ascomycota including the *Torpedospora/Bertia/Melanospora* (TBM) clade, Koralionastetales, Lulworthiales, Magnaporthales and Phyllachorales, while the taxonomic position of many marine taxa is uncertain. Jones et al. (2009) included only 12 marine Basidiomycota, mostly isolated from lignocellulosic materials, but yeasts were not included in their review. If yeasts, endophytic isolates and environmental sequences, are included in the estimate of marine basidiomycetes, the current number stands at 67 (Jones and Fell 2012), see Chapter 4. Previous reviews have never taken into account the true zoosporic fungi in the calculation of the number of marine fungi and 12 taxa in eight genera are included in this treatise (Table 1.1). This number does not reflect the true diversity of fungi in the marine environment, but until more intensive collection of these fungi, with improved techniques, our knowledge on this group will remain fragmentary.

Table 1.1 Taxonomic classification of obligate marine fungi.

Phylum	Class	Order	Habitat	Number of species
Ascomycota	Arthoniomycetes	Arthoniales		2 genera, 2 species
	Dothideomycetes	Capnodiales		3 genera, 9 species
		Dothideales		1 genus, 1 species
		Dothideales *incertae sedis*		10 genera, 14 species
		Pleosporales		25 genera, 65 species
		Pleosporales *incertae sedis*		11 genera, 16 species
		Dothideomycetes *incertae sedis*		3 genera, 3 species
		Jahnulales		2 genera, 3 species
	Eurotiomycetes	Onygenales		1 genus, 1 species
		Eurotiales		1 genus, 1 species
		Chaetothyriales		1 genus, 1 species
		Pyrenulales		4 genera, 9 species
		Verrucariales		2 genera, 25 species
	Lecanoromycetes	Lecanorales		1 genus, 3 species
	Leotiomycetes	Helotiales		3 genera, 3 species
	Lichinomycetes	Lichinales		1 genus, 2 species
	Orbiliomycetes			
	Arthoniomycetes	Arthoniales		1 genus, 1 species
		Melaspileaceae *incertae sedis*		1 genus, 1 species
	TBM clade			4 genera, 10 species
	Sordariomycetes	Hypocreales		7 genera, 9 species
		Coronophorales		1 genus, 1 species
		Microsporales		58 genera, 148 species
		Diaporthales		4 genera, 5 species
		Diaporthales *incertae sedis*		1 genus, 1 species
		Sordariales		3 genera, 4 species
		Sordariales *incertae sedis*		1 genus, 1 species
		Savoryellales		1 genus, 5 species

Table continued on next page.

Table 1.1 (Continued)

Phylum	Class	Order	Habitat	Number of species
		Chaetosphaeriales		1 genus, 1 species
		Ophiostomatles		1 genus, 1 species
		Xylariales		16 genera, 24 species
		Xylariales *incertae sedis*		6 genera, 17 species
		Lulworthiales		8 genera, 31 species
		Koralionastetales		2 genera, 13 species
		Magnaporthales		3 genera, 3 species
		Phyllachorales		2 genera, 2 species
		Phyllachorales *incertae sedis*		3 genera, 3 species
	Ascomycetes *incertae sedis*	Ascomycota order *incertae sedis*		16 genera, 20 species
Chytridiomycota				6 genera, 10 species
Neocallimastigomycota				1 genus, 1 species
Blastocladiomycota				1 genus, 1 species
Basidiomycota	Ustilaginomycetes	Urocystales		1 genus, 1 species
		Ustilaginales		1 genus, 1 species
	Cystobasidiomycetes	Cystobasidiales		1 genus, 1 species
		Erythrobasidiales		2 genera, 2 species
	Microbotryomycetes	Sporidiales		1 genus, 4 species
		Leucosporidiales		1 genus, 2 species
	Tremellomycetes	Cystofilobasidiales		1 genus, 3 species
		Filobasidiales		2 genera, 2 species
		Tremellales		3 genera, 6 species
		Trichosporonales		1 genus, 1 species
	Exobasidiomycetes	Georgefischeriales		1 genus, 1 species
		Microstromatales		1 genus, 1 species
		Malasseziales		1 genus, 3 species
	Agariomycetes	Agaricales		5 genera, 7 species
	Agariomycetes *incertae sedis*			2 genera, 3 species
	Wallemiomyces	Wallemiales		1 genus, 1 species

Estimated number of marine fungi

Currently, the estimated number of marine fungi stands at 530 and includes Ascomycota, Basidiomycota and anamorphic fungi (Jones et al. 2009; Figure 1.3). Jones (2011) estimated that there could be as many as 10,000 marine fungi and suggested areas for finding the missing fungi: (1) unidentified/misidentified species; (2) marine-derived (facultative) fungi; (3) other groups of fungi (e.g., Chytridiomycota); (4) wider geographical sampling; (5) non-culturable fungi; (6) cryptic species or species with similar morphology; and (7) characterization of host species specific parasites (Table 1.2). Recent molecular techniques have suggested a great diversity of marine fungi in various environments and advanced pyrosequencing techniques could even reveal a greater diversity of marine fungi.

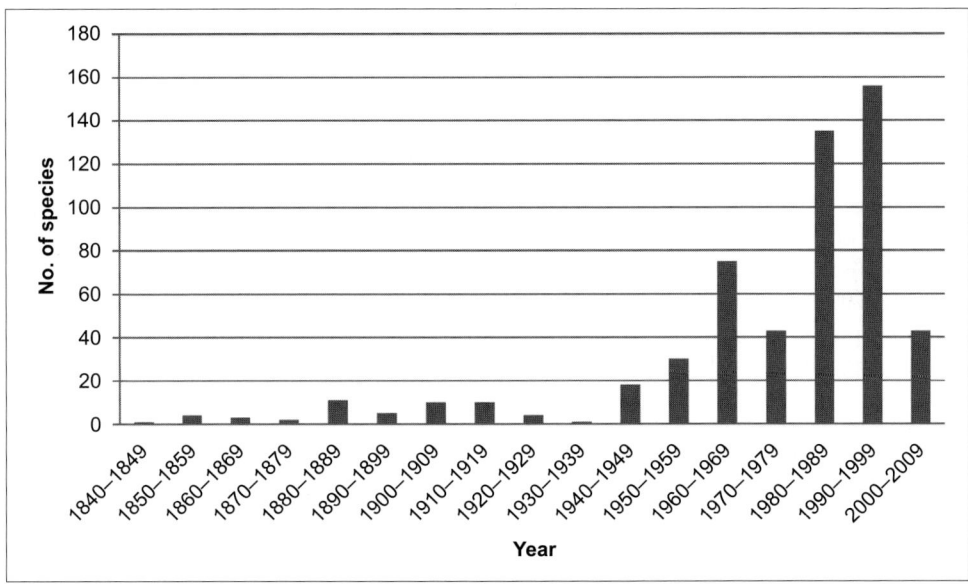

Figure 1.3 Histogram of numbers of obligate marine fungi from Jones et al. (2009).

Table 1.2 Estimated number of marine fungi.

Group	Number
Marine fungi	560
Facultative marine fungi	100
Marine yeasts	1,500
Misidentified fungi	100
Marine-derived	1,500
Deep-sea fungi	300
Planktonic fungi	500
Endophytes, algicolous and cryptic species	7,500
Total	**12,060**

World distribution of marine fungi

Marine fungi have a distinct geographical distribution, generally being referred to tropical/subtropical and temperate, although Jones and Pang (2012) listed seven cosmopolitan marine fungi, including *Aniptodera cheasapeakensis, Ceriosporopsis halima, Corollospora maritima, Lignincola laevis, Savoryella lignicola, Torpedospora radiata* and *Zalerion maritima*. Using ordination analysis, Schmit and Shearer (2004) found that marine microfungal communities are more similar within a single ocean basin than between different ocean basins, although many mangrove-associated fungal species have wide longitudinal distributions (Alias and Jones 2009; Pang et al. 2010b). A number of factors—biological, chemical and physical—can affect the diversity of marine fungi, including the availability of substrata for colonization, nature of the substratum, interference competition, water temperature, pH and salinity (Jones 2000).

Water temperature is regarded as the most important physical factor controlling the occurrence of marine fungi at different geographical locations (Booth and Kenkel 1986; Hughes 1986). Booth and Kenkel (1986) also divided marine fungi into 6 groups based on temperature and salinity: (1) warm water homeothermic euryhaline; (2) warm water euryhalothermic; (3) cool water eurythermic homohaline; (4) cool water euryhalothermic; (5) cool water homeothermic euryhaline; and (6) eurytolerant species.

Jones (1993) preferred to categorize marine fungi into three groups according to their biogeographical distribution: tropical to subtropical, temperate to arctic and cosmopolitan. Marine fungi restricted to subtropical and tropical areas include many mangrove taxa, e.g., *Aigialus* spp., *Bathyascus* spp. *Halorosellinia oceanica, Lulworthia grandispora* (Jones 1993; Jones and Pang 2012). Temperate marine fungi include *Toriella tubulifera, H. appendiculata, Lautosporopsis circumvestita* and *Marinospora* spp. (Koch and Petersen 1996). Wood blocks exposed to sea water at South Georgia, Antarctica revealed that marine fungi were typical temperate water species, including *L. circumvestita* and *T. tubulifera* (Pugh and Jones 1986). Similar results were obtained on trapped wood collected at Tromsø, northern mainland Norway; however, the diversity of marine lignicolous fungi at Longyearbyen inside the Arctic Circle was different, as it included mainly new taxa (*Remispora spitsbergenensis, Havispora longyearbyenensis*; Pang et al. 2011). In the same study, *Sablecola chinensis* was identified from the wood. This fungus had only previously been reported in tropical locations (Pang et al. 2004). This suggests that wider geographical sampling is required before a definitive picture of the distribution of marine fungi can be created. Also, distribution based on fruiting structures may not be entirely accurate and molecular techniques might help to resolve this issue.

In Figure 1.4, we have plotted the world distribution of marine fungi based on published data. This may, however, give a false picture on the distribution of marine fungi. While some locations have been intensively studied: Friday Harbor, USA; Hong Kong; India; Malaysia; and Thailand, other geographical areas have received scant attention: Africa (Steinke and Jones 1993), China (Sun et al. 2008) and Russia (Bagry-Shakhmatova 1983; Bubnova 2010; Zaitsev et al. 2010).

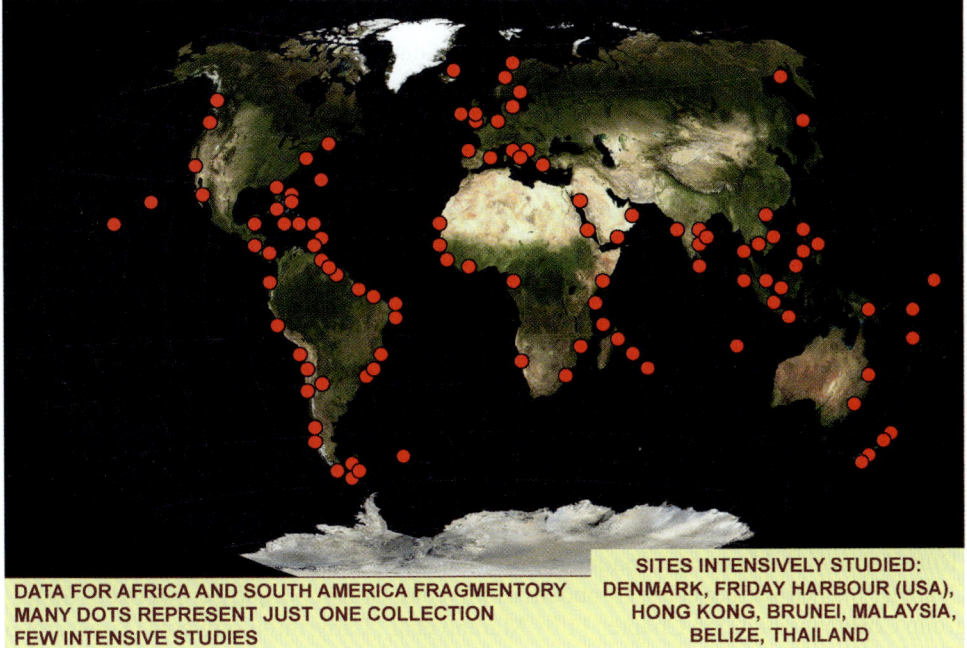

Figure 1.4 World distribution of marine fungi.

Objectives and outline of the volume

The primary objective of this volume is to review the most recent information on the occurrence, distribution, phylogeny, ecology and application of marine fungi. Although there have been a number of review articles recently on the subject, there is no comprehensive analysis of the group as a whole (Jones 2011). For example, chytridiomycetes, the ascomycetous and basidiomycetous yeasts are rarely included in books on marine fungi (Jones et al. 2009). Furthermore, other zoosporic organisms (Cryptomycota, Hyphochytriomycota, Oomycota, Perkinozoa, Labyrinthulomycota and Phytomyxea), often studied by mycologists in the past, are also ignored in symposia and edited volumes on marine mycology. Fungi and fungal-like organisms play a vital role in the ecology of our oceans, yet are rarely given the coverage in books on marine biology.

In marine biology it is vital that all fungal-like organisms are detailed in order to obtain a better understanding of their overall role in the biology of the world's oceans. This volume hopes to address this issue by their inclusion.

The volume is in four parts. The first and second sections will detail the immense progress made in our knowledge of the phylogeny and classification of true marine fungi and fungal-like organisms, respectively. Biodiversity and ecological studies of marine fungi on different substrates are covered in section three, while the final section details the application of these organisms and their commercial potential. Reviews and original research articles are combined to summarize recent research and to give an overview of current investigations. The final chapter will deal with gaps in our current knowledge and suggestions for further areas for research.

Phylogeny of true marine fungi

Phylogenetic studies of marine fungi have exploded with the application of molecular techniques, which has led to a radical change in their classification, a topic dealt with in section one. This section is devoted to Dothideomycetes (Chapter 2 by Jones et al.), Sordariomycetes (Chapter 3 by Pang), Basidiomycota (Chapter 4 by Jones and Fell), anamorphic fungi (Chapter 5 by Abdel-Wahab and Bahkali), marine yeasts (Chapter 6 by Fell), zoosporic true fungi (Chapter 7 by Gleason et al.) and morphology and ultrastructure of marine fungi (Chapter 8 by Hsieh and Jones).

Phylogeny of fungal-like organisms

Traditionally, marine fungi were categorized into higher and lower fungi. The higher marine fungi included filamentous basidiomycetes and ascomycetes, and their phylogeny is introduced in section one. Zoosporic organisms, predominantly chytrids, thraustochytrids and oomycetes, made up the lower marine fungi. A general introduction to fungal-like organisms is given in Chapter 9 (by Neuhauser et al.) and individual groups based on the current classification are introduced in subsequent chapters: Cryptomycota (Rozellida) and Mesomycetoaoea (Ichthyosporae) (Supergroup Opisthokonta) in Chapter 10 by Glockling et al.; Hyphochytriomycota, Oomycota and Perkinsozoa (Supergroup Chromalveolata) in Chapter 11 by Marano et al.; Labyrinthulomycota (Supergroup Chromalveolata) in Chapter 12 by Leaño and Damare; and Phytomyxea (Supergroup Rhizaria) in Chapter 13 by Neuhauser et al.

Biodiversity of marine fungi

Marine fungi can be categorized into different ecological groups based on their colonization and growth on substrata/environments. Early studies focused on the fungi on algae and on driftwood/trapped wood (Sutherland 1916; Barghoorn and Linder 1944). Over the past 30 years, many new marine fungi have been discovered on decayed mangrove substrates (wood, *Nypa*) in tropical/sub-tropical locations and on dead submerged salt marsh plants, especially *Juncus roemerianus* (Kohlmeyer and Volkmann-Kohlmeyer 2001). Unique marine environments and niches have recently been explored for fungi, including the endophytic fungi of marine plants and animals and fungi in deep-sea sediments. Fungi associated with these substrata are reviewed in various chapters in this volume: mangrove substrata in Chapter 14 by Sridhar et al.; *Nypa fruticans* in Chapter 15 by Loilong et al.; endophytic fungi of mangrove plants and seagrasses in Chapter 16 by Sakayaroj et al.; algae in Chapter 17 by Jones et al.; salt marsh plants in Chapter 18 by Calado and Barata; and marine animals and deep-sea sediments in Chapter 19 by Singh et al.

Application of marine fungi

Marine fungi produce wood-modifying enzymes, including cellulases, hemicellulases and laccases (Bucher et al. 2004; Luo et al. 2005) and this aspect is reviewed in Chapter 22 by Sridhar. Other than wood-degrading enzymes, marine fungi produce novel enzymes that could be of industrial value due to their occurrence in the salt water environments (see Chapter 21 by Velmurugan and Lee). Over the past two decades, a surge of studies on secondary metabolites (over 10,000) from marine fungi, especially those from marine animals, has been seen with the hope of finding new pharmaceutical drugs for various human diseases (see Chapter 20 by Ebel). Maintenance of cultures is essential, especially for those of industrial/medical value, since many of them do not survive for many generations of subculturing on agar media at room temperature. Methods for long-term preservation of filamentous fungi and zoosporic organisms are detailed in Chapter 23, written by Nakagiri.

This volume is concludes with a chapter in which the future of marine mycology is discussed and considers topics that have been neglected to date. As outlined above, knowledge of the world distribution of marine fungi is incomplete, detailed studies of their physiology and biochemistry are limited, while the identification and the role of marine derived fungi require a more detailed investigation (see Chapter 24 by Pang and Jones).

Acknowledgements

We are particularly grateful to Drs Frank Gleason and Sally Glockling for their suggestions and contributions to this introduction and for their major contributions to the section on fungal-like organisms. Figure 1.1 is reproduced from the journal *Science*.

References

Abdel-Wahab MA, Pang KL, Nagahama T, Abdel-Aziz FA, Jones EBG. Phylogenetic evaluation of anamorphic species of *Cirrenalia* and *Cumulospora* with the description of eight new genera and four new species. Mycol Progr. 2010;9:537–558.

Adl SM, Simpson AGB, Farmer MA, Andersen RA, Anderson OR, Barta JR, Bowser SS, Brugerolle G, Fensome RA, Fredericq S, James TY, Karpov S, Kugrens P, Krug J, Lane CE, Lewis LA, Lodge J, Lynn DH, Mann DG, Mccourt RM, Mendoza L, Moestrup O, Mozley-Standridge SE, Nerad TA, Shearer CA, Smirnov AV. The new higher level classification of Eukaryotes with emphasis on the taxonomy of protists. J Eukaryot Microbiol. 2005;52:399–451.

Alias SA, Jones EBG. Marine fungi from mangroves of Malaysia. Institute of Ocean and Earth Studies, No. 8, Kuala Lumpur, 2009.

Bagry-Shakhmatova LM. Toward studying higher marine fungi of the Sea of Japan. Biologia Morya 1983;5:51–56.

Baldauf SL. The deep roots of Eukaryotes. Science 2003;300:1703–1706.

Barghoorn ES, Linder DH. Marine fungi: Their taxonomy and biology. Farlowia 1944;1:395–467.

Bass D, Howe A, Brown N, Barton H, Demidova M, Michelle H, Li L, Sanders H, Watkinson SC, Willcock S. Yeast forms dominate fungal diversity in the deep oceans. Proc Biol Sci. 2007;274:3069–3077.

Berbee ML, Taylor JW. Dating the evolutionary radiations of the true fungi. Can J Bot. 1993;71:1114–1127.

Booth T, Kenkel N. Ecological studies of lignicolous marine fungi: a distribution model based on ordination and classification. In: Moss ST, ed. The biology of marine fungi. Cambridge: Cambridge University Press, 1986:297–310.

Bubnova EN. Fungal diversity in bottom sediments of the Kara Sea. Bot Mar. 2010;53:595–600.

Bucher VVC, Hyde KD, Pointing SB, Reddy CA. Production of wood decay enzymes, mass loss and lignin solubilization in wood by marine ascomycetes and their anamorphs. Fungal Divers. 2004;15:1–14.

Cotton AD. Notes on marine Pyrenomycetes. Trans Br Mycol Soc. 1909;31:92–99.

Hariot P. Algues. In: Botanique TV, ed. Mission Scientifique du Cap Horn. 1882–1883. Paris: Gauthier-Villars et fils, 1889:3–109.

Hughes GC. Biogeography and the marine fungi. In: Moss ST, ed. The biology of marine fungi. Cambridge: Cambridge University Press, 1986:275–295.

Hyde KD, Goh TK. Adaptations for dispersal in filamentous freshwater fungi. Fungal Divers. 2003; 10:231–258.

James TY, Kauff F, Schoch CL, Matheny PB, Hofstetter V, Cox CJ, Celio G, Gueidan C, Fraker E, Miadlikowska J, Lumbsch HT, Rauhut A, Reeb V, Arnold AE, Amtoft A, Stajich JE, Hosaka K, Sung G-H, Johnson D, O'Rourke B, Crockett M, Binder M, Curtis JM, Slot JC, Wang Z, Wilson AW, Schüßler A, Longcore JE, O'Donnell K, Mozley-Standridge S, Porter D, Letcher PM, Powell MJ, Taylor JW, White MM, Griffith GW, Davies DR, Humber RA, Morton JB, Sugiyama J, Rossman AY, Rogers JD, Pfiser DH, Hewitt D, Hansen K, Hambleton S, Sjoemaker RA, Kohlmeyer J, Volkmann-Kohlmeyer B, Spotts RA, Serdani M, Crous PW, Hughes KW, Matsuura K, Langer E, Langer G, Untereiner WA, Lücking R, Büdel B, Geiser DM, Aptroot A, Diederich P, Schmitt I, Schultz M, Yahr R, Hibbett DS, Lutzoni F, McLaughlin DJ, Spatafora JW, Vilgalys R. Reconstructing the early evolution of Fungi using a six-gene phylogeny. Nature 2006a;443:818–822.

James TY, Letcher PM, Longcore JE, Mozley-Standridge SE, Powell MJ, Griffith GW, Vilgalys R. A molecular phylogeny of the flagellated fungi (Chytridiomycota) and description of a new phylum (Blastocladiomycota). Mycologia 2006b;989:860–871.

Jones EBG. Tropical marine fungi. In: Isaac S, Frankland JC, Watling R, Whalley AJS, eds. Aspects of tropical mycology. Cambridge: Cambridge University Press, 1993:73–89.

Jones EBG. Ultrastructure and taxonomy of the aquatic ascomycetous order Halosphaeriales. Can J Bot. 1995;73:S790–S801.

Jones EBG. Marine fungi: some factors influencing biodiversity. Fungal Divers. 2000;4:53–73.

Jones EBG. Are there more marine fungi to be described? Bot Mar. 2011;54:343–354.

Jones EBG, Pang KL. Tropical aquatic fungi. Biodivers Conserv. 2012; DOI: 10.1007/s10531-011-0198-6.

Jones EBG, Moss ST, Cuomo V. Spore appendage development in the lignicolous marine Pyrenomycetes *Chaetosphaeria chaetosa* and *Halosphaeria trullifera*. Trans Br Mycol Soc. 1983; 80:93–200.

Jones EBG, Sakayaroj J, Suetrong S, Somrithipol S, Pang KL. Classification of marine Ascomycota, anamorphic taxa and Basidiomycota. Fungal Divers. 2009;35:1–187.

Koch J, Jones EBG. The identity of *Crinigera maritima* and three new genera of marine cleistothecial ascomycetes. Can J Bot. 1989;67:1183–1197.

Koch J, Petersen KRL. A check list of higher marine fungi on wood from Danish coasts. Mycotaxon 1996;15:397–414.

Kohlmeyer J. Tropical marine fungi. PSZNI Mar Ecol. 1984;5:329–378.

Kohlmeyer J, Spatafora JW, Volkmann-Kohlmeyer B. Lulworthiales, a new order of marine Ascomycota. Mycologia 2000;92:453–458.

Kohlmeyer J, Volkmann-Kohlmeyer B. The biodiversity of fungi on *Juncus roemerianus*. Mycol Res. 2001;105:1411–1412.

Luo W, Vrijmoed LLP, Jones EBG. Screening of marine fungi for lignocellulose-degrading enzyme activities. Bot Mar. 2005;48:379–386.

Malloch D, Blackwell M. Dispersal of fungal diaspores. In: Carroll GC, Wicklow DT, eds. The fungal community: its organisation and role in the ecosystem. New York: Marcel Dekker, 1992:147–172.

Pang KL, Alias SA, Chiang MWL, Vrijmoed LLP, Jones EBG. *Sedecimiella taiwanensis* gen. *et* sp. nov., a marine mangrove fungus in the Hypocreales (Hypocreomycetidae, Ascomycota). Bot Mar. 2010a;53:493–498.

Pang KL, Chow RKK, Chan CW, Vrijmoed LLP. Diversity and physiology of marine lignicolous fungi in Arctic waters: a preliminary account. Polar Res. 2011;30:1–5.

Pang KL, Jones EBG, Vrijmoed LLP. Two new marine fungi from China and Singapore, with the description of a new genus, *Sablecola*. Can J Bot. 2004;82:485–490.

Pang KL, Sharuddin SS, Alias SA, Nor NAM, Awaluddin HH. Diversity and abundance of lignicolous marine fungi from the east and west coasts of Peninsular Malaysia and Sabah (Borneo Island). Bot Mar. 2010b;53:515–523.

Pugh GJF, Jones EBG. Antarctic marine fungi: a preliminary account. In: Moss ST, ed. The biology of marine fungi. Cambridge: Cambridge University Press, 1986:323–330.

Sakayaroj J, Pang KL, Jones EBG. Multi-gene phylogeny of the Halosphaeriaceae: its ordinal status, relationships between genera and morphological character evolution. Fungal Divers. 2011; 46:87–109.

Schmit JP, Shearer CA. Geographical and host distribution of lignicolous mangrove microfungi. Bot Mar. 2004;47:496–500.

Shearer CA. The freshwater ascomycetes. Nova Hedw. 1993;56:1–33.

Spatafora JW, Blackwell M. The polyphyletic origins of ophiostomatoid fungi. Mycol Res. 1994; 98:1–9.

Spatafora JW, Volkmann-Kohlmeyer B, Kohlmeyer J. Independent terrestrial origins of the Halosphaeriales (marine Ascomycota). Am J Bot. 1998;85:1569–1580.

Steinke TD, Jones EBG. Marine and mangrove fungi from the Indian Ocean coast of South Africa. South Afr J Bot. 1993;59:385–390.

Suetrong S, Boonyuen N, Pang KL, Ueapattanakit J, Klaysuban A, Sri-indrasutdhi V, Sivichai S, Jones EBG. A taxonomic revision and phylogenetic reconstruction of the Jahnulales (Dothideomycetes), and the new family Manglicolaceae. Fungal Divers. 2011;51:163–188.

Suetrong S, Sakayaroj J, Phongpaichit S, Jones EBG. Morphological and molecular characteristics of a poorly known marine ascomycete, *Manglicola guatemalensis*. Mycologia 2009; 102:83–92.

Sun SL, Jin J, Li BD, Lu BS. Wood-inhabiting marine fungi from the coast of Shandong, China III. Mycosystema 2008;27:66–74.

Sutherland GK. Additional notes on marine pyrenomycetes. Trans Br Mycol Soc. 1916;5: 257–263.

Taylor TN, Hass H, Remy W. Devonian fungi: interactions with the green alga *Palaeonitella*. Mycologia 1992;84:901–910.

Taylor TN, Remy W, Hass H, Kerp H. Fossil arbuscular mycorrhizae from the early Devonian. Mycologia 1995;87:560–573.

Vijaykrishna D, Jeewon R, Hyde KD. Molecular taxonomy, origins and evolution of freshwater ascomycetes. Fungal Divers. 2006;23:351–390.

Zaitsev Y, Kopytina N, Garkusha O, Serbinova I. Preliminary observations on the Samsun Bay splash zone biodiversity. J Black Sea/Mediter Environ. 2010;16:245–252.

Phylogeny of marine fungi

2 Phylogeny of the Dothideomycetes and other classes of marine Ascomycota

E.B. Gareth Jones, Kevin D. Hyde, Ka-Lai Pang and Satinee Suetrong

Jones et al. (2009) documented 530 (in 321 genera) obligate marine fungi, which include 424 ascomycete species (in 251 genera), 94 anamorphic species (in 61 genera) and 12 basidiomycete species (in nine genera). The marine Ascomycota belong to the classes Dothideomycetes, Eurotiomycetes, Laboulbeniomycetes, Lecanoromycetes, Leotiomycetes, Lichinomycetes, Arthoniomycetes and Sordariomycetes, with Dothideomycetes and Sordariomycetes being the largest groups. The Sordariomycetes are detailed in Chapter 3. Marine asexual (anamorphic) fungi are reviewed in Chapter 5, which includes data on their molecular phylogeny.

In the Dothideomycetes ascomata develop in multilocular ascostromata, or may develop singly or rarely be cleistothecial; asci are bitunicate with or without an ocular chamber and rarely a conspicuous apical ring; and ascospores are hyaline, but generally brown and have various septation (Suetrong et al. 2009). In recent years a number of studies have examined the molecular phylogeny of the Dothideomycetes (Khashnobish and Shearer 1996; Inderbitzin et al. 2002; Tam et al. 2003; Jones et al. 2008; Suetrong et al. 2009; Zhang et al. 2009a, 2009b, 2009c, 2012b), with 50 species and 95 isolates sequenced.

Dothideomycetes

Marine Dothideomycetes comprise 108 species in 64 genera (Jones et al. 2009), the majority belonging in the order Pleosporales. Two orders are referred to the Dothideomycetidae (Capnodiales, Dothideales), one to the Pleosporomycetidae (Pleosporales), while three are classified in Dothideomycetes *incertae sedis* (Hysteriales, Patellariales and Jahnulales) (Jones et al. 2009; Suetrong et al. 2009, 2010, 2011a). Sequence data are available for a large number of marine Dothideomycete species, however a number of taxa cannot be taxonomically assigned until they are freshly collected, cultured and sequenced (e.g., *Belizeana, Bicrouania* and *Lautitia*). In this chapter, major lineages of marine Dothideomycetes – already with sequence data supporting their phylogeny – will be introduced and discussed, with particular emphasis on new families within the Pleosporales (Suetrong et al. 2009, 2011b; Zhang et al. 2011).

Dothideomycetidae

Fourteen genera are assigned to this subclass, but the taxonomic position of only a few species is resolved including *Mycosphaerella eurypotami* and *Scirrhia annulata*, while the position of *Heleiosa barbatula* is unresolved, as it did not cluster with any other taxon, and forming a sister group to the Dothideales and Capnodiales clades with weak support (Suetrong et al. 2009).

Capnodiales

Jones et al. (2009) list three genera of the Capnodiales with marine species: *Mycosphaerella, Sphaerulina* and *Pharcidia*, but only two *Mycosphaerella* species have been sequenced. *Mycosphaerella eurypotami* (isolated from the salt marsh plant *Juncus roemerianus*) is grouped as a sister taxon with other *Mycosphaerella* species (Suetrong et al. 2009), while a single isolate of *M. pneumatophorae* did not group in the genus (Schoch et al. 2006). Although *Scirrhia annulata* was tentatively referred to the Dothideales (Kohlmeyer et al. 1996), it groups with *Mycosphaerella* species with moderate support in the Capnodiales (Suetrong et al. 2009). Crous et al. (2011) referred a new *Scirrhia* species (*S. brasiliensis*) in the family Mycosphaerellaceae, thus confirming the observations of Suetrong et al. (2009), and it is the first confirmed genus in the family to have well-developed pseudoparaphyses and prominent hypostroma in which the ascomata are arranged in parallel rows.

Other marine *Mycosphaerella* species found on seaweeds including *M. apophlaeae* (a mycobiont of *Apophlaea lyalii* and *A. sinclarii*) and *M. ascophylli* (a mycobiont of *Ascophyllum nodosum* and *Pelvetia canaliculata*) were transferred to the new genus *Mycophycias* by Kohlmeyer and Volkmann-Kohlmeyer (1998a). This was based on morphological differences from other *Mycosphaerella* species and the mycophycobiosis habit of the two species. The new genus was referred to the Verrucariales, which also includes lichens in marine habitats (Kohlmeyer and Volkmann-Kohlmeyer 1998b; Gueidan et al. 2007). However, a phylogeny-based study placed *M. ascophylli* in the Capnodiales (Toxopeus et al. 2011). Although sequence data resolve the taxonomic position of *M. ascophylli, M. apophlaeae* requires further study at the molecular level (Toxopeus et al. 2011).

Dothideales and Dothideomycetidae *incertae sedis*

Eleven genera with marine species are referred to this group, of which sequences are only available for *Heleiosa barbatula*. Its taxonomic position remains unresolved, as it did not cluster with any other taxon, and just formed a sister group to the Dothideales and Capnodiales clades with weak support (Suetrong et al. 2009).

Many unusual species are placed in this group, e.g., the large (circa 180×150 µm), thick-walled (4–10 µm), hyaline, muriform (19–24 trans-septa) spored *Lautospora gigantea* and *L. simillima* referred to the new family Lautosporaceae (Kohlmeyer et al. 1995), and three *Thalassoascus* species parasitic on various brown seaweeds (Kohlmeyer 1981). However, phylogenetic assignment is not possible due to lack of recent collections and cultures of these fungi. These comprise unique marine lineages and their collection and sequencing is required to resolve their taxonomic position.

Pleosporomycetidae

Pleosporales

This order has been intensively investigated in recent years and many new families and marine lineages have been identified. Although many genera have been sequenced, the phylogenetic relationship of many remains unresolved, and further taxon sampling with a wider range of genes being sequenced is required. Schoch et al. (2006) identified seven families in the order supported by molecular data: Leptosphaeriaceae, Lophiostomataceae, Phaeosphaeriaceae, Pleosporaceae, Sporormiaceae, Testudinaceae and Trematosphaeriaceae. Further families have since been supported by sequence data: Massarinaceae, Montagnulaceae, and the new families Aigialaceae, Amniculicolaceae, Lentitheciaceae, Morosphaeriaceae, and Trematosphaeriaceae (Zhang et al. 2008, 2009a, 2009b, 2011; Suetrong et al. 2009, 2011b).

Aigialaceae

This family was proposed by Suetrong et al. (2009) to accommodate the genera *Aigialus* (five species) and the monotypic genera *Ascocratera* (one species) and *Rimora* (Suetrong et al. 2009), all marine taxa. More recently two terrestrial genera have been referred to this family: *Fissuroma* (two species) and *Neoastrosphaeriella* (one species) (Liu et al. 2011). The family is characterized by ascomata with cleft-like ostioles, black, carbonaceous to coriaceous, trabeculate pseudoparaphyses embedded in a gelatinous matrix, and hyaline to brown, septate to muriform ascospores with a gelatinous sheath (Suetrong et al. 2009; Liu et al. 2011). Liu et al. (2011) provide a key to the genera placed in the Aigialaceae. No anamorphs have been reported in the family.

Amniculicolaceae

Zhang et al. (2009a) erected the family to accommodate the genus *Amniculicola* (three species, Zhang et al. 2008), *Murispora* (*M. rubicunda*) and *Neomassariosphaeria* (*N. typhicola*, and *N. grandispora*); most are aquatic ascomycetes in freshwater or salt marshes (Zhang et al. 2009a). *Neomassariosphaeria typhicola* is regarded as being a marine species and is known from *Spartina* spp., *Juncus roemerianus*, *Phragmites communis* and *Typha latifolia* (Kohlmeyer and Kohlmeyer 1979). This is a species referred to various genera, most recently *Massariosphaeria* (Leuchtmann 1984; Jones et al. 2009) but molecular data place it in this family with high bootstrap support.

Didymellaceae

Few marine species are to be found in this family: *Didymella* (four species) and *Leptosphaerulina* (one species) (Jones et al. 2009). The only marine species sequenced is *Didymella fucicola*. It forms a well-supported sister group to other taxa in the family, in the analysis by Suetrong et al. (2009). The anamorphs of Didymellaceae are generally *Phoma*-like species, and these are common in the marine environment, with eight species listed by Jones et al. (2009). Therefore further marine species can be expected in this family (Aveskamp et al. 2011).

Lentitheciaceae

In a phylogenetic study of *Lophiostoma* and *Massarina* species, a new genus *Lentithecium* was proposed based on *Lophiostoma* (*Massarina*) *fluvitale* (Zhang et al. 2009a). Subsequently a well-supported clade, the family Lentitheciaceae (Zhang et al. 2009b) with the genera *Lentithecium*, *Stagonospora*, *Keissleriella*, *Wettsteinina*, *Ophiosphaerella* and *Katumotoa*, was introduced. Two marine species nest in this family: *Lentithecium*

(*Massarina*) *phragmiticola*, isolated from *Phragmites australis* and *Keissleriella rarum* growing on the salt marsh sedge *Juncus roemerianus* (Poon and Hyde 1998; Suetrong et al. 2009).

Lophiostomataceae

Although *Lophiostoma* and *Massarina* species have often been confused based on morphological features, molecular data show that the Lophiostomataceae are distally placed from the Massarinaceae in the Pleosporales. Two marine species are placed in this family: *Lophiostoma (Platystomum) scabridisporum*, which was isolated from driftwood in sand dunes (Abdel-Wahab and Jones 2000); and *Paraliomyces lentiferus*, which occurs on mangrove wood (Suetrong et al. 2009). Liu et al. (2011) show that the Lophiostomataceae forms a well-supported sister clade to the Sporormiaceae.

Massarinaceae

Most of the marine *Massarina* species that have been sequenced have been referred to other families, while other species including *M. beaurivagea*, *M. cystophorae*, *M. lacertensis*, *M. mauritiana*, *M. ricifera* and *M. rhizophorae* require study to confirm their position in the family (Jones et al. 2009).

Montagnulaceae

Tremateia halophila occurs on the senescent culms of *Juncus roemerianus* and groups in the Montagnulaceae with high bootstrap support (Suetrong et al. 2009). The species is characterized by an apical cap to the ascus, I-ocular chamber and muriform ascospores with a wide, mucilaginous sheath and a *Phoma*-like anamorph (Kohlmeyer et al. 1995).

Morosphaeriaceae

This is a new family erected by Suetrong et al. (2009) for two *Massarina* species that did not group in the Massarinaceae: *Morosphaeria velataspora* and *M. ramunculicola*. The taxa *Helicascus kanaloanus* and *H. nypae* form a sister group to *Morosphaeria* species with high bootstrap support. *Kirschsteiniothelia elaterascus*, a freshwater ascomycete, also groups with *Helicascus* spp. in this family (Shearer 1993; Liu et al. 2011; Suetrong et al. 2011a; Zhang et al. 2012). Based on morphology and phylogenetic analyses of LSU and SSU rDNA, *K. elaterascus* was transferred to the genus *Helicascus* (Zhang et al. 2012). *Helicascus elaterascus* is comparable to *H. kanaloanus*, in having immersed lenticular ascomata clustered beneath pseudoclypeus, clavate asci with long, narrow, coiled endoascus and brown, unequal two-celled ascospores.

Phaeosphaeriaceae

Jones et al. (2009) list three genera with marine species in this family. These are *Carinispora*, *Lautitia* and *Phaeosphaeria*. The latter contains ten species, of which four have been sequenced: *Ph. albopunctata*, *Ph. olivacea*, *Ph. spartinicola* and

Ph. typharum (Suetrong et al. 2009; Zhang et al. 2009b). *Loratospora aestuarii*, which occurs on *J. roemerianus* (Kohlmeyer and Volkmann-Kohlmeyer 1993) also nests in the Phaeosphaeriaceae forming a sister group to *Ph. elongata* (Zhang et al. 2009b). This species had previously been referred to the Planistromellaceae by Barr (1996) because its locules opened schizogenously by a periphysate ostiole.

Pleosporaceae

Jones et al. (2009) included five genera in this family; however a phylogenetic study places *Helicascus* in the Morosphaeriaceae, *Falciformispora* in the Trematosphaeriaceae and *Tremateia* in Montagnulaceae (Suetrong et al. 2009). Sequence data are available for three marine taxa: *Decorospora gaudefroyi* (Inderbitzin et al. 2002) and the anamorphic species *Alternaria maritima* (Zhang et al. 2009b) and *Dendryphiella arenaria* and *D. salina* (Jones et al. 2008). In all analyses, *D. arenaria* and *D. salina* form a sister group with *Pleospora herbarum* and *P. sedicola* with strong support (Suetrong et al. 2009).

Sporormiaceae

Mantle et al. (2006) described the anamorphic species *Amorosia littoralis* isolated from the littoral zone in the Bahamas. Based on molecular data, it was referred to the Sporormiaceae. However, in the phylogenetic study by Suetrong et al. (2009) this has weak support and further studies are required to confirm its taxonomic position.

Testudinaceae

Two marine species are referred to this family: *Verruculina enalia* (Schoch et al. 2006) and *Massarina ricifera*, the latter with weak support and a long branch (Suetrong et al. 2009). Members of the Testudinaceae form a monophyletic clade and are characterized by ascospores that are 1-septate, brown without germ slits and with or without ornamentation (Kruys et al. 2006). However, *V. enalia* and *M. ricifera* share few characters with members of this family, and further collections are required to verify its position in the family. Another mangrove species, *Quintaria lignatilis*, groups in the Testudinaceae, as a sister group to the genera *Verruculina, Ulospora* and *Neotestudina* with weak support (Schoch et al. 2009; Suetrong et al. 2009; Zhang et al. 2011).

Trematosphaeriaceae

Various phylogenetic studies have referred to the family Trematosphaeriaceae (Schoch et al. 2009; Suetrong et al. 2009; Zhang et al. 2009b); however, it has not formally been proposed (Cannon and Kirk 2007). Suetrong et al. (2011a) therefore introduced the family Trematosphaeriaceae to accommodate the genera *Falciformispora, Halomassarina* and *Trematosphaeria* (Figures 2.1 and 2.3). The main distinguishing characters of the family are medium-sized ascomata with apapillate ostioles, a relatively wide, coriaceous peridium, cellular pseudoparaphyses and cylindro-clavate asci. Ascospores range from two-celled to many celled and are hyaline or brown. *Falciformispora lignatilis* and *Halomassarina thalassiae* are both mangrove species; however, three marine *Trematosphaeria* species have yet to be sequenced to confirm their position in the family.

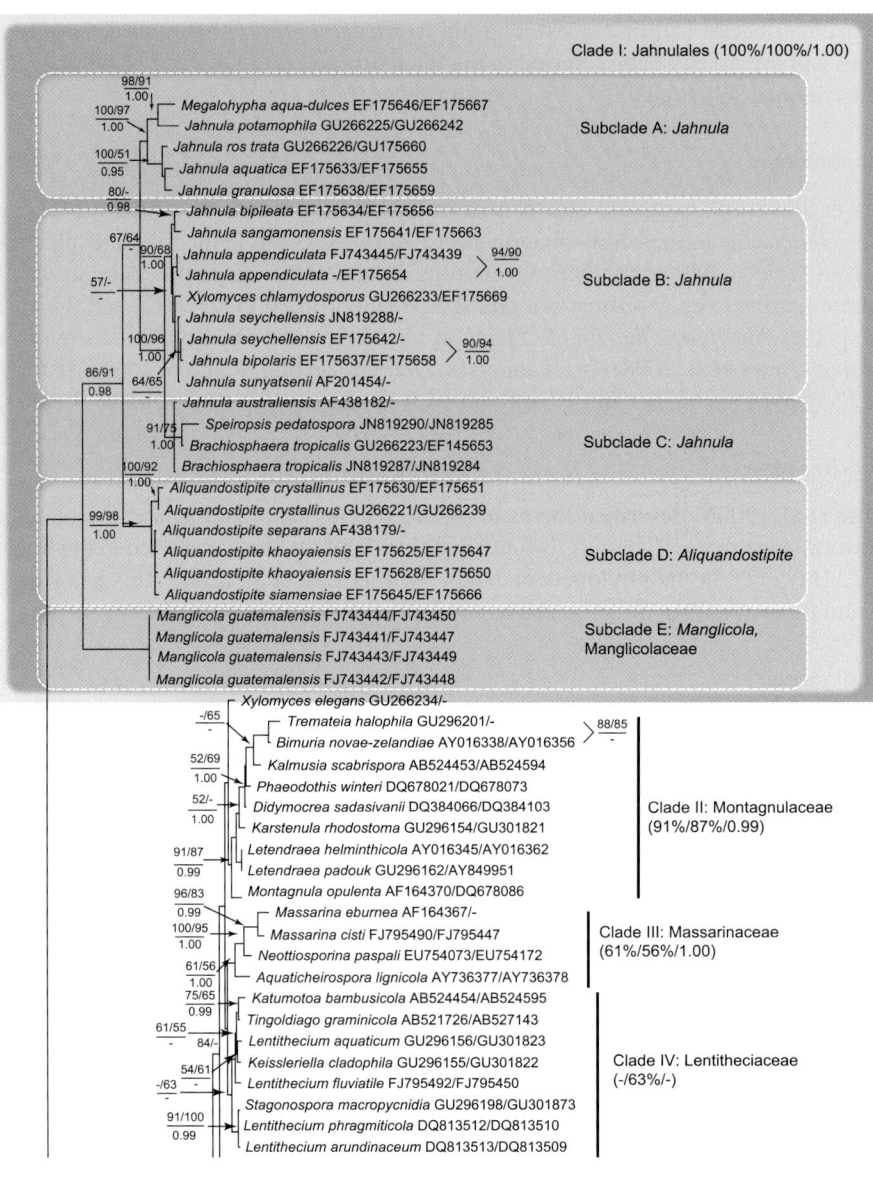

Figure 2.1 (Continued)

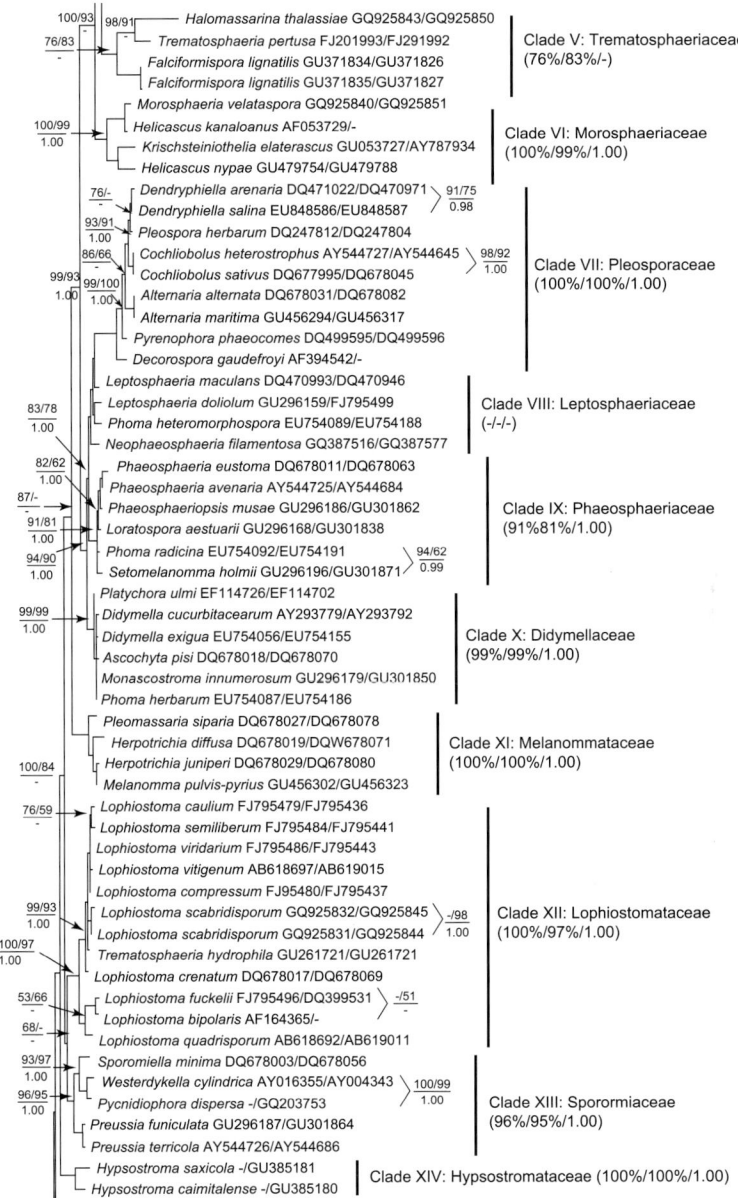

Figure 2.1 (Continued)

Figure 2.1 (Continued)

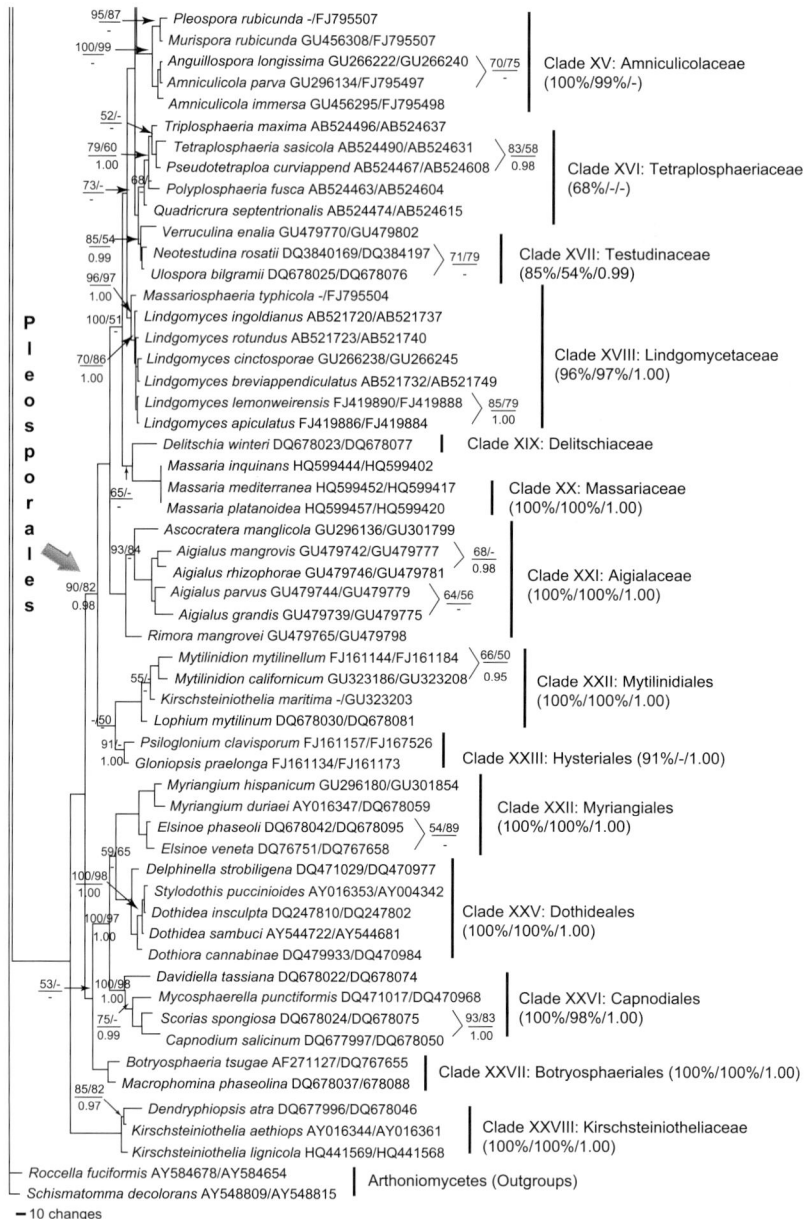

Figure 2.1 One of 10 MPTs inferred from combined SSU and LSU rDNA sequences of Jahnulales and other taxa of the Dothideomycetes, generated with maximum parsimony (tree length = 4337, C.I. = 0.362, R.I. = 0.794, R.C. = 0.288). Maximum likelihood (BSML, left) and parsimony (BSMP, right) bootstrap values >50% are given above the node. Bayesian posterior probabilities >0.95 are given below each node (BYPP). The internodes that are highly supported by all bootstrap (100%) and posterior probabilities (1.00) are shown as a thicker line.

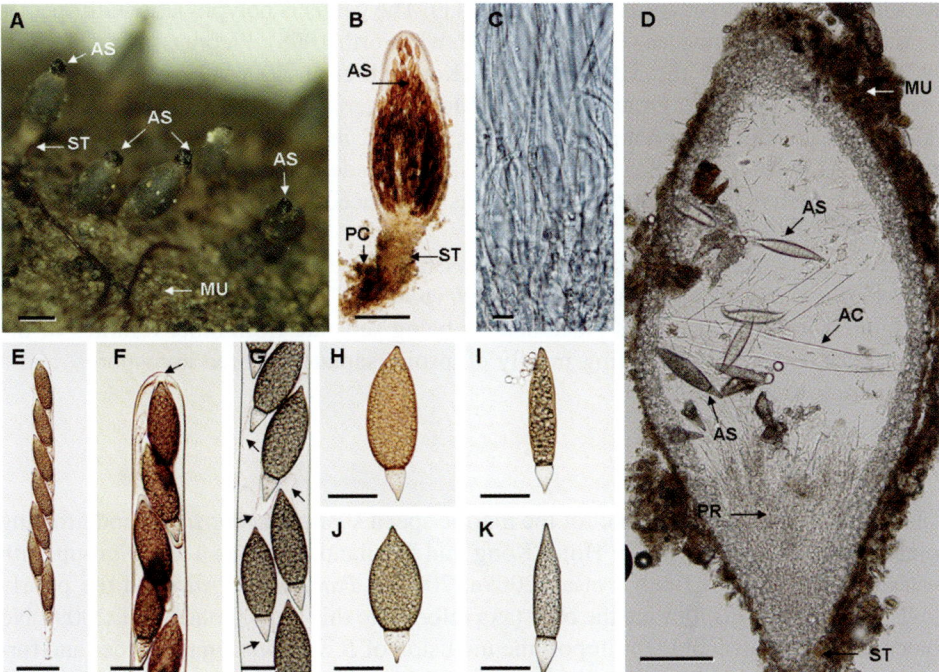

Figure 2.2 *Manglicola guatemalensis*. (A, B) Mature ascomata with stalk (ST) on the surface of brackish water palm *Nypa fruticans*, partially immersed in mud (MU), composed of pseudoparenchymatous cells (PC) and ascospores (AS) exuded at ostiole. (C) Trabeculate pseudoparaphyses numerous, septate simple. (D) Longitudinal section of ascoma. Ascoma with stalk (ST), asci (AC), pseudoparaphyses (PR) and ascospores (AS) covered with mud (MU). (E) Cylindrical ascus. (F) Ascus tip with an apical apparatus (arrowed) comprising a lens-shape disk. (G) Ascospores with appendages (arrowed) in ascus. (H, I, J, K) Bicelled ascospores. Scale bars: A–B = 500 μm; C = 10 μm; D = 200 μm; E = 100 μm; F–K = 50 μm.

Pleosporales, Family *incertae sedis*

Mauritiana rhizophorae, *Halotthia posidoniae* and *Pontoporeia biturbinata* form a monophyletic group with moderate support (Suetrong et al. 2009; Zhang et al. 2011) in basal clade to the families Sporormiaceae and Lophiostomataceae. These three genera share few common features with *H. posidoniae* and *P. biturbinata* which grow on the submerged rhizomes of seagrasses, while *M. rhizophorae* is described from the mangrove tree *Rhizophora mucronata*; morphologically they differ especially in ascospore characteristics (Zhang et al. 2012b). *Pontoporeia biturbinata* has a long branch length in published phylograms (Suetrong et al. 2009; Zhang et al. 2011), which suggests deviation from the other taxa in the clade.

Mytilinidiales

Hawksworth (1985) introduced the genus *Kirschsteiniothelia* for *K. aethiops* based on *Sphaeria aethiops* and currently 18 species are recorded in Index Fungorum, although only seven are recognized by Zhang et al. (2011). The genus is polyphyletic with species referred to the Pleosporaceae (Eriksson and Hawksworth 1998), Pleomassariaceae (Barr 1993) and questionably the Massarinaceae (Kodsueb et al. 2006). *Kirschsteiniothelia elaterascus* was shown to group within the Morosphaeriaceae (Pleosporales) (Suetrong et al. 2009; Liu et al. 2011; Zhang et al. 2011), while *K. maritima* clustered with *Mytilinidion* spp. as a sister group to the Mytilinidiaceae clade (Suetrong et al. 2009). More recently the genus *Halokirschsteiniothelia* was introduced to accommodate *K. maritima* (Boonmee et al. 2012).

A new family, Kirschsteiniotheliaceae was introduced to accommodate the genus *Kirschsteiniothelia* and its anamorph *Dendryphiopsis* (Boonmee et al. 2012). The high support index indicates its placement as being separate from other families of the Pleosporales and differs in having mostly ellipsoid, septate, colored ascospores, and a *Dendryphiopsis* anamorph.

Patellariales

Molecular data are only available for the marine species *Patellaria atrata* found growing on mangrove wood collected in Hong Kong and Thailand. It forms a sister group with *Hysteropatella* species (Boehm et al. 2009a, 2009b). *Patellaria atrata* and the poorly known *Banhegyia setispora* are the only taxa referred to this order (Jones et al. 2009). No molecular data are available to support the inclusion of *B. setispora* in this order and further collections are required to enable a re-evaluation of its phylogenetic relationship.

Jahnulales

The order Jahnulales was proposed by Pang et al. (2002) to accommodate ascomycetes with stalked/sessile and dimorphic ascomata, hyphal stalk cells that are about 40 μm wide and ascospores that are unequally two-celled with or without various types of appendages or sheaths. Molecular data were presented for six species: *Aliquandostipite sunyatsenii* (referred to *Jahnula*), *A. khaoyaiensis*, *Jahnula bipolaris*, *J. australiensis*, *J. siamensiae* (a new species), and *Patescospora separans* (new genus and species) and referred to the family Aliquandostipitaceae (Inderbitzin et al. 2001).

All taxa assigned to the order were freshwater species, mostly growing on submerged wood, and lacking anamorphs. The order was then emended by Campbell et al. (2007) to include wide, brown hyphae and a wider variation in ascospores characters, not known when the Jahnulales was proposed. Hibbett et al. (2007) did not accept the order and referred it to Dothideomycetes *incertae sedis*. Suetrong et al. (2011a) have reviewed the phylogenetic relationship of the order and provided a key to the genera.

Aliquandostipitaceae

The family Aliquandostipitaceae was introduced by Inderbitzin et al. (2001) based on the genus *Aliquandostipite* (type species *A. khaoyaiensis*), to which most other genera have been referred. Most species in this family are freshwater taxa, although the anamorphs *Xylomyces rhizophorae* and *X. chlamydosporus* have been reported to be found in mangrove habitats (Kohlmeyer and Volkmann-Kohlmeyer 1998c).

program Biodiversity Research and Training Grants BRT R_251006 and R_325015 to study marine fungi in Thailand. For their continued interest and support, we also thank BIOTEC including Prof Morakot Tanticharoen, Dr Kanyawim Kirtikara and Dr Lily Eurwilaichitr.

References

Abdel-Wahab MA, Jones EBG. Three new marine ascomycetes from driftwood in Australian sand dunes. Mycoscience 2000; 41:379–388.
Aveskamp MM, Gruyter de J, Woudenberg JHC, Verkley GJM, Crous PW. Highlights of the Didymellaceae: A polyphasic approach to characterize *Phoma* and related pleosporalean genera. Stud Mycol. 2011; 65:1–60.
Barr M. Teichosporaceae, another family in the Pleosporales. Mycotaxon 2002; 72:373–389.
Barr ME. Notes on the Pleomassariaceae. Mycotaxon 1993; 49:129–142.
Barr ME. Planistromellaceae, a new family in the Dothideales. Mycotaxon 1996; 60:433–442.
Boehm EW, Schoch CL, Spatafora JW. On the evolution of the Hysteriaceae and Mytilinidiaceae (Pleosporomycetidae, Dothideomycetes, Ascomycota) using four nuclear genes. Mycol Res. 2009a; 113:461–479.
Boehm EWA, Mugambi GK, Miller AN, Huhndorf SM, Marincowitz S, Spatafora JW, Schoch CL. A molecular phylogenetic reappraisal of the Hysteriaceae, Mytilinidiaceae and Gloniaceae (Pleosporomycetidae, Dothideomycetes) with keys to world species. Stud Mycol. 2009b; 64: 49–83.
Boonmee S, KoKo TW, Chuleatirote E, Hyde KD, Chen H, Cai L, McKenzie EHC, Jones EBG, Kodsueb R, Hassan BA. Two new *Kirschsteiniothelia* species with *Dendryphiopsis* anamorphs cluster in Kirschsteiniotheliaceae fam. nov. Mycologia 2012; 104:698–714.
Campbell J, Ferrer A, Raja HA, Sivichai S, Shearer CA. Phylogenetic relationships among taxa in the Jahnulales inferred from 18S and 28S nuclear ribosomal DNA sequences. Can J Bot. 2007; 85:873–882.
Cannon PF, Kirk PM. Fungal families of the world. Egham, UK: CABI UK Centre, 2007.
Crous P, Minnis AM, Pereira OL, Alfenas AC, Alfenas RF, Rossman AY, Groenwald JZ. What is *Scirrhia*? IMA Fungus 2011; 2:127–133.
Eriksson OE, Hawksworth DL. Outline of the ascomycetes. Syst Ascomycet. 1998; 16:83–296.
Gueidan C, Roux C, Lutzoni F. Using a multigene phylogenetic analysis to assess generic delineation and character evolution in Verrucariaceae (Verrucariales, Ascomycota). Mycol Res. 2007; 111:1145–1168.
Gueidan C, Savíc S, Thüs H, Roux C, Tibell L, Prieto M, Heiëmarsson S, Breuss O, Orange A, Fröberg L, Wynns A, Navarro-Rosinés P, Krzewicka B, Pykälä J, Grube K, Lutzoni F. Generic classification of the Verrucariaceae (Ascomycota) based on molecular and morphological evidence: recent progress and remaining challenges. Taxon 2009; 58:184–208.
Hambleton S, Sigler L. *Meliniomyces*, a new anamorph for root-associated fungi with phylogenetic affinities to *Rhizoscyphus ericae* (=*Hymenoscyphus ericeae*), Leotiomycetes. Stud Mycol. 2005; 53:1–27.
Hawksworth DL. *Kirschsteiniothelia*, a new genus for the *Microthelia incrustans*-group (Dothideales). Bot J Linn Soc. 1985; 91:181–202.
Hibbett DS, Binder M, Bischoff JF, Blackwell M, Cannon PF, Eriksson OE, Huhndorf S, James ST, Kirk PM, Lucking R, Lumbsch HT, Lutzoni F, Matheny PB, Mclaughlin DJ, Powell MJ, Redhed S, Scoch CL, Spatafora JF, Stalpers JA, Vilgalys R, Aime MC, Aptroot AA, Bauer R, Begerow D, Benny GL, Castelbury LA, Crous PW, Dai C, Gams W, Geiser DM, Griffith GW, Gueidan C, Hawksworth DL, Hestmark G, Hosaka K, Humber RA, Hyde KD, Ironisde JE,

Koljag U, Kurtzman CP, Larsson KH, Lichtwardt R, Longcore J, Miadlikowska J, Miller A, Moncalvo JM, Mozley-Standridge S, Oberwinkler F, Parmasto E, Reeb V, Rogers JD, Roux C, Ryvarden L, Sampaio JP, Schßler A, Sugiyama J, Thorn RG, Tibell L, Untereiner WA, Walker C, Wang Z, Weir A, Weiss M, White MM, Winka K, Yao YJ, Zhang N. A higher-level phylogenetic classification of the fungi. Mycol Res. 2007; 111:509–547.

Inderbitzin P, Kohlmeyer J, Volkmann-Kohlmeyer B, Berbee ML. *Decorospora*, a new genus for the marine ascomycetes *Pleospora gaudefroyi*. Mycologia 2002; 91:651–659.

Inderbitzin P, Landvik S, Abdel-Wahab MA, Berbee ML. Aliquandostipitaceae, A new family for two new tropical ascomycetes with unusually wide hyphae and dimorphic ascomata. Am J Bot. 2001; 88:52–61.

Janso JE, Bernan VS, Greenstein M, Bugni TS, Ireland CM. *Penicillium dravuni*, a new marine-derived species from an alga in Fiji. Mycologia 2005; 97:444–453.

Jones EBG, Klaysuban A, Pang KL. Ribosomal DNA phylogeny of marine anamorphic fungi: *Cumulospora varia*, *Dendryphiella* species and *Orbimyces spectabilis*. Raff Bull Zool. 2008; suppl 19:11–18.

Jones EBG, Sakayaroj J, Suetrong S, Somrithipol S, Pang KL. Classification of marine Ascomycota, anamorphic taxa and Basidiomycota. Fungal Divers. 2009; 35, pp. 1–187.

Khashnobish A, Shearer CA. Phylogenetic relationships in some *Leptosphaeria* and *Phaeosphaeria* species. Mycol Res. 1996; 100:1355–1363.

Kodsueb R, Dhanasekaran V, Aptroot A, Lumyong S, Mckenzie EHC, Hyde KD, Jeewon R. The family Pleosporaceae: intergeneric relationships and phylogenetic perspectives based on sequence analyses of partial 28S rDNA. Mycologia 2006; 98:571–583.

Kohlmeyer J. Marine fungi from Easter Island and notes on *Thalassoascus*. Mycologia 1981; 73:833–843.

Kohlmeyer J, Hawksworth DL, Volkmann-Kohlmeyer B. Observations on two "borderline" lichens: *Mastodia tessellata* and *Collemopsidium pelvetiae*. Mycol Prog. 2004; 3:51–56.

Kohlmeyer J, Kohlmeyer E. Marine Mycology. The higher fungi. New York: Academic Press, 1979.

Kohlmeyer J, Volkmann-Kohlmeyer B. *Atrotorquata* and *Loratospora*: New ascomycete genera on *Juncus roemerianus*. Syst Ascomycet. 1993; 12:7–22.

Kohlmeyer J, Volkmann-Kohlmeyer B. *Mycophycias*, a new genus for the mycobionts of *Apophlaea*, *Ascophyllum* and *Pelvetia*. Syst Ascomycet. 1998a; 16:1–7.

Kohlmeyer J, Volkmann-Kohlmeyer B. *Dactylospora canariensis* sp. nov. and notes on *D. haliotrepha*. Mycotaxon 1998b; 67:247–250.

Kohlmeyer J, Volkmann-Kohlmeyer B. A new marine *Xylomyces* on *Rhizophora* from the Caribbean and Hawaii. Fungal Divers. 1998c; 1:159–164.

Kohlmeyer J, Volkmann-Kohlmeyer B, Eriksson OE. Fungi on *Juncus roemerianus*. 2. New dictyosporous ascomycetes. Bot Mar. 1995; 38:165–174.

Kohlmeyer J, Volkmann-Kohlmeyer B, Eriksson OE. Fungi on *Juncus roemerianus*. 8. New bitunicate ascomycetes. Can J Bot. 1996; 74:1830–1840.

Kruys A, Eriksson OE, Wedin M. Phylogenetic relationships of coprophilous Pleosporales (Dothideomycetes, Ascomycota), and the classification of some bitunicate taxa of unknown position. Mycol Res. 2006; 110:527–536.

Landvik S, Shaile NFJ, Eriksson OE. SSU rDNA sequence support for a close relationship between the Elaphomycetales and the Eurotiales and Onygenales. Mycoscience 1996; 37:237–241.

Leuchtmann A. Über *Phaeosphaeria* Miyake und andere bitunicate Ascomyceten nit mehrfach querseptierten ascosporen. Sydowia 1984; 37:75–194.

Lindemuth R, Wirtz N, Lumbsch HT. Phylogenetic analysis of nuclear and mitochondrial rDNA sequences supports the view that loculoascomycetes (Ascomycota) are not monophyletic. Mycol Res. 2001; 105:1176–1181.

Liu FK, Phookamsak R, Jones EBG, Zhang Y, Ko-Ko T, Hu H, Boonmee S, Doilom M, Chukeatirote E, Bahkali AH, Wang Y, Hyde KD. *Astrosphaeriella* is polyphyletic, with species in *Fissuroma* gen. nov., and *Neoastosphaeriella* gen. nov. Fungal Divers. 2011; 51:135–154.

Lumbsch HT, Schmitt I, Lindemuth R, Miller A, Mangold A, Fernandez F, Huhndorf S. Performance of four ribosomal DNA regions to infer higher-level phylogenetic relationships of inoperculate euascomycetes (Leotiomyceta). Mol Phy Evol. 2005; 34:512–524.

Mantle PG, Hawksworth DL, Pazoutova S, Collinson LM, Rassing BR. *Amorosia littoralis* gen. et sp. nov., a new genus and species name for scorpinone and caffeine producing hyphomycetes from the littoral zone in The Bahamas. Mycol Res. 2006; 110:1371–1378.

Mohr F, Ekman S, Heegaard E. Evolution and taxonomy of the marine *Collemopsidium* species (lichenized Ascomycota) in north-west Europe. Mycol Res. 2004; 108:515–532.

Pang KL, Abdel-Wahab MA, El-Sharouney HM, Sivichai S, Jones EBG. Jahnulales (Dothideomyces, Ascomycota) a new order of lignicolous freshwater ascomycetes. Mycol Res. 2002; 106:1031–1042.

Pérez-Ortega S, De los Rios A, Crespo A, Sancho LG. Symbiotic lifestyle and phylogenetic relationships of the bionts of *Mastodia tessellata* (Ascomycota, *Incertae sedis*). Am J Bot. 2010; 97:738–752.

Poon MOK, Hyde KD. Biodiversity of intertidal estuarine fungi on *Phragmites* at Mai Po marshes, Hong Kong. Bot Mar. 1998; 41:141–155.

Schatz S. Taxonomic revision of two Pyrenomycetes associated with littoral-marine green algae. Mycologia 1980; 72:110–117.

Schoch CL, Shoemaker RA, Seifert KA, Hambleton S, Spatafora JW, Crous PW. A multigene phylogeny of the Dothideomycetes using four nuclear loci. Mycologia 2006; 98:1041–1052.

Schoch CL, Sung GH, López-Giráldez F, Towsend JP, Miadlikowska J, Hofstetter V, Robbertse B, Matheny PB, Kauff F, Wang Z, Gueidan C, Andrie RM, Trippe K, Ciufetti LM, Wynns A, Fraker E, Hodkinson BP, Bonito G, Groenewald JZ, Arzanlou M, de Hoog GS, Crous PW, Hewitt D, Pfister DH, Peterson K, Gryenhout M, Wingfield MJ, Aptroot A, Suh SO, Blackwell M, Hillis DM, Griffith GW, Castlebury LA, Rossman AY, Lumbusch HT, Lückunbg R, Büdel B, Rauht A, Diederich P, Erta D, Geiser DM, Hosaka K, Inderbitzin P, Kohlmeyer J, Volkmann-Kohlmeyer B, Mostert L, O'Donnell K, Sipman J, Rogers JD, Shoemaker RA, Sugiyama J, Summmerbell RC, Hohnston PR, Stenroos S, Dyer PS, Crittenden PD, Cole PD, Hansen K, Trappe JM, Yahr R, Lutzoni F, Spatafora JW. The Ascomycota tree of life: A phylum-wide phylogeny clarifies the origin and evolution of fundamental reproductive and ecological traits. Syst Biol. 2009; 58:211–223.

Shearer CA. A new species of *Kirschsteiniothelia* (Pleosporales) with an unusual fissitunicate ascus. Mycologia 1993; 85:963–969.

Suetrong S, Boonyuen N, Pang KL, Ueapattanakit J, Klaysuban A, Sri-Indrasutdhi V, Sivichai S, Jones EBG. A taxonomic revision and phylogenetic reconstruction of the Jahnhuales (Dothideomycetes), and the new family Manglicolaceae. Fungal Divers. 2011a; 51:163–188.

Suetrong S, Hyde KD, Zhang Y, Bahkali AH, Jones EBG. Trematosphaeriaceae fam. nov. (Dothideomycetes, Ascomycota). Crypt Mycol. 2011b; 32:343–358.

Suetrong S, Sakayaroj J, Phongpaichit S, Jones EBG. Morphological and molecular characteristics of a poorly known marine ascomycete, *Manglicola guatemalensis* (Jahnulales: Pezizomycotina; Dothideomycetes, *Incertae sedis*): a new lineage of marine ascomycetes. Mycologia 2010; 102:83–92.

Suetrong S, Schoch CL, Spatafora JW, Kohlmeyer J, Volkmann-Kohlmeyer B, Sakayaroj J, Phongpaicht S, Tanaka K, Hairayama K, Jones EBG. Molecular systematics of the marine Dothideomycetes. Stud Mycol. 2009; 64:155–173.

Tam WY, Pang KL, Jones EBG. Ordinal placement of selected marine Dothideomycetes inferred from SSU ribosomal DNA sequence analysis. Bot Mar. 2003; 4:487–494.

Toxopeus J, Kozera CJ, O'Leary SJB, Garbary DJ. A reclassification of *Mycophycias ascophylli* (Ascomycota) based on nuclear large ribosomal subunit DNA sequences. Bot Mar. 2011; 54:325–334.

Zhang H, Hyde KD, Abdel-Wahab MA, Abdel-Aziz FA, KoKo TW, Zhao R, Ali H, Bahkali AH, Zhou D. Morphological and molecular characterization of the genus *Helicascus* with a description of a new species. Mycologia 2012.

Zhang Y, Crous PW, Schoch CL, Hyde KD. Pleosporales. Fungal Divers. 2011; 53:1–221.

Zhang Y, Fournier J, Crous PW, Pointing SB, Hyde KD. Phylogenetic and morphological assessment of two new species of *Amniculicola* and their allies (Pleosporales). Persoonia 2009a; 23:48–54.

Zhang Y, Jeewon R, Fournier J, Hyde KD. Multi-gene phylogeny and morphotaxonomy of *Amniculicola lignicola*: a novel freshwater fungus from France and its relationships to the Pleosporales. Mycol Res. 2008; 112:1186–1194.

Zhang Y, Schoch CL, Fournier J, Crous PW, De Gruyter J, Woudenberg JHC, Hirayama K, Tanaka K, Pointing SB, Hyde, KD. Multi-locus phylogeny of the Pleosporales: a taxonomic, ecological and evolutionary re-evaluation. Stud Mycol. 2009b; 64:85–102.

Zhang Y, Wang HK, Fournier J, Crous PW, Jeewon R, Pointing SB, Hyde KD. Towards a phylogenetic clarification of *Lophiostoma/Massarina* and morphologically similar genera in the Pleosporales. Fungal Divers. 2009c; 38:225–251.

3 Phylogeny of the marine Sordariomycetes

Ka-Lai Pang

Jones et al. (2009) documented 530 (in 321 genera) obligate marine fungi, which include Ascomycota 424 species (in 251 genera), asexual (anamorphic) fungi 94 species (in 61 genera) and Basidiomycota 12 species (in nine genera). The marine Ascomycota belong to the classes Dothideomycetes, Eurotiomycetes, Laboulbeniomycetes, Lecanoromycetes, Leotiomycetes, Lichinomycetes, Arthoniomycetes and Sordariomycetes, while the Dothideomycetes and the Sordariomycetes are among the largest. While the phylogeny of the Dothideomycetes is discussed elsewhere in this volume, the phylogeny of the Sordariomycetes is the main topic of this chapter.

Taxa of the Sordariomycetes possess perithecial, or less frequently, cleistothecial ascomata and inoperculate unitunicate or prototunicate asci (Alexopoulos et al. 1996). Spatafora and Blackwell (1994) was the first published study to include a sequence of a marine Sordariomycetes (*Halosphaeriopsis mediosetigera*) in a phylogenetic analysis and it was found to be related to members of the Microascaceae, Microascales. Since then, there have been numerous studies on the phylogeny of marine fungi to resolve different taxonomic issues, including the origin of marine fungi (Spatafora et al. 1998), classification of taxonomically problematic taxa (Sakayaroj et al. 2011), and delineation of morphologically related taxa (Campbell et al. 2003; Pang et al. 2003b).

A total of 283 species of marine fungi belong to the Sordariomycetes (Abdel-Wahab et al. 2009; Jones et al. 2009; Pang et al. 2010; Abdel-Wahab and Nagahama 2011), and have been classified into several orders predominantly based on morphology: Hypocreales, Coronophorales, Microascales, Diaporthales, Sordariales, Chaetosphaeriales, Ophiostomatales, Xylariales, Lulworthiales, Koralionastetales, Magnaporthales and Phyllachorales (Jones et al. 2009). Nevertheless, the correct taxonomic placement of a number of marine sordariomycetous taxa is still unknown and many have been tentatively placed in a class until sequence data are available to confirm their placement. In this chapter, major ordinal lineages of marine Sordariomycetes that already have sequence data supporting their phylogeny are introduced: Sordariales (Cai et al. 2006), Microascales (formerly Halosphaeriales; Spatafora et al. 1998), Lulworthiales (Kohlmeyer et al. 2000), TBM (*Torpedospora/Bertia/Melanospora*) clade (Schoch et al. 2007), Hypocreales (Rossman et al. 2001; Pang et al. 2010), Xylariales (Smith et al. 2003), Ophiostomatales (Spatafora et al. 2006), Koralionastetales (Campbell et al. 2009), Magnaporthales (Thongkantha et al. 2009) and Savoryellales (Boonyuen et al. 2011). A more detailed account of the Halosphaeriaceae (Microascales) is provided, since most marine Ascomycota belong to this family.

Hypocreomycetidae

Hypocreales

Members of the Hypocreales generally produce brightly-colored ascomata, ovoid to cylindrical asci with an apical pore and spherical to needle-like ascospores (Samuels and Seifert 1987; Alexopoulos et al. 1996) and they are not common inhabitants of the marine environment. Most marine Hypocreales belong to the Bionectriaceae: *Heleococcum japonense, Kallichroma tethys, K. glabrum, Emericellopsis maritima* (Rossman et al. 2001; Zuccaro et al. 2003). *Heleococcum japonense* forms a group with *Hydropisphaera* spp., *Roumegueriella rufula* and *Selinia pulchra* while *Kallichroma* spp. constitute a separate clade (Rossman et al. 2001). Recently, Pang et al. (2010) discovered a new marine fungus *Sedecimiella taiwanensis* from a mangrove in Taiwan and it is phylogenetically related to the Niessliaceae. Other taxa, e.g., *Halonectria* and *Payosphaeria*, have been referred to other families in the Hypocreales based on morphology; however a molecular study is required to confirm their taxonomic placement.

Microascales (formerly Halosphaeriales)

Most marine Sordariomycetes belong to the Halosphaeriaceae, Microascales. The ordinal status of the Halosphaeriales was rejected and reduced to a family, Halosphaeriaceae, with the Microascaceae in the Microascales (Hibbett et al. 2007). However, both families differ significantly in terms of morphology and ecology (Sakayaroj et al. 2011). Currently, there are 148 species (58 genera) of aquatic fungi in the family, and freshwater taxa are present in *Aniptodera, Ascosacculus, Fluviatispora, Lignincola, Luttrellia, Magnisphaera, Nais, Natantispora, Oceanitis, Panorbis* and *Phaeonectriella*.

Morphology has traditionally been used to identify and classify the taxa in the Halosphaeriaceae; however, it does not infer intergeneric relationships or give any indications on the evolution of the group. A rapid radiation in this group of fungi may be the cause for such wide variation of morphology. As a consequence, classification of taxa in the Halosphaeriaceae has been problematic, with the various types of ascoma, ascus, ascospore and ascospore appendage morphology being present. Among these structures, ascospore appendage morphology and ontogeny have been the major characters for the delineation of genera in the family and Jones (1995) categorized these into 12 ontogenic groups. Early studies using DNA sequence data to infer the phylogeny and intergeneric relationship of genera in the Halosphaeriaceae were on the complex of species with unfurling ascospore appendages based on the 18S and 28S rRNA genes (Kong et al. 2000; Campbell et al. 2003; Pang et al. 2003b). Subsequently, many other genera in the family have also been studied and reviewed: *Corollospora* (Campbell et al. 2002), *Lignincola, Nais* and *Saagaromyces* (Pang et al. 2003a), *Aniptodera, Halosarpheia, Phaeonectriella* and *Tirispora* (Campbell et al. 2003; Pang et al. 2003b), *Halosphaeria* (Pang et al. 2004), *Haligena* (Sakayaroj et al. 2005b), and *Antennospora* and *Arenariomyces* (Pang et al. 2008). Recently, Sakayaroj et al. (2011) resolved the taxonomy of *Ceriosporopsis* and *Remispora* using phylogenetic analysis of 18S, 28S rRNA and *rpb2* genes. New genera and combinations were made in these studies, mostly supported by ascospore and ascospore appendage morphology. In some cases, distinctive generic morphological characteristics are lacking for the new genera. For instance, *Halosarpheia aquatica* and *H. heteroguttulatus* were phylogenetically unrelated to *Halosarpheia sensu stricto* and transferred to the new genus *Ascosacculus* (Campbell et al. 2003). The main characteristics of *Ascosacculus* include periphysate necks, saccate asci without apical structures, fusiform to cylindrical ascospores with heavy oil guttulation and unfurling ascospore appendages. These morphological characteristics do not differentiate this genus from other related genera with unfurling ascospore appendages, such as *Aniptodera, Halosarpheia, Natantispora* and *Panorbis*.

One problem with interpreting information from phylogenetic trees resides in how to define the clades for a genus. Current practice is based primarily on a clade, supplemented with significant morphological differences with the adjacent taxon. For example, Pang et al. (2008) investigated the phylogeny of *Antennospora* and *Halosphaeria* and found that *Antennospora salina* forms a well-supported monophyletic clade with *Arenariomyces trifurcatus*, the type species of the genus. A new genus, *Haiyanga*, was established for *A. salina* based on the morphological differences between *A. trifurcatus* and *A. salina*. A consensus tree of the 38 most parsimonious trees based on the partial nuclear large subunit (LSU) rRNA gene and the mean pairwise distance (adjusted for missing data) of closely related taxa in comparatively well-supported terminal clades [consistent with the study by Sakayaroj et al. (2011)] are shown in Figure 3.1. The pairwise distance between clade A consisting of *Arenariomyces, Gesasha* and *Haiyanga* is between 7.8% and 14.8%, while that in clades D (*Kochiella, Ocostaspora, Morakotiella*) and G (*Oceanitis, Ophiodeira*) is below 7.3%, and this justifies the creation of *Haiyanga* for *A. salina*. However, the pairwise distance within each of the genera *Gesasha* and *Remispora* is 12.8% and 0.7–7.4%, implying that they may represent more than one genus. Likewise, the pairwise distances between isolates of *Saagaromyces abonnis, Morakotiella salina* and *Magnisphaera spartinae* are 4.0%, 6.7% and 4.8%, respectively. These results may indicate cryptic species within the circumscribed species in the Halosphaeriaceae. The presence of cryptic species has also been suspected based on morphology in *Saagaromyces abonnis*, together with *Aniptodera chesapeakensis, Ceriosporopsis halima, Corollospora maritima* and *Haiyanga salina* (Jones 2011). All of these suggest inconsistency in the circumscription of genera and species in the family based on sequence data. A taxonomical review is warranted based on pairwise distance calculation but this will require a robust phylogenetic tree, ideally from a multi-gene analysis.

Savoryellales

Savoryella species are characterized by dark brown to black ascomata, clavate to cylindrical asci with a comparatively flattened apical ring and versicolorous septate ascospores, brown central cells and hyaline end cells (Jones and Eaton 1969). Eleven species are now referred to *Savoryella* from both freshwater and marine habitats (Jones et al. 2009). Taxonomically, *Savoryella* has been referred to the Sphaeriales *incertae sedis* (Kohlmeyer and Kohlmeyer 1979), ascomycetes *incertae sedis* (Eriksson and Hawksworth 1986; Kohlmeyer 1986), Amphisphaeriaceae (Eriksson and Hawksworth 1987), Sordariales (Jones and Hyde 1992) and Halosphaeriales (now Microascales; Barr 1990; Read et al. 1993), based on morphology and ultrastructure. Vijaykrishna et al. (2006) and Cai et al. (2006) referred *S. elongata* and *S. longispora* to the order Hypocreales based on LSU rDNA data, however, the type species *S. lignicola* was not included in the analysis. Recently, Boonyuen et al. (2011) studied the phylogeny of *Savoryella, Ascotaiwania, Ascothailandia* and the anamorphic *Canalisporium* based on four genes and these genera

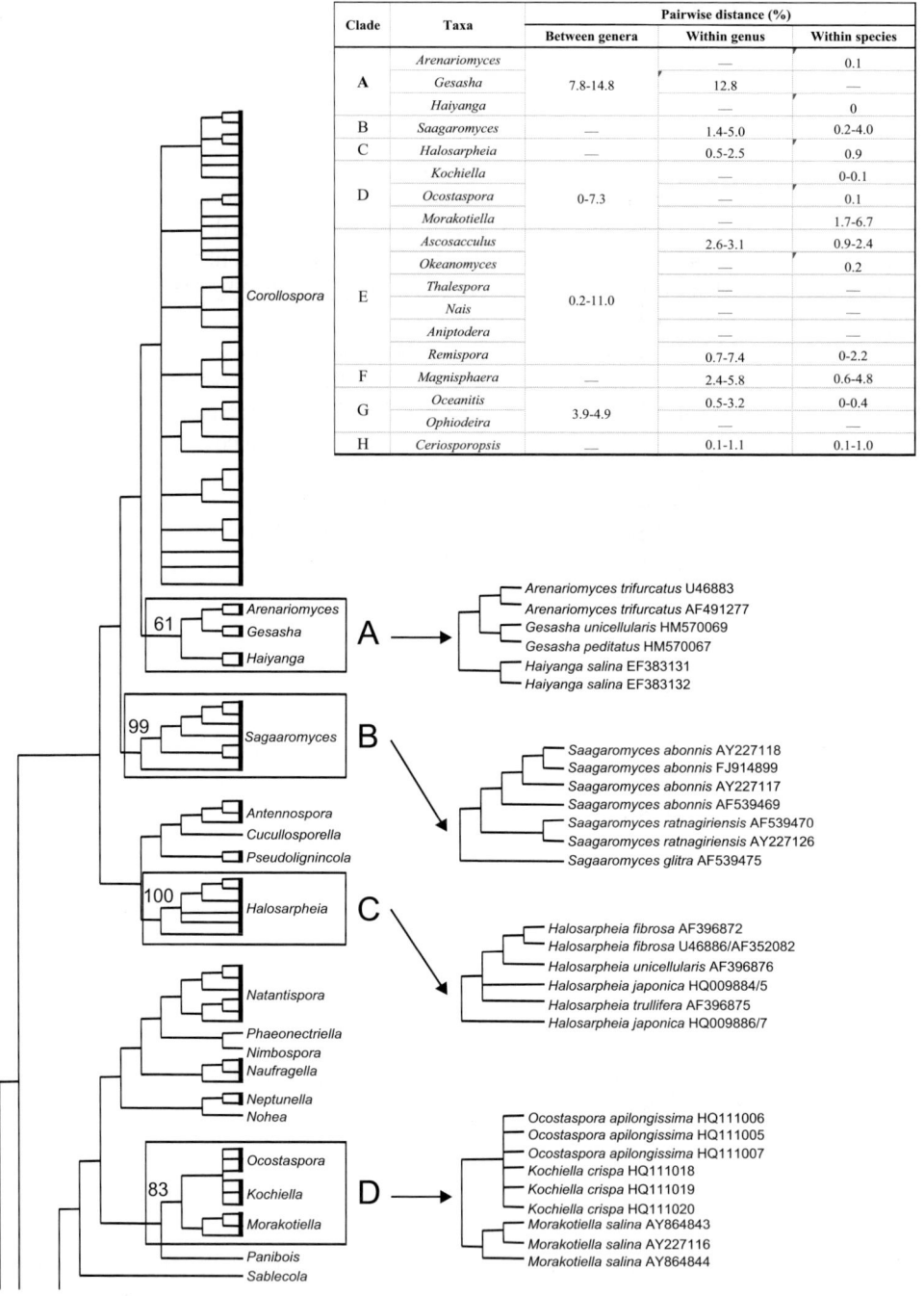

Figure 3.1 (Continued)

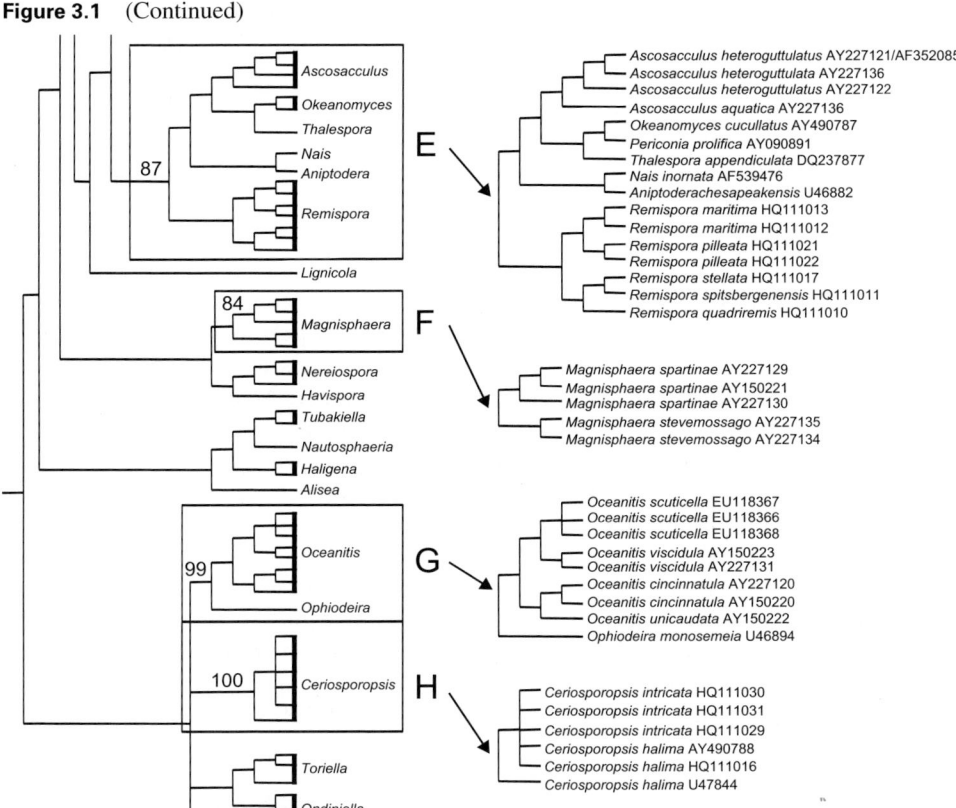

Figure 3.1 The consensus tree of 38 MPTs (tree length = 3075, CI = 0.380, RI = 0.746) with bootstrap values for selected clades based on a maximum parsimony analysis (heuristic search, gaps as missing data, starting tree(s) obtained via stepwise addition, random sequence addition of 10,000 replicas, a tree-bisection-reconnection branch-swapping algorithm, and MULTREES off) of a data set of the partial nuclear large subunit sequences of taxa in Halosphaeriaceae. The same data set was used for the calculation of mean pairwise distance (adjusted for missing data).

form a well-supported monophyletic group within the Hypocreomycetidae. As a result, a new order, Savoryellales, has been introduced.

Sordariomycetidae

Magnaporthales

Magnaporthales is a recently-established order of the Sordariomycetes, and is closely related to the Diaporthales and Ophiostomatales (Thongkantha et al. 2009). Morphological characteristics of this order include a lack of stromata, immersed, black ascomata with long hairy necks, the presence of paraphyses, cylindrical asci with a large apical pore mostly surrounded by a ring that stains blue in iodine (J+), filiform, and septate ascospores with a sheath (Cannon 1994; Thongkantha et al. 2009). *Gaeumannomyces, Magnaporthe, Pseudohalonectria* and *Ophioceras* were included in the order, while *Buergenerula spartinae* was the only marine species sequenced in the study and was found to be related to *Gaeumannomyces* species (Thongkantha et al. 2009). *Gaeumannomyces medullaris*, isolated from *Juncus roemerianus*, groups with *Magnaporthe grisea* (Schoch et al. 2009) and thus, can also be referred to the Magnaporthales.

Ophiostomatales

Lanspora coronata is a lignicolous marine fungus described from driftwood in the Seychelles (Hyde and Jones 1986). It is characterized by perithecial ascomata, deliquescing asci and ascospores with crown-like appendages at both ends and these characteristics fit into the criteria of the Halosphaeriaceae, Microascales. Spatafora et al. (2006) have shown that this fungus is phylogenetically unrelated to the Halosphaeriaceae, however, and groups with *Ophiostoma*. Members of the Ophiostomatales are characterized by perithecial ascomata, globose to ovoid evanescent asci, the absence of both periphyses and paraphyses and the presence of *Sporothrix* and *Leptographium* asexual states for some species (Alexopoulos et al. 1996). Whether *L. coronata* belongs to the Ophiostomatales remains unknown, since periphyses are present in this species (Hyde and Jones 1986). Collections of two related marine fungi (Figure 3.2A–B) with similar morphology to *L. coronata* (Figure 3.2C) have been made in northern Taiwan and phylogenetically they are related to *L. coronata* based on a combined analysis of the nuclear 18S and 28S rRNA and *rpb2* genes (unpublished results).

Sordariales

Members of the Sordariales are characterized by: dark or pallid perithecia or cleistothecia with periphyses, sometimes embedded in a subiculum; ovoid to cylindrical asci with a

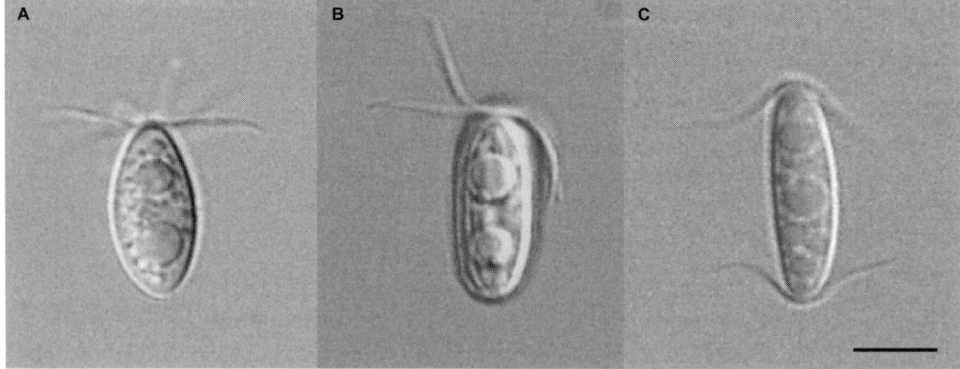

Figure 3.2 Marine taxa of the Ophiostomatales. (A) An unknown fungus collected at Shih-Ti-Ping, Taiwan. (B) A similar fungus to (A) but with a striated ascospore wall collected at Ying-Ke-Shih, Taiwan. (C) *Lanspora coronata*. Scale bar = 5 μm.

thin refractile apical ring; the presence of paraphyses and pseudoparenchyma; and hyaline to dark ascospores that may have ornamentations, sheaths or germ pores (Alexopoulos et al. 1996). Only four marine species are represented in the Sordariales, including *Biconiosporella corniculata, Chaetomium heteropilum, Zopfiella latipes* and *Zopfiella marina* (Jones et al. 2009). Cai et al. (2006) investigated the phylogeny of *Zopfiella* spp. using 3 genes (partial 28S rDNA, ITS/5.8S rDNA and partial b-tubulin) and included *Z. latipes*. *Zopfiella* was inferred to be a polyphyletic genus with *Z. latipes* grouping in the Sordariales but did not group with *Z. tabulata*, the type species of the genus (Cai et al. 2006). *Zopfiella marina* was later found to be phylogenetically related to *Z. latipes* based on ITS sequences (Vicente et al. 2009).

Xylariomycetidae

Xylariales

Taxa of the Xylariales are common inhabitants of the upper tidal area of mangroves or are found on the intertidal parts of the palm *Nypa fruticans* (Jones et al. 2009). Common morphological characteristics are dark-colored ascomata in a stroma, persistent cylindrical asci with an amyloid apical ring, the presence of paraphyses, and hyaline or dark ascospores that may possess germ slits or pores (Alexopoulos et al. 1996). Few marine species from the Xylariales have been sequenced and many have been assigned to various families based on morphology. Many taxa in the Xylariaceae have been sequenced and confirmed their inclusion in the family: *Anthostomella torosa* (Zhang et al. 2006), *Halorosellinia oceanica* (Smith et al. 2003), *Nemania maritima* (Pinnoi et al. 2010) and *Pedumispora rhizophorae* (Klaysuban et al. 2011). Many taxa await a molecular study to confirm their taxonomic placement in the order, e.g., *Adomia* and *Oxydothis*.

Sordariomycetes *incertae sedis*

Koralionastetales

Pontogeneia is a genus containing eight species that are parasitic on brown and green algae (Jones et al. 2009). Distinctive characteristics of *Pontogeneia* species include the presence of septate paraphyses, mostly filamentous and multi-septate ascospores and parasitic occurrence on marine algae (Kohlmeyer 1975; Kohlmeyer and Kohlmeyer 1979). Koralionastetaceae was established to accommodate *Koralionaste*, a genus of ascomycetes with thick-walled ascospores that germinate into hyphae bearing phialidic antheridia with enteroblastic spermatia and associated with corals (Kohlmeyer and Volkmann-Kohlmeyer 1987a). Phylogenetically, *Koralionaste ellipticus* and *Pontogeneia microdictyi* form a sister group to the Lulworthiales (Campbell et al. 2009), but differ in that they possess paraphyses and periphyses and lack the apical mucus-containing chambers or gelatinous sheaths (except *Lindra*) found in the Lulworthiales (Campbell et al. 2009). A new order, Koralionastetales, was established to accommodate these genera based on morphological and molecular data (Campbell et al. 2009).

Lulworthiales

Lulworthia fucicola was described by Sutherland (1916) on *Fucus* in Dorset, UK, and was included in the Halosphaeriaceae. Together with *Lindra*, they differed from other genera in the Halosphaeriaceae in having filamentous ascospores. Spatafora et al. (1998) sequenced the 18S and 28S of the rRNA genes and discovered that *Lindra marinera*, *Lindra thalassiae* and *Lulworthia grandispora* are phylogenetically unrelated to the Halosphaeriaceae. Lulworthiales was introduced to accommodate these two genera (Kohlmeyer et al. 2000).

Campbell et al. (2005) investigated the phylogeny of a number of *Lulworthia* and *Lindra* species and uncovered the polyphyly of both genera. *Lulworthia crassa* was transferred to *Kohlmeyeriella* and *L. marinera* was reduced to synonymy with *L. thalassiae*. New genera, *Lulwoana* and *Lulwoidea*, were introduced to accommodate *Lulworthia lignoarenaria* and *Lulworthia uniseptata*, respectively. Unfortunately, both *Lindra* and *Lulworthia* are still not monophyletic genera after these taxonomic revisions. Part of the reason lies to the fact that the type species *Lindra inflata* has not been sequenced and *L. fucicola* has never been collected on *Fucus versiculosus* since Sutherland (1916). Campbell et al. (2005) neotypified *L. fucicola* with a collection on wood from Chile but significant morphological differences are discernible between the two types, including the carbonaceous, neckless ascomata on collections on *Fucus*.

Other marine genera have also been found to be related to the order including *Spathulospora* (Inderbitzin et al. 2004), *Kohlmeyeriella* (Campbell et al. 2002), *Haloguignardia* (Inderbitzin et al. 2004) and *Rostrupiella* (Koch et al. 2007). Many marine anamorphic species also belong to this order: *Cumulospora marina, Matsusporium tropicale, Hydea pygmea, Zalerion maritima, Moromyces varius, Halazoon melhae, H. fuscus, Moleospora maritima, Anguillospora marina* and *Orbimyces spectabilis* (Abdel-Wahab et al. 2010).

Torpedospora/Bertia/Melanospora (TBM) clade

Swampomyces is characterized by immersed ascomata with a clypeus, cylindrical asci with an apical ring and hyaline ellipsoidal ascospores without appendages and has been referred to Phyllachorales (Kohlmeyer and Volkmann-Kohlmeyer 1987b) or Ascomycota *incertae sedis* (Read et al. 1995). *Torpedospora* has dark-colored ascomata, deliquescing asci, paraphyses and ascospores with radiating appendages at one or both ends (Kohlmeyer and Kohlmeyer 1979).

Torpedospora was tentatively placed in the Halosphaeriaceae but was transferred to Sphaeriales *incertae sedis* (Kohlmeyer and Kohlmeyer 1979). Sakayaroj et al. (2005a) investigated the phylogeny of *Swampomyces* and *Torpedospora* and showed that these taxa do not group with any of the known orders in the Sordariomycetes, but instead form an independent lineage of marine Ascomycota in the Hypocreomycetidae. Schoch et al. (2006) confirmed the placement of these taxa, although *Swampomyces* species split into two clades. *Juncigena adarca, Etheirophora blepharospora* and *Etheirophora unijubata* also form part of this clade. *Juncigena adarca* is an obligate marine fungus on *Juncus roemerianus* with fuscous ascomata, pseudoparaphyses, fusiform to cylindrical asci and three-septate ascospores and was placed to Magnaporthaceae (Kohlmeyer et al. 1997). *Etheirophora* species have dark-colored ascomata, cylindrical asci and ascospores with bristle-like appendages (Kohlmeyer and Volkmann-Kohlmeyer 1989).

Taxa in this clade have been described from marine habitats but morphologically have little in common.

Concluding remarks

Sequence analysis of nuclear ribosomal genes has greatly advanced our knowledge of the phylogeny of a number of marine fungi, especially genera in the Halosphaeriaceae. Such studies have clarified the marine origin of the Halosphaeriaceae, resolved the phylogenetic relationships between taxa and consolidated the higher classification of a number of marine fungi that were previously of unknown taxonomic affinity. Significant DNA sequence data, especially the nuclear rRNA genes, have been accumulated for marine fungi and made available in NCBI, which will enable future study of marine fungal diversity by the direct extraction, amplification and sequencing of DNA from marine substrata.

In the Sordariomycetes, marine taxa so far sequenced have been referred to ten different clades corresponding to nine orders and one unnamed clade, but some of these clades are only represented by a few marine species. The taxonomic ambiguity in the Sordariomycetes lies within the Hypocreomycetidae, where the genera *Etheirophora, Juncigena, Swampomyces* and *Torpedospora* have not been referred to any order and sequencing of extra genes may provide support for the creation of a new order for these taxa. In addition, many marine taxa in the Sordariomycetes remain to be sequenced to confirm their current morphology-based taxonomic placement, but many of them have not been collected since their original description, e.g., *Argentinomyces* and *Trailia*.

The Halosphaeriaceae contains most marine Ascomycota and generic delineation within the family has been problematic. The partial nuclear 28S rRNA gene has revealed significant divergence in pairwise distance in the current circumscription of genera and species in the family and this may be a supplementary criterion for the future delineation of taxa, in addition to morphology and phylogeny. Delineation of species has been problematic in *Lulworthia*, and the key identifiable character is the length of the ascospores. Due to the overlapping of ascospore length between different species, *Lulworthia* collections – especially those from the mangrove environment – have been difficult to identify. Pairwise distance of sequences of different *Lulworthia* species can be calculated and used to delimit species.

Although at least ten ordinal lineages of fungi in the Sordariomycetes have been proven to include marine taxa, many questions remain unanswered. When did fungi migrate into the marine environment? How did they invade the sea? Answers to these questions will require an accurate estimation of divergence dates at nodes leading to various marine fungal taxa based on nucleic acid variation and the fungal fossil record, in particular the node leading to the Microascales, i.e., the time when Halosphaeriaceae separated from the Microascaceae. However, fossil records of fungi are rare and, when available, proper identification of the fossilized sample is important in order to give an accurate calibration of divergence time (Taylor and Berbee 2006).

Acknowledgements

K.L. Pang would like to thank National Science Council of Taiwan for a research grant (NSC 98–2621-B-019–002-MY3) for providing financial support.

References

Abdel-Wahab MA, Nagahama T. *Gesasha* (Halosphaeriales, Ascomycota), a new genus with three new species from the Gesashi mangroves in Japan. Nova Hedw. 2011;92:497–512.

Abdel-Wahab MA, Nagahama, T, Abdel-Aziz FA. Two new *Corollospora* species and one new anamorph based on morphological and molecular data. Mycoscience 2009;50:147–155.

Abdel-Wahab MA, Pang KL, Nagahama T, Abdel-Aziz FA, Jones EBG. Phylogenetic evaluation of anamorphic species of *Cirrenalia* and *Cumulospora* with the description of eight new genera and four new species. Mycol Prog. 2010;9:537–558.

Alexopoulos CJ, Mims CW, Blackwell M. Introductory mycology. 4th ed. New York: John Wiley & Sons, 1996.

Barr ME. Prodromus to nonlichenized, pyrenomycetous members of class Hymenoascomycetes. Mycotaxon 1990;39:43–184.

Boonyuen N, Chuaseeharonnachai C, Suetrong S, Sri-indrasutdhi V, Sivichai S, Pang KL, Jones EBG. Savoryellales (Hypocreomycetidae, Sordariomycetes): a novel lineage of aquatic ascomycetes inferred from multiple-gene phylogenies of the genera *Ascotaiwania, Ascothailandia*, and *Savoryella*. Mycologia 2011;103:1351–1371.

Cai L, Jeewon R, Hyde KD. Molecular Systematics of *Zopfiella* and allied genera: evidence from multi-gene sequence analyses. Mycol Res. 2006;110:359–368.

Campbell J, Shearer CA, Mitchell JI, Eaton RA. *Corollospora* revisited: a molecular approach. In: Hyde KD, ed. Fungi in marine environments. Hong Kong: Hong Kong University Press, 2002;15–33.

Campbell J, Anderson JL, Shearer CA. Systematics of *Halosarpheia* based on morphological and molecular data. Mycologia 2003;95:530–552.

Campbell J, Volkmann-Kohlmeyer B, Gräfenhan T, Spatafora JW, Kohlmeyer J. A reevaluation of Lulworthiales: relationships based on 18S and 28S rDNA. Mycol Res. 2005;109:556–568.

Campbell J, Inderbitzin P, Kohlmeyer J, Volkmann-Kohlmeyer B. Koralionastetales, a new order of marine Ascomycota in the Sordariomycetes. Mycol Res. 2009;113:373–380.

Cannon P. The newly recognized family Magnaporthaceae and its relationships. Syst Ascomycet. 1994;13:25–42.

Eriksson OE, Hawksworth DL. An alphabetical list of the generic names of ascomycetes. Syst Ascomycet. 1986;5:3–111.

Eriksson OE, Hawksworth DL. Notes on ascomycete systematics. Nos 225–463. Syst Ascomycet. 1987;6:111–165.

Hibbett DS, Binder M, Bischoff JF, Blackwell M, Cannon PF, Eriksson OE, Huhndorf S, James T, Kirk PM, Lücking R, Lumbsch HT, Lutzoni F, Matheny PB, McLaughlin DJ, Powell MJ, Redhead S, Schoch CL, Spatafora JW, Stalpers JA, Vilgalys R, Aime MC, Aptroot A, Bauer R, Begerow D, Benny GL, Castlebury LA, Crous PW, Dai Y-C, Gams W, Geiser DM, Griffith GW, Gueidan C, Hawksworth DL, Hestmark G, Hosaka K, Humber RA, Hyde KD, Ironside JE, Kõljalg U, Kurtzman CP, Larsson K-H, Lichtwardt R, Longcore J, Miadlikowska J, Miller A, Moncalvo J-M, Mozley-Standridge S, Oberwinkler F, Parmasto E, Reeb V, Rogers JD, Roux C, Ryvarden L, Sampaio JP, Schüßler A, Sugiyama J, Thorn RG, Tibell L, Untereiner WA, Walker C,

Wang Z, Weir A, Weiss M, White MM, Winka K, Yao YJ, Zhang N. A higher-level phylogenetic classification of the Fungi. Mycol Res. 2007;111:509–547.

Hyde KD, Jones EBG. Marine fungi from Seychelles. II. *Lanspora coronata* gen. *et* sp. nov. from driftwood. Can J Bot. 1986;64:1581–1585.

Inderbitzin P, Lim SR, Volkmann-Kohlmeyer B, Kohlmeyer J. The phylogenetic position of *Spathulospora* based on DNA sequences from dried herbarium material. Mycol Res. 2004;108:737–748.

Jones EBG. Ultrastructure and taxonomy of the aquatic ascomycetous order Halosphaeriales. Can J Bot. 1995;73(suppl 1):S790–S801.

Jones EBG. Are there more marine fungi to be described? Bot Mar. 2011;54:391–402.

Jones EBG, Eaton RA. *Savoryella lignicola* gen. *et* sp. nov. from water-cooling towers. Trans Br Mycol Soc. 1969;52:161–174.

Jones EBG, Hyde KD. Taxonomic studies on *Savoryella* Jones *et* Eaton (Ascomycotina). Bot Mar. 1992;35:83–91.

Jones EBG, Sakayaroj J, Suetrong S, Somrithipol S, Pang KL. Classification of marine Ascomycota, anamorphic taxa and Basidiomycota. Fungal Divers. 2009;35:1–187.

Klaysuban A, Sakayaroj J, Jones EBG. Morphological characterization and molecular phylogeny of a marine fungus *Pedumispora rhizophorae*. In: Abstracts, Asian Mycological Congress 2011 & the 12th International Marine and Freshwater Mycology Symposium, 7–11 August 2011, Incheon, Korea: University of Incheon Convention Center, 2011:488.

Koch J, Pang KL, Jones EBG. *Rostrupiella danica* gen. *et* sp. nov., a *Lulworthia*-like marine lignicolous species from Denmark. Bot Mar. 2007;50:294–301.

Kohlmeyer J. Revision of algicolous *Zigonella* spp. and description of *Pontogenia* gen. nov. (Ascomycetes). Botanische Jahrbücher für Systematik. 1975;96:200–211.

Kohlmeyer J. Taxonomic studies of the marine Ascomycotina. In: Moss ST, ed. The biology of marine fungi. Cambridge: Cambridge University Press, 1986:234–257.

Kohlmeyer J, Kohlmeyer E. Marine mycology – the higher fungi. New York: Academic Press, 1979.

Kohlmeyer J, Volkmann-Kohlmeyer B. Koralionastetaceae fam. nov. (Ascomycetes) from coral rocks. Mycologia 1987a;79:764–778.

Kohlmeyer J, Volkmann-Kohlmeyer B. Marine fungi from Belize with a description of two new genera of Ascomycetes. Bot Mar. 1987b;30:195–204.

Kohlmeyer J, Volkmann-Kohlmeyer B. Hawaiian marine fungi, including two new genera of Ascomycotina. Mycol Res. 1989;92:410–421.

Kohlmeyer J, Volkmann-Kohlmeyer B, Eriksson OE. Fungi on *Juncus roemerianus* 9. New obligate and facultative marine Ascomycotina. Bot Mar. 1997;40:291–300.

Kohlmeyer J, Spatafora J, Volkmann-Kohlmeyer B. Lulworthiales, a new order of marine Ascomycota. Mycologia 2000;92:453–458.

Kong RYC, Chan JYC, Mitchell JI, Vrijmoed LLP, Jones EBG. Relationships of *Halosarpheia*, *Lignincola* and *Nais* inferred from partial 18S rDNA. Mycol Res. 2000;104:336–343.

Pang KL, Vrijmoed LLP, Kong RYC, Jones EBG. *Lignincola* and *Nais*, polyphyletic genera of the Halosphaeriales (Ascomycota). Mycol Prog. 2003a;2:29–36.

Pang KL, Vrijmoed LLP, Kong RYC, Jones EBG. Polyphyly of *Halosarpheia* (Halosphaeriales, Ascomycota): implications on the use of unfurling ascospore appendage as a systematic character. Nova Hedw. 2003b;77:1–18.

Pang KL, Jones EBG, Vrijmoed LLP, Vikineswary S. *Okeanomyces*, a new genus to accommodate *Halosphaeria cucullata* (Halosphaeriales, Ascomycota). Bot J Linn Soc. 2004;146:223–229.

Pang KL, Jones EBG, Vrijmoed LLP. Autecology of *Antennospora* (Ascomycota, Fungi) and its phylogeny. Raffles Bull Zool. 2008;19:1–10.

Pang KL, Alias SA, Chiang MWL, Vrijmoed LLP, Jones EBG. *Sedecimiella taiwanensis* gen. *et* sp. nov., a marine mangrove fungus in the Hypocreales (Hypocreomycetidae, Ascomycota). Bot Mar. 2010;53:493–498.

Pinnoi A, Phongpaichit P, Jeewon R, Tang AMC, Hyde KD, Jones EBG. Phylogenetic relationships of *Astrocystis eleiodoxae* sp. nov. (Xylariaceae). Mycosphere 2010;1:1–9.

Read SJ, Jones EBG, Moss, ST. Taxonomic studies of marine Ascomycotina: ultrastructure of the asci, ascospores and appendages of *Savoryella* species. Can J Bot. 1993;71:273–283.

Read SJ, Jones EBG, Moss ST, Hyde KD. Ultrastructure of asci and ascospore of two mangrove fungi: *Swampomyces armeniacus* and *Marinosphaera mangrovei*. Mycol Res. 1995;12:1465–1471.

Rossman AY, McKemy JM, Pardo-Schultheiss RA, Schroers HJ. Molecular studies of the Bionectriaceae using large subunit rDNA sequences. Mycologia 2001;93:100–110.

Sakayaroj J, Pang KL, Jones EBG, Vrijmoed LLP, Abdel-Wahab MA. A systematic reassessment of marine ascomycetes *Swampomyces* and *Torpedospora*. Bot Mar. 2005a;48:395–406.

Sakayaroj J, Pang KL, Phongpaichit S, Jones EBG. A phylogenetic study of the genus *Haligena* (Halosphaeriales, Ascomycota). Mycologia 2005b;97:804–811.

Sakayaroj J, Pang KL, Jones EBG. Multi-gene phylogeny of the Halosphaeriaceae: its ordinal status, relationships between genera and morphological character evolution. Fungal Divers. 2011;46:87–109.

Samuels GJ, Seifert KA. Taxonomic implications of variation among hypocrealean anamorphs. In: Sugiyama J, ed. Pleomorphic fungi: the diversity and its taxonomic implications. New York: Elsevier Science Ltd, 1987:29–56.

Schoch CL, Sung GH, Volkmann-Kohlmeyer B, Kohlmeyer J, Spatafora JW. Marine fungal lineages in the Hypocreomycetidae. Mycol Res. 2006;111:154–162.

Schoch CL, Sung GH, Lopez-Giraldez F, Townsend JP, Miadlikowska J, Hofstetter V, Robbertse B, Matheny BP, Kauff F, Wang Z, Gueidan C, Andrie RM, Trippe K, Ciufetti L, Wynns A, Fraker E, Hodkinson B, Bonito G, Yahr R, Groenewald JZ, Arzanlou M, De Hoog GS, Crous P, Hewitt D, Pfister DH, Peterson K, Gryzenhout M, Wingfield MJ, Aptroot A, Suh SO, Blackwell M, Hillis DM, Griffith GW, Castlebury LA, Rossman AY, Lumbsch HT, Lücking R, Büdel B, Rauhut A, Diederich P, Ertz D, Geiser DM, Hosaka K, Inderbitzin P, Kohlmeyer J, Volkmann-Kohlmeyer B, Mostert L, O'Donnel K, Sipman H, Rogers JD, Shoemaker RA, Sugiyama J, Summerbell RC, Untereiner W, Johnston P, Stenroos S, Zuccaro A, Dyer P, Crittenden P, Cole MS, Hansen K, Trappe JM, Lutzoni F, Spatafora JW. The Ascomycota tree of life: A phylum-wide phylogeny clarifies the origin and evolution of fundamental reproductive and ecological traits. Syst Biol. 2009;58:224–239.

Smith GJD, Liew ECY, Hyde KD. The Xylariales: a monophyletic order containing 7 families. Fungal Divers. 2003;13:175–208.

Spatafora JW, Blackwell M. The polyphyletic origins of ophiostomatoid fungi. Mycol Res. 1994;98:1–9.

Spatafora JW, Volkmann-Kohlmeyer B, Kohlmeyer J. Independent terrestrial origins of the Halosphaeriales (marine Ascomycota). Am J Bot. 1998;85:1569–1580.

Spatafora JW, Sung GH, Johnson D, Hesse C, O'Rourke B, Serdani M, Spotts R, Lutzoni F, Hofstetter V, Miadlikowska J, Reeb V, Gueidan C, Fraker E, Lumbsch T, Lucking R, Schmitt I, Hosaka K, Aptroot A, Roux C, Miller AN, Geiser DM, Hafellner J, Hestmark G, Arnold AE, Budel B, Rauhut A, Hewitt D, Untereiner WA, Cole MS, Scheidegger C, Schultz M, Sipman H, Schoch CL. A five-gene phylogeny of Pezizomycotina. Mycologia 2006;98:1018–1028.

Sutherland GK. Additional notes on marine pyrenomycetes. Trans Br Mycol Soc. 1916;5:257–263.

Taylor JW, Berbee ML. Dating divergences in the fungal tree of life: review and new analyses. Mycologia 2006;98:838–849.

Thongkantha S, Jeewon R, Vijaykrishna D, Lumyong S, McKenzie EHC, Hyde KD. Molecular phylogeny of Magnaporthaceae (Sordariomycetes) with a new species *Ophioceras chinadaoense* from *Dracaena loureiroi* in Thailand. Fungal Divers. 2009;34:155–171.

Vicente F, Basilio A, Platas G, Collado J, Bills GF, González Del Val A, Martín J, Tormo JR, Harris GH, Zink DL, Justice M, Kahn JN, Peláez F. Distribution of the antifungal agents sordarins across filamentous fungi. Mycol Res. 2009;113:754–770.

Vijaykrishna D, Jeewon R, Hyde KD. Molecular taxonomy, origins and evolution of freshwater ascomycetes. Fungal Divers. 2006;23:351–390.

Zhang N, Castlebury LA, Miller AN, Huhndorf S, Schoch CL, Seifert K, Rossman AY, Rogers JD, Kohlmeyer J, Volkmann-Kohlmeyer B, Sung GH. Sordariomycetes systematics: an overview of the systematics of the Sordariomycetes based on a four-gene phylogeny. Mycologia 2006;8:1076–1087.

Zuccaro A, Schulz B, Mitchell JI. Molecular detection of ascomycetes associated with *Fucus serratus*. Mycol Res. 2003;107:1451–1466.

4 Basidiomycota

E.B. Gareth Jones and Jack W. Fell

Few filamentous basidiomycetes have been reported from marine and estuarine habitats, an ecological but taxonomically diverse group (Table 4.1). Morphologically basidiomycetous fungi can be categorized into three groups:

(i) single-celled yeasts free floating in the sea or growing attached to or within their hosts/substrates (Vogel et al. 2007; Kutty and Philip 2008; Boekhout et al. 2011);
(ii) filamentous species that may form basidiomes visible to the naked eye and grow predominantly on woody substrates (Jones and Choeyklin 2008); and
(iii) those isolated as endophytes (Sakayaroj et al. 2010).

Others include marine-derived fungi from various substrates (Jones 2011), or are known only as sequences recovered from marine sediments in the deep sea, and hydrothermal vents (see Chapter 19).

Marine filamentous basidiomycetes can be found on a variety of substrata: feathers in sand (e.g., *Nia epidermoidea*), culms of salt marsh grasses (*Nia globospora*), seaweeds (*Mycaureola dilseae*), wood (e.g., *Digitatispora lignicola, D. marina, Nia vibrissa*), and mangrove wood (*Calathella mangrovei, Halocyphina villosa, Physalacria maipoensis*) (Table 4.1, Figures 4.1 and 4.2). Many species remain to be described, as illustrated by the observations of Statzell-Tallman et al. (2008) who reported 55 and 58 species of ascomycetous and basidiomycetous yeasts, respectively, from three mangrove habitats, 50% of which are un-described. Gao et al. (2010) employed DGGE to sample planktonic fungal communities and identified 124 clones including 46 fungal species that belonged to the Ascomycota (four species) and the majority belonging to the Basidiomycota (42 species). Of the 42 basidiomycetes, 39 were likely to be new fungal phylotypes. Ryvarden (2010) also reported on 20 wood-inhabiting basidiomycetes collected on coniferous driftwood from the coast of Finnmark, northernmost Norway, most of which were corticoid (16). Terrestrial basidiomycetes can be active in estuarine habitats, as shown by the occurrence of butt-rot of *Xylocarpus granatum* in Thai mangroves (Figures 4.4 and 4.5). The decay of the *Xylocarpus* trunks extends to the roots system, which is regularly subject to inundation by brackish water. Mwangi (2001) also reported butt and heart rot of *Rhizophora apiculata* in Kenya, and identified *Phellinus pachyphloeus* and *Ph. rimosus* as the causative species.

Currently there is no definitive estimate of the number of species of marine basidiomycetes: while some species can be shown to be endemic to the marine environment, others have only been isolated once or twice. A compilation of the above accounts for 67 taxa, including those known from occasional collections or isolations, environmental sequences and endophytic fungal isolates. The number may be even higher, however, with more intensive studies. This observation is supported by molecular data (Bass et al. 2007; Statzell-Tallman et al. 2008; Nagano et al. 2010; Sakayaroj et al. 2010; Preedanon et al. 2011).

Table 4.1 Selective examples of basidiomycetes isolated from marine habitats.

Classification	Species	Substratum	Reference
Group 1: Filamentous fungi			
Agaricales: *Nia* clade	*Calathella mangrovei*	Mangrove wood, tropical	Jones and Agerer (1992)
Agaricales: *Nia* clade	*Halocyphina villosa*	Mangrove wood, tropical	Kohlmeyer and Kohlmeyer (1965)
Agaricales: *Nia* clade	*Nia vibrissa*	Wood, cosmopolitan	Moore and Meyers (1959)
Agaricales: *Nia* clade	*Nia globospora*	*Spartina* culms, temperate	Barata et al. (1997)
Agaricales: *Nia* clade	*Nia epidermoidea*	Feathers in contact with sand, temperate	Rosello et al. (1993)
Agaricales: Physalacriaceae	*Mycaureola dilseae*	Red alga *Dilsea carnosa*	Porter and Farnham (1986)
Agaricales: Physalacriaceae	*Physalacria maipoensis*	Mangrove substrates, temperate	Inderbitzin and Desjardin (1999)
Atheliales: *Digitatispora* clade	*Digitatispora lignicola*	Wood, temperate	Jones (1986)
Atheliales: *Digitatispora* clade	*Digitatispora marina*	Wood, temperate	Doguet (1962)
Russulales: Peniophoraceae	*Haloaleurodiscus mangrovei*	Mangrove wood, tropical	Maekawa et al. (2005)
Urocystidiales	*Flamingomyces ruppiae*	On *Ruppia maritima* sea grass	Bauer et al. (2007)
Ustilaginales	*Parvulago marina*	On *Eleocharis parvula* sedge	Bauer et al. (2007)
Group 2: Yeasts			
Cystobasidiales	*Occultifur externus*	Mangrove, sea water	Fell et al. (2011); Sampaio (2011)
Cystofilobasidiales	*Cystofilobasidium bisporidii*	Sea water	Fell et al. (1973, 2001)
Cystofilobasidiales	*Cystofilobasidium capitatum*	Sea water	Fell et al. (1973, 2001)
Cystofilobasidiales	*Cystofilobasidium infirmominiatum*	Sea water	Fell et al. (1973, 2001)
Tremellales	*Kwoniella mangroviensis*	Mangroves	Statzell-Tallman et al. (2008)
Tremellomycetes *incertae sedis*	*Cryptococcus marinus*	Sea water, Great Barrier Reef in Australia	van Uden and ZoBell (1962)
Filobasidiales	*Filobasidium* sp.	Arabian Sea	Jebaraj et al. (2010)
Sporidiobolales	*Rhodosporidium babjevae*	Mangrove	Fell et al. (2011)

Sporidiobolales	*Rhodosporidium fluviale*	Brackish water	Fell et al. (1988)
Sporidiobolales	*Rhodosporidium diobovatum*	Sea water, mangrove detritus	Fell et al. (1973, 2011)
Sporidiobolales	*Rhodosporidium paludigenum*	Mangrove *Juncus roemerianus* marshes	Fell et al. (1973, 2001)
Sporidiobolales	*Rhodosporidium sphaerocarpum*	Sea water	Fell et al. (1973, 2001)
Agaricostilbales	*Sterigmatomyces halophilus*	Sea water	Fell et al. (2001), Kurtzman and Fell (2006)
Erythrobasidiales	*Sakaguchia dacryoidea*	Sea water	Fell et al. (1973, 2001), Kurtzman and Fell (2006)
Tremellales: *Bulleromyces* clade	*Cryptococcus mangaliensis*	Sea grass	Fell et al. (2011)
Erythrobasidiales	*Rhodotorula cladiensis*	Saw grass marsh	Fell et al. (2011)
Erythrobasidiales	*Rhodotorula marina*	Mangroves, seawater	Fell et al. (2011), Fonseca et al. (2011)
Sporidiobolales	*Rhodotorula evergladiensis*	Sea grass	Fell et al. (2011)
Georgefischeriales	*Tilletiopsis* sp.	Deep sea	Singh et al. (2010)
Ustilaginales	*Pseudozyma abaconensis*	Coral reef waters	Statzell-Tallman et al. (2010)
Microstromatales	*Jaminaea lanaiensis*	Driftwood	Mahdi et al. (2009), Wei et al. (2011)
Microstromatales	*Sympodiomycopsis kandeliae*	Mangrove flower	Wei et al. (2011)
Group 3 Endophytes, Phylotypes			
Russulales	*Peniophora* sp. (KH302)	*Enhalus acoroides*, sea grass	Sakayaroj et al. (2010)
Trichosporonales	*Cryptococcus curvatus*	Sediments	Takishita et al. (2006)
Cystobasidiales	*Occultifur externus*	Mangrove	Fell et al. (2011)
Erythrobasidiales	*Erythrobasidium hasegawianum*	Mangrove	Fell et al. (2011)
Malasseziales	*Malassezia*	Deep sea	Edgcomb et al. (2010)
Tremellales	*Habbaella surugaebsis*	Deep sea	Nagahama et al. (2003)
Tremellales	*Cryptococcus* spp.	Hydrothermal vent fauna	Le Calvez et al. (2009), Burgaud et al. (2010)
Wallemiales	*Wallemia sebi*	Deep sea	Singh et al. (2011a, 2011b)

Table continued on next page.

Table 4.1 (Continued)

Classification	Species	Substratum	Reference
Agaricales	*Schizophyllum commune*	Sea water, marine sponges	Gao et al. (2008)
Corticiales	*Phlebia* sp.	Sea water, marine sponges	Gao et al. (2008)
Polyporales	Unidentified strain	Sea water, marine sponges	Gao et al. (2008)
Group 4 Species collected or isolated from marine habitats, but not considered endemic			
Ustilaginales	*Pseudozyma aphidis*, *P. hubeiensis*	Mangrove	Fell et al. (2011)
Sporidiobolales	*Sporidiobolus pararoseus*, *Sporobolomyces blumeae*, *S. carnicolor, S. japonicus*	Mangroves	Fell et al. (2011)
Tremellales	*Cryptococcus flavescens*, *C. laurentii, C. liquefaciens*, *C. magnus, C. tephrensis*, *C. victoriae*	Mangroves	Fell et al. (2011)
Erythrobasidiales	*Erythrobasidium hasegawianum*	Mangrove	Fell et al. (2011)
Hymenochaetales	*Fulviformes* spp.	But-rot of *Xylocarpus granatum*	Preedanon et al. (2011)
Polyporales	*Grammothele fuligo*	*Nypa* palm trunk	Jones and Choeyklin (2008)
Agaricales	*Schizophyllum commune*	*Nypa* fronds, cosmopolitan	Unpublished data
Agaricales	*Henningsomyces* sp.	*Nypa* fronds, tropical	Unpublished data
Agaricales	*Coprinus* sp.	*Nypa* fronds	Unpublished data
Polyporales	*Hyphoderma* sp.	*Nypa* fronds	Unpublished data
Agaricales	*Marasmiellus* sp.	*Nypa* fronds	Unpublished data
Agaricales	*Psathyrella* sp.	*Nypa* fronds	Unpublished data

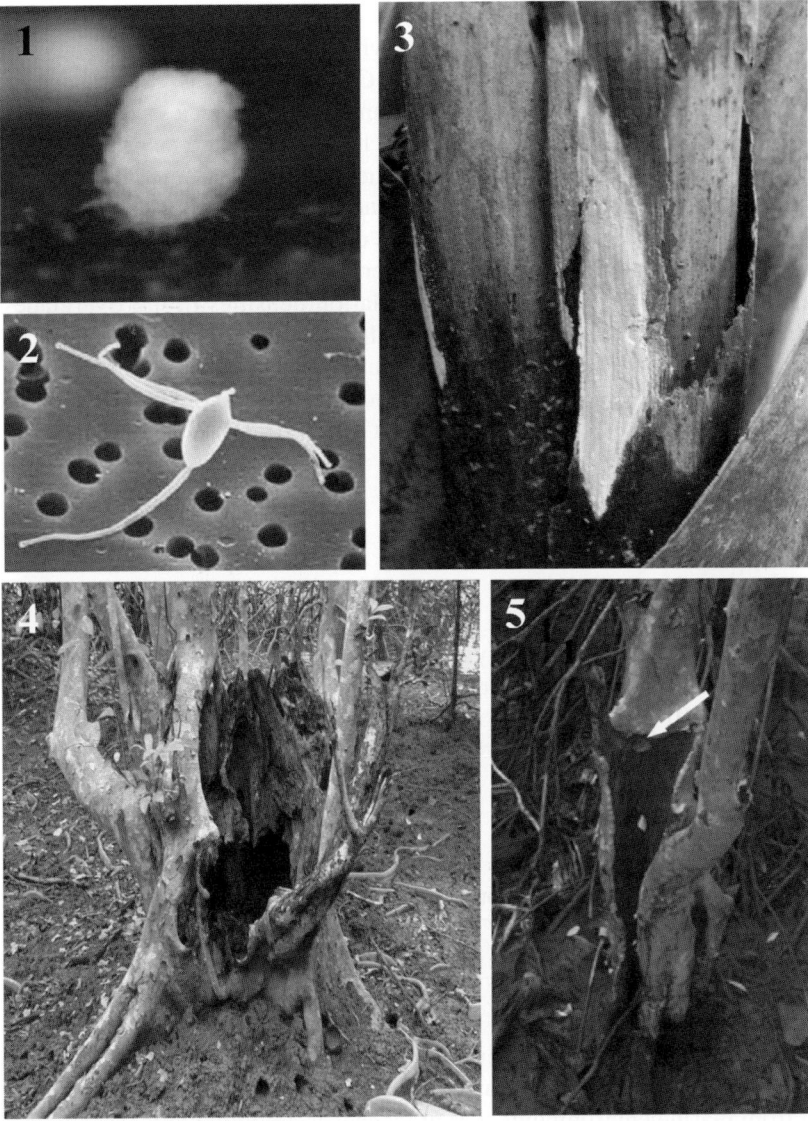

Figures 4.1–4.5 Figure 4.1. Basidiome of *Calathella mangrovei* on frond of *Nypa* palm. Figure 4.2 Appendaged basidiospore of *Nia vibrissa*. Figure 4.3 Immature fruit body of *Grammothele fuligo* on *Nypa* frond in brackish water. Figure 4.4 Butt-rot of the mangrove tree *Xylocarpus granatum*. Figure 4.5 Butt-rot with the basidiocarp shown by an arrow.

Numerous basidiomycetous yeasts have been recovered from oceanic and mangrove waters, but how many are indigenous remains to be determined. Fell et al. (2011) isolated yeasts from three tidal stations in the Shark River Slough, Florida Everglades, which yielded 44 basidiomycetes, of which eight and 17 were from sites with salinities ranging from 0 to 30‰ (mangrove site 5) and 12–33‰ (mangrove site 6), respectively.

Many of these were only collected once, e.g., *Auriculibuller fuscus* and *Erythrobasidium hasegawianum*, from the mangrove stations, while those in the sawgrass and mangrove stations were more frequently isolated, e.g., *Cryptococcus laurentii, Kwoniella mangroviensis* and *Rhodosporidium babjevae*. Many of these species were described from terrestrial habitats but are repeatedly isolated from the marine environment. However, others are known only from mangroves or other marine environments, e.g., *Kwoniella mangroviensis* and *Rhodosporidium paludigenum*. Fell et al. (2011) concluded that the role of yeasts in the Everglades ecosystem "is open to speculation". Some species may be associated with the mangrove plants, whereas other species may degrade organic materials in the organically-rich water. Population densities in the Everglades waters ranged from 70 to 1300 cells/liter (Statzell-Tallman et al. 2008). Although we know that the yeasts are present in high numbers and taxonomically diverse, their roles require in-depth study. For a detailed account of the morphology and phylogeny of basidiomycetous yeasts, the reader is referred to Boekhout et al. (2011).

Agaricomycotina

Tremellomycetes

Cystofilobasidiales

Nine genera are included in the Cystofilobasidiales, with marine representatives in *Cystofilobasidium* (*C. bisporidii, C. capitatum* and *C. infirmominiatum*; Fell et al. 1973; Boekhout et al. 2011). They are mostly pink-colored yeasts with teliospores that are holobasidia and hyphal septa with dolipores lacking parenthesomes (Fell et al. 2000; Boekhout et al. 2011).

Filobasidiales

A well supported monophyletic group with the genera *Carcinomyces, Christiansenia, Syzygospora* (Carcinomycetaceae), *Filobasidium* and some *Cryptococcus* species (Sampaio 2004). This is an order accepted in the higher-level phylogenetic classification of the fungi by Hibbett et al. (2007).

Most are terrestrial species but a *Filobasidium* sp. was reported as one of the most common phylotypes from hydrothermal vents by Le Calvez et al. (2009). Occasionally, species of *Cryptococcus* that are members of this order, such as *C. albidus*, are isolated from marine environments. *Cryptococcus* is a polyphyletic, asexual genus and members are found within all five major lineages of the order (Fonseca et al. 2011). Consequently, a species labeled *Cryptococcus* does not refer to a specific phylogenetic lineage.

Tremellales

This order is characterized by dimorphic life cycles, having sacculate membranous caps to the dolipore septum, and tremelloid basidia consisted of four cells with longitudinal to oblique walls. Swann and Taylor (1995) have confirmed that the order is an early diverging lineage within the hymenomycetes.

basidiospores discharged actively per basidium, cylindrical to ellipsoidal and in an apical cluster (Cannon and Kirk 2007).

The yeast genus *Tilletiopsis* is polyphyletic, comprising five lineages, with some species grouping in the Entylomatales and Microstromatales (Sampaio 2004). The genus includes ballistoconidia-producing anamorphs, with most species being filamentous (Sampaio 2004). Singh et al. (2010) isolated and identified 28 fungi (16 filamentous fungi and 12 yeasts) from deep-sea sediments from the Central Indian Basin. These species included two isolates of a *Tilletiopsis* sp. that grouped with 100% bootstrap support with *T. albescens*. The *Tilletiopsis* sp. was among the three species recovered from the deepest part of the sediment core (Singh et al. 2010).

Microstromatales

Two marine species are referred to this order: *Sympodiomycopsis lanaiensis* and *S. kandeliae*. *Sympodiomycopsis lanaiensis* was isolated from driftwood collected on a beach at Lana'i, Hawai'I (Mahdi et al. 2009); however, Wei et al. (2011) transferred the fungus to *Jaminaea* based on sequence data. The species has spheroid to oval cells and reproduces by polar budding. Conidia form at the internodes of hyphal strands, while ballistoconidia and sexual structures have not been observed (Mahdi et al. 2009). *Sympodiomycopsis kandeliae* was isolated from flowers of *Kandelia obovata* in a mangrove forest in Taiwan (Wei et al. 2011).

Malasseziales

Members of this order are lipophilic yeasts with thick cell walls, that are multilamellate with a unique substructure of electron-opaque helicoid band, anamorphic with no known sexual phase, and primarily pathogens of the skin of warm-blooded animals, causing pityriasis versicolor for example (Weiss et al. 2004).

This order comprises one genus (*Malassezia*) with 13 species (Guého-Kellermann et al. 2011). In the marine environment *Malassezia* spp. have been identified from DNA from seawater and sediments and from the sponges *Suberites zeteki* and *Mycale armata* (Gao et al. 2008; Singh et al. 2011a, 2011b). Singh et al. (2011a) investigated the fungal diversity of deep-sea sediments by amplification of environmental DNA with fungal or eukaryote-specific primer sets. Eight operational taxonomic units were recovered spreading across seven different classes of the Ascomycota and Basidiomycota, the majority belonging to the latter. Three operational taxonomic units (09, 21 and 27) were referable to *Malassezia* species, and may well represent novel taxa.

Concluding remarks

Jones et al. (2009) list only 12 marine basidiomycetes (in nine genera) in their monograph on the classification of marine fungi, while basidiomycetous yeasts were not included. This monograph is similar to other texts on marine fungi as the techniques for the study differ from those for yeasts. Jones (2011), however, indicates that many basidiomycetes are still to be discovered and described.

The true domain of marine yeasts, like filamentous fungi, has been open to question. The restrictive use of the terms obligate and facultative marine fungi has also hindered progress in determining the marine mycoflora. Do all species recovered from seawater and marine habitats play an active role in the ecology of the sea? This should be the primary criterion in deciding the origin of marine species.

The use of sequence data has greatly increased our knowledge of the occurrence of marine basidiomycetes, especially from marine deep-sea sediments, hydrothermal vent environments, and anoxic sites (see Chapter 19). However, many of the sequences recovered are of human pathogens and terrestrial-like species e.g., *Malassezia* spp., *Tilletiopsis* sp. and mycoparasites of terrestrial mushrooms, *Occultifur internus*. What is the substrate/host for these taxa in the deep sea sediments? The use of pyrosequencing techniques is going to dramatically increase our knowledge of the occurrence and distribution of fungi in marine habitats and opens up new frontiers of discovery.

Acknowledgements

EBG Jones thanks University Malaya for a visiting professorship, Dr. Siti Aisyah Alias and Norlailatul Asikin binti Mohamad Nor for logistic support, and research funding from research grants (UMRG: RG057–09SUS and FRGS: FP087–2010A). Dr. Rattaket Choeyklin is thanked for Figures 4.1 and 4.3.

References

Agerer R. *Lachnella-Crinipellis, Stigmatolemina-Fistulina*: zwei verwandtschafsreihen? Z Mykol. 1978;44:51–70.
Aime MC, Patheny PB, Henk DA, Frieders EM, Nilsson RH, Piepenbring M, McLaughlin DJ, Szabo LJ, Begerow D, Sampaio JP, Bauer R, Weiß M, Oberwinkler F, Hibbett D. An overview of the higher level classification of Pucciniomycotina based on combined analysis of nuclear large and small subunit rDNA sequences. Mycologia 2006;98:896–905.
Alexopoulus CJ, Mims CW, Blackwell M. Introductory mycology. New York: Wiley, 1996.
Barata M, Basilio MC, Baptista-Ferreira JL. *Nia globospora*, a new marine gasteromycete on baits of *Spartina maritima* in Portugal. Mycol Res. 1997;101:687–690.
Bass D, Howe A, Brown N, Barton H, Demidova M, Michelle H, Li L, Sander H, Watkinson SC, Willcock S, Richards TA. Yeast forms dominate fungal diversity in the deep oceans. Proc R Soc B. 2007;274:3069–3077.
Bauer R, Begerow D, Sampaio JP, Weiß M, Oberwinkler F. The simple-septate basidiomycetes: a synopsis. Mycol Prog. 2006;5:41–66.
Bauer R, Stoll M, Piatek M, Vanky, Oberwinkler F. *Flamingomyces* and *Parvulago*, new genera of marine smut fungi (Ustilaginomycotina). Mycol Res. 2007;111:1199–1206.
Binder M, Hibbett DS. Higher-level phylogenetic relationships of Homobasidiomycetes (mushroom-forming fungi) inferred from four rDNA regions. Mol Phylogenet Evol. 2001;22:76–90.
Binder M, Hibbett DS, Molitoris HP. Phylogenetic relationships of the marine gasteromycetes *Nia vibrissa*. Mycologia 2001;93:679–688.
Binder M, Hibbett DS, Wang Z, Farnham WF. Evolutionary relationships of *Mycaureola dilsea* (Agaricales), a basidiomycetes pathogen of a subtidal Rhodophyte. Am J Bot. 2006;93:547–556.
Boekhout T, Fonseca Á, Sampaio JP, Bandoni RJ, Fell JW, Kwong-Chung. Discussion of teleomorphic and anamorphic basidiomycetous yeasts. In: Kurtzman CP, Fell JW, Boekhout T, eds. The yeasts, a taxonomic study. 5th ed. Amsterdam: Elsevier, 2011:1339–1372.

Burgaud G, Arzu D, Durand L, Cambon-Bonavita M-A, Barbier G. Marine culturable yeasts in deep-sea hydrothermal vents: species richness and association with fauna. FEMS Microbio Ecol. 2010;73:121–133.

Cannon PF, Kirk PM. Fungal families of the world. London: CABI, 2007.

Doguet G. *Digitatispora marina*, n.g., n.sp., Basidiomycéte marin. C R Hebd Séances Acad Sci. 1962;254:4336–4338.

Edgcomb VP, Beaudoin D, Gast R, Biddle JF, Teske A. Marine subsurface eukaryotes: the fungal majority. Environ Microbiol. 2010;13:172–183.

Fazzani K, Jones EBG. Spore release and dispersal in marine and brackish water fungi. Mater Organismen 1977;12:235–248.

Feldmann G. Une ustilaginale marine, paraite du *Ruppia maritima* L. Rev Gen Bot. 1959; 66:35–39.

Fell JW. Distribution of yeasts in water masses of the Southern Ocean. In: Colwell R, Morita R, eds. Effect of the ocean environment on microbial activities. Baltimore: University Park Press, 1974:510–523.

Fell JW, Statzell AC, Hunter IL, Phaff HJ. *Leucosporidium* gen. n., the heterobasidiomycetous stage of several yeasts of the genus *Candida*. Antonie Leeuwenhoek 1969;35:433–462.

Fell JW, Hunter I, Tallman AS. Marine basidiomycetous yeasts (*Rhodosporidium* spp. n.) with tetrapolar and multiple allelic bipolar mating systems. Can J Microbiol. 1973;19:643–657.

Fell JW, Kurtzman CP, Tallman AS, Buck JD. *Rhodosporidium fluviale* sp. n. a homokaryotic red yeast from estuarine water. Mycologia 1988;80:560–564.

Fell JW, Boekhout T, Fonesca A, Scorzetti G, Statzell-Tallman A. Biodiversity and systematics of basidiomycetous yeasts as determined by large-subunit rDNA D1/D2 domain sequence analysis. Int J Syst Evol Microbiol. 2000;50:1351–1371.

Fell JW, Boekhout T, Fonseca A, Sampaio JP. Basidiomycetous yeasts. In: McLaughlin DJ, McLaughlin EG, Lemke PA, eds. Mycota VII, Part B: systematics and evolution. Berlin Heidelberg, New York: Springer, 2001:3–35.

Fell JW, Statzell-Tallman A, Scorzetti G, Gutiérrez MH. Five new species of yeasts from fresh water and marine habitats in the Florida Everglades. Antonie Leeuwenhoek 2011;99:533–549.

Fonseca A, Boekhout T, Fell JW. *Cryptococcus* Vuillemin (1901). In: Kurtzman CP, Fell JW, Boekhout T, eds. The yeasts, a taxonomic study. 5th ed. Amsterdam: Elsevier, 2011:1661–1738.

Gao Z, Li BL, Zheng CC, Wang GY. Molecular detection of fungal communities in the Hawaiian marine sponges *Suberites zeteki* and *Mycale armata*. Appl Environ Microbiol. 2008;74: 6091–6101.

Gao Z, Hohnson ZI, Wang GL. Molecular characterization of the spatial diversity and novel lineages of mycoplankton in Hawaiian coastal waters. ISME J. 2010;4:111–120.

Guého-Kellermann E, Batra R, Boekhout T. *Malassezia* Baillon (1889). In: Kurtzman CP, Fell JW, Boekhout T, eds. The yeasts, a taxonomic study. 5th ed. Amsterdam: Elsevier, 2011:1807–1832.

Hibbett DS. After the gold rush, or before the flood? Evolutionary morphology of mushroom forming fungi (Agaricomycetes) in the early 21st century. Mycol Res. 2007;111:1001–1018.

Hibbett DS, Binder M. Evolution of marine mushrooms. Biol Bull. 2001;201:319–322.

Hibbett DS, Pine EM, Langer E, Langer G, Donoghue MJ. Evolution of gilled mushrooms and puffballs inferred from ribosomal DNA sequences. PNAS. 1997;94:12002–12006.

Hibbett DS, Binder M, Bischoff JF, Blackwell M, Cannon PF, Eriksson OE, Huhndorf S, James ST, Kirk PM, Lucking R, Lumbsch HT, Lutzoni F, Matheny PB, Mclaughlin DJ, Powell MJ, Redhed S, Scoch CL, Spatafora JF, Stalpers JA, Vilgalys R, Aime MC, Aptroot AA, Bauer R, Begerow D, Benny GL, Castelbury LA, Crous PW, Dai YC, Gams W, Geiser, DM, Griffith GW, Gueidan C, Hawksworth DL, Hestmark G, Hosaka K, Humber RA, Hyde KD, Ironisde JE, Koljag U, Kurtzman CP, Larsso, KH, Lichtwardt R, Longcore J, Miadlikowska J, Miller A, Moncalvo JM, Mozley-Standridge S, Oberwinkler F, Parmasto E, Reeb V, Rogers JD., Roux C, Ryvarden L,

Sampaio JP, Schßler A, Sugiyama J, Thorn RG, Tibell L, Untereiner WA, Walker C, Wang Z, Weir A, Weiss M, White MM, Winka K, Yao YJ, Zhang N. A higher-level phylogenetic classification of the fungi. Mycol Res. 2007;111:509–547.

Inderbitzin P, Desjardin DE. A new halotolerant species of *Physalacria* from Hong Kong. Mycologia 1999;91:666–668.

Jebaraj CS, Raghukumar C, Behnke A, Stoeck T. Fungal diversity in oxygen-depleted regions of the Arabian Sea revealed by targeted environmental sequencing combined with cultivation. FEMS Microb Ecol. 2010;71:399–412.

Jones EBG. *Digitatispora lignicola* sp. nov. A new marine lignicolous Basidiomycotina. Mycotaxon 1986;27:155–159.

Jones EBG. Are there more marine fungi to be described? Bot Mar. 2011;54:343–354.

Jones EBG, Agerer R. *Calathella mangrovei* sp. nov. and observations on the mangrove fungus *Halocyphina villosa*. Bot Mar. 1992;35:259–265.

Jones EBG, Choeyklin R. Ecology of marine and freshwater basidiomycetes. In: Boddy L, Frankland JC, van West P, eds. Ecology of saprotrophic Basidiomycetes. London: Elsevier, 2008:301–324.

Jones EBG, Sakayaroj J, Suetrong S, Somrithipol S, Pang KL. Classification of marine Ascomycota, anamorphic taxa and Basidiomycota. Fungal Divers. 2009;35:1–187.

Kohlmeyer J, Kohlmeyer E. New marine fungi from mangroves and trees along eroding shorelines. Nova Hedw. 1965;9:89–104.

Kurtzman CP, Fell JW. Yeast systematics and phylogeny-implications of molecular identification methods for studies in ecology. In: Carlos R, Gabor P, eds. Biodiversity and ecophysiology of yeasts. Germany: Springer, 2006:11–30.

Kutty SN, Philip R. Marine yeasts – a review. Yeast 2008;25:465–483.

Le Calvez T, Burgaud G, Mahe S, Barbier G, Vandenkoornhuyse P. Fungal diversity in deep-sea hydrothermal ecosystems. Appl Environ Microbiol. 2009;75:6415–6421.

Maekawa N, Suhara H, Kinjo K, Kondo R, Hashi Y. *Haloaleurodiscus mangrovei* gen. sp. nov. (Basidiomycota) from mangrove forests in Japan. Mycol Res. 2005;109:825–832.

Mahdi LE, Statzell-Tallman A, Fell JW, Brown MV, Donachie SP. *Sympodiomycopsis lanaensis* sp. nov., a basidiomycetous yeast (Ustilaginomycotina: Microstromatales) from marine driftwood in Hawaii. FEMS Yeast Res. 2009;8:1357–1363.

Moore RT. Taxonomic proposals for the classification of marine yeasts and other yeast-like fungi including the smuts. Bot Mar. 1980;23:361–373.

Moore RT, Meyers S. Thalassiomycetes I. Principles of delimitation of the marine mycota with a description of a new aquatically adapted Deuteromycete. Mycologia 1959;51:871–876.

Mwangi JG. (2001). A new pest causing decline of mangrove forests in Kenya. Eastern Arc Mountains Information Source. (Accessed April 10, 2012 at www.easternarc.org.)

Nagahama T, Hamamoto M, Nakase T, Takaki Y, Horikoshi K. *Cryptococcus surugaensis* sp. nov., a novel yeast species from sediment collected on the deep-sea floor of Suruga Bay. Int J Syst Evol Microbiol. 2003;53:2095–2098.

Nagano Y, Nagahama T, Hatada Y, Nunoura T, Takami H, Miyazaki J, Takai K, Horikoshi K. Fungal diversity in deep-sea sediments-the presence of novel fungal groups. Fungal Ecol. 2010;3:316–325.

Porter D, Farnham WF. *Mycaureola edulis*, a marine basidiomycetes parasite of the red alga, *Dilsea carnosa*. Trans Br Mycol Soc. 1986;87:575–582.

Preedanon S, Sakayaroj J, Suetrong S, Klaysuban A, Jones EBG. Molecular characterization of Butt Rot-associated Basidiomycetes occurring on *Xylocarpus granatum* in Thailand. In: Abstract 12th International Marine and Freshwater Mycological Symposium, Incheon, Korea, 2011. Korean Mycological Society.

Rossello MA, Descals E, Cabrer R. *Nia epidermoidea*, a new marine gasteromycete. Mycol Res. 1993;97:68–70.

Ryvarden L. Basidiomycetes on driftwood in Finnjmark, Norway. Agarica 2010;29:2–4.

Sakayaroj J, Preedanon S, Supaphon O, Jones EBG, Phongpaichit S. Phylogenetic diversity of endophyte assemblages associated with the tropical sea grass *Enhalus acoroides* in Thailand. Fungal Divers. 2010;42:27–45.

Sampaio JP. Diversity, phylogeny and classification of basdiomycetous yeasts. In: Agerer R, Piepenbring M, Blanz P, eds. Frontiers in Basidiomycete mycology. Eching: IHW-Verlag, 2004:49–80.

Sampaio JP. *Rhodotorula* Harrison (1928). In: Kurtzman CP, Fell JW, Boekhout T, eds. The yeasts, a taxonomic study. 5th ed. Amsterdam: Elsevier, 2011:1873–1928.

Singer R. The Agaricales in modern taxonomy. 4th ed. Köngstein. Germany: Koeltz Sci Books, 1986.

Singh P, Raghukumar C, Verma P, Shouche Y. Phylogenetic diversity of culturable fungi from the deep-sea sediments of the Central Indian Basin and their growth characteristics. Fungal Divers. 2010;40:89–102.

Singh P, Raghukumar C, Verma P, Shouche Y. Fungal community analysis in the deep-sea sediments of the Central Indian Basin by culture-independent approach. Microb Ecol. 2011a;61:507–517.

Singh P, Raghkumar C, Verma P, Schouche Y. Assessment of fungal diversity in deep-sea sediments by multiple primer approach. World J Microbiol Biotechnol. 2011b;28:659–667.

Spatafora JW, Volkmann-Kohlmeyer B, Kohlmeyer J. Independent terrestrial origins of the Halosphaeriales (marine Ascomycota). Am J Bot. 1998;85:1569–1998.

Statzell-Tallman A, Belloch C, Fell JW. *Kwoniella mangroviensis* gen. nov., sp. nov. (Tremellales, Basidiomycota) a teleomorphic yeast from mangrove habitats in the Florida Everglades and Bahamas. FEMS Yeast Res. 2008;8:1-3-113.

Statzell-Tallman A, Scorzetti G, Fell JW. *Candida spencermartinsiae* sp. nov., *Candida taylorii* sp. nov. and *Pseudozyma abaconensis* sp. nov., novel yeasts from mangrove and coral reef ecosystems. Inter J Syst Evol Microbiol. 2010;60:1978–1984.

Suh SO, Hirata A, Sugiyama J, Komagata K. Septal ultrastructure of basidiomycetous yeasts and their taxonomic implications with observations on the ultrastructure of *Erythrobasidium hasegawianum* and *Sympodiomycopsis papioedili*. Mycologia 1993;85:30–66.

Swann EC, Taylor JW. Phylogenetic perspectives on basidiomycetes systematics: evidence from the 18S rRNA gene. Can J Bot. 1995;73:862–868.

Takishita K, Tsuchiya M, Reimer JD, Maruyama T. Molecular evidence demonstrating the basidiomycetous fungus *Cryptococcus curvatus* is the dominant microbial eukaryote in sediment at the Kuroshima Knoll methane seep. Extremophiles 2006;10:165–169.

van Uden N, Zobell CE. *Candida marina* nov. spec., *Torulopsis torresii* nov. spec. and *T. maris* nov. spec., three yeasts from the Torres Strait. Antonie Leeuwenhoek 1962;28:275–283.

Vogel C, Rogerson A, Scatz S, Laubach H, Tallman A, Fell JW. Prevalence of yeasts in beach sand at three bathing beaches in South Florida. Water Res. 2007;41:1915–1930.

Wei YH, Liou GY, Liu HY, Lee FL. *Sympodiomycopsis kandeliae* sp. nov., a basidiomycetous anamorphic fungus from mangroves, and reclassification of *Sympodiomycopsis lanaiensis* as *Jaminaea lanaiensis* comb. nov. Int J Syst Evol Microbiol. 2011;61:469–469.

Weiss M, Baucr R, Begerow D. Spotlights on heterobasidiomycetes. In: Agerer R, Piepenbring M, Blanz P, eds. Frontiers in Basidiomycete mycology. Eching: IHW-Verlag, 2004:7–48.

5 Taxonomy of filamentous anamorphic marine fungi: morphology and molecular evidence

Mohamed A. Abdel-Wahab and Ali H.A. Bahkali

Anamorphic fungi are an artificial assemblage of asexual ascomycetes and basidiomycetes that no longer have formal taxonomic status (Barnett and Hunter 1998). Their natural classification is possible when their teleomorphic stages are known by cultural studies or by molecular phylogenetics. In the past two decades, almost half of the marine anamorphic fungi described have been sequenced and their phylogenetic placement has been verified. Eleven new lineages of marine fungi have been established based on molecular phylogenetic studies that include two orders (Lulworthiales, Koralionastetales), one unnamed clade (TBM clade) and eight new families (Jones 2011b). Many genera were shown to be polyphyletic, e.g., *Cirrenalia*, *Cumulospora* and *Zalerion*, and this resulted in the creation of several new genera (Bills et al. 1999; Jones et al. 2008; Abdel-Wahab et al. 2010). Molecular taxonomy also confirms the teleomorphs of many anamorphic genera and this will be discussed below.

The number of anamorphic filamentous marine fungi has increased from 60 (Kohlmeyer and Volkmann-Kohlmeyer 1991), to 94 (Jones et al. 2009b) and to 143 (in this chapter). However, filamentous anamorphic marine fungi far exceed this number, as suggested by Jones (2011a). Hundreds of fungi have been recorded from marine sediments (from coastal areas, mangrove sediments and the deep sea), endophytes of marine algae, sea grasses, marine animals, submerged parts of mangrove trees and dead sea and other high-saline waters (Jones 2011a), and this needs serious consideration.

Anamorphs of Dothideomycetes

Forty-four marine anamorphic fungi belonging to the class Dothideomycetes were recorded from diverse habitats and are listed in Tables 5.1 and 5.2. These fungi belong to three orders – Botryosphaeriales, Capnodiales and Pleosporales – and 12 families. Of the 44 dothideomycetous fungi, the taxonomic placement of 16 species remains unresolved (Table 5.1) and requires further morphological and molecular studies.

Anamorphs of Eurotiomycetes

We list one species, *Aspergillus sydowii* (Trichocomaceae, Eurotiales), that is widely distributed and causes devastating disease in sea fans (*Gorgonia* spp.) in the Bahamas, Caribbean, Costa Rica, Cuba, Mexico, USA and Venezuela (Nieves-Rivera 2002). However, the anamorphs of this class are very frequently isolated in large numbers from marine habitats and are often referred to as "marine derived species" (see Chapter 19). Substrata that support members of this group are: coastal and deep sea sediments, epiphytes and endophytes of marine plants and animals (see Chapters 16 and 19). Clearly

Table 5.1 Anamorphic fungi recorded from marine habitats.

Ascomycota
Dothideomycetes
Botryosphaeriales
Botryosphaeriaceae
Diplodia orae-maris Linder, *Farlowia* 1: 403 (1944)
Teleomorph: unknown
Substrate: driftwood

*Macrophoma spp.
Teleomorph: unknown
Substrate: intertidal wood, marsh plants, seawater and sediment

Dothideomycetes *incertae sedis*
Botryosphaeriales *incertae sedis*
Camarosporium palliatum Kohlm. *et* E. Kohlm., *Marine Mycology, the Higher Fungi* 519 (1979)
Teleomorph: unknown
Substrate: salt marsh plants

Camarosporium roumeguerei Sacc., *Michelia* 2(6):112 (1880)
Teleomorph: unknown
Substrate: salt marsh plants

Capnodiales
Davidiellaceae
Cladosporium algarum Cooke *et* Massee in Cooke, *Grevillea* 16(79): 80 (1888)
Teleomorph: unknown
Substrate: drift algae and leaves and dead rhizomes of *Zostera marina*

Mycosphaerellaceae
Rhabdospora avicenniae Kohlm. *et* E. Kohlm., *Mycologia* 63(4): 851 (1971)
Teleomorph: *Leptosphaeria avicenniae*?
Substrate: bark and pneumatophores of *Avicennia* and prop roots of *Rhizophora*

Septoria ascophylli Melnik *et* M. Petrov, *Nov Sist Niz Rast* 211 (1966)
Teleomorph: *Mycophycias ascophylli* (Cotton) Kohlm. *et* Volkm.-Kohlm., *Syst Ascomycet* 16(1–2): 3 (1998)
Substrate: symbiotic forming mycophycobiosis on *Ascophyllum nodosum*

Pleosporales
Didymellaceae
Phoma marina Lind, *Nordisk Forlag Copenhagen,* 214 (1913)
Teleomorph: *Lautitia danica* (Berl.) S. Schatz, *Can J Bot* 62(1): 31 (1984)
Substrate: parasitic on the red alga *Chondrus crispus*

Phoma laminariae Cooke *et* Massee, *Grevillea* 18(87): 53 (1890)
Teleomorph: unknown
Substrate: decaying thalli of the brown alga *Laminaria*

Table 5.1 (Continued)

Phoma suaedae Jaap, *Schr Naturw Ver Schles-Holst* 14(1): 27 (1907)
Teleomorph: unknown
Substrate: dead stems and leaves of *Suaeda maritima*

Leptosphaeriaceae
*Coniothyrium obiones Jaap, *Schr Naturw Ver Schles-Holst* 14(1): 29 (1907)
Teleomorph: unknown
Substrate: salt marsh plant *Halimione portulacoides*

Montagnulaceae
Montagnula perforans (Roberge ex Desm.) Aptroot, *Mycosphaerella and its anamorphs: 2. Conspectus of Mycosphaerella* 150 (2006)
Teleomorph: *Amarenomyces ammophilae* (Lasch) O.E. Erikss., *Op Bot* 60: 124 (1981)
Substrate: Saprobic on *Ammophila arenaria*

Phaeosphaeriaceae
Hendersonia typhae Oudem., *Ned kruidk Archf*, 3sér. 1: 255 (1873)
Teleomorph: *Phaeosphaeria typharum* (Desm.) L. Holm, *Symb Bot Upsal* 14: 126 (1957)
Substrate: Saprobic on *Spartina* spp.

Stagonospora haliclysta Kohlm., *Bot Mar* 16(4): 213 (1973)
Teleomorph: unknown
Substrate: brown seaweed *Pelvetia canaliculata*

Stagonospora spp.
Teleomorph: unknown
Substrate: salt marsh plants

Pleosporaceae
Amarenographium metableticum (Trail) Eriksson, *Mycotaxon* 15: 199 (1982)
Teleomorph: *Amarenomyces ammophilae* (Lasch) O.E. Erikss.
Substrate: bark, maritime grasses and salt marsh plants

*Dendryphiella arenariae Nicot, *Rev Mycol*, Paris 23: 93 (1958)
Teleomorph: unknown
Substrate: decaying marine or estuarine plants, sea foam

*Dendryphiella salina (G.K. Sutherl.) Pugh *et* Nicot, *Trans Br Mycol Soc* 47: 266 (1964)
Teleomorph: unknown
Substrate: decaying marine or estuarine plants, driftwood and wood panels

*Dictyosporium elegans Corda, *Weitenweber's Beitr Nat*: 87 (1836)
Teleomorph: unknown
Substrate: *Rhizophora stylosa* wood

*Stemphylium gracilariae E.G. Simmons, *Mem N Y Bot Gdn* 49: 305 (1989)
Teleomorph: *Pleospora gracilariae* E.G. Simmons *et* S. Schatz
Substrate: red alga *Gracilaria*

Table continued on next page.

Table 5.1 (Continued)

Stemphylium maritimum T.W. Johnson, *Mycologia* 48(6): 844 (1957)
Teleomorph: unknown
Substrate: submerged wood panels

*Stemphylium triglochinicola B. Sutton *et* Piroz., *Trans Br Mycol Soc* 46(4): 519 (1963)
Teleomorph: *Pleospora triglochinicola* J. Webster
Substrate: old leaves and inflorescences of *Triglochin maritima*

Sporormiaceae
*Amorosia littoralis Mantle *et* D. Hawksw., *Mycol Res* 110: 1373 (2006)
Teleomorph: unknown
Substrate: driftwood and sediment in the intertidal zone

Pleosporales *incertae sedis*
*Alternaria maritima G.K. Sutherl., *New Phytol* 15: 46 (1916)
Teleomorph: unknown
Substrate: seawater, sediment, driftwood, intertidal wood, wood panels.

*Ascochyta obiones (Jaap) P.K. Buchanan, *Mycol Pap* 156: 28 (1987)
Teleomorph: unknown
Substrate: salt marsh plants

Ascochyta salicorniae Magnus, in Jaap, *Schr naturw Ver Schles-Holst* 12: 30 (1902)
Teleomorph: unknown
Substrate: salt marsh plants

Dictyosporium pelagicum (Linder) G.C. Hughes ex Johnson *et* Sparrow, *Trans Br Mycol Soc* 46(1): 137 (1963)
Teleomorph: unknown
Substrate: submerged wood

Periconia abyssa Kohlm., *Rev Mycol*, Paris 41(2): 202 (1977)
Teleomorph: unknown
Substrate: submerged wood in the deep sea

*Phialophorophoma litoralis Linder, *Farlowia* 1: 403 (1944)
Teleomorph: unknown
Substrate: drift and intertidal wood

*Pseudorobillarda phragmitis (Cunnell) M. Morelet, *Bull Soc Sci Nat Arch Toulon et du Var* 175: 6 (1968)
Teleomorph: unknown
Substrate: wood panels

Sporidesmium salinum E.B.G. Jones, *Trans Br Mycol Soc* 46(1): 135 (1963)
Teleomorph: unknown
Substrate: wood panels

Table 5.1 (Continued)

Dothideomycetes *incertae sedis*
Bactrodesmium linderi (J.L. Crane *et* Shearer) M.E. Palm *et* E.L. Stewart, *Mycotaxon* 15: 319 (1982)
Teleomorph: unknown
Substrate: intertidal and drift wood

Epicoccum spp.
Teleomorph: unknown
Substrate: washed-up *Laminaria* sp., *Spartina alterniflora*, wood panels, sea foam

Eurotiomycetes
Eurotiales
Trichocomaceae
*Aspergillus sydowii (Bainier *et* Sartory) Thom *et* Church, *Aspergilli*: 47 (1926)
Teleomorph: unknown
Substrate: pathogen of sea fans *Gorgonia* spp.

Leotiomycetes
Leotiales
Leotiaceae
*Halenospora varia (Anastasiou) E.B.G. Jones, *Fungal Divers* 35: 154 (2009)
Teleomorph: unknown
Substrate: intertidal and drift wood

Helotiales
Sclerotiniaceae
Botryophialophora marina Linder in Barghoorn *et* Linder, *Farlowia* 1: 404 (1944)
Teleomorph: unknown
Substrate: wood panels, sediments

Helotiales *incertae sedis*
*Dactylaria sp., *Bot Mar* 54: 218 (2011)
Teleomorph: unknown
Substrate: driftwood

Orbiliomycetes
Orbiliales
Orbiliaceae
Arthrobotrys arthrobotryoides (Berl.) Lindau, *Rabenh Krypt-Fl*, Edn 2: 371 (1906)
Teleomorph: in the genus *Orbilia*
Substrate: decaying mangrove leaves and wood baited with nematodes

*Arthrobotrys brochopaga (Drechsler) S. Schenck, W.B. Kendr. *et* Pramerr, *Can J Bot* 55(8): 982 (1977)
Teleomorph: in the genus *Orbilia*
Substrate: decaying mangrove leaves baited with nematodes

Table continued on next page.

Table 5.1 (Continued)

Arthrobotrys dactyloides Drechsler, *Mycologia* 29(4): 486 (1937)
Teleomorph: in the genus *Orbilia*
Substrate: decaying mangrove leaves baited with nematodes

Arthrobotrys javanica (Rifai *et* R.C. Cooke) Jarow, *Acta Mycologica* 6(2): 373 (1970)
Teleomorph: in the genus *Orbilia*
Substrate: decaying mangrove leaves baited with nematodes

Arthrobotrys mangrovispora Swe, Jeewon, Pointing *et* K.D. Hyde, *Bot Mar* 51: 332 (2008)
Teleomorph: in the genus *Orbilia*
Substrate: decaying wood of marine mangrove, baited with *Panagrellus redivivus*

Arthrobotrys musiformis Deschler, *Mycologia* 29(4): 481 (1937)
Teleomorph: in the genus *Orbilia*
Substrate: decaying mangrove leaves and wood baited with nematodes

Arthrobotrys oligospora Fresen., *Beitr Mykol* 1: 18 (1850)
Teleomorph: in the genus *Orbilia*
Substrate: decaying mangrove leaves and wood baited with nematodes

Arthrobotrys polycephala (Drechsler) Rifai, *Reinwardtia* 7(4): 371 (1968)
Teleomorph: in the genus *Orbilia*
Substrate: nematode trapping fungi

Arthrobotrys pyriformis (Juniper) Schenk, W.B. Kendr. *et* Pramer, *Can J Bot* 55(8): 984 (1977)
Teleomorph: in the genus *Orbilia*
Substrate: nematode trapping fungi

Arthrobotrys superba Corda, Pracht-Fl. *Eur Schimmelbild*: 43 (1839)
Teleomorph: in the genus *Orbilia*
Substrate: decaying mangrove wood

Arthrobotrys vermicola (R.C. Cooke *et* Satchuth.) Rifai, *Reinwardtia* 7(4): 371 (1968)
Teleomorph: in the genus *Orbilia*
Substrate: decaying mangrove leaves and wood

Dactylellina spp.
Teleomorph: in the genus *Orbilia*
Substrate: decayed mangrove leaves and wood baited with nematodes

Monacrosporium eudermatum (Drechsler) Subram., *J Indian bot Soc* 42: 293 (1964)
Teleomorph: in the genus *Orbilia*
Substrate: decaying mangrove leaves and wood baited with nematodes

Monacrosporium huisuniana (Drechsler) Subram., *J Indian bot Soc* 42: 293 (1964)
Teleomorph: in the genus *Orbilia*
Substrate: decaying mangrove leaves and wood baited with nematodes

Table 5.1 (Continued)

Monacrosporium thaumasium (Drechsler) Subram., *J Indian bot Soc* 42: 293 (1964)
Teleomorph: in the genus *Orbilia*
Substrate: decaying mangrove leaves and wood baited with nematodes

Sordariomycetes
Diaporthales
Diaporthaceae

Phomopsis sp., *Bot Mar* 54: 218 (2011)
Teleomorph: *Diaporthe* sp., *Bot Mar* 54: 218 (2011)
Substrate: driftwood

Valsaceae

Cytospora rhizophorae Kohlm. *et* E. Kohlm., *Mycologia* 63(4): 847 (1971)
Teleomorph: unknown
Substrate: intertidal *Rhizophora*

Gnomoniaceae

Gloeosporidina cecidii (Kohlm.) B. Sutton, *The Coelomycetes* (Kew): 517 (1980)
Teleomorph: unknown
Substrate: on *Sargassum natans*

Microascales
Halosphaeriaceae

Cirrenalia macrocephala (Kohlm.) Meyers *et* Moore, *Am J Bot* 47: 347 (1960)
Teleomorph: unknown
Substrate: driftwood, intertidal wood, wood panels

Cirrenalia basiminuta Raghu-Kumar *et* Zainal in Raghu-Kumar, Zainal *et* Jones, *Mycotaxon* 31(1): 163 (1988)
Teleomorph: unknown
Substrate: wood panels

Cirrenalia pseudomacrocephala Kohlm., *Mycologia* 60(2): 266 (1968)
Teleomorph: unknown
Substrate: driftwood

Clavatospora bulbosa (Anast.) Nakagiri *et* Tubaki, *Bot Mar* 28(11): 489 (1985)
Teleomorph: *Corollospora pulchella* Kohlm., I. Schmidt *et* N.B. Nair
Substrate: submerged and intertidal wood and rotting algae

Halosigmoidea luteola (Nakagiri *et* Tubaki) Nakagiri, K.L. Pang *et* E.B.G. Jones, *Bot Mar* 52(4): 349–359 (2009)
Teleomorph: *Corollospora luteola* Nakagiri *et* Tubaki
Substrate: sea foam

Table continued on next page.

Table 5.1 (Continued)

Halosigmoidea marina (Haythorn *et* E.B.G. Jones) Nakagiri, K.L. Pang *et* E.B.G. Jones, *Bot Mar* 52(4): 349–359 (2009)
Teleomorph: in the genus *Corollospora*
Substrate: cast algae, *Laminaria saccharina*

Halosigmoidea parvula Zuccaro, J.I. Mitch. *et* Nakagiri, *Bot Mar* 52(4): 349–359 (2009)
Teleomorph: in the genus *Corollospora*
Substrate: sea foam, *Fucus serratus*

Monodictys pelagica (T.W. Johnson) E.B.G. Jones, *Trans Br Mycol Soc* 46(1): 138 (1963)
Teleomorph: *Nereiospora cristata* (Kohlm.) E.B.G. Jones, R.G. Johnson *et* S.T. Moss
Substrate: drift and intertidal wood, wood panels and submerged leaves

Periconia prolifica Anastasiou, *Nova Hedw* 6: 260 (1963)
Teleomorph: *Okeanomyces cucullatus* (Kohlm.) K.L. Pang *et* E.B.G. Jones
Substrate: intertidal and drift wood

Trichocladium achrasporum (Meyers *et* Moore) Dixon in Shearer *et* Crane, *Mycologia* 63(2): 244 (1971)
Teleomorph: *Halosphaeriopsis mediosetigera* (Cribb *et* J.W. Cribb) T.W. Johnson
Substrate: intertidal and drift wood

Trichocladium melhae E.B.G. Jones, Abdel-Wahab *et* Vrijmoed, *Fungal Divers* 7: 50 (2001)
Teleomorph: unknown
Substrate: intertidal and drift wood, wood panels.

Varicosporina anglusa Abdel-Wahab *et* Nagahama, *Mycoscience* 50(3): 150 (2009)
Teleomorph: *Corollospora anglusa* Abdel-Wahab *et* Nagahama
Substrate: on balsa wood, derived from teleomorph

Varicosporina prolifera Nakagiri, *Trans Mycol Soc Japan* 27(2): 198 (1986)
Teleomorph: *Corollospora intermedia* I. Schmidt
Substrate: sea foam

Varicosporina ramulosa Meyers *et* Kohlm., *Can J Bot* 43: 916 (1965)
Teleomorph: in the genus *Corollospora*
Substrate: drift and washed-up algae, sea grasses and synthetic sponge

Hypocreales
Bionectriaceae
Acremonium fuci Summerb., Zuccaro *et* W. Gams, *Stud Mycol* 50(1): 288 (2004)
Teleomorph: in the genus *Emericellopsis*
Substrate: *Fucus* thalli

Acremonium potronii Vuillemin, *Bull Séanc Soc Sci Nancy*, Sér. 3 11: 147 (1910)
Teleomorph: in the genus *Emericellopsis*
Substrate: skin lesion in dolphin

Table 5.1 (Continued)

Clavicipitaceae
Pochonia sp., *Bot Mar* 54: 218–220 (2011)
Teleomorph: unknown
Substrate: driftwood

Nectriaceae
Tubercularia pulverulenta Speg., *Anal Soc cient argent* 12(1): 32 (1881)
Teleomorph: unknown
Substrate: saprobic on various *Salicornia* species

Hypocreales *incertae sedis*
Stachybotrys atra Corda, *Icon fung* (Prague) 1: 21 (1837)
Teleomorph: unknown
Substrate: submerged twigs of *Tamarix aphylla*

Stachybotrys mangiferae P.C. Misra *et* S.K. Srivast., *Trans Br Mycol Soc* 78(3): 556 (1982)
Teleomorph: unknown
Substrate: submerged wood from *Rhizophora stylosa*

Fusarium sp.
Teleomorph: unknown
Substrate: submerged twigs of *Tamarix aphylla*, sediments and sand dunes

Hypocreomycetidae *incertae sedis*
Plectosphaerellaceae
Plectosporium oratosquillae Duc, Yaguchi *et* Udagawa, *Mycopathologia* 167(5): 237 (2009)
Teleomorph: unknown
Substrate: diseased mantis shrimp (*Oratosquilla oratoria*)

Lulworthiales
Lulworthiaceae
Anguillospora marina Nakagiri *et* Tubaki, *Mycologia* 75(3): 488 (1983)
Teleomorph: *Lindra obtusa* Nakagiri *et* Tubaki
Substrate: sea foam

Cumulospora marina I. Schmidt, *Mycotaxon* 24: 421 (1985)
Teleomorph: unknown
Substrate: intertidal and drift wood

Halazoon fuscus (I. Schmidt) Abdel-Wahab, K.L. Pang, Nagahama, Abdel-Aziz *et* E.B.G. Jones, *Mycol Progr* 9(4): 547 (2010)
Teleomorph: unknown
Substrate: intertidal and drift wood

Halazoon melhae Abdel-Aziz, Abdel-Wahab *et* Nagahama, *Mycol Progr* 9(4): 547 (2010)
Teleomorph: unknown
Substrate: drift wood

Table continued on next page.

Table 5.1 (Continued)

Hydea pygmea (Kohlm.) K.L. Pang *et* E.B.G. Jones, *Mycol Progr* 9(4): 547 (2010)
Teleomorph: unknown
Substrate: intertidal and drift wood

Matsusporium tropicale (Kohlm.) E.B.G. Jones *et* K.L. Pang *Mycol Progr* 9(4): 547 (2010)
Teleomorph: unknown
Substrate: intertidal and drift wood

Moleospora maritima Abdel-Wahab, Abdel-Aziz *et* Nagahama, *Mycol Progr* 9(4): 547 (2010)
Teleomorph: unknown
Substrate: intertidal and drift wood

Moromyces varius (Chatmala *et* Somrith.) Abdel-Wahab, K.L. Pang, Nagahama, Abdel-Aziz *et* E.B.G. Jones, *Mycol Progr* 9(4): 547 (2010)
Teleomorph: unknown
Substrate: intertidal and drift wood

Orbimyces spectabilis Linder in Barghoorn *et* Linder, *Farlowia* 1: 404 (1944)
Teleomorph: unknown
Substrate: intertidal and drift wood

Zalerion maritima (Linder) Anastasiou, *Can J Bot* 41: 1136 (1963)
Teleomorph: *Lulwoana uniseptata* (Nakagiri) Kohlm., Volkm.-Kohlm., J. Campb., Spatafora *et* Gräfenhan, *Mycol Res* 109(5): 562 (2005)
Substrate: intertidal and drift wood

Microascales
Microascaceae
Scopulariopsis halophilica Tubaki, *Trans Mycol Soc Japan* 14(4): 367 (1973)
Teleomorph: unknown
Substrate: seaweed *Undaria pinnatifida* and algae

TBM clade
Glomerulispora mangrovis Abdel-Wahab *et* Nagahama, *Mycol Progr* 9(4): 547 (2010)
Teleomorph: unknown
Substrate: driftwood

Moheitospora adarca (Kohlm., Volkm.-Kohlm. *et* O.E. Erikss.) Abdel-Wahab, Abdel-Aziz *et* Nagahama, *Mycol Progr* 9(4): 547 (2010)
Teleomorph: *Juncigena adarca* Kohlm., Volkm.-Kohlm. *et* O.E. Erikss.
Substrate: *Juncus roemerianus*

Moheitospora fruticosae Abdel-Wahab, Abdel-Aziz *et* Nagahama, *Mycol Progr* 9(4): 547 (2010)
Teleomorph: unknown
Substrate: driftwood

Table 5.1 (Continued)

Trichosphaeriales
Trichosphaeriaceae
Brachysporium helgolandicum Schaumann, *Helgl Wiss Meeresunters* 25(1): 26 (1973)
Teleomorph: unknown
Substrate: driftwood

Koorchaloma galateae Kohlm. *et* Volkm.-Kohlm., *Bot Mar* 44(2): 147 (2001)
Teleomorph: unknown
Substrate: senescent culms of *Juncus roemerianus*

*Koorchaloma spartinicola V.V. Sarma, S.Y. Newell *et* K.D. Hyde, *Bot Mar* 44(4): 321 (2001)
Teleomorph: in the genus *Kananascus*
Substrate: leaf sheaths of *Spartina alterniflora*

Xylariales
Amphisphaeriaceae
Robillarda rhizophorae Kohlm., *Can J Bot* 47: 1483 (1969)
Teleomorph: unknown
Substrate: *Rhizophora* wood

Xylariales *incertae sedis*
Dinemasporium marinum Sv. Nilsson, *Bot Not* 110: 321 (1957)
Teleomorph: unknown
Substrate: driftwood

Sordariomycetidae *incertae sedis*
Magnaporthaceae
Trichocladium medullare Kohlm. *et* Volkm.-Kohlm., *Mycotaxon* 53: 349 (1995)
Teleomorph: *Gaeumannomyces medullaris* Kohlm., Volkm.-Kohlm. *et* O.E. Eriks.
Substrate: dead culms of *Juncus roemerianus*

Apiosporaceae
Arthrinium algicola (N.J. Artemczuk) E.B.G. Jones, Sakay., Suetrong, Somrith. *et* K.L. Pang, *Fungal Divers*: 150 (2010)
Teleomorph: unknown
Substrate: brown alga *Cystoseira barbata*

Pezizomycotina *incertae sedis*
Asteromyces cruciatus Moreau *et* M. Moreau ex. Hennebert, *Rev Mycol* Paris 6(3–4): 79 (1941)
Teleomorph: unknown
Substrate: drift algae, wood panels, sea foam

Blodgettia confervoides Harvey, Smithson. *Contr Bot* 10: 48 (1858)
Teleomorph: unknown
Substrate: symbiotic with *Cladophora caespitosa* and *C. fuliginosa*

Table continued on next page.

Table 5.1 (Continued)

Cytoplacosphaeria phragmiticola Poon *et* K.D. Hyde, *Bot Mar* 41(2): 148 (1998)
Teleomorph: unknown
Substrate: intertidal *Phragmites australis*

Helicorhoidion nypicola K.D. Hyde *et* Goh, *Mycol Res* 103(11): 1420 (1999)
Teleomorph: unknown
Substrate: intertidal *Nypa fruticans*

Heliscella stellatacula (P.W. Kirk ex Marvanová *et* Sv. Nilsson) Marvanová, *Trans Br Mycol Soc* 75(2): 224 (1980)
Teleomorph: unknown
Substrate: submerged wood in estuary

Hymenopsis chlorothrix Kohlm. *et* Volkm.-Kohlm., *Mycol Res* 105(4): 504 (2001)
Teleomorph: unknown
Substrate: dead standing culms of *Juncus roemerianus*

Mycoenterolobium platysporum Goos, *Mycologia* 62(1): 172 (1970)
Teleomorph: unknown
Substrate: submerged *Rhizophora stylosa* wood

Nypaella frondicola K.D. Hyde *et* B. Sutton, *Mycol Res* 96(3): 210 (1992)
Teleomorph: unknown
Substrate: intertidal fronds of *Nypa fruticans*

Octopodotus stupendus Kohlm. *et* Volkm.-Kohlm., *Mycologia* 95(1): 117 (2003)
Teleomorph: unknown
Substrate: dead leaves of *Spartina alterniflora*

Phialophorophoma litoralis Linder, *Farlowia* 1: 403 (1944)
Teleomorph: unknown
Substrate: dead leaves of *Spartina alterniflora*

Phragmospathula phoenicis Subram. *et* N.G. Nair, *Antonie van Leeuwenhoek* 32: 384 (1966)
Teleomorph: unknown
Substrate: intertidal wood of *Rhizophora stylosa*

Plectophomella nypae K.D. Hyde *et* B. Sutton, *Mycol Res* 96(3): 211 (1992)
Teleomorph: unknown
Substrate: intertidal fronds of *Nypa fruticans*

Pleurophomopsis nypae K.D. Hyde *et* B. Sutton, *Mycol Res* 96(3): 213 (1992)
Teleomorph: unknown
Substrate: intertidal fronds of *Nypa fruticans*

Trichocladium alopallonellum (Meyers *et* R.T. Moore) Kohlm. *et* Volkm.-Kohlm., *Mycotaxon* 53: 352 (1995)
Teleomorph: unknown
Substrate: intertidal and driftwood, wood panels, submerged leaves

Table 5.1 (Continued)

Trichocladium constrictum I. Schmidt, *Natur Naturschutz Mecklenberg* 12: 114 (1974)
Teleomorph: unknown
Substrate: submerged wood

Trichocladium lignicola I. Schmidt, *Natur Naturschutz Mecklenberg* 12: 116 (1974)
Teleomorph: unknown
Substrate: submerged wood

Trichocladium nypae K.D. Hyde *et* Goh, *Mycol Res* 103(11): 1420 (1999)
Teleomorph: unknown
Substrate: intertidal *Nypa fruticans*

Xylomyces rhizophorae Kohlm. *et* Volkm.-Kohlm., *Fungal Divers* 1: 160 (1998)
Teleomorph: unknown
Substrate: prop root of *Rhizophora mangle*

Basidiomycota
Agaricomycetes
Cantharellales
Botryobasidiaceae
Allescheriella bathygena Kohlm., *Rev Mycol* Paris 41(2): 199 (1977)
Teleomorph: *Botryobasidium*?
Substrate: submerged wood in the deep sea

*Species that have been sequenced for one or more genes.

Table 5.2 Teleomorphs with unnamed anamorphs.

Ascomycotina
Dothideomycetes
Botryosphaeriales
Botryosphaeriaceae
**Botryosphaeria* sp., *Bot Mar* 54: 216–218 (2011)
Anamorph: pycnidial coelomycete
Substrate: driftwood

Capnodiales
Mycosphaerellaceae
Mycosphaerella staticicola (Pat.) Dias, *Mém Soc Brot* 21: 72 (1971)
Anamorph: pycnidial coelomycete
Substrate: drying inflorescences of Plumbaginaceae

Mycosphaerella suaedae-australis Hansf., *Proc Linn Soc N S W* 79(3–4): 122 (1954)
Anamorph: pycnidial coelomycete (*Septoria* sp.)
Substrate: dead stems of *Suaeda australis*

Table continued on next page.

Table 5.2 (Continued)

Pleosporales

Arthopyreniaceae

Leiophloea pelvetiae (G.K. Sutherl.) Kohlm. *et* E. Kohlm., *Marine Mycology, the Higher Fungi* (London): 376 (1979)
Anamorph: pycnidial coelomycete
Substrate: symbiotic epiphytic blue-green algae on *Pelvetia canaliculata*

Leptosphaeriaceae

Leptosphaeria albopunctata (Westend.) Sacc., *Syll. fung.* (Abellini) 2: 72 (1883)
Anamorph: pycnidial coelomycete
Substrate: driftwood, *Juncus, Phragmites* and *Spartina*

Massarinaceae

Massarina cystophorae (Cribb *et* J.W. Herb.) Kohlm. *et* E. Kohlm., *Marine Mycology, the Higher Fungi* (London): 427 (1979)
Anamorph: spermogonia
Substrate: parasitic on *Cystophora retroflexa*

Montagnulaceae

Tremateia halophila Kohlm., Volkm.-Kohlm. *et* O.E. Erikss., *Bot Mar* 38(2): 166 (1995)
Anamorph: *Phoma* sp.
Substrate: senescent culms of *Juncus roemerianus*

Pleosporaceae

*Pleospora spartinae (J. Webster *et* M.T. Lucas) Apinis *et* Chesters, *Trans Br Mycol Soc* 47(3): 432 (1964)
Anamorph: pycnidial coelomycete
Substrate: saprobic on *Spartina townsendii*

Zopfiaceae

Halotthia posidoniae (Durieu *et* Mont.) Kohlm., *Nova Hedw* 6: 9 (1963)
Anamorph: spermogonia
Substrate: parasitic on living rhizomes of *Posidonia oceanica*

Pleosporales *incertae sedis*

Passeriniella obiones (P. Crouan *et* H. Crouan) K.D. Hyde *et* Mouzouras, *Trans Br Mycol Soc* 91(1): 183 (1988)
Anamorph: spermogonia
Substrate: parasitic on living rhizomes of *Posidonia oceanica*

*Massariosphaeria typhicola (P. Karst.) Leuchtm., *Sydowia* 37: 168 (1984)
Anamorph: pycnidial coelomycete
Substrate: intertidal and driftwood, dead culms of *Spartina* spp.

Dothideomycetes *incertae sedis*

Mastodiaceae

Turgidosculum complicatulum (Nyl.) Kohlm. *et* E. Kohlm., *Marine Mycology, the Higher Fungi* (London): 361 (1979)

Table 5.2 (Continued)

Anamorph: spermogonia
Substrate: symbiotic with *Prasiola borealis* and *P. tessellata*

Turgidosculum ulvae (M. Reed) Kohlm. *et* E. Kohlm., *Bot Jb* 92(2–3): 429 (1972)
Anamorph: spermogonia
Substrate: symbiotic with *Blidingia minima*

Dothideomycetes *incertae sedis*

Thalassoascus tregoubovii Ollivier, C. r. hebd. *Séanc Acad Sci*, Paris 182: 1348 (1926)
Anamorph: spermogonia
Substrate: parasitic on *Aglaozonia* spp. and *Cystoseira* spp.

Didymosphaeria danica (Berl.) I.M. Wilson *et* Knoyle, *Trans Br Mycol Soc* 44(1): 57 (1961)
Anamorph: *Phoma* sp.
Substrate: parasitic on *Chondrus crispus*

Sordariomycetes
Microascales
Halosphaeriaceae

*Halosarpheia japonica Abdel-Wahab *et* Nagahama, *Mycol Progr* (2011)
Anamorphic: helicoid conidia
Substrate: driftwood

Hypocreales
Bionectriaceae

*Heleococcum japonense Tubaki, *Trans Mycol Soc Japan* 8: 5 (1967)
Anamorph: *Trichothecium* type
Substrate: wood panels

Hypocreales *incertae sedis*

Halonectria milfordensis E.B.G. Jones, *Trans Br Mycol Soc* 48(2): 287 (1965)
Anamorph: pycnidial coelomycete
Substrate: intertidal wood, test panels

Melanopsamma balani (G. Winter) Meyers, *Mycologia* 49: 485 (1957)
Anamorph: pycnidial coelomycete
Substrate: saprobic or symbiotic with algae on shells of mollusks and cirripedes

Lulworthiales *incertae sedis*

Haloguignardia decidua Cribb *et* J.W. Cribb, *Pap Univ Qd* 3: 97 (1956)
Anamorph: spermogonia
Substrate: parasitic on *Sargassum* spp.

Haloguignardia irritans (Setch. *et* Estee) Cribb *et* J.W. Cribb, *Pap Univ Qd* 3: 98 (1956)
Anamorph: spermogonia
Substrate: parasitic on *Cystoseira osmundacea* and *Halidrys dioica*

Table continued on next page.

Table 5.2 (Continued)

Haloguignardia oceanica (Ferd. *et* Winge) Kohlm., *Mar Biol* 8: 344 (1971)
Anamorph: spermogonia
Substrate: parasitic on *Sargassum fluitans* and *S. natans*

Haloguignardia tumefaciens (Cribb *et* J.W. Cribb) Cribb *et* J.W. Cribb, *Pap Univ Qd* 3: 98 (1956)
Anamorph: spermogonia
Substrate: parasitic on *Sargassum decipiens*, *S. fallax*, *S. globulariaefolium* and *S. sinclairii*

Magnaporthales
Magnaporthaceae
Buergenerula spartinae Kohlm. et R.V. Gessner, *Can J Bot* 54(15): 1764 (1976)
Anamorph: conidia 5 × 2 μm produced in pure culture
Substrate: parasitic on *Spartina alterniflora*

Sordariales
Chaetomiaceae
Zopfiella marina Furuya *et* Udagawa, *J Jap Bot* 50(8): 249 (1975)
Anamorph: hyaline, one-celled conidia that produced in culture
Substrate: marine sediment at depth of 120 m

Sordariomycetes *incertae sedis*
Phyllachorales *incertae sedis*
Phycomelaina laminariae (Rostr.) Kohlm., *Phytopath Z* 63: 350 (1968)
Anamorph: spermogonia
Substrate: parasitic on *Laminaria* spp.

---

*Species that have been sequenced for one or more genes.

this group of fungi plays an ecological role in marine habitats and this needs further ecological, physiological and molecular studies to be carried out.

Anamorphs of Leotiomycetes

Three species belonging to two orders were reported from marine habitats, of which *Halenospora varia* (Leotiaceae, Leotiales) is the most frequently recorded. *H. varia* was originally described as a *Zalerion* species and transferred to *Halenospora* based on a molecular study (Bills et al. 1999; Jones et al. 2009b). *H. varia* is a cosmopolitan and very common lignicolous species in marine habitats (Jones et al. 2009b).

Anamorphs of Orbiliomycetes

Johnson and Autery (1961) reported the first nematode-trapping fungus (*Arthrobotrys dactyloides*) from pine panels submerged in brackish habitats in the USA. Seventeen nematode-trapping fungal species were recorded from decaying wood and leaves collected at eight different mangrove sites in Hong Kong (Swe et al. 2009). The 17 species belong to three genera – namely *Arthrobotrys, Dactylella* and *Monacrosporium* – that have *Orbilia* teleomorphs. There is only one record of the genus *Orbilia*, *O. marina* Phillips ex Boyd in Smith (Smith 1909), which is transferred to the genus *Laetinaevia* (Kirk and Spooner 1984).

Anamorphs of Sordariomycetes

Fifty-seven anamorphic fungi belong to this class (Tables 5.1 and 5.2), and the most frequent species on various substrata in marine habitats are found in the Halosphaeriaceae and Lulworthiaceae families. Marine anamorphic fungi belonging to the following orders/groups have been reported: Diaporthiales (three species), Hypocreales (11 species), Lulworthiales (14 species), Magnaporthales (one species), Microascales (16 species), Phyllachorales (one species), TBM clade (three species), Trichosphaeriales (three species), Xylariales (two species) and Sordariomycetes *incertae sedis* (two species).

Anamorphs of Halosphaeriaceae

The Halosphaeriaceae comprises 126 species in 53 genera (Sakayaroj et al. 2011), Chapter 3. Fifteen anamorphic fungi are currently included in the family, with 12 species sequenced and their taxonomy verified. Ten teleomorph/anamorph connections have been supported by molecular taxonomy and they are listed in Tables 5.1 and 5.2. Additionally, fifteen connections include anamorphs of the genera: *Corollospora* [*Clavatospora, Halosigmoidea* (three species), *Varicosporina* (two species); Abdel-Wahab et al. 2009; Jones et al. 2009a], *Halosarpheia japonica* – un-named helicoid conidia (Abdel-Wahab and Nagahama 2010), *Nereiospora cristata – Monodictys pelagica* (Campbell et al. 2002), *Okeanomyces cucullatus – Periconia prolifica* (Pang et al. 2004) and *Halosphaeriopsis mediosetigera – Trichocladium achrasporum* (Abdel-Wahab 2011b). Phylogenetic analyses of LSU rDNA sequence of *Trichocladium melhae* proved it is a member of Halosphaeriaceae (Figures 5.1 and 5.2–5.6).

Anamorphs of Lulworthiaceae

The Lulworthiaceae comprise eight genera: *Haloguignardia, Kohlmeyeriella, Lulworthia, Lulwoana, Lulwoidea, Lindra, Rostrupiella, Spathulospora* and their anamorphs: *Anguillospora, Cumulospora, Halazoon, Hydea, Matsusporium, Moleospora, Moromyces, Orbimyces* and *Zalerion* (Jones et al. 2008; Abdel-Wahab et al. 2010). Currently ten anamorphic species are included in the family, all of which have been sequenced. Two teleomorph/anamorph connections have been supported by molecular data, namely *Lindra obtusa – Anguillospora marina* and *Lulwoana uniseptata – Zalerion maritima* (Campbell et al. 2005). The genus *Cirrenalia* is polyphyletic (Abdel-Wahab et al. 2010) with the type species *C. macrocephala* nested within Halosphaeriaceae, three species belonging to Lulworthiaceae and one to the TBM clade. Four new genera were named to accommodate the displaced species (Abdel-Wahab et al. 2010). Both *Cumulospora* species have teleomorphs in the Lulworthiaceae; however they were distantly related to

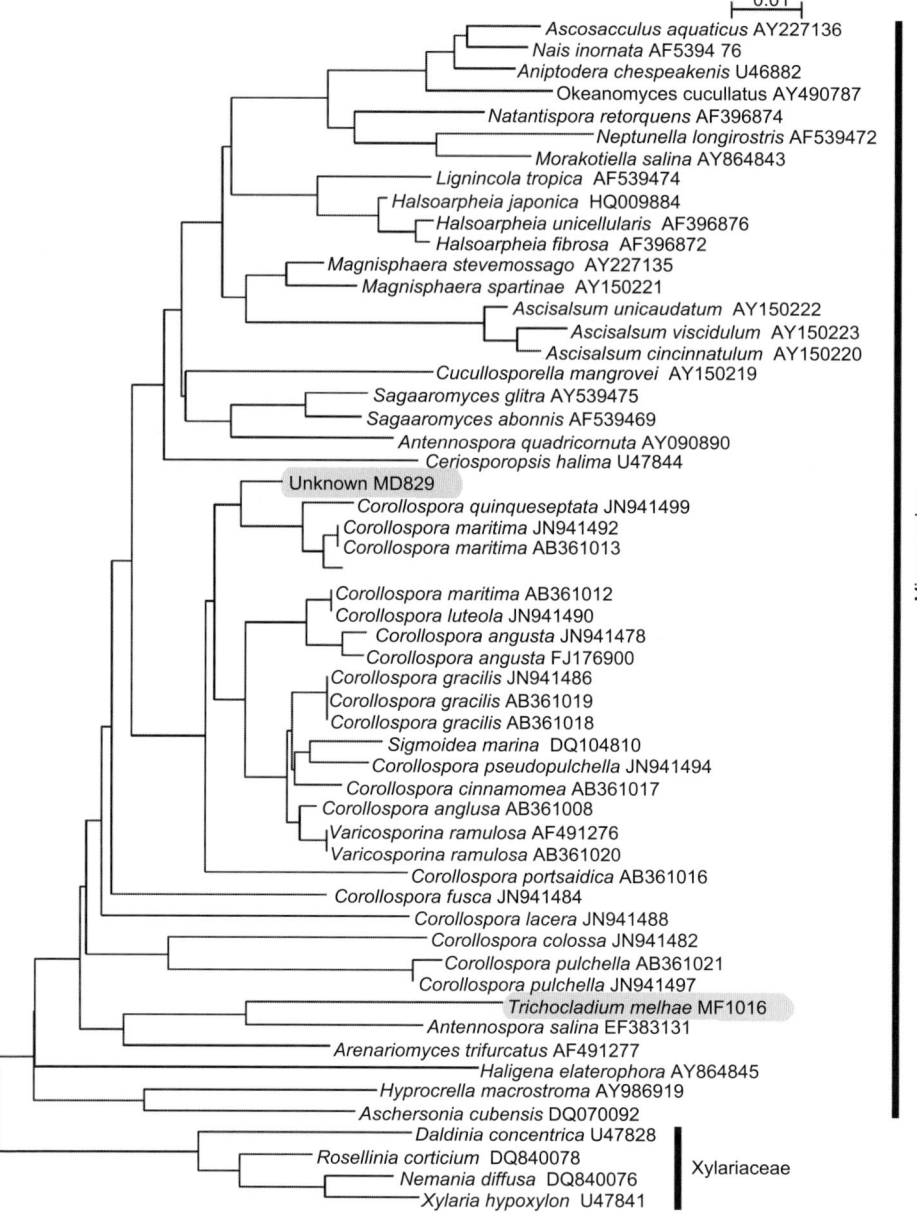

Figure 5.1 Neighbor-joining tree based on the LSU rDNA data set showing the placement of *Trichocladium melhae* and the chlamydospore-producing fungus MD829 in the Halosphaeriaceae.

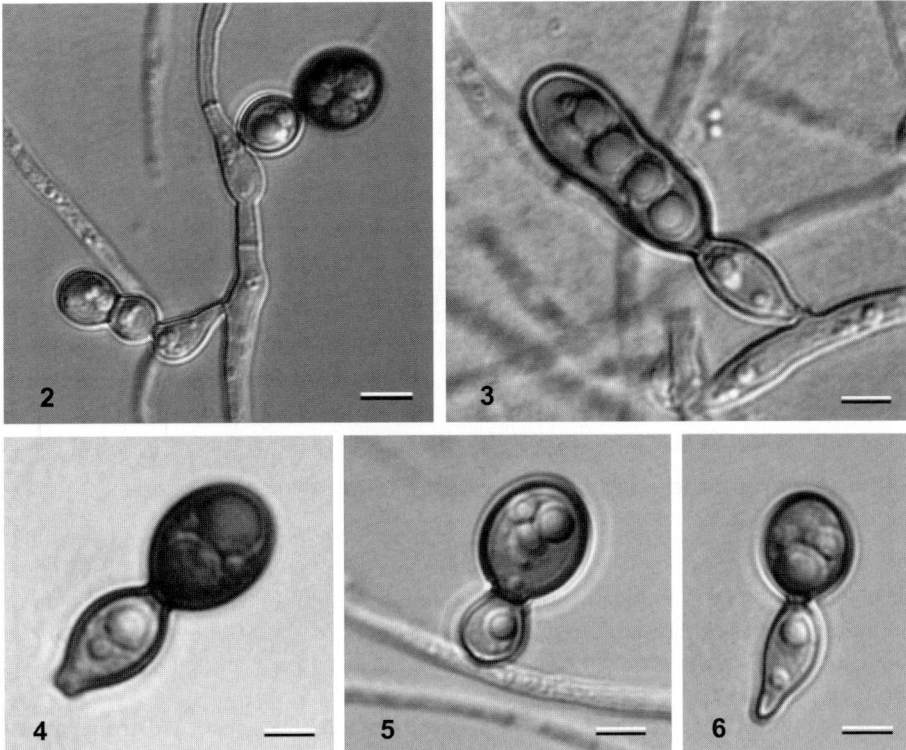

Figures 5.2–5.6 Conidia of *Trichocladium melhae* recorded from intertidal wood in Gesashi mangrove, Okinawa, Japan. Scale bars: 4–8 = 10 μm.

each other and a new genus, *Moromyces*, was proposed for *Cumulospora varia* (Abdel-Wahab et al. 2010).

Anamorphs of TBM clade

The TBM clade is a lineage of marine genera with some terrestrial genera that was first recognized by Sakayaroj et al. (2005) and named by Schoch et al. (2006). The name of the clade is the initials of the three genera, *Torpedospora*, *Bertia* and *Melanospora*. The clade also includes the marine genera *Etheirophora*, *Juncigena* and *Swampomyces* and their anamorphs *Glomerulispora* and *Moheitospora* (Abdel-Wahab et al. 2010).

Anamorphic ascomycetes with uncertain higher taxonomic placement

We list 18 fungi (Table 5.1) that need further morphological and molecular studies in order to resolve their taxonomical placement.

Anamorphs of Basidiomycota

Allescheriella bathygena is the only anamorphic filamentous fungus reported from marine habitats that was described from wood collected at an ocean depth of 1722 meters near the islands of the Bahamas (Kohlmeyer 1977).

Substrata for anamorphic marine fungi

Substrata that supported anamorphic species were: intertidal wood including mangroves (supported the growth of 34 species), driftwood (30 species), salt marsh plants (26 species), submerged wood panels (19 species), living and decaying algae (16 species), nematode trapping fungi (15 species), sea foam (seven species), sea grasses (four species), deep sea samples (three species) and parasites of marine animals (three species) and decaying leaves (two species).

Chlamydospores-producing fungi in marine habitats: cryptic species

Many marine fungi complete their lifecycles as chlamydospores (e.g., *Xylomyces rhizophorae*; Kohlmeyer and Volkmann-Kohlmeyer 1998) and most of these fungi go undetected. A common fungus in marine habitats in Egypt (MD829) produces chlamydospores as a mass of dark thick-walled cells (Figures 5.7–5.10). Phylogenetic analyses of the LSU rDNA sequence of the fungus proved it as an undescribed species of *Corollospora* (Figure 5.1).

Another chlamydospore-producing *Xylomyces* species was recorded from intertidal wood from Gesashi mangroves in Okinawa, Japan (Figures 5.11–5.14). Phylogenetic analyses of its SSU rDNA sequence suggest it to be a member of the Lulworthiales (Figure 5.15). The genus *Xylomyces* is polyphyletic with most species in the Jahnulales (Suetrong et al. 2011). Further morphological and molecular studies are needed to estimate the actual diversity of chlamydospore-producing species in marine habitats and to establish a universal system to name them.

Conclusion

One hundred forty-four anamorphic fungal species were recorded from marine habitats. All but one belong to Ascomycota. Anamorphic fungi in the phylum Ascomycota belong to five classes, namely Dothideomycetes, Eurotiomycetes, Leotiomycetes, Orbiliomycetes and Sordariomycetes. Of the 143 species, 66 were sequenced for one or more genes and the sequences were deposited in the GenBank. Sixty-six teleomorph/anamorph connections have been established based on morphology, of which 38 connections have been supported by molecular phylogenetics. Of the established 66 teleomorph/anamorph connections, 26 teleomorphic fungi have unnamed anamorphs.

Most of the published ecological and taxonomical studies about marine fungi are based on single short visits. Numbers of anamorphic marine fungi will increase when those sites are extensively surveyed and new locations and substrata are examined. Four undescribed anamorphic marine fungi were recorded from driftwood collected at a

Taxonomy of filamentous anamorphic marine fungi | 85

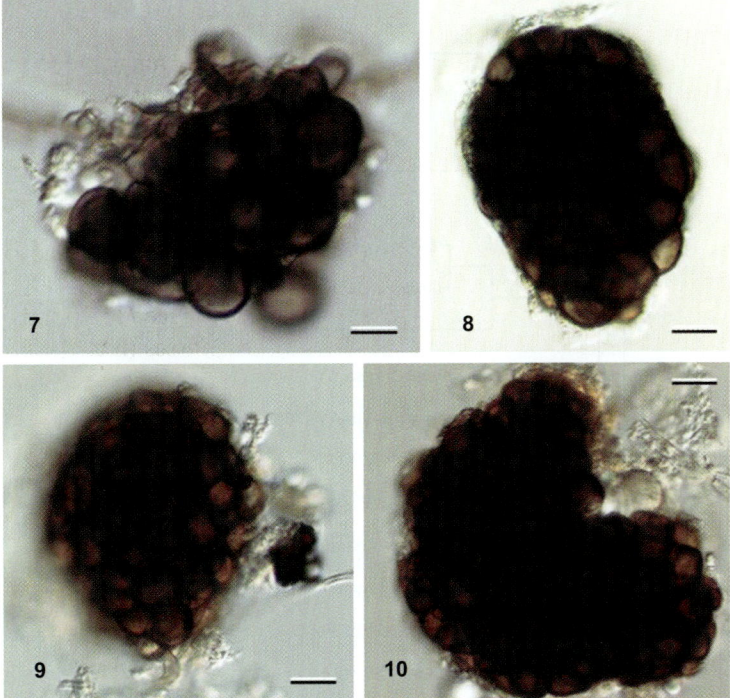

Figures 5.7–5.10 Chlamydospores of an undescribed species of *Corollospora*. Scale bars = 25 μm.

Figures 5.11–5.14 Chlamydospores of *Xylomyces* sp. belonging to the Lulworthiales. Scale bars: 13–14 = 25 μm, 15–16 = 10 μm.

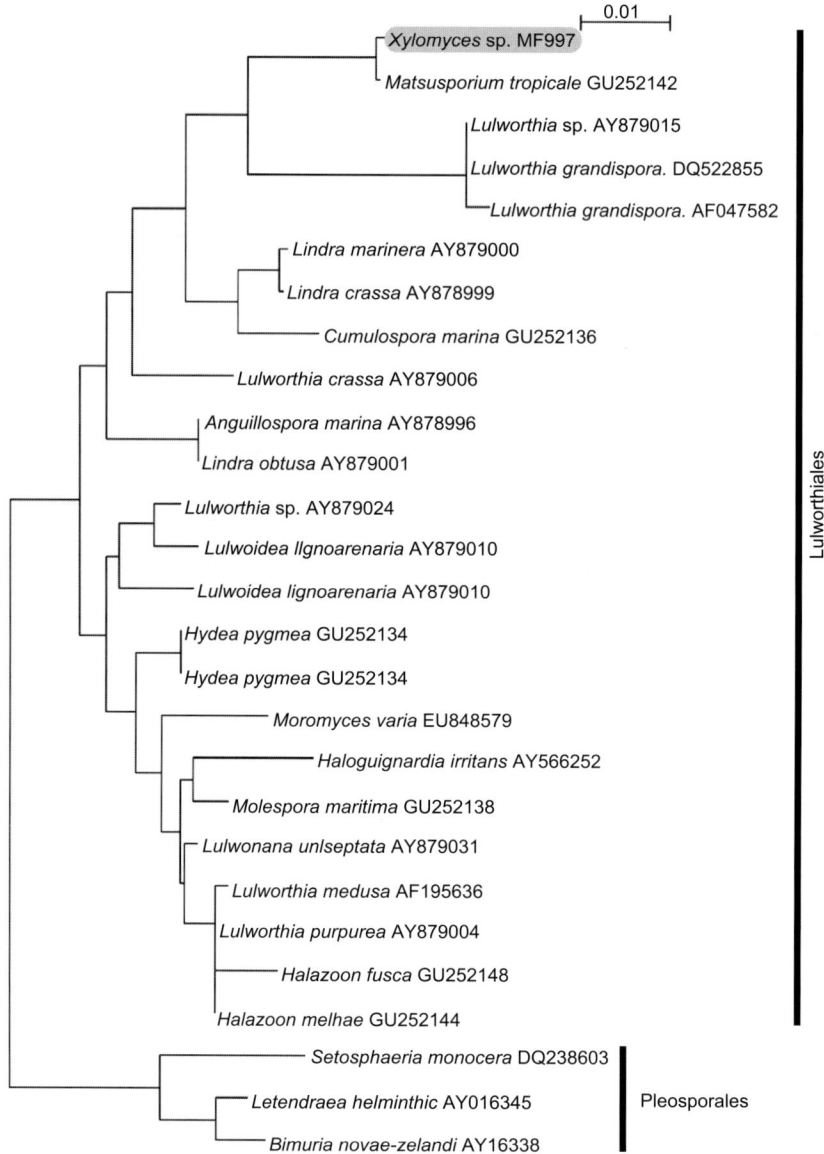

Figure 5.15 Phylogenetic tree based on a maximum parsimony analysis of the SSU rDNA data set showing the phylogenetic placement of *Xylomyces* sp. in the Lulworthiales.

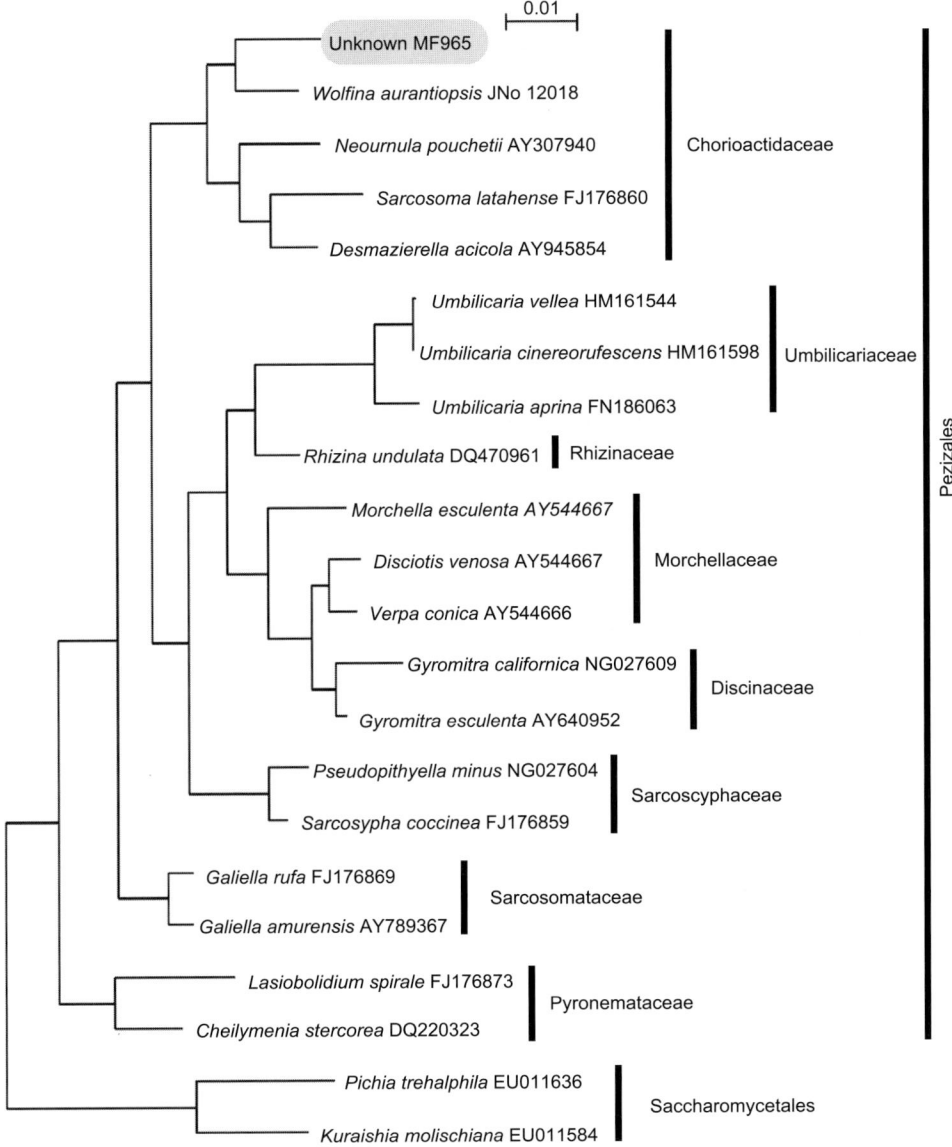

Figure 5.16 Phylogenetic tree based on a maximum parsimony analysis of the LSU rDNA data set showing the phylogenetic placement of an unknown marine anamorphic fungus in the Pezizales.

Figures 5.17–5.20 An undescribed marine fungus belonging to the Pezizales recorded from driftwood from Hakkeijima beach, Yokosuka, Japan. Figure 5.17. Branched conidiophores. Scale bar = 50 μm. Figure 5.18. Base of aggregated conidiophores. Scale bar = 25 μm. Figures 5.19 and 5.20. Conidial heads. Scale bars = 10 μm.

small sandy beach in Japan (Abdel-Wahab 2011a), of which a member of the Pezizales (Figures 5.16 and 17–20) is illustrated here.

Acknowledgements

This project was supported by King Saud University, Deanship of scientific research, College of Science Research Centre.

References

Abdel-Wahab MA. Lignicolous marine fungi from Yokosuka, Japan. Bot Mar. 2011a;53: 283–289.

Abdel-Wahab MA. Marine fungi from Sarushima Island, Japan, with a phylogenetic evaluation of the genus *Naufragella*. Mycotaxon 2011b;115:443–456.

Abdel-Wahab MA, Nagahama T. *Halosarpheia japonica* sp. nov. (Halosphaeriales, Ascomycota) from marine habitats in Japan. Mycol Prog. 2010;11:85–92.

Abdel-Wahab MA, Nagahama T, Abdel-Aziz FA. Two new *Corollospora* species and one anamorph based on morphological and molecular data. Mycoscience 2009;50:147–155.

Abdel-Wahab MA, Pang KL, Nagahama T, Abdel-Aziz FA, Jones EBG. Phylogenetic evaluation of anamorphic species of *Cirrenalia* and *Cumulospora* with the description of eight new genera and four new species. Mycol Prog. 2010;9:537–558.

Barnett HL, Hunter BB. Illustrated genera of imperfect fungi. 4th ed. APS Press: St. Paul, MN, 1998.

Bills GF, Platas G, Peelaez F, Masurekar P. Reclassification of a pneumocandin-producing anamorph, *Glarea lozoyensis* gen. et sp. nov., previously identified as *Zalerion arboricola*. Mycol Res. 1999;103:179–192.

Campbell J, Shearer CA, Mitchell JI, Eaton RA. *Corollospora* revisited: a molecular approach. In: Hyde KD, ed. Fungi in marine environments. Hong Kong University Press: Hong Kong, 2002:15–33.

Campbell J, Volkmann-Kohlmeyer B, Gräfenhan T, Spatafora JW, Kohlmeyer J. A re-evaluation of Lulworthiales: relationships based on 18S and 28S rDNA. Mycol Res. 2005;109:556–568.

Johnson TW, Autery GL. An *Arthrobotrys* from brackish water. Mycologia 1961;53:432–433.

Jones EBG. Are there more marine fungi to be described? Bot Mar. 2011a;54:343–354.

Jones EBG. Fifty years of marine mycology. Fungal Divers. 2011b;50:73–112.

Jones EBG, Klaysuban A, Pang KL. Ribosomal DNA phylogeny of marine anamorphic fungi: *Cumulospora varia*, *Dendryphiella* species and *Orbimyces spectabilis*. Raff Bull Zool Suppl. 2008;19:11–18.

Jones EBG, Zuccaro A, Nakagiri A, Mitchell JL, Pang KL. Phylogenetic position of freshwater and marine *Sigmoidea* species: introducing a marine hyphomycete *Halosigmoidea* gen. nov. (Halosphaeriales). Bot Mar. 2009a;52:349–359.

Jones EBG, Sakayaroj J, Suetrong S, Somrithipol S, Pang KL. Classification of marine Ascomycota, anamorphic taxa and Basidiomycota. Fungal Divers. 2009b;35:1–187.

Kirk PM, Spooner BM. An account of the fungi of Arran, Gigha and Kintyre. Kew Bull. 1984;38:503–597.

Kohlmeyer J. New genera and species and species of higher fungi from the deep sea (1615–5315 m). Rev Mycol. 1977;41:189–206.

Kohlmeyer J, Volkmann-Kohlmeyer B. Illustrated key to the filamentous marine fungi. Bot Mar. 1991;34:1–61.

Kohlmeyer J, Volkmann-Kohlmeyer B. A new marine *Xylomyces* on *Rhizophora* from the Caribbean and Hawaii. Fungal Diver. 1998;1:159–164.

Nieves-Rivera AM. Sea fan aspergillosis. What is it? Inoculum 2002;53:10–13.

Pang KL, Jones EBG, Vrijmoed LLP, Vikineswary S. *Okeanomyces*, a new genus to accommodate *Halosphaeria cucullata* (Halosphaeriales, Ascomycota). Bot J Linn Soc. 2004;146:223–229.

Sakayaroj J, Pang KL, Jones EBG, Vrijmoed LLP, Abdel-Wahab MA. A systematic reassessment of marine Ascomycetes *Swampomyces* and *Torpedospora*. Bot Mar. 2005;48:395–406.

Sakayaroj J, Pang KL, Jones EBG. Multi-gene phylogeny of the Halosphaeriaceae: its ordinal status, relationships between genera and morphological character evolution. Fungal Divers. 2011;46:87–109.

Schoch CL, Sung GH, Volkmann-Kohlmeyer B, Kohlmeyer J, Spatafora JW. Marine fungal lineages in the Hypocreomycetidae. Mycol Res. 2006;110:257–263.

Suetrong S, Boonyuen N, Pang KL, Ueapattanakit J, Klaysuban A, Sri-indrasutdhi V, Sivichai S, Jones EBG. A taxonomic revision and phylogenetic reconstruction of the Jahnulales (Dothideomycetes), and the new family Manglicolaceae. Fungal Divers. 2011;51:163–188.

Smith AL. New or rare microfungi. Trans Br Mycol Soc. 1909;3:111–123.

Swe A, Jeewon R, Pointing SB, Hyde KD. Diversity and abundance of nematode-trapping fungi from decaying litter in terrestrial, freshwater and mangrove habitats. Biodivers Conserv. 2009;18:1695–1714.

6 Yeasts in marine environments

Jack W. Fell

Marine-occurring yeasts were first reported by Fischer and Brebeck (1894); subsequently yeasts have been observed in all oceans of the world, ranging from nearshore environments to oceanic surface waters and deep sea sediments. The early work emphasized the occurrence and distribution of yeasts in the world's ocean with the somewhat difficult task of determining the nomenclature of these species. The taxonomic puzzle became easier to resolve with the advent of molecular systematics. Currently marine yeast research is discovering a vast diversity of unknown genera and species, with a focus toward understanding their biology, ecology and role(s) in marine ecosystems. Reviews of marine yeast research have been presented by Kriss et al. (1967), Morris (1968), Fell (1976), Hagler and Ahearn (1987) and Kutty and Phillip (2008).

Systematics

Yeasts are a phylogenetically diverse group of ascomycetous and basidiomycetous yeasts that occur in virtually all environments. There are approximately 1500 known species in 149 genera. Yeasts can be defined by their distinguishing trait of asexual reproduction by budding or fission, which results in growth by single cells. Some species of yeasts produce limited to extensive hyphae; however the yeast growth phase is their predominant life form.

There is extensive but growing information on yeasts in terrestrial environments; yeasts from marine habitats fit within this framework of knowledge. As will be discussed, yeasts that occur in marine environments may be either terrestrial species that were introduced into coastal waters or endemic marine species whose life histories reside in specific marine habitats. Generally, the introduced species are able to survive for short to long periods of time and some species appear to have successfully taken up residence in both habitats.

Research with yeasts in marine environments will undoubtedly provide many surprises and new avenues of research. We anticipate that the majority of yeasts from marine habitats will fit into the present, and evolving, phylogenetic scheme for yeasts. This phylogenetic diversity within the ascomycetous and basidiomycetous yeasts is extensive. Yeasts and yeast-like species in the phylum Ascomycota are included in two subphyla: the Taphrinomycotina and Saccharomycotina with 72 teleomorphic and 14 anamorphic genera, respectively (Kurtzman 2011). The Basidiomycota includes three subphyla, the Pucciniomycotina, Agaricomycotina and Ustilaginomycotina, with 34 teleomorphic and 28 anamorphic genera (Boekhout et al. 2011). There is an extensive diversity of marine yeast species within the two phyla. Diversity within the basidiomycetes is discussed in Chapter 4. The majority of the ascomycetes from marine environments are terrestrially-associated species with a wide phylogenetic diversity. To date, the few endemic (or apparently endemic) species fall within five different families (Table 6.1).

There are close relationships between some of the species within these families, for example: the four species of *Metschnikowia* (Figure 6.1), which are often associated with marine animals, and several species of *Candida* (Figure 6.2) which belong to the Yamadazyma clade (Kurtzman 2011; Lachance et al. 2011) of the Debaryomycetaceae.

Table 6.1 Representative yeasts in the Saccharomycetales that occur mainly in marine habitats.

Family	Species	Habitat	Reference
Trichomonascaceae	*Blastobotrys parvus*	Oceanic water Southern Ocean	Fell and Statzell (1971), Fell (1976)
Debaryomycetaceae	*Candida atlantica*	Shrimp eggs and seawater, North Atlantic Ocean	Siepmann and Hohnk (1962), Gadanho et al. (2003)
Debaryomycetaceae	*Candida atmospherica*	Hydrothermal vents	Gadanho and Sampaio (2005)
Incertae sedis-*Ogataea* clade	*Candida maris*	Seawater Pacific	Van Uden and ZoBell (1962)
Debaryomycetaceae	*Candida oceani*	Hydrothermal vent	Burgaud et al. (2011)
Metschnikowiaceae	*Candida rhizophoriensis*	Florida Everglades	Fell et al. (2011)
Metschnikowiaceae	*Candida sharkiensis*	Florida Everglades	Fell et al. (2011)
Debaryomycetaceae	*Candida spencermartinsiae*	Coral reefs Florida and Belize; hydrothermal vent	Statzell-Tallman et al. (2010), Gadanho and Sampaio (2005)
Debaryomycetaceae	*Candida taylori*	Mangrove swamps and coral reefs, Florida	Statzell-Tallman et al. (2010)
Metschnikowiaceae	*Candida torressi*	Seawater Torres Strait, Australia	Van Uden and ZoBell (1962)
Saccharomycetaceae	*Kluyveromyces aestuarii*	Mangroves, Florida, Bahamas and Brazil	Fell (1961), Lachance (2011a)
Saccharomycetaceae	*Lachancea meyersii*	Mangroves: Bahamas	Fell et al. (2004)
Metschnikowiaceae	*Metschnikowia australis*	Seawater, South Shetland Islands; krill, Antarctica	Fell and Hunter (1968), Donachie and Zadanowski (1998)
Metschnikowiaceae	*Metschnikowia bicuspidata*	Pathogen of marine invertebrates, seawater, Pacific	Lachance (2011b)
Metschnikowiaceae	*Metschnikowia krissii*	Seawater, pathogen of copepods, Pacific Ocean	Lachance (2011b)
Metschnikowiaceae	*Metschnikowia zobellii*	Seawater and fish guts Pacific, seawater Atlantic Ocean	Van Uden and Castello-Branco (1961), Miller and Phaff (1998)

Yeasts in marine environments | 93

Figure 6.1 Marine yeasts in the ascomycetous genus *Metschnikowia*. Phylogenetic tree based on likelihood analysis of LSU D1/D2 rDNA gene sequences. See Lachance (2011b) for the relationship between these species within the genus.

Figure 6.2 Marine yeasts in the Yamadazyma clade of the Debaryomycetaceae. Phylogenetic tree based on the likelihood analysis of LSU D1/D2 rDNA gene sequences. See Kurtzman (2011) for the relationship between these species within the ascomycetous yeast.

These four *Metschnikowia* species were isolated from cold-water habitats in the Pacific and Southern Oceans. There are also marine species of *Candida*, which are assigned to the Metschnikowiaceae (Table 6.1), however, their habitats range from cold to subtropical waters and there is no evidence to suggest that they are animal associated. The marine members of the Debaryomycetaceae (Figure 6.2) occur in diverse environments ranging from mangroves to deep sea vents. The diversity of habitats and the presence of undescribed species (Mary 92, Mary 101) from hydrothermal vents (Gadanho and Sampaio 2005) suggest that many new species in this cluster will be discovered.

The systematics of these phylogenetic groups of yeasts is detailed in the 5th edition of *The Yeasts, A Taxonomic Study* (Kurtzman et al. 2011). The numbers of species and genera described in that treatise have risen dramatically from the 4th edition (Kurtzman and Fell 1998), where 100 genera and ~700 species were described. Yeast diagnostics has, for many years, relied on phenotypic characteristics, particularly morphology and the abilities to utilize various carbon and nitrogen compounds. The advent of nucleotide sequence analysis, particularly in the ITS and D1/D2 regions of the large subunit rDNA, is largely responsible for this rapid increase in the numbers of taxa described. Diagnostic sequences for all known yeasts were published by Kurtzman and Robnett (1998), Fell et al. (2000) and Scorzetti et al. (2002), which provided a method for rapid and accurate species identifications. The ability to ascertain the phylogenetic relationships between species led to the rapid discovery of unknown species and the description of new genera.

Yeast systematics is continuing to rapidly evolve due, in part, to our meager knowledge of the ecology of yeast species that inhabit the various habitats on earth. The currently described ~1500 species possibly represents as few as 1% of the yeasts in nature, an estimate that seems to be confirmed by the rapid rate of publications of new species. Consequently, as unknown species are discovered, we can anticipate a greater understanding of their biology, ecology and phylogeny.

Yeast ecology

Yeast ecology follows the general precepts of fungal ecology. Yeasts inhabit the majority of the ecological habitats throughout the world, where they are active decomposers in terrestrial, aquatic and marine food webs. Some species have distinct habitats and may act in close relationships with other micro- and macro-organisms as mutualists, parasites and pathogens. For example, members of the genus *Trichosporon* are generally associated with soils throughout most of the world, although some species can be serious pathogens of man and other animals (Gueho-Kellerman et al. 2011). In contrast, *Cyniclomyces guttulatus* is confined to rabbits and a few other herbivores (Boundy-Mills and Miller 2011). A comprehensive review of yeast ecology has been presented by Starmer and Lachance (2011). In the following discourse, we will discuss the occurrence of facultative, endemic and transient yeasts in coastal and oceanic environments.

Yeasts in marine environments: coastal regions

Nearshore areas are subject to terrestrial influxes due to natural drainage and human-induced activities ranging from swimming to land runoff, garbage dumping and sewer outfalls. Terrestrial and human-associated yeasts are introduced into these waters and may persist due, in part, to their tolerance to the saline conditions in sea water. Swimming beaches are a prime example of the direct pollution of marine environments where clinically-important yeasts have been reported. Vogel et al. (2007) isolated *Candida tropicalis, Candida albicans* and *Trichosporon aashii* in Florida USA beaches and Papadakis et al. (1997) identified *C. tropicalis, C. albicans* and *Trichosporon* spp. from bathing beaches in Greece. Human-associated species have been found in coastal waters of Greece (Velegraki-Abel et al. 1987), polluted shrimp and sediments in Brazil

(Hagler et al. 1982; Pagnocca et al. 1989; Soares et al. 1997) and beaches in Portugal (Sabino et al. 2011). Boguslawska-Was and Dabrowski (2001) suggested that the introduced species may have contributed to the pollution of Szczecin Lagoon in Poland via the degradation of industrial and domestic wastes. These pollution indicator species may extend into nearshore waters. In particular, Chen et al. (2009) reported that *C. tropicalis* is the most frequently recovered yeast in coastal waters of northeastern Taiwan.

There is an extensive diversity of yeasts that inhabit nearshore waters; a list of species that can and have been found would possibly include a majority of the known genera. For example, a study of unpolluted waters in the Florida Everglades demonstrated the presence of 64 described species (ten genera, 33 species of ascomycetes and 15 genera, 41 species of basidiomycetes), with an approximately equal number of undescribed species. Many of these species were undoubtedly introduced from the surrounding terrestrial vegetation. Some species, such as *Kwoniella mangroviensis, Candida sharkiensis, C. rhizophoriensis* and *Cryptococcus mangaliensis*, appear to be specifically associated with mangrove habitats (Statzell-Tallman et al. 2008). An important point is that studies in the Everglades natural environment have not shown the presence of pollution-indicator species (Lazarus et al. 1974; Fell et al. 2011).

Not all species are visitors to marine environments: an ever-increasing number of species are being found that appear to be endemic to shoreline and nearshore habitats. Among the earliest known endemic species is the basidiomycete *Cryptococcus marinus*, which was isolated by van Uden and ZoBell (1962) from water adjacent to algae and corals in the Torres Strait, Australia. There have not been any confirmed additional isolations of the species (Fonseca et al. 2011). The ascomycete *Kluyveromyces aestuarii* was first isolated from sediment in Biscayne Bay in Florida, USA (Fell 1961). It was subsequently isolated from mangrove areas in the Bahamas, sea water in the Torres Strait, and was found to be associated with marine invertebrates in Brazil (reviewed by Lachance 2011a), suggesting that the species is a subtropical/tropical marine inhabitant.

The ascomycetous genus *Metschnikowia* has significant endemic marine representatives. *Metschnikowia australis* was isolated from seawater in the South Shetland Islands (Fell and Hunter 1968) and from Antarctic krill (Donachie and Zdanowski 1998). *Metschnikowia bicuspidata* is a pathogen of marine invertebrates and fishes. The species caused milky disease in cultured crabs (*Portunus trituberculatus*) in China, with a high mortality of crabs and a significant economic loss (Wang et al. 2007). Similarly, the species has been implicated as a virulent pathogen of *Daphnia magna* (Ebert et al. 2000a, 2000b), Chinook salmon (Moore and Strom 2003) and freshwater prawns (Chen et al. 2007). Lachance (2011b) reported that one possible mechanism of predation was through the forceful ejection of spear-like spores. *Metschnikowia krissii*, a pathogen of copepods, was isolated off the coast of California and offshore in the Straits of Georgia, Canada, to water depths of 200 m. Pathogenicity by the species may be due to the production of extracellular enzymes that hydrolyze chondroitin sulfate and proteins (Seki and Fulton 1969).

There are species of yeasts that are capable of inhabiting both terrestrial and nearshore marine environments. They may be generalists, such as *Debaryomyces hansenii* and *Candida parapsilosis*, that are capable of utilizing a wide assortment of nutrients, or they may be niche-associated. An interesting example of the latter is the finding, via molecular sequence analysis, of *Malassezia* spp. associated with sponges in the coastal waters of Hawaii (Gao et al. 2008). This ascomycetous genus, which consists of 13 known species, is usually associated with the skin of humans and various animals such as cats, dogs, goats and cows (Gueho-Kellerman et al. 2011), see Chapter 19. Some species may cause diseases. Members of the genus have been isolated or identified by molecular analysis from terrestrial soils, with an association (Duarte et al. 2002; Renker et al. 2003) or presumed association (Fell et al. 2006) with nematodes. The molecular identification of the genus from the marine environment is significant as the data open the door for a transition from molecular parataxonomy to systems ecology and biology. Studies of the potential association of *Malassezia* with marine nematodes or other invertebrates should provide knowledge on the role and lifecycles of these yeasts in the environment and, in turn, give us an insight into the pathogenicity of the species. Molecular information has been paramount in recognizing the presence of these species, which would not be isolated by standard culturing conditions. Most species of *Malassezia* are lipophilic or lipid-dependent and require the supplement of a long-chain fatty acid (e.g., oleic acid) in the isolation medium for growth.

Offshore environments

Research in the open ocean, particularly the deep sea, faces two important problems: expensive ship time and the need to acquire samples aseptically, in conjunction with environmental parameters and from specific depths in the water column. Consequently, oceanographically-based studies are limited in number. Methods of collection and isolation of strains are detailed in Fell (1976) and do not need to be repeated here.

Early oceanographic studies in our laboratory examined the distribution of species in water masses of the Indian and Antarctic Oceans, via sampling from surface waters to the deep sea in longitudinal transects (Fell 1967, 1974, 1976). The studies confirmed the presence of yeasts in the open ocean, as reported by Kriss et al. (1967). Significantly, certain species were found to inhabit specific water masses or oceanographic regions. For example, *Blastobotrys (Sympodiomyces) parvus* was restricted to warmed Antarctic waters in the vicinity of the polar front and northward into the Subantarctic and Antarctic Intermediate waters (Fell and Statzell 1971). *Candida natalensis* was isolated in a narrow longitudinal zone in the Indo-Pacific (149E–177E, 50S–64S) from the polar front southward 245–3755 m deep in the water column. Population densities were 1–6 cells/liter. Similarly, *Candida norvegica* was restricted to a narrow geographical zone (151E–175W) southward from the polar front (53S–77S). The species was found in water masses with temperatures below 5°C. Population densities ranged from 1 to 200 cells/liter with an average of 30 cells/liter.

A basic problem with the early studies was that methods of identification were based on phenotypic characters; consequently no precise comparisons with known species were possible. Based on present molecular systematics, *C. natalensis* has only been studied from one strain, which was isolated from soils in Africa (van der Walt and Tscheuschner 1957). *C. norvegica* is a widely disseminated strain from soils, plants and clinical sources (Lachance et al. 2011). Whether or not the marine and terrestrial strains of the two species are related has not been examined. Fortunately, many of these strains, which were collected from the Indian and Antarctic Oceans, have been preserved at Agricultural Research Service (Northern Regional Research Laboratory) collection at the National Center for Agricultural Utilization Research, Peoria, IL, USA and the Centraalbureau voor Schimmelcultures, Utrecht, Netherlands. The strains are available for study via contact with the curators at either institution: the strains are not listed on the institutions' websites. A species, whose identity is not questioned due to the presence of a sexual cycle, is *Leucosporidium antarcticum*. This basidiomycete was found in Antarctic waters adjacent to the ice pack in concentrations of 1–22 cells/liter and at depths to 3000 m (Fell 1974). These few examples show that certain species are restricted in habitat, although the reasons for the ecological associations are not known. Clearly, the association of the psychrophilic species *L. antarcticum* with the ice pack would appear to be temperature dependent. The role of *L. antarcticum* may be associated with the high productivity of plankton and other members of the food chain due to the coastal upwelling that brings nutrient-rich deep waters to the surface.

Deep sea sediments

There has been considerable research into fungi and other microbes in deep sea sediments, including hydrothermal vents. Recent studies, using cloning and sequencing of DNA extracted directly from sediments or sequence analysis of isolated strains, have emphasized the occurrence of unknown species. Nagano et al. (2010) investigated deep sea sediments at five locations off the coast of Japan that ranged in depth from 1200 m to 10,000 m. The data, which were obtained by culture-independent polymerase chain reaction of the ITS region, documented a diversity of unknown species in addition to sequences similar to *Candida parapsilosis, Cryptococcus skinneri* and three species of *Metschnikowia (M. colocasiae, M. continentalis* and *M. kamakouana)*. These three species are generally associated with terrestrial plants and beetles.

Hydrothermal vents are biologically active areas due to the chemicals and hot water that issue from the vents. Le Calvez et al. (2009) examined vents in the Pacific and Atlantic Ocean and reported a diversity of unknown fungal groups, including species of *Cryptococcus* and *Filobasidium*. Unfortunately for yeast identification, the direct polymerase chain reaction method employed the SSU region, which is a broad spectrum approach and not conducive to precise identification of yeasts. Bass et al. (2007), who also employed the SSU region, found a diversity of unknown species and concluded that yeasts, rather than filamentous fungi, were the prevalent form of fungi in the deep sea. Gadanho and Sampaio (2005) explored thermal systems of the Mid-Atlantic Rift using culture based methods and the D1/D2 region of the LSU rDNA. They reported numbers of colonies ranging from 0 to 5940 cells/liter and species identifications included both marine- and terrestrially-associated species. They reported two unusual species: the clinically important *Exophiala dermatitidis* and *Trichosporon dermatis*. Species often found in marine waters included *Candida atlantica, Candida atmospherica, C. parapsilosis, Rhodosporidium diobovatum* and *R. sphaerocarpum*. Unknown species represented 33% of the yeast taxa. Similar results were reported by Burgaud et al. (2010) from a series of studies at vents in the Mid-Atlantic Ridge, South Pacific Basins and East Pacific Rise.

Summary

Yeasts occur in nearshore waters, in conjunction with filamentous fungi, with a high diversity of species and genera and a population density that can be as high as thousands of cells per liter. These inshore waters are susceptible to the influx of pollution indicator species, such as *C. tropicalis, C. albicans* and *Trichosporon* spp. These species, due to factors such as ocean currents and weather conditions, can extend their presence into offshore waters.

In offshore regions there is a reduction in the diversity of species and population densities. However, yeasts are the dominant fungal form, as the unicellular yeast cell is better adapted to the aqueous environment than fungal hyphae. Numerically, yeast cells in open ocean waters may range from 0 to 10 cells/liter, although regions of high organic activity will result in intensive yeast activity. For example, Meyers et al. (1967) reported cell numbers of up to 3000 cells/liter in the North Sea and speculated on there being a relationship to *Noctiluca* blooms.

Cell counts should be taken into the context that the numbers represent a standing crop of those species that will grow on the culture medium. Feeding rates of marine invertebrates, varied growth rates on culture media and other factors must be taken into consideration. Not all species will grow on any one specific isolation medium; however, molecular methods are demonstrating the presence of additional species. Neither cultures, nor molecular methods are completely reliable in their ability to discern the complete yeast community structure; therefore both methods should be used in concert. Direct polymerase chain reaction of marine materials has the advantage of the rapid detection of many known and unknown species; however, culture studies are necessary to develop knowledge of their biology and ecology and to exploit their potentially commercially- and clinically-important resources.

Molecular methods used in the identification and phylogeny of yeasts is presently based on the ITS and D1/D2 LSU regions of the rDNA. The SSU region is not used, due to the lack of species discrimination. Consequently, the SSU database only contains data on a few species. These methods will change as whole-genome sequence data becomes more universally available. The speculation that the known species of yeasts may represent only 1% of yeasts in nature provides the potential for rapid increases in our knowledge of marine yeast biology and ecology. This assumption must be tempered with species losses due to climate change and other moderate to severe impacts on the environment as a result of commercial and industrial shoreline to oceanic activities. Consequently, there is a need for rapid responses in marine yeast research.

Research is not only important for academic knowledge, but also because of its industrial and clinical significance. In biotechnology, yeasts are considered to be the most important group of organisms as they are, on a worldwide basis, the major producers of biotechnical products (see the review by Johnson and Echavarri-Erasun 2011). A literature search on the exploitation of marine yeasts for industrial processes reveals an impressive level of research, such as the use of their bioactive substances in mariculture (Chi et al. 2006), production of extracellular enzymes (Chi et al. 2009), bioethanol (Kathiresan et al. 2011) and unsaturated fatty acids (Cui et al. 2011). The intensity of research is undoubtedly higher than is publically known due to commercial research, which is rarely published. Undoubtedly, research on marine yeasts is a critically important endeavor that requires increased effort and expertise.

References

Bass D, Howe A, Brown N, Barton H, Demidova M, Michelle H, Li L, Sanders H, Watkinson SC, Willcock S, Richards TA. Yeast forms dominate fungal diversity in the deep oceans. Proc R Soc B. 2007;274:3069–3077.

Boekhout T, Fonseca A, Sampaio JP, Bandoni RJ, Fell JW, Kwon-Chung KJ. Discussion of teleomorphic and anamorphic basidiomycetous yeasts. In: Kurtzman CP, Fell JW, Boekhout T, eds. The yeasts, a taxonomic study. 5th ed. Amsterdam: Elsevier, 2011:1339–1374.

Boguslawska-Was E, Dabrowski W. The seasonal variability of yeasts and yeast-like organisms in water and bottom sediment of the Szczecin Lagoon. Int J Hyg Environ Health. 2001;203: 451–458.

Boundy-Mills K, Miller MW. *Cyniclomyces* van der Walt *et* Scott (1971). In: Kurtzman CP, Fell JW, eds. The yeasts, a taxonomic study. 5th ed. Amsterdam: Elsevier, 2011:357–361.

Burgaud G, Arzur D, Durand L, Cambon-Bonavita M, Barbier G. Marine culturable yeasts in deep-sea hydrothermal vents: species richness and association with fauna. FEMS Microbiol Ecol. 2010;73:121–133.

Burgaud G, Arzur D, Sampaio JP, Barbier G. *Candida oceani* sp. nov., a novel yeast isolated from a Mid-Atlantic Ridge hydrothermal vent (-2300 meters). Antonie Leeuwenhoek 2011;100:75–82.

Chen SC, Chen YC, Kwang JM, Manopo I, Wang PC, Chaung HC, Liaw LL, Chiu SH. *Metschnikowia bicuspidata* dominates in Taiwanese cold-weather yeast infections of a *Macrobrachium rosenbergii*. Dis Aquat Organ. 2007;75:191–199.

Chen Y, Yanagida F, Chen L. Isolation of marine yeasts from coastal waters of northeastern Taiwan. Aquatic Biol. 2009;8:55–60.

Chi Z, Liu Z, Gao L, Fong F, Ma C, Wang X, Li H. Marine yeasts and their applications in mariculture. J Ocean Univ China. 2006;3:251–256.

Chi Z, Chi Z, Zhang T, Liu G, Li J, Wang X. Production, characterization and gene cloning of the extracellular enzymes from marine-derived yeasts and their potential applications. Biotech Adv. 2009;27:236–255.

Cui Y, Fraser C, Gardner G, Huang D, Reith M, Windust AJ. Isolation and optimization of the oleaginous yeast *Sporobolomyces roseus* for biosynthesis of 13C isotopically labeled 18-carbon unsaturated fatty acids and trans 18:1 and 18:2 derivatives through synthesis. J Ind Microbiol Biotechnol. 2011;39:153–161.

Donachie SP, Zdanowski MK. Potential digestive function of bacteria in krill *Euphausia superba* stomach. Aquat Microb Ecol. 1998;14:129–136.

Duarte ER, Lachance M-A, Hamdan JS. Identification of atypical strains of *Malassezia* spp. from cattle and dog. Can J Microbiol. 2002;48:749–752.

Ebert D, Lipsitch M, Mangin KL. The effect of parasites on host population density and extinction: experimental epidemiology with *Daphnia* and six micro-parasites. Am Nat. 2000a;156: 459–477.

Ebert D, Zschokke-Rohringer CD, Carius HJ. Dose effects and density-dependent regulation of two microparasites of *Daphnia magna*. Oecologia 2000b;122:200–209.

Fell JW. A new species of *Saccharomyces* isolated from a subtropical estuary. Antonie Leeuwenhoek 1961;27:27–30.

Fell JW. Distribution of yeasts in the Indian Ocean. Bull Mar Sci. 1967;17:454–470.

Fell JW. Distribution of yeasts in water masses of the Southern Ocean. In: Colwell R, Morita R, eds. Effect of the ocean environment on microbial activities. Baltimore: University Park Press, 1974:510–523.

Fell JW. Yeasts in oceanic regions. In: Jones EBG, ed. Recent advances in aquatic mycology. London: Elek Science, 1976:93–124.

Fell JW, Hunter IL. Isolation of heterothallic yeast strains of *Metschnikowia kamienski* and their mating reactions with *Chlamydozyma wickerham* spp. Antonie Leeuwenhoek 1968;34:365–376.

Fell JW, Statzell A. *Sympodiomyces* gen. n., a yeast-like organism from southern marine waters. Antonie Leeuwenhoek 1971;37:359–367.

Fell, JW, Boekhout T, Fonseca A, Scorzetti G, Statzell-Tallman A. Biodiversity and systematics of basidiomycetous yeasts as determined by large subunit rD1/D2 domain sequence analysis Int J Syst Evol Microbiol. 2000;50:1351–1371.

Fell, JW, Statzell-Tallman A, Kurtzman CP. *Lachancea meyersii* sp. nov., an ascosporogenous yeast from mangrove regions in the Bahama Islands. Stud Mycol. 2004;50:359–363.

Fell JW, Scorzetti G, Connell L, Craig S. Biodiversity of micro-eukaryotes in Antarctic Dry Valley soils with <5% soil moisture. Soil Biol Biochem. 2006;38:3107–3115.

Fell JW, Statzell-Tallman A, Scorzetti G, Gutiérrez MH Five new species of yeasts from fresh water and marine habitats in the Florida Everglades. Antonie Leeuwenhoek 2011;99:533–549.

Fischer B, Brebeck C. Zur Morphologie, Biologie und Systematik der Kahmpilze, der *Monilia candida* Hansen und des Soorerregers C. Jena: Fischer, 1894:1–52.

Fonseca A, Boekhout T, Fell JW. *Cryptococcus* Vuillemin (1901). In: Kurtzman CP, Fell JW, Boekhout T, eds. The yeasts, a taxonomic study. 5th ed. Amsterdam: Elsevier, 2011:1661–1738.

Gadanho M, Sampaio J. Occurrence and diversity of yeasts in the Mid-Atlantic Ridge hydrothermal fields near the Azores archipelago. Microb Ecol. 2005;50:408–417.

Gadanho M, Almeida JMF, Sampaio JP. Assessment of yeast diversity in a marine environment in the south of Portugal by microsatellite-primed PCR. Antonie Leeuwenhoek 2003;84:217–227.

Gao Z, Li B, Zheng C, Wang G. Molecular detection of fungal communities in the Hawaiian marine sponges *Suberites zeteki* and *Mycale armata*. Appl Environ Microbiol. 2008;74:6091–6101.

Gueho-Kellerman E, Batra R, Boekhout T. *Malassezia* Baillon (1889). In: Kurtzman CP, Fell JW, eds. The yeasts, a taxonomic study. 4th ed. Amsterdam: Elsevier, 2011:1807–1832.

Hagler AN, Ahearn DG. Ecology of aquatic yeasts. In: Rose AH, Harrison JS, eds. The yeasts. 2nd ed. Vol. 1. London: Academic Press, 1987:181–205.

Hagler AN, Oliveira RB, Mendonca-Hagler LC. Yeasts in the intertidal sediments of a polluted estuary in Rio de Janeiro, Brazil. Antonie Leeuwenhoek 1982;48:53–56.

Johnson EA, Echavarri-Erasun C. Yeast biotechnology. In: Kurtzman CP, Fell JW, eds. The yeasts, a taxonomic study. 4th ed. Amsterdam: Elsevier, 2001:21–44.

Kathiresan K, Saravanakumar K, Senthilraja. Bio-ethanol production by marine yeasts isolated from coastal marine mangrove sediment. Internat Multidiscipl Res J. 2011;1:19–24.

Kriss AE, Mishustina IE, Mitskevich IN, Zemtsoval EV. Microbial populations of oceans and Seas. New York: St Martin's Press, 1967.

Kurtzman CP. Discussion of teleomorphic and anamorphic ascomycetous yeasts and yeast-like taxa. In: Kurtzman CP, Fell JW, eds. The yeasts, a taxonomic study, 4th ed. Elsevier: Amsterdam, 2011:293–307.

Kurtzman CP, Fell JW. The yeasts, a taxonomic study. 4th ed. Amsterdam: Elsevier, 1998:1055.

Kurtzman CP, Robnett CJ. Identification of clinically important ascomycetous yeasts based on nucleotide divergence in 5' end of the large-subunit (26S) ribosomal DNA gene. J Clin Microbiol. 1997;35:1216–1223.

Kurtzman CP, Fell JW, Boekhout T. The yeasts, a taxonomic study. 5th ed. Vols 1–3. Amsterdam: Elsevier, 2011:2080.

Kutty SN, Philip R. Marine yeasts – a review. Yeast. 2008;24:465–483.

Lachance M-A. *Kluyveromyces* van der Walt (1971). In: Kurtzman CP, Fell JW, eds. The yeasts, a taxonomic study. 5th ed, Amsterdam: Elsevier, 2011a:471–482.

Lachance MA. *Metschnikowia* Kamienski (1899). In: Kurtzman CP, Fell JW, eds. The yeasts, a taxonomic study. 5th ed. Amsterdam: Elsevier, 2011b:575–621.

Lachance MA, Boekhout T, Scorzetti G, Fell JW, Kurtzman CP. *Candida* Berkhout (1923). In: Kurtzman CP, Fell JW, eds. The yeasts, a taxonomic study. 5th ed. Amsterdam: Elsevier, 2011:987–1278.

Lazarus CR, Koburger JA. Identification of yeasts from the Suwannee River Florida Estuary. Appl Environ Microbiol. 1974;27:1108–1111.

Le Calvez T, Burgaud G, Mahé S, Barbier G, Vandenkoornhuyse P. Fungal diversity in deep-Sea hydrothermal ecosystems. Appl Environ Microbiol. 2009;75:6415–6421.

Meyers SP, Ahearn DG, Gunkel W, Roth FJ. Yeasts from the North Sea. Mar Biol. 1967;1:118–123.

Miller MW, Phaff HJ. *Metschnikowia* Kamienski. In: CP Kurtzman, JW Fell, eds. The yeasts, a taxonomic study. 4th ed. Amsterdam: Elsevier, 1998:256–267.

Moore MM, Strom MS. Infection and mortality by the yeast *Metschnikowia bicuspidate* var. *bicuspidate* in Chinook salmon fed live adult brine shrimp (*Artemia franciscana*). Aquaculture 2003;220:43–57.

Morris EO. Yeasts of marine origin. Oceanogr Mar Biol Ann Rev. 1968;6:201–230.

Nagano Y, Nagahma T, Hatada Y, Nunoura T, Takami H, Miyazaki J, Takai K, Horikoshi K. Fungal diversity in deep-sea sediments – the presence of novel fungal groups. Fungal Ecol. 2010;3:316–325.

Pagnocca FC, Mendonca-Hagler LC, Hagler AN. Yeasts associated with the white shrimp *Penaeus schmitti*, sediment and water of Sepetiba Bay, Rio de Janeiro, Brazil. Yeast, 1989;5:5479–5483.

Papadakis JS, Mavridou A, Richardson SC, Lamprini M, Marcelou U. Bather-related microbial and yeast populations in sand and seawater. Water Res. 1997;31:799–804.

Renker C, Alphei J, Buscot F. Soil nematodes associated with the mammal pathogenic fungal genus *Malassezia* (Basidiomycota: Ustilginomycetes) in Central European forests. Biol Fertil Soils. 2003;37:70–72.

Sabino R, Verissimo C, Cunha MA, Wergikoski B, Ferreira FC, Rodrigues R, Prada H, Falcao L, Rosado L, Pinheiro C, Paixao E, Brandao J. Pathogenic fungi: an unacknowledged risk at coastal resorts? New insights on microbiological sand quality in Portugal. Mar Poll Bull. 2011;62:1506–1511.

Scorzetti G, Fell JW, Fonseca A, Statzell-Tallman A. Systematics of basidiomycetous yeasts: a comparison of large sub-unit D1D2 and internal transcribed spacer rDNA regions. FEMS Yeast Res. 2002;2:495–517.

Seki H, Fulton J. Infection of marine copepods by *Metschnikowia* sp. Mycopath Mycol Appl. 1969;38:61–70.

Siepmann R, Hohnk W. Uber Hefen und einige Pilze (Fungi imp., Hyphales) aus dem Nordatlantik. Veroeff Inst Meeresforsch Bremerhaven. 1962;8:79–97.

Soares CAG, Maury M, Pagnocca FC, Araujo FV, Mendonca-Hagler LC, Hagler AN. Ascomycetous yeasts from tropical intertidal dark mud of southeast Brazilian estuaries. J Gen Appl Microbiol. 1997;43:265–272.

Starmer WT, Lachance MA. Yeast ecology. In: Kurtzman CP, Fell JW, eds. The yeasts, a taxonomic study. 5th ed. Amsterdam: Elsevier, 2011:65–86.

Statzell-Tallman A, Belloch, Fell JW. *Kwoniella mangroviensis* gen. nov., sp. nov. a tremellaceous yeast from mangrove habitats in the Florida Everglades and Bahamas. FEMS Yeast Res. 2008;8:103–113.

Statzell-Tallman A, Scorzetti G, Fell JW. *Candida spencermartinsiae* sp. nov., *Candida taylorii* sp. nov. and *Pseudozyma abaconensis* sp. nov., novel yeasts from mangrove and coral reef ecosystems. Inter J Syst Evol Microbiol. 2010;60:1978–1984.

van der Walt JP, Tscheuschner IT. *Pichia vanriji* sp. nov. isolated from soil. J Gen Microbiol. 1956;15:459–461.

van Uden N, Casgtelo-Branco. *Metschnikowiella zobellii* sp. nov. and *M. krissii*, sp. nov. two yeasts from the Pacific Ocean pathogenic for *Daphnia magna*. J Gen Microbiol. 1961;26:141–148.

van Uden N, ZoBell CE. *Candida marina* nov. spec., *Torulopsis torresii* nov. spec. and *T. maris* nov. spec., three yeasts from the Torres Strait. Antonie Leeuwenhoek 1962;28:275–283.

Velegraki-Abel V, Marselou-Kinti U, Richardson D. Incidence of yeasts in coastal sea water of the Attica Peninsula, Greece. Water Res. 1987;21:1363–1369.

Vogel C, Rogerson A, Schatz S, Laubach H, Tallman A, Fell J. Prevalence of yeasts in beach sand at three bathing beaches in South Florida. Water Res. 2007;41:1915–1920.

Wang X, Chi Z, Yue L, Li J, Li M, Wu L. A marine killer yeast against the pathogenic yeast strain in crab (*Portunus trituberculatus*) and an optimization of the toxin production. Microbiol Res. 2007;162:77–85.

7 Zoosporic true fungi

Frank H. Gleason, Frithjof C. Küpper and Sally L. Glockling

The zoosporic true fungi include a very large and diverse group of microorganisms that have often historically been referred to as chytrids. These fungi are frequently observed in fresh water and soil ecosystems (Powell 1993; Barr 2001; Shearer et al. 2007). Powell (1993) called for wider recognition of the zoosporic true fungi which, despite their environmental and economic importance, had either been excluded from many biology text books or had been erroneously discussed within the kingdom Protista. Fossil records have shown that the zoosporic true fungi were in existence over 400 million years ago (Taylor et al. 1992). Although the literature on zoosporic true fungi from soil and freshwater environments is extensive, there has been very little research on the marine species. In this chapter, we discuss some of the general characteristics of the zoosporic true fungi. Then we focus on the morphology and ecology of several well-documented marine species.

Taxonomy and phylogeny

Zoosporic true fungi were historically considered to be 'lower fungi' or Phycomycetes along with the Oomycota because of their zoosporic mode of reproduction (Sparrow 1960; Barr 1990). Bartniki-Garcia (1970) considered the zoosporic true fungi to be true fungi based largely on their biochemistry. Early phylogenetic analysis of the ribosomal RNA genes of a few species of zoosporic true fungi confirmed that this group was indeed basal within the kingdom Fungi (Forster et al. 1990) and unrelated to the biflagellate Oomycota.

The classification of the zoosporic true fungi was previously based entirely on morphological characteristics observed under the light microscope (Sparrow 1960; Karling 1977). However, the systematics of this group has recently undergone considerable reorganization based on the ultrastructure of zoospores (Barr 1981) and, more recently, on sequences of ribosomal RNA genes (James et al. 2000, 2006; Letcher et al. 2008; Mozley-Standridge et al. 2009; Simmons et al. 2009; Wakefield et al. 2010).

Most of the zoosporic true fungi that were formerly considered to be in the phylum Chytridiomycota, as defined by Barr (2001), are currently placed into three phyla – Blastocladiomycota, Chytridiomycota and Neocallimastigomycota (James et al. 2006). A fourth phylum including the Monoblepharidales has recently been proposed (M.J. Powell pers. comm.). Species in the genus *Olpidium* have not yet been assigned to any phylum (James et al. 2006) and the phylogeny of *Olpidium* will be considered later in this chapter. Species in the genus *Rozella* are no longer considered to be zoosporic true fungi (Lara et al. 2010; Jones et al. 2011) and are currently placed in a new clade, the Cryptomycota, which falls close to the point where animals and fungi diverge in phylogenetic trees (see Chapter 1 and Chapter 10). Molecular studies of zoosporic true fungi are limited by the numbers of species that can be maintained in culture and by the current database of sequences available in GenBank. The aquatic and obligate parasitic niche of many species renders them more difficult to isolate into pure culture and to acquire molecular data.

There are approximately 1,250 species described in the zoosporic true fungi (James et al. 2006; Shearer et al. 2007). A comprehensive molecular analysis of zoosporic true fungi based on ribosomal RNA genes indicated that this group, under its present classification, is polyphyletic (James et al. 2000). Later molecular analyses of SSU rRNA genes of 35 species showed that zoosporic true fungi fall into many clades (Küpper et al. 2006; Letcher et al. 2006, 2008). The same studies also confirmed that many traditional genera of zoosporic true fungi, including *Chytridium* and *Rhizophydium*, are polyphyletic in origin. A more accurate assignment of taxa, particularly of those from aquatic environments, may become possible in the future following further ultrastructural studies of zoospores. In addition to the molecular evaluation of known species of zoosporic true fungi, environmental sequencing techniques are now revealing the presence of further species, particularly from marine environments, in clades that are not morphologically known (Küpper et al. 2006; Le Calvez et al. 2009; Nagano et al. 2010).

Morphology

The zoosporic true fungi produce zoospores in asexual reproduction. These are characterized by a single posteriorly-directed whiplash flagellum (Sparrow 1960). The mature sporangia are thin-walled, saccate or globose structures that develop exit papilla that may be plugged with mucoid material prior to zoospore release or may have an operculum to allow the zoospores to exit. Zoospore cleavage is intrasporangial and the zoospores are flagellate and fully motile when they exit the sporangium. In eucarpic species, rhizoids are produced around the sporangium and these may form a network. Many species produce thick-walled resting spores in addition to sporangia, and these may be formed either sexually or asexually. The resting spore wall may be multi-layered and either smooth or sculptured.

Ultrastructure and biochemistry

Ultrastructural details of the zoosporic true fungi have been well documented and recognized as having phylogenetic and taxonomic implications (Barr 1981, 1986; Beakes et al. 1988; Longcore 1992; James et al. 2006). The mitochondria of the Chytridiomycota and the Blastocladiomycota have plate-like cristae similar to other members of the super group Opisthokonta (Barr 2001; Lara et al. 2010). However, the species in the order Neocallimastigales, which are obligate anaerobes in the digestive systems of some mammals and reptiles, possess hydrogenosomes in the place of mitochondria (Trinci et al. 1994).

Ultrastructural features, particularly of the zoospore kinetosome and associated flagellar rootlet system, were considered to be well-conserved features of phylogenetic significance (Barr 1986; Longcore 1992). However, James et al. (2006) have shown that, while these detailed ultrastructural features are consistent within a lineage, some features have been repeatedly lost in separate lineages and others have repeatedly evolved. The mapping of these ultrastructural features onto phylogenetic trees has shown that our concept of the importance of ultrastructural details in classification needs to be reevaluated (James et al. 2006). The zoospore body of fungi in the Chytridiomycota contains a conspicuous microbody-lipid globule complex (Powell 1978). Ultrastructural differences have been observed between species belonging to the Blastocladiomycota and those of the Chytridiomycota. In particular, a closed mitotic pole has been reported for species in the Blastocladiomycota, whereas this is open in species belonging to the Chytridiomycota (James et al. 2006).

The thallus walls of the zoosporic true fungi contain chitin, as do the mycelia and nonmotile spores of other true fungi (Barr 2001; James et al. 2006). Carbohydrate is stored as glycogen and an α-aminoadipic acid lysine synthetic pathway is present in zoosporic true fungi (James et al. 2006).

General ecology

Zoosporic true fungi include a diverse group of ubiquitous microorganisms that have frequently been observed as saprobes and parasites in fresh water and soil ecosystems (Sparrow 1960; Powell 1993; Barr 2001; Kagami et al. 2007; Shearer et al. 2007). A few species have also been isolated from marine habitats, but the paucity of zoosporic true fungi in saline environments may be due to poor sampling. Sparrow (1960) showed that zoosporic true fungi utilize an extensive range of substrates, which demonstrates their diverse nutritional capabilities. These fungi are also important biodegraders of cellulose, chitin and keratin (Powell 1993).

Most species of zoosporic true fungi are saprotrophs but many are parasites of algae and higher plants (Sparrow 1960; Powell 1993; Barr 2001; Kagami et al. 2007; Shearer et al. 2007), some causing important economic crop diseases such as black wart of potatoes (Powell 1993). *Olpidium brassicae*, an obligate root parasite, has been shown to be a vector of several plant viruses (Powell 1993). A few species parasitize the eggs of invertebrates, such as nematodes and rotifers (Karling 1946; Barron and Szijarto 1986) and some species are known to infect dipteran larvae (Gleason et al. 2010a). The lifecycle of these parasites can be complex and *Coelomomyces*, an important parasite of mosquitoes, requires a second host, a crustacean, to complete its reproduction. This fungal parasite also ingeniously utilizes the reproductive system of the mosquito host for propagation and dispersal of its own resting spores (Powell 1993). Only one species of zoosporic true fungi, *Bactrachochytrium dendrobatidis*, is known to infect vertebrates, and this causes serious population declines in frogs (Longcore et al. 1999).

Zoosporic true fungi have been isolated from soil and fresh water ecosystems from a range of climatic zones, from the tropics to the Arctic (Powell 1993). Experimental studies have shown they can survive harsh environmental conditions (Powell 1993) and they have often been observed in extreme natural environments (Gleason et al. 2010b). For example, zoosporic true fungi have been shown to represent about three-quarters of the total fungal diversity in soil at high elevations despite the requirement for free water for zoospore dispersal (Freeman et al. 2009). The resting spores of zoosporic true fungi have been reported to have extended longevity (Laidlaw 1985), which may enable repeated outbreaks of infection or disease.

Some species of zoosporic true fungi are obligate parasites that rely on their zoospores to locate and penetrate a specific host in order to propagate. The zoospores are thought to be chemotactic and to respond to chemical cues released from the host (Powell 1993). Chemotaxis would be particularly important where infection is host-specific, the host is mobile or the environment is fluid, such as in the intertidal zone. The cellular basis for signal reception and transduction in chemotactic zoospores has been reported in an aquatic species infecting diatoms (Canter and Jaworski 1980).

Marine and brackish environments

Only a few species of zoosporic true fungi have been reported in brackish and marine ecosystems (Sparrow 1936, 1960; Johnson and Sparrow 1961; Porter and Kirk 1987; Shearer et al. 2007; Alias and Jones 2009). In particular, a few species belonging to four genera, *Rhizophydium*, *Chytridium*, *Thalassochytrium* and *Phlyctochytrium* (now *Spizellomyces*), have been characterized as estuarine or marine ecotypes (Sparrow 1960; Johnson and Sparrow 1961; Ulken 1978; Amon 1984; Raghukumar 1986a; Shields 1990; Küpper and Müller 1999; Nyvall et al. 1999; Alias and Jones 2009). Most of these species are thought to be parasites; however, many of the species found in marine ecosystems may not be correctly identified and their descriptions may be incomplete.

Little is known about the distribution of zoosporic true fungi in brackish and marine environments because only about a dozen ecotypes have been properly identified and characterized and their global distribution has not been thoroughly studied (Ulken 1978; Amon 1984; Raghukumar 1986a; Shields 1990; Müller et al. 1999; Nyvall et al. 1999; Gleason et al. 2011). However, analysis of environmental samples using molecular techniques has revealed unknown clades of zoosporic true fungi along with other groups of heterotrophic eukaryotic microorganisms in many diverse marine ecosystems (Le Calvez et al. 2009; Lara et al. 2010; Nagano et al. 2010; Jones et al. 2011). This suggests that marine ecotypes are more prevalent than previously thought.

Marine and estuarine species

Chytridiomycota

Rhizophydium littoreum

Rhizophydium littoreum is an epibiotic, monocentric species that has been observed growing on the surface of two common genera of siphonous green algae, *Bryopsis* and *Codium* (Chlorophyta), in estuaries along the east coast of North America. This fungus has never been documented on other genera of estuarine or marine algae. Four morphotypes of this fungus have been isolated (Kazama 1972; Amon 1976; Porter and Smiley 1980; Amon 1984) but little is known about the pathology of these estuarine morphotypes on their green algal hosts.

Another morphotype of *R. littoreum* was observed growing primarily as a saprobe on dead eggs of the subtidal yellow rock crab, *Cancer anthonyi* (Shields 1990). This fungus also attacks and kills live eggs, although significantly more dead eggs are infected. The prevalence of infection ranged from 14 to 52% throughout the year at sites along the west coast of North America.

Since all five morphotypes grow well in pure culture on chemically-defined media, they are probably facultative parasites with a low degree of virulence, or more likely saprobes. These fungi produce large numbers of zoospores that range in size from 3 to 4 µm in diameter. Sequence analysis of two ribosomal RNA genes from one of the estuarine isolates confirms that this fungus belongs in the order Rhizophydiales (GenBank accession numbers DQ485540, DQ485604; Letcher et al. 2006). However, since molecular data are only available for one isolate, nothing is known about the genetic diversity within these morphotypes.

Thalassochytrium gracilariopsis

Thalassochytrium gracilariopsis is an endobiotic, polycentric species that infects *Gracilariopsis* spp. (Rhodophyta; Nyvall et al. 1999). This fungus has never been reported as a parasite on other species of estuarine or marine algae, and only one morphotype has been studied.

Nyvall et al. (1999) noted that *T. gracilariopsis* released zoospores in such large numbers that they caused a yellowing of the sea water. The zoospores are ovoid and measure about 3 µm in diameter. Intensely orange and large endobiotic sporangia (50–300 µm in diameter) are embedded in algal tissues with only the tips of the discharge tubes protruding. Multinucleate, septate hyphae penetrate between the cells in the tissues of the algal host and form haustoria, which mainly penetrate the medullary cells.

The fungal infection does not kill the thalli of the host but causes the thylakoid membranes to become disorganized. The starch reserves are also degraded in infected medullary cells during sporulation of the fungus. The impact of *T. gracilariopsis* on the growth of the host is not known. Whether this is a true parasitic or mutualistic relationship has not been determined, but the relationship appears to be host specific.

This fungus has never been grown in pure culture, and molecular data are not yet available.

Chytridium polysiphoniae

Chytridium polysiphoniae is an epibiotic, monocentric species that infects *Pylaiella littoralis* and other species of brown algae (Phaeophyta; Raghukumar 1987; Küpper and Müller 1999; Müller et al. 1999; Küpper et al. 2006). This fungus has also been reported as a parasite of members of three species of red algae (Rhodophyta) in the genera *Callithamnion* (Martin 1922), *Polysiphonia* (Sparrow 1936; Johnson 1966) and *Centroceras* (Raghukumar 1986a). In addition, Sparrow (1936) reported *C. polysiphoniae* as a saprotroph on *Ceramium*. However, it is not clear whether the reports by Küpper and Müller (1999), Raghukumar (1986a, 1987), Johnson (1966) and Sparrow (1936) all relate to the same biological species. This is because this species has few morphological features, sequence data are unavailable, and cross-infection studies were not carried out by most of these investigators.

The sporangia of *C. polysiphoniae* are attached to the surface of the algal host cells while the haustoria penetrate into the cytoplasm of individual cells. The zoospores of *C. polysiphoniae* are 3–4 µm in diameter. These parasites of brown and red algae have not yet been grown in pure culture, but they can be grown in co-culture with the brown alga *P. littoralis* (Müller et al. 1999) or with the red alga *Centroceras clavulatum* (Raghukumar 1986a), and therefore are considered to be obligate parasites or biotrophs.

Zoospores of *C. polysiphoniae* attack the uniseriate filaments of filamentous brown algae, the distal uniseriate parts of polystichous thalli and the filamentous microthalli of brown algae with heteromorphic lifecycles, such as gametophytes (Müller et al. 1999).

In "culture" the cytoplasm of individual infected host cells disintegrates and the cells eventually die, but the remaining parts of the algal host will continue to grow. This fungus prefers to infect actively-growing uniseriate filaments. It is not known whether *C. polysiphoniae* can attack cells in the sheet-like thalli of larger brown algae such as kelps. Infection of filamentous brown algae by *C. polysiphoniae* and other eukaryotic parasites such as the oomycete *Eurychasma* (Wilce et al. 1982) causes fragmentation of the filaments. These fragments are carried around in water currents along with attached sporangia belonging to the parasites (Müller et al. 1999). This is probably an effective method for dispersal of the parasites of brown algae. Zoospores of *C. polysiphoniae* also attack the filaments and seta of the red alga *Centroceras clavulatum* (Raghukumar 1986a), but details of the progression of this disease in red algae are not clearly understood.

Only one isolate of *C. polysiphoniae*, pathogenic on brown algae, has been sequenced, and it clustered with that of a novel clade of the Chytridiomycota (GenBank accession number AY032608; Küpper et al. 2006). Thus *C. polysiphoniae* should probably not be placed within the genus *Chytridium*. This clade is composed of environmental sequences of uncultured and morphologically-unknown taxa. As more DNA samples are collected and ribosomal and other genes are sequenced, new clades are likely to be described.

Phlyctochytrium mangrovii (now in the genus *Spizellomyces*)

Ulken (1978) observed a variety of zoosporic true fungi growing on pine pollen baits in marine mud samples mixed with diluted sea water. In particular, *Phlyctochytrium mangrovii* was isolated into pure culture from mud collected from four mangrove swamps (Cananeia, Brazil; Pearl Harbor, HI, USA; He'eia, HI, USA; and Vera Cruz, Mexico). This fungus was described as a new species based on morphological studies. Ulken (1978) investigated the effects of several physical factors, including the concentration of sodium chloride, on the growth of the four isolates of *P. mangrovii* in culture. Although these isolates grew in a wide range of salt concentrations, they appeared to be better adapted to environments with salinity values below that of pure sea water; therefore *P. mangrovii* is probably an estuarine saprotroph.

Nowakowskiella spp.

Anastasiou and Churchland (1969) and Newell and Fell (1980) frequently observed *Nowakowskiella* species growing on leaves used as bait. The leaf baits were placed either in seawater samples collected from the ocean or in leaf litter bags incubated directly in the ocean at two sites in south eastern Florida, one in Venezuela and three in British Columbia. The trophic status of *Nowakowskiella* species has not yet been accurately determined, but it is likely that they are saprotrophs.

Other species

Several other species of chytridiomycetes, which have been considered to be parasites in marine ecosystems, are listed in Table 7.1. Finally, Alias and Jones (2009) listed a few more species reported from marine ecosystems, but the identity of these species and their range of tolerance to salinity have not yet been accurately determined.

Table 7.1 Some species of parasitic and saprotrophic zoosporic true fungi isolated from estuarine and marine ecosystems.

Species name	Ecotype	Mode of nutrition	Host genus	Reference
Rhizophydium littoreum	Probably estuarine	Facultative parasite	*Bryopsis, Codium*	Amon (1984)
Rhizophydium littoreum	Marine	Facultative parasite	*Cancer*	Shields (1990)
Thalassochytrium gracilariopsis	Marine	Probable biotroph	*Gracilariopsis*	Nyvall et al. (1999)
Chytridium polysiphoniae	Marine	Probable biotroph	*Pyaiella*, other genera of brown algae	Küpper and Müller (1999), Müller et al. (1999)
Chytridium polysiphoniae	Marine	Probable biotroph	*Centroceras, Polysiphonia*	Johnson (1966), Raghukumar (1986a)
Olpidium rostriferum, Olpidium sp.	Marine	Probable biotroph	*Cladophora, Pseudo-nitzschia*	Raghukumar (1987), Elbrächter and Schnepf (1998)
Rhizophydium sp.	Marine	Probable biotroph	*Pseudo-nitzschia, Bellerochea, Cylindrotheca*	Elbrächter and Schnepf (1998), Hanic et al. (2009)
Rhizophydium sp.	Probably estuarine	Probable biotroph	*Rhizoctonium, Ulothrix*	Huth (1974)
Coenomyces sp.	Probably estuarine	Probable biotroph	*Cladophora, Rhizoclonium*	Raghukumar (1986b)
Phlyctochytrium mangrovis (now *Spizellomyces*)	Probably estuarine	Saprotroph	Pine pollen	Ulken (1978)
Unknown (obligate anaerobes)	Marine	Mutualist	*Endocardium, Amblyrhynchus*	Thorsen (1999), Mackie et al. (2004)

Blastocladiomycota

There has been only one report of a species of Blastocladiomycota from marine or estuarine ecosystems (Ulken 1977). This ecotype of *Catenaria* was observed growing on baits incubated with sediment collected from a mangrove swamp in New Zealand and mixed with distilled water. Ulken (1977) isolated this fungus into pure culture, but never accurately tested its range of tolerance to salinity. Her data suggest that it is an estuarine ecotype. However, in fresh water ecosystems, many species of *Catenaria, Coelomycidium* and *Coelomomyces* are common parasites of dipteran larvae, especially midges, black flies and mosquitoes. A few species of *Catenaria* also infect nematodes (Gleason et al. 2010a). Some species of dipteran larvae and other invertebrates live in brackish water and in intertidal ecosystems, but the parasites present in these habitats have not been properly investigated and their tolerance to salinity is not known.

Neocallimastigales

There are two reports of fungi that are probably members of the phylum Neocallimastigales in the digestive systems of sea urchins (Thorsen 1999) and marine iguanas (Mackie et al. 2004). The taxonomic placement of these fungi needs to be confirmed with molecular data.

Olpidium clade

Olpidium

Olpidium species are morphologically simple parasites that infect the host with uniflagellate zoospores and form sporangia and sometimes resting spores inside the host cells. Dick (2001) lists over 80 species that have been assigned to *Olpidium*, but many of these taxa have synonyms with other genera of zoosporic true fungi, such as *Chytridium*, and others have been redescribed in Oomycete genera (Dick 2001).

The number of recognized *Olpidium* species is around 50 (Sekimoto et al. 2011). Several species are ubiquitous obligate parasites of the roots of flowering plants in soil ecosystems (Herrera-Vásquez et al. 2009; Hartwright et al. 2010) and others have been described as parasites of fungi and pollen. *Olpidium* species are also known to parasitize Zygnematales (Charophyceae) and other fresh-water algae (Dasgupta and John 1988). About 12 species of marine parasites of algae have been described in *Olpidium*, although several of these reports date from the 19th and early 20th centuries and type material cannot be verified (Dick 2001). Raghukumar (1987) reported *Olpidium rostriferum* as a parasite of *Cladophora* in marine environments in southern India. *Olpidium* species have also been described as parasites of protoctists (Scherffel 1926) and a few species parasitize invertebrates, especially the eggs of rotifers and nematodes in soil and freshwater (Karling 1946; Barron and Szijarto 1986; Glockling 1994; Barron 2004).

The tolerance of these species to salinity is not known. *Olpidium* species found in invertebrates, diatoms, and green algae have not yet been properly characterized by molecular methods, and their relationships to species infecting plant roots are not known. A multigene sequence analysis of *O. brassicae* (now *O. virulentus*), a parasite of *Brassica* roots, however, unexpectedly placed this species not with the phyla of true zoosporic fungi but among the Zygomycota (James et al. 2006). The credibility of this result was subsequently strengthened as 18S sequencing of *O. vermicola*, a parasite of nematode eggs, positioned it at the same place as *O. brassicae* (T.Y. James pers. comm.). A more recent multigene phylogeny of *O. virulentus* and *O. bornovanus* also placed *Olpidium* in a clade among the Zygomycota (Sekimoto et al. 2011). This phylogenetic research suggests that, although *Olpidium* species have morphological characteristics that are typical of zoosporic true fungi, they are most closely related to terrestrial fungi. Sekimoto et al. (2011) postulate that the simple thallus of *Olpidium* may have evolved as a reduced hyphal condition. Two putative marine species of *Olpidium* are listed in Table 7.1.

Ecology of marine and estuarine species

Chytridium polysiphoniae can infect a wide range of species of brown algae that grow in the intertidal and subtidal zones of cold temperate and polar oceans. Infection experiments in the laboratory using 48 species of cultivated brown algae showed that 23 species were susceptible to infection by *C. polysiphoniae* (Müller et al. 1999). Müller et al. (1999) reported that, in co-culture, *C. polysiphoniae* infected species of brown algae belonging to nine orders: Ectocarpales, Dictyosiphonales, Sphacelariales, Tilopteridales, Chordariales, Sporochnales, Scytosiphonales, Desmarestiales and Laminariales.

Massive epidemics of *C. polysiphoniae* have been observed along the European coast. The prevalence of infection in thalli of the algal host, *Pylaiella*, can often exceed 10% at sites in the Shetlands (Küpper and Müller 1999). Since this fungus can infect such a large number of species from the colder oceans of the world, it may be considered a ubiquitous (cosmopolitan) pathogen in these ecosystems (even though this is currently not well supported by field records outside the European North Atlantic). This fungus exists in a narrow temperature range of between 4 and 15°C in co-culture with its host (Müller et al. 1999), which may explain these patterns of distribution. Furthermore *C. polysiphoniae* has not been found in the Baltic or Mediterranean Seas, which suggests that it has a narrow range of tolerance to salinity.

Although the survey of susceptible hosts was not extended by Müller et al. (1999) to red algae, several authors have reported that some species of red algae are also hosts for *C. polysiphoniae* (Raghukumar 1986a). However, it is not known whether the same ecotypes of this fungus can infect both brown and red algal hosts. Zoospores of *C. polysiphoniae* attacked *Centroceras clavulatum* (a red alga) but were unable to infect a species of *Ceramium* (another red alga) in culture (Raghukumar 1986a). Populations of *C. clavulatum* were infected at several sites near Goa and in the Lakshadweep Islands, India, but were not infected at other sites. Although infection was detected throughout the year, the highest incidence occurred during the summer months. In the laboratory, better infection was achieved at 30°C than at 24°C.

These data suggest differences between the ecotypes of *C. polysiphoniae* found in cold-temperate regions and those found in the tropics. Therefore, it can be concluded that *C. polysiphoniae* infects a wide range of algal hosts in the marine environment, but the extent of its distribution needs further investigation. In contrast, only a few hosts are known for species in *Rhizophydium, Thalassochytrium* and *Olpidium*.

The effect of parasitic zoosporic true fungi on the reproductive rates of their hosts in the marine environment is unknown and only one study has investigated this topic. Gachon et al. (2006) used spatially-resolved (imaging) microscopic measurements of chlorophyll fluorescence kinetics to follow the fate of individual cells of *P. littoralis* when infected by *C. polysiphoniae*. Gachon et al. (2006) concluded that parasitism negatively impacted photosynthetic rates, but the significance at the population level has not yet been determined.

The effect of salinity on the growth of zoosporic true fungi in general has been reviewed by Gleason et al. (2010b) and Gleason et al. (2011). None of the species of zoosporic true fungi isolated from soil were able to grow, but some species survived in a dormant state in full-strength sea water. The data from Ulken (1978), Amon (1984), and Raghukumar (1986a) suggest that some species isolated from marine environments, such as certain ecotypes of *P. mangrovis, R. littoreum* and *Coenomyces*, have a tolerance to salinity but grow better in environments with reduced salinity. Therefore, these species are probably better adapted to estuaries than to the ocean. It appears that *C. polysiphoniae, T. gracilariopsis* and one ecotype of *R. littoreum* are atypical of most zoosporic true fungi because they have adapted well to oceanic environments. However, Harrison and Jones (1974) found two species of *Phlyctochytrium* (*Spizellomyces*) that can also grow in full-strength seawater media.

The large number of zoospores released by the zoosporic true fungal parasites of marine algae may provide important food resources for grazing and filter-feeding zooplankton and metazoans in coastal marine environments, where marine algae form dense communities and where the prevalence of infection is high. No data on the biomass of zoosporic true fungi in marine ecosystems are currently available.

DNA sequences attributed to zoosporic true fungi have been detected in environmental samples collected from various sites at deep-sea hydrothermal vents in the Mid-Atlantic Ridge and the East Pacific Rise (Le Calvez et al. 2009) and from deep-sea sediments from five sites near Japan (Nagano et al. 2010). Previously unknown clades of zoosporic true fungi were detected in these environments, suggesting that many more unknown species of zoosporic true fungi are likely to occur in marine ecosystems. Some of these DNA sequences found in the ocean may belong to known species for which sequences have not yet been recorded in public databases.

Conclusion

The thalli of zoosporic true fungi growing on the surface of marine algae are often difficult to observe under the light microscope without the use of fluorescent staining techniques. In many cases, due to a lack of distinctive morphological features, molecular techniques are necessary for an accurate identification. However, data available in public databases such as GenBank do not include many sequences from zoosporic true fungi collected from the marine environment. Published records of environmental sampling worldwide indicate that little attention has been paid to this group of microorganisms. Available data suggest that, as a group, zoosporic true fungi have not adapted as well as many other groups of eukaryotic microorganisms to marine ecosystems, but that evolutionary success has been achieved in terrestrial and fresh water habitats instead. Nonetheless, from the few studies conducted to date, it can be concluded that at least some zoosporic true fungi are important components of microbial populations in marine and estuarine ecosystems. The small number of species of zoosporic true fungi reported from marine ecosystems is probably due to poor sampling and a lack of knowledge necessary for the accurate identification of these fungi.

References

Alias SA, Jones EBG. Marine fungi from mangroves of Malaysia. IOES monograph series 8. Institute of Ocean and Earth Sciences. Kuala Lumpur: University of Malaya, 2009:109.

Amon JP. An estuarine species of *Phlyctochytrium* (Chytridiales) having a transient requirement for sodium. Mycologia 1976;68:470–480.

Amon JP. *Rhizophydium littoreum*: a chytrid from siphonaceous marine algae – an ultrastructural examination. Mycologia 1984;76:132–139.

Anastasiou CJ, Churchland LM. Fungi on decaying leaves in marine habitats. Can J Bot. 1969;47:251–257.

Barr DJS. The phylogenetic and taxonomic implications of flagellar rootlet morphology among zoosporic fungi. BioSystems 1981;14:359–370.

Barr DJS. *Allochytrium expandens* rediscovered: morphology, physiology and zoospore ultrastructure. Mycologia 1986;78:439–448.
Barr DJS. Phylum Chytridiomycota. In: Margulis L, Corliss JO, Melkonian D, Chapman DJ, eds. Handbook of Protoctista. Boston, MA: Jones & Bartlett, 1990:454–466.
Barr DJS. Chytridiomycota. In: McLaughlin DL, McLaughlin EG, Lemke PA, eds. The Mycota. vol. VII, Part A. New York: Springer-Verlag, 2001:93–112.
Barron GL. Fungal parasites and predators of rotifers, nematodes, and other invertebrates. In: Mueller GM, Bills GF, Foster MS, eds. Biodiversity of fungi, inventory and monitoring methods. Amsterdam: Elsevier Academic Press, 2004:435–450.
Barron GL, Szijarto E. A new species of *Olpidium* parasitic in nematode eggs. Mycologia 1986;78:972–975.
Bartnicki-Garcia S. Cell wall composition and other biochemical markers in fungal phylogeny. London: Academic Press, 1970:81–103.
Beakes GW, Canter HM, Jaworski GHM. Zoospore ultrastructure of *Zygorhyzidium affluens* and *Z. planktonicum*, two chytrids parasitizing the diatom *Asterionella formosa*. Can J Bot. 1988;66:1054–1067.
Canter HM, Jaworski GHM. Some general observations on zoospores of the chytrid *Rhizophydium planktonicum* Canter emend. New Phytol. 1980;84:515–531.
Dasgupta SN, John R. A contribution to our knowledge of the zoosporic fungi. Bull Bot Survey of India. 1988;30:1–82.
Dick MW. Straminipilous fungi. Dortrecht: Kluwer Academic Publishers, 2001.
Elbrächter M, Schnepf E. Parasites of harmful algae. In: Anderson DM, Cembella AD, Hallegraeff GM, eds. Physiological ecology of harmful algal blooms. Berlin: Springer-Verlag, 1998:351–363.
Forster H, Coffey MD, Elwood H, Sogin ML. Sequence analysis of the small subunit ribosomal RNAs of three zoosporic fungi and implications for fungal evolution. Mycologia 1990;82:306–312.
Freeman KR, Martin AP, Karki D, Lynch RC, Mitter MS, Meyer AF, Longcore JE, Simmons DR, Schnidt SK. Evidence that chytrids dominate fungal communities in high elevation soils. Proc Nat Acad Sci. 2009;106:18315–18320.
Gachon CMM, Küpper H, Küpper FC, Šetlik I. Single-cell chlorophyll fluorescence kinetic microscopy of *Pylaiella littoralis* (Phaeophyceae) infected by *Chytridium polysiphoniae* (Chytridiomycota). Eur J Phycol. 2006;41:395–403.
Gleason FH, Marano AV, Johnson P, Martin WW. Blastocladian parasites of invertebrates. Fungal Biol Rev. 2010a;24:56–67.
Gleason FH, Schmidt SK, Marano AV. Can zoosporic true fungi grow or survive in extreme or stressful environments? Extremophiles 2010b;14:417–425.
Gleason FH, Küpper FC, Amon JP, Picard K, Gachon CMM, Marano AV, Sime-Ngando T, Lilje O. Zoosporic true fungi in marine environments: a review. Mar Freshw Res. 2011;62:383–393.
Glockling SL. Predacious and parasitiodal fungi in association with herbivore dung in deciduous woodlands. PhD thesis. Department of Botany, University of Reading: UK, 1994.
Hanic LA, Sekimoto S, Bates SS. Oomycete and chytrid infections of the marine diatom *Pseudonitzchia pungens* (Bacillariophyceae) from Prince Edward Island, Canada. Botany 2009;87:1096–1105.
Harrison JL, Jones EBG. Patterns of salinity tolerance displayed by the lower fungi. Veröff Inst Meeresforsch Bremerh. 1974;Suppl 5:197–220.
Hartwright LM, Hunter PJ, Walsh JA. A comparison of *Olpidium* isolates from a range of host plants using internal transcribed spacer sequence analysis and host range studies. Fungal Biol. 2010;114:26–33.

Herrera-Vásquez JA, Cebrián MdC, Alfaro-Fernández A, Córdoba-Sellés MdC, Jordá C. Multiplex PCR assay for the simultaneous detection and differentiation of *Olpidium bornovanus, O. brassicae*, and *O. virulentus*. Mycol Res. 2009;113:602–610.

Huth K. Chytridineen als Parasiten benthischer Algen im Weserästuar. Veröff Inst Meeresforsch Bremerh. 1974;Suppl 5:105–113.

James TY, Porter D, Leander CA, Vilgalys R, Longcore JE. Molecular phylogenetics of the Chytridiomycota supports the utility of ultrastructure data in chytrid systematics. Can J Bot. 2000;78:336–350.

James TY, Letcher PM, Longcore JE, Mozley-Standridge SE, Porter D, Powell MJ, Griffith GW, Vilgalys R. A molecular phylogeny of the flagellated fungi (Chytridiomycota) and description of a new phylum (Blastocladiomycota). Mycologia 2006;98:860–871.

Johnson TW Jr. *Rozella marina* in *Chytridium polysiphoniae* from Icelandic waters. Mycologia 1966;58:490–494.

Johnson TW Jr, Sparrow FK Jr. Fungi in oceans and estuaries. Weinheim, Germany: J. Cramer Publisher, 1961.

Jones MDM, Forn I, Gadelha C, Egan MJ, Bass D, Massana R, Richards TA. Discovery of novel intermediate forms redefines the fungal tree of life. Nature 2011;474:200–203.

Kagami M, de Bruin A, Ibelings BW, van Donk E. Parasitic chytrids: their effects on phytoplankton communities and food-web dynamics. Hydrobiol. 2007;578:113–129.

Karling JS. Brazilian Chytrids. Additional parasites of rotifers and nematodes. Lloydia 1946;9:1–12.

Karling JS. Chytridiomycetarum iconographia. Vaduz: J. Cramer Publisher, 1977.

Kazama FY. Development and morphology of a chytrid isolated from *Bryopsis plumosa*. Can J Bot. 1972;50:499–505.

Küpper FC, Müller DG. Massive occurrence of the heterokont parasites *Anisolpidium, Eurychasma* and *Chytridium* in *Pylaiella littoralis* (Ectocarpales, Phaeophyceae). Nova Hedw. 1999;69:381–389.

Küpper FC, Maier I, Müller DG, Loiseaux-De Goer S, Guillou L. Phylogenetic affinities of two eukaryotic pathogens of marine macroalgae, *Eurychasma dicksonii* (Wright) Magnus and *Chytridium polysiphoniae* Cohn. Crypt Algol. 2006;27:165–184.

Laidlaw WMR. A method for the detection of the resting sporangia of potato wart disease (*Synchytrium endobioticum*) in the soil of old outbreak sites. Potato Res. 1985;28:223–232.

Lara E, Moreira D, Lopez-Garcia P. The environmental clade LKM11 and *Rozella* from the deepest branching clade of fungi. Protist 2010;161:116–121.

Le Calvez T, Burgaud G, Mahe S, Barbier G, Vandenkoornhuyse P. Fungal diversity in deep-sea hydrothermal ecosystems. Appl Environ Microbiol. 2009;75:6415–6421.

Letcher PM, Powell MJ, Churchill PF, Chambers JG. Ultrastructural and molecular phylogenetic delineation of a new order, the Rhizophydiales (Chytridiomycota). Mycol Res. 2006;110:898–915.

Letcher PM, Vélez CG, Barrantes ME, Powell MJ, Churchill PF, Wakefield WS. Ultrastructural and molecular analysis of Rhizophydiales (Chytridiomycota) isolates from North America and Argentina. Mycol Res. 2008;112:759–782.

Longcore JE. Morphology and zoospore ultrastructure of *Chytriomyces angularis* sp. nov. (Chytridiales). Mycologia 1992;84:442–451.

Longcore JE, Pessier AP, Nicolas DK. *Bactrachochytrium dendrobatidis* gen. *et* sp. nov., a chytrid pathogenic to amphibians. Mycologia 1999;91:219–227.

Mackie RI, Rycyk M, Ruemmler RL, Aminov JI, Wikelski M. Biochemical and microbiological evidence for fermentative digestion in free-living land iguanas (*Conolophus pallidus*) and marine iguanas (*Amblyrhynchus cristatus*) on the Galapagos Archipelago. Physiol Biochem Zool. 2004;77:127–138.

Martin GW. *Rhizophydium polysiphoniae* in the United States. Bot Gaz. 1922;73:236–238.

Mozley-Standridge SE, Letcher PM, Longcore JE, Porter D, Simmons R. Cladochytriales – a new order in Chytridiomycota. Mycol Res. 2009;113:489–507.

Müller DG, Küpper FC, Küpper H. Infection experiments reveal broad host ranges of *Eurychasma dicksonii* (Oomycota) and *Chytridium polysiphoniae* (Chytridiomycpota), two eukaryotic parasites in marine brown algae (Phaeophyceae). Phycol Res. 1999;47:217–223.

Nagano Y, Nagahama T, Hatada Y, Nunoura T, Takami H, Miyazaki J, Takai K, Horikoshi K. Fungal diversity in deep-sea sediments – the presence of novel fungal groups. Fungal Ecol. 2010;3:316–325.

Newell SY, Fell JW. Mycoflora of turtle grass (*Thalassia testudinum* König) as recorded after seawater incubation. Bot Mar. 1980;23:265–275.

Nyvall P, Pedersén M, Longcore J. *Thalassochytrium gracilariopsidis* (Chytridiomycota), gen. et sp. nov., endosymbiotic in *Gracilariopsis* sp. (Rhodophyceae). J Phycol. 1999;35:176–185.

Porter D, Smiley R. Development of the sporangium and discharge apparatus in a marine chytrid, *Phlyctochytrium* sp. Bot Mar. 1980;23:99–116.

Porter D, Kirk PW Jr. Marine fungi: taxonomic and ecological considerations II. Lower fungi. In: Mulherj KG, Singh VP, eds. Frontiers in applied microbiology. Lucknow, India: Print House, 1987:235–256.

Powell MJ. Phylogenetic implications of the microbody-lipid globule complex in zoosporic fungi. BioSystems 1978;10:167–180.

Powell MJ. Looking at mycology with a Janus face: a glimpse at Chytridiomycetes active in the environment. Mycologia 1993;85:1–20.

Raghukumar C. *Chytridium polysiphoniae* – a fungal pathogen on the red alga *Centroceras clavulatum* (C. Agardh) Montagne, from Goa. Ind J Mar Sci. 1986a;15:42–44.

Raghukumar C. Fungal parasites of the marine green algae, *Cladophora* and *Rhizoclonium*. Bot Mar. 1986b;29:289–297.

Raghukumar C. Fungal parasites of marine algae from Mandapam (South India). Dis Aquat Organ. 1987;3:137–146.

Scherffel A. Beiträge zur Kenntnis der Chytridineen. Teil III. Archiv. Prostistenkd. 1926;54:510–528.

Sekimoto S, Rochon D, Long JE, Dee JM, Berbee ML. A multigene phylogeny of *Olpidium* and its implications for early fungal evolution. BMC Evol Biol. 2011;11:331.

Shearer CA, Descals E, Kohlmeyer B, Kohlmeyer J, Marvanová L, Padgett D, Porter D, Raja HA, Schmit JP, Thornton HA, Voglmayr H. Fungal biodiversity in aquatic habitats. Biodiv Conserv. 2007;16:49–67.

Shields JD. *Rhizophydium littoreum* on the eggs of *Cancer anthonyi:* Parasite or saprobe? Biol Bull. 1990;179:201–206.

Simmons DR, James TY, Meyer AF, Longcore JE. Lobulomycetales, a new order in the Chytridiomycota. Mycol Res. 2009;113:450–460.

Sparrow FK Jr. Biological observations on the marine fungi of Woods Hole Waters. Biol Bull. 1936;70:236–263.

Sparrow FK Jr. Aquatic Phycomycetes. 2nd ed. Ann Arbor, MI, USA: University of Michigan Press, 1960.

Taylor TN, Remy W, Hass H. Fungi from the Lower Devonian Rhynic chert: Chytridiomycetes. Am J Bot. 1992;79:1233–1241.

Thorsen MS. Abundance and biomass of the gut-living microorganisms (bacteria, protozoa and fungi) in the irregular sea urchin *Echinocardium cordatum* (Spatangoida: Echinodermata). Mar Biol. 1999;133:353–360.

Trinci APJ, Davies DR, Gull K, Lawrence M, Nielsen BB, Rickers A, Theodorou MK. Anaerobic fungi in herbivorous animals. Mycol Res. 1994;98:129–152.

Ulken A. Phycomyceten aus der Laguma de Mandinga, Veracruz, Mexico. Veröff Inst Meeresforshch Bremerh. 1977;16:177–189.

Ulken A. Growth experiments with different isolates of *Phlyctochytrium mangrovis* (Phycomycetes). Veröff Inst Meeresforshch Bremerh. 1978;17:21–31.

Wakefield WS, Powell MJ, Letcher PM, Barr DJS, Churchill PF, Longcore JE, Chen S-F. A molecular phylogenetic evaluation of the Spizellomycetales. Mycologia 2010;102:596–604.

Wilce RT, Schneider CW, Quinlan AV, van den Bosch K. The life history and morphology of free-living *Pilayella littoralis* (L.) Kjellman (Ectocarpaceae, Ectocarpales) in Nahant Bay, Massachusetts. Phycologia 1982;21:336–354.

8 Morphology and ultrastructure of marine fungi with special reference to the origin of the membrane complex in the marine ascomycete *Corollospora gracilis*

Sung-Yuan Hsieh and E.B. Gareth Jones

Marine fungi show a broad range in their form and function, with many specially adapted to life in an aquatic habitat. All taxonomical groups are to be found in the marine environment and their basic structure conforms to those of terrestrial taxa (Kohlmeyer and Kohlmeyer 1979; Alexopoulos et al. 1996).

Basidiomycota

Marine basidiomycetes are generally small in size (<1 cm), ranging from basidiomata that are gasteroid (*Nia vibrissa*), cyphellaceous (*Halocyphina villosa* and *Henningsomyces* sp.), resupinate (*Digitatispora marina* and *Haloaleurodiscus mangrovei*), stipitate (*Physalacria maipoensis*) to the unicellular basidiomycetous yeasts (*Cystofilobasidium capitatum*), see also Chapters 4 and 6.

The compact size of fruit bodies of most marine basidiomycetes is considered to be an adaptation to an aquatic habitat (Jones 2000; Hibbett and Binder 2001; Jones and Choeyklin 2008). Many species also produce tetra-radiate basidiospores, which are thought to aid dispersal and entrapment to substrates (Jones 1994). Hibbett and Binder (2001) speculated as to the possible evolution of the marine filamentous basidiomycetes and considered *Nia vibrissa* as the most derived member, with a shift from terrestrial to periodically-immersed or fully-submerged substrata, loss of basidiospory, evolution of appendaged spores and an enclosed fruit body. They hypothesized that the mangrove-inhabiting *Halocyphina villosa* may be morphologically and ecologically intermediate between *N. vibrissa* and terrestrial cyphelloid forms, such as the genera *Cyphellopsis*, *Calathella*, and *Henningsomyces*, the latter two containing species also occurring in mangrove habitats (Alias and Jones 2009; Pang et al. 2011). In less saline waters, terrestrial species may also colonize substrates, e.g., *Grammothele fuligo* and *Schizophyllum commune* (see Chapter 15).

Ascomycota

Cleistothecial, perithecoid, and apothecial ascomata occur in marine ascomycetes as well as unicellular marine yeasts (see also Chapters 2, 3 and 6). The most common forms are those with solitary perithecial ascomata (Dothideomycetes and Sordariomycetes), while none produce stromata with embedded perithecia (Kohlmeyer and Kohlmeyer 1979). Ascomycetes in oceanic waters often have deliquescing asci and elaborate appendaged ascospores, again adaptations to a marine habitat (Jones 1995). Ascomycetes on mangrove substrata are generally adapted to their intertidal location with active ejection of ascospores that lack elaborate appendages, but may possess gelatinous sheaths (Suetrong et al. 2009; Sakayaroj et al. 2011).

Asexual fungi

Few asexual (anamorphic) fungi appear to be morphologically adapted to the marine environment (e.g., tetra-radiate conidia: *Clavatospora bulbosa, Heliscella stellatacula* and *Varicosporina ramulosa*), with most sharing the common features of terrestrial genera (e.g., *Acremonium, Arthrobotrys, Dendryphiella* and *Zalerion*; Jones et al. 2009; Seifert et al. 2011).

Chytridiomycota

The morphology of this group is detailed in Chapter 7, with the possession of zoospores produced in variously formed zoosporangia. Few species have been documented from marine habitats and this topic requires closer study.

Introduction to the ultrastructure of marine fungi

In the 1980s and 1990s, scanning and transmission electron microscopy (SEM and TEM, respectively) was widely used to provide information on the morphology and structure of marine fungi. For example, spore wall ornamentation was extensively used to distinguish species: *Dactylospora haliotrepha* (longitudinal striations) could be distinguished from *D. mangrovei* (verrucose ascospore wall); *Halophytophthora mycoparasitica* (zoosporangial walls with spines from which mucilage spreads out over the wall) from *H. masteri* (smooth walled), see Figures 8.1 and 8.2 (Nakagiri et al. 1998; Alias and Jones 2009). Topics researched included:

  (i) searching for diagnostic features for the characterization of species;
  (ii) the development of ascospores and their appendages;
  (iii) observation of the attachment of spores to various substrata; and
  (iv) following the stages in the development of reproductive structures and the formation of ascomata.

The earliest study of the ultrastructure of marine fungi was by Lutley and Wilson (1972), who illustrated the ascus and ascospore development in *Ceriosporopsis halima, Halosphaeria appendiculata* and *Torpedospora radiata*. They showed the thin-walled nature of the ascus wall in *C. maritima,* and the differentiated ascospore wall layers, using a freeze-etched technique.

Species identification at the ultrastructural level

Ultrastructure micrographs were commonly used to characterize new species, highlighting the unique features of reproductive structures (Table 8.1). Many marine ascomycetes possess ascospores with elaborate appendages and SEMs were used to illustrate the differences between genera and species (Jones and Moss 1978; Jones 1995). For example: fragmentation of an exosporic sheath (*Halosphaeriopsis mediosetigera*);

Subsequently, ensheathed hyphae are formed by mucilage that further attaches the spore to the substratum (Hyde et al. 1986b). Mucilaginous hyphal sheaths have been known for some time (Szaniszlo et al. 1968; Palmer et al. 1983) but their function was speculative. Hyde et al. (1986a) followed the germination of 15 marine fungi on wood veneers and polycarbonate membranes and reported the development of hyphal sheaths in all species. It was postulated that these sheaths serve in the adhesion of the developing hyphae prior to penetration of the substratum.

Development of reproductive structures in the Ascomycota

Few detailed studies of ascomata and ascus development of marine fungi have been published, and mostly of sections observed at the light microscope level (Wilson 1954, 1956; Lloyd and Wilson 1962; Kohlmeyer and Kohlmeyer 1966; Pang et al. 2011). Au et al. (1999b) showed at the TEM level the ascoma wall of *Capronia ciliomaris* comprised two to three layers with setae arising from the outer wall layer. Initially the upper third of the centrum was filled with rounded cells embedded in an extracellular matrix, but broke down at maturity with the extension of the asci. Periphysoidal elements arose from the inner upper third of the ascomal wall and ostiolar canal, extended through the ostiole and merged with a crown of apical setae. Hsieh et al. (2007) followed ascoma development in *Corollospora gracilis* from ascogonium initiation, antheridium-ascogonium conjugation, production of ascogenous hyphae to peridium and ostiole development. Pit-connections of the centrum pseudoparenchyma have been reported for a number of marine ascomycetes, e.g., *Haiyanga salina, Arenariomyces trifurcatus* and *Kohlmeyeriella tubulata* (Kohlmeyer and Kohlmeyer 1979; Kohlmeyer and Volkmann-Kohlmeyer 1989), but their origin has not been determined.

At the TEM level, pit-connections in *C. gracilis* were found to be modified ascomycetous septal pores, fluorescing with Calcofluor white, which stains material rich in 1, 4-β-glucans. Another observation made in this study was the occurrence of a plug of thick-walled non-melanized cells at the base of the ostiole that separated the rest of the centrum tissue by a thin melanized separation layer. Similar structures have been reported by Kohlmeyer and Volkmann-Kohlmeyer (1987, 1989) in *Corollospora cinnamomea* and *C. armoricana*, and may occur in other species in the genus (Nakagiri and Tokura 1987). This plug may play a role in preventing the entry of seawater into the centrum until ascospore release is imminent.

Ascosporogeneis in marine ascomycetes

Ascosporogenesis in all ascomycetes examined has been shown to be similar and comprises two major processes (Beckett 1981):

(i) uninucleate portions of ascus cytoplasm are delimited by a pair of unit membranes; and
(ii) ascospore wall material is deposited between these two unit membranes that separate as the spore matures.

Depending on the position of the ascospore-delimiting membrane relative to the haploid nuclei, two basic types of ascospore delimitation have been characterized (Reeves 1967; Beckett 1981). First, the Hemiascus-type (yeast and yeast-like fungi) where the ascospore-delimiting membrane forms adjacent to the spindle pole bodies and envelopes individual nuclei. In *Taphrina* sp. and *Tuber* sp., however, delimitation is by invagination of the ascus plasmalemma. Second, the Euascus-type when the ascospore-delimiting membrane is formed initially as a single double-membrane cylinder – the ascus vesicle – in close proximity to the ascus wall and which surrounds all haploid nuclei. The ascus vesicle then constricts around each haploid nucleus to delimit the ascospores. This type is characteristic of most filamentous ascomycetes.

Lutley and Wilson (1972) were the first to investigate ascosporogenesis in the marine ascomycete *Ceriosporopsis halima* by TEM when ascospores were delimited from the ascus cytoplasm by a pair of unit membranes with the wall laid down between them. The origin of the delimiting membrane and the presence of an ascus vesicle were not reported. Subsequent studies of several species of the Halosphaeriaceae showed a similar sequence of delimitation (Shearer and Miller 1977; Johnson 1982; Farrant 1988; Baker 1991), with an ascus vesicle observed in *Halosphaeriopsis mediosetigera* (Johnson 1982). These ultrastructure studies showed that ascospore delimitation in the marine ascomycetes studied is similar to that reported for terrestrial ascomycetes.

Several researchers (Johnson 1982; Farrant 1988; Baker 1991) have noted the presence of a "membrane complex" in the developing asci of members of the Halosphaeriaceae. This complex is formed by elaboration of the outer delimiting membrane and is thought to result from the coalescence of epiplasmic vesicles (Johnson 1982; Farrant 1988; Baker 1991).

Once the ascospore wall is formed, a membrane complex is observed whose function is to prevent the premature expansion of the appendages until they are released from the ascomata. For example, in the ascospores of *Halosarpheia, Cucullosporella* and *Saagaromyces* spp., once released from the ascomata and in water, the membrane complex deliquesces and the hamate polar appendages unfurl to form long thread-like appendages (Alias et al. 2001; Baker et al. 2001; Jones 2006). Similarly, the ascospore appendages of *Corollospora* species only expand when the membrane complex ruptures in water and the exosporic layer peels away from the spore wall and undergoes a fragmentation process (Jones et al. 1983a; Jones 2006; Hsieh et al. 2007).

Origin of the membrane complex in *Corollospora gracilis*

The preparation of material for scanning and TEM is fully outlined by Hsieh et al. (2007) with grids examined in a JEOL 100S TEM at 80 kV or a Hitachi H-600 TEM at 75 kV.

Ascus vesicle and ascospore delimitation

Ascospore-delimiting membranes in *Corollospora gracilis* were first observed at the four-nucleate stage of the ascus (Figure 8.3). At prophase III, numerous variously-sized membrane vesicles formed laterally along the periphery of the ascus adjacent to the ascus plasmalemma. Initially these vesicles were rather flattened and discontinuous (Figure 8.4). At metaphase III, these peripheral vesicles started to fuse to form a continuous double-membrane sac, the ascus vesicle (Figure 8.5). The ascus vesicle was close to the ascus wall, discrete from the ascus plasmalemma, and formed a cylinder that encompassed most of the ascus cytoplasm (Figure 8.6).

When the ascus was at the eight-nucleate stage, the ascus vesicle started to invaginate to delimit the ascospores (Figure 8.7), with mitochondria and other organelles condensing around the nuclei. The cytoplasm between the ascus plasmalemma and the ascus vesicle became highly vesiculated. Ascospore delimitation was completed when the eight ascospore initials were bounded by the inner and outer delimiting membranes derived from the ascus

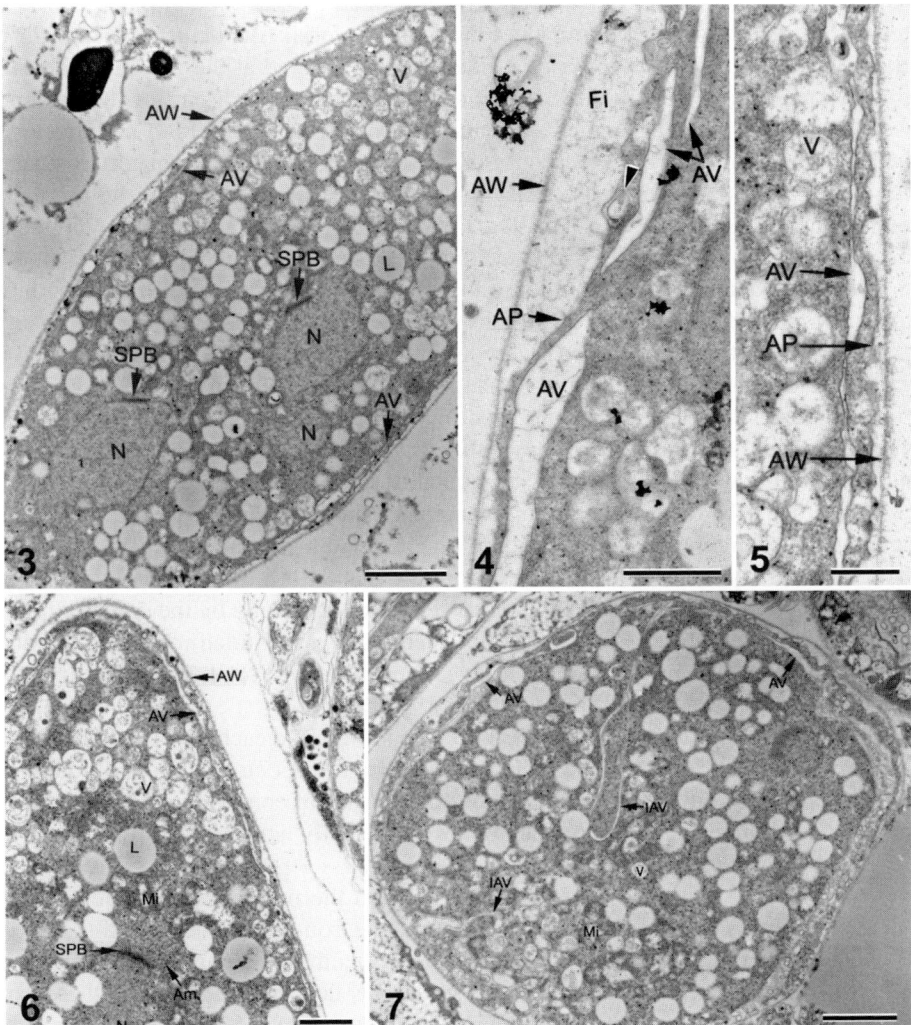

Figures 8.3–8.7 TEM micrographs *Corollospora gracilis*. Figure 8.3 Oblique section of an ascus showing nuclei at the prophase III stage (prior to the post-meiotic mitosis). Several regions of the ascus vesicle (AV) are present adjacent to the ascus wall (AW). Bar scale = 2 μm. Figure 8.4 Ascus vesicle (AV) regions prior to their fusion. Bar scale = 1 μm. Figure 8.5 Ascus vesicle (AV) adjacent to and parallel with the ascus cell wall (AW) of a tetranucleate ascus. Bar scale = 0.5 μm. Figure 8.6 At metaphase III. The ascus vesicle (AV) is almost complete. Bar scale = 1 μm. Figure 8.7 Transverse section of a young ascus, showing invaginations of the ascus vesicles (IAV) that will cleave the cytoplasm and delimit ascospores. Bar scale = 2 μm.

vesicle (Figures 8.8 and 8.9). The ascospore initials were uninucleate, irregular in shape and lacked the polar spines of mature spores. Prior to ascospore wall formation, the ascospores became ellipsoid to sub-fusiform with the nucleus positioned centrally.

Membrane complex

Concurrent with episporium deposition changes occurring in the epiplasm and to the outer delimiting membrane, some electron dense material (or inclusions, ~15–25 nm in diameter) appeared in the lomasome-like structure located at the margin of developing ascospore. This lomasome-like structure was associated with part of the ascospore plasmalemma, but was also connected with numerous tubes (30–60 nm in diameter) or lamellae in the epiplasm. Electron-dense inclusions appeared to be transported from the lomasome-like structure to the ends of the epiplasmic lamellae through the channel of lamellae (Figure 8.10).

The tips of the epiplasmic lamellae enlarged and become detached to produce many vesicles containing electron-dense inclusions, (~100–300 nm in diameter) in the epiplasm (Figure 8.11). The electron-dense bodies appeared to fuse with evaginations of the outer delimiting membrane (Figure 8.12). Upon fusion, the membrane around the electron-dense material lost its integrity and the electron-dense material was deposited on to the outer delimiting membrane. Continued deposition of electron-dense material formed a continuous electron-dense layer that was external but contiguous with the outer delimiting membrane (Figure 8.13). Subsequently, a unit membrane formed external to the electron-dense layer, the latter being bounded on the inner surface by the outer delimiting membrane (Figure 8.14). The origin of the external membrane was not determined, although it is likely to have been derived from the membrane around the electron-dense body. The electron-dense deposited material, delimited internally by the outer delimiting membrane and externally by a newly-formed membrane, constituted what has previously been referred to as the membrane complex (Johnson 1980; Porter 1982; Farrant 1988; Baker 1991; Johnson et al. 1991; Yusoff et al. 1994a).

The membrane complex formed a separate "sheath" around each developing ascospore (Figures 8.15 and 8.16) and was 25–50 nm thick (Figure 8.17). Occasionally the membrane complex was further elaborated by the coalescence of epiplasmic membranes to its surface. Formation of the membrane complex was usually completed before the formation of the exosporium (Figure 8.18).

Transition of the outer delimiting membrane into the membrane complex coincided with a reduction in the number of epiplasmic electron-dense bodies and the loss of detectable organelles in the epiplasm. The mature membrane complex was not closely adpressed to the episporium and the exosporium was deposited external to the episporium but bounded externally by the membrane complex (Figure 8.19).

Abbreviations in legends: AP: ascus plasmalemma; AS: ascospore; ASP: ascospore plasmalemma; AV: ascus vesicle; AW: ascus wall; DM: delimiting membrane; Eb: electron-dense body; EL: epiplasmic lamella; EP: epiplasm; Ep: episporium; ER: endoplasmic reticulum; Fi: fibrillar material; IAV: invagination of ascus vesicle; IDM: inner delimiting membrane; L: lipid body; LEA: equatorial exosporial appendage; Lo: lomasome-like body; MC: membrane complex; Mi: mitochondrion; MT: marginal thickenings of equatorial exosporial appendage; Mt: microtubule; N: nucleus; ODM: outer

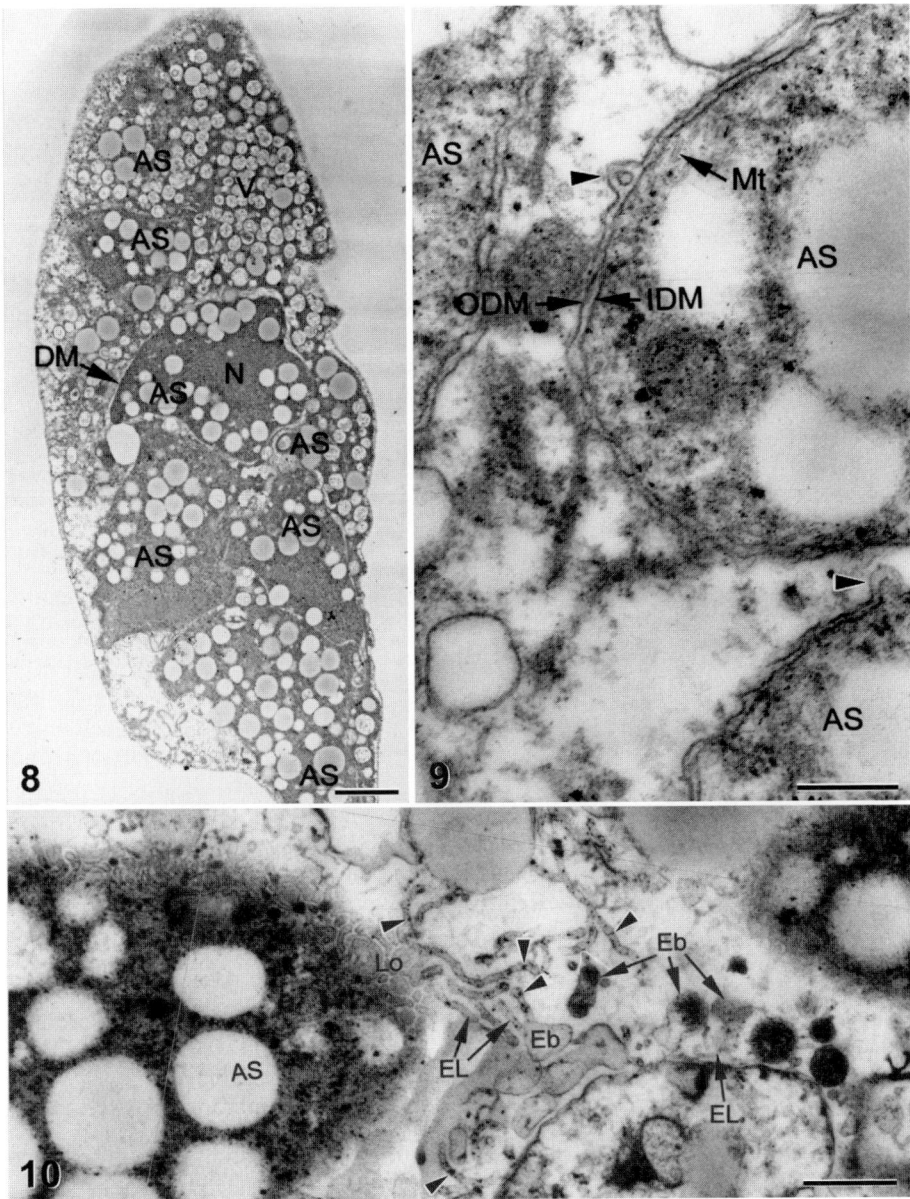

Figures 8.8–8.10 TEM micrographs *Corollospora gracilis*. Figure 8.8 Oblique section through an ascus with recently-delimited irregularly-shaped ascospore initials (AS). Delimiting membranes (DM) surround ascospore initials and separate them from the epiplasm. Bar scale = 2 μm. Figure 8.9 Transverse section through three adjacent ascospores (AS) within an ascus. The delimiting membranes consist of outer (ODM) and inner (IDM) delimiting membranes. Bar scale = 0.3 μm. Figure 8.10 The peripheral region of a partly delimited ascospore (AS) showing lomasome-like structures (Lo) associated with the ascospore plasmalemma and continuous with epiplasmic lamellae (30–60 nm in diameter). The lamellae (EL) contain electron-dense inclusions (arrow heads) and terminally form the electron-dense bodies (Eb). Bar scale = 0.5 μm.

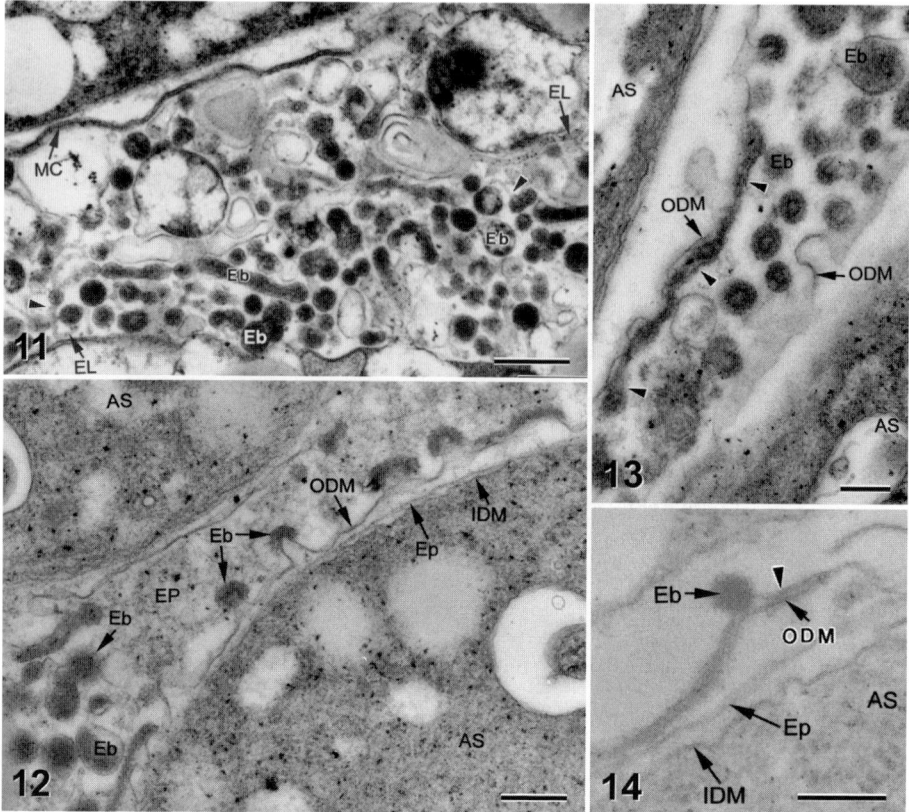

Figures 8.11–8.14 TEM micrographs *Corollospora gracilis*. Figure 8.11 Electron-dense-bodies (Eb) in the epiplasm, some still continuous with the lamellae (arrow heads). Note another unit membrane (arrow head) present external to the electron dense layer. Bar scale = 0.5 µm. Figure 8.12 Transverse section of two adjacent developing ascospores (AS). The membrane-bounded electron-dense bodies (Eb) in the epiplasm (EP) are closely associated or fused with evaginations of the outer delimiting membrane (ODM). Bar scale = 0.3 µm. Figure 8.13 Progressive deposition of electron-dense bodies (Eb) onto the outer delimiting membrane (ODM) forming a continuous layer (arrow heads). Bar scale = 0.2 µm. Figure 8.14 Electron-dense bodies (Eb) fused with evaginations of the outer delimiting membrane (ODM). Bar scale = 0.2 µm.

delimiting membrane; PE: protrusion of episporium into exosporium; PS: primary polar spine; Sm: spindle microtubule; SPB: spindle pole body; V: vesicle; Va: vacuole.

Ascospore delimitation in other marine fungi

The major ontogenetic stages in ascospore delimitation are characteristic of *Corollospora gracilis*:

   (i) formation of an ascus vesicle;
   (ii) invagination of the ascus vesicles to delimit ascospores; and
   (iii) formation of the inner and outer delimiting membrane, the former forming the spore plasmalemma and the latter the spore investing membrane.

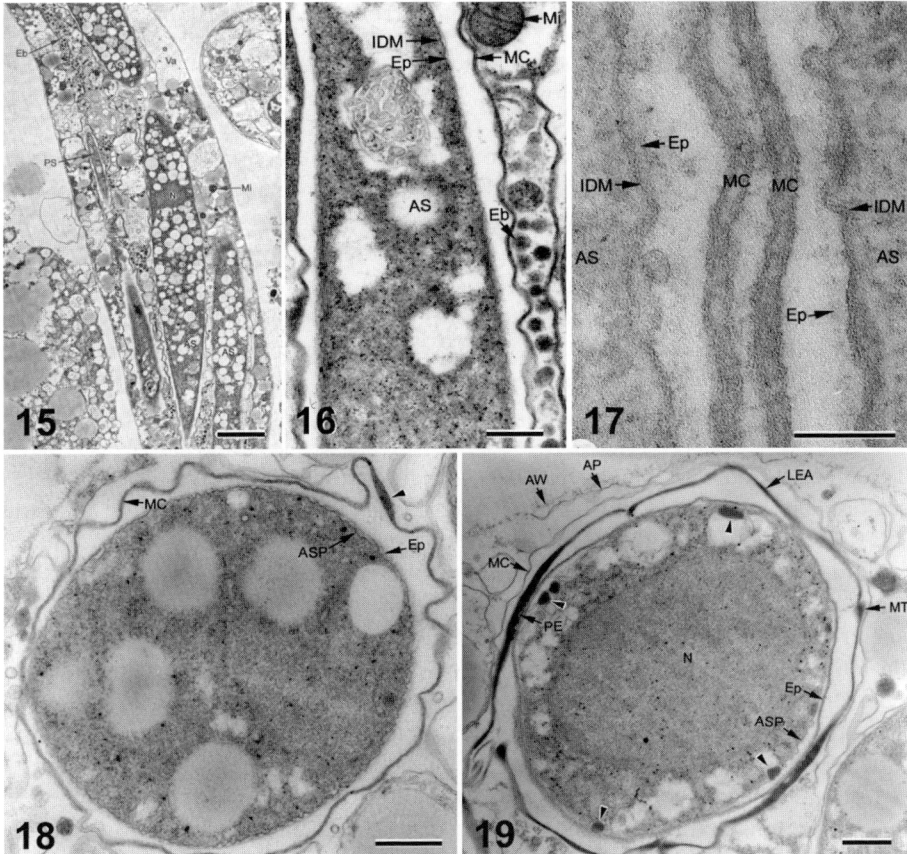

Figures 8.15–8.19 TEM micrographs of *Corollospora gracilis*. Figure 8.15 Longitudinal section of an ascus with developing ascospores. The ascospores (AS) are elongated, fusiform and extended at both ends to form the developing polar spines (PS). The sporoplasm contains one nucleus (N) and many lipid bodies. The epiplasm contains lipid bodies, numerous electron-dense bodies (Eb), mitochondria (Mi), and fibrillar vacuoles (Va). Bar scale = 3 μm. Figure 8.16 Longitudinal section of part of an immature elongated ascospore showing that the outer delimiting membrane has become electron-dense and thicker to form the membrane complex (MC), and is separated from the episporium (Ep). Bar scale = 0.5 μm. Figure 8.17 Higher magnification of the two adjacent membrane complexes (MC) of adjacent ascospores (AS). Each complex consists of two unit membranes between which there is electron-dense material. Bar scale = 0.1 μm. Figure 8.18 Transverse section of an ascospore with a fully-developed electron-dense membrane complex (MC). Bar scale = 0.5 μm. Figure 8.19 Transverse section of a developing ascospore showing four equatorial exosporial appendages (LEA) with thickened margins (MT) and are surrounded by the membrane complex (MC). Bar scale = 0.5 μm.

Ascospore delimitation in *C. gracilis* is similar to that described for most terrestrial Euascomycetes and confirms that ascospore delimitation is a stable process throughout the Euascomycetes (Beckett 1981).

Membrane complex in marine fungi – the formation and function

Previous studies of marine ascomycetes have established the presence of an elaborated outer delimiting membrane. However, no studies have demonstrated the pattern of development for the membrane complex (Johnson 1982; Farrant 1988; Baker 1991). In *C. gracilis*, the ultrastructural evidence indicates that the epiplasm plays an important role in the formation of the complex. The most prominent stage is the occurrence of electron-dense bodies in the epiplasm and their subsequent fusion with the outer delimiting membrane. The origin of the electron-dense bodies has not been resolved, although it appears that they and their contents are from lomasome-like structures present at the poles of delimited ascospores. At this stage, the development of the polar regions of the spore lacks cell-wall layers and the plasmalemma is associated with lomasome-like bodies that are continuous with epiplasmic lamellae. The lamellae extend into the regions surrounding the ascospore and give rise to the electron-dense bodies.

The precise nature of the electron-dense bodies requires histochemical analysis, but based on their morphology, position and occurrence it is suggested that the "lomasome-epiplasmic lamellae-electron-dense body complex" is functionally an endoplasmic reticulum–Golgi body equivalent. Material synthesized in the sporoplasm is transported via the lomasome-like body to the epiplasmic lamellae. Within the lamellae, 15–25 nm diameter electron-dense inclusions occur and these aggregate in the more distal regions of the epiplasmic lamellae to form the electron-dense bodies that are budded-off and later fuse to form the membrane complex.

The structure of the membrane complex in *C. gracilis* is more complex than that reported for other marine ascomycetes. Previous studies have shown the membrane complex to be a thickened electron-dense outer delimiting membrane, with the suggestion that the thickening derives from the coalescence of epiplasmic membranes. In *C. gracilis*, the mature membrane complex comprises two unit membranes between which there is an electron-dense amorphous matrix.

The role of the epiplasm in the development of ascospore delimiting structures has been a subject of debate, as the epiplasm lacks nuclei and functional organelles. However, the sequence of events leading to the development of the membrane complex in *C. gracilis* involves continuity between the spore cytoplasm and the epiplasm, thus allowing sporoplasmic control of the process. This continuity has not been observed in any other species of the marine ascomycetes, although the membrane complex appears to be ubiquitous in this ecological group of fungi. It is likely that the association between the sporoplasm and epiplasm is ephemeral and only occurs during the early stages of ascosporogenesis, and thus may not have been observed by scientists in previous works.

Acknowledgements

SYH is grateful to the late Dr Steven Moss for all his assistance and patience in teaching him electron microscopy while at the University of Portsmouth.

References

Alexopoulos CJ, Mims CW, Blackwell M. Introductory mycology. 4th ed. New York: John Wiley & Sons, 1996.
Alias SA, Jones EBG. Marine fungi from mangroves of Malaysia. Institute of Ocean and Earth Studies, No. 8, Kuala Lumpur, 2009;109.
Alias SA, Moss ST, Jones EBG. *Cucullosporella mangrovei*, ultrastructure of ascospores and their appendages. Mycoscience 2001;42:405–411.
Au DWT, Jones EBG, Vrijmoed LLP. Ultrastructure of asci and ascospores of the mangrove ascomycete *Dactylospora haliotrepha*. Mycoscience 1996;37:129–135.
Au DWT, Jones EBG, Vrijmoed LLP. Observations on the biology and ultrastructure of the asci and ascospores of *Julella avicenniae* from Malaysia. Mycol Res. 1999a;103:865–872.
Au DWT, Jones EBG, Vrijmoed LLP. The ultrastructure of *Capronia ciliomaris*, an intertidal marine fungi from San Juan Island. Mycologia 1999b;91:326–333.
Baker TA. Taxonomic studies of the Halosphaeriaceae with special reference to ultrastructure of spore ontogeny. Ph.D. Thesis, Council for National Academic Awards, Portsmouth Polytechnic: Portsmouth, UK, 1991.
Baker TA, Jones EBG, Moss ST. Ultrastructure of ascus and ascospore appendages of the mangrove fungus *Halosarpheia ratnagiriensis* (Halosphaeriales, Ascomycota). Can J Bot. 2001;79:1–11.
Bauer R, Luta M, Piatek M, Vanky K, Oberwinkler F. *Flamingomyces* and *Parvulago*, new genera of marine smut fungi (Ustilaginomycotina). Mycol Res. 2007;111:1199–1206.
Beckett A. Ascospore formation. In: Turian G, Hohl H, eds. The fungal spore: morphogenetic controls. London: Academic Press, 1981:107–129.
Brooks RD. The presence of dolipore septa in *Nia vibrissa* and *Digitatispora marina*. Mycologia 1975;67:172–174.
Cole GT. Conidium ontogeny in marine hyphomycetous fungi: *Asteromyces cruciatus* and *Zalerion maritimum*. Mar Biol. 1976;38:147–158.
Farrant CA. Ultrastructure studies in the Halosphaeriaceae with special reference to *Halosarpheia* Kohlmeyer and *Aniptodera* Shearer *et* Miller. Ph.D. Thesis, Council for National Academic Awards, Portsmouth Polytechnic: Portsmouth, UK, 1988.
Hale MS, Eaton RA. The ultrastructure of soft rot fungi. II. Cavity-forming hyphae in wood cell walls. Mycologia 1985a;77:594–605.
Hale MS, Eaton RA. The oscillatory growth of fungal hyphae on wood cell walls. Trans Br Mycol Soc. 1985b;84:227–288.
Hibbett DS, Binder M. Evolution of marine mushrooms. Biol Bull. 2001;201:319–322.
Hsieh SY, Moss ST, Jones EBG. Ascoma development in the marine ascomycete *Corollospora gracilis* (Halosphaeriales, Hypocreomycetidae, Sordariomycetes). Bot Mar. 2007;50:1–12.
Hyde KD. Frequency of occurrence of lignicolous marine fungi in the tropics. In: Moss ST, ed. The biology of marine fungi. Cambridge: Cambridge University Press, 1986:311–322.
Hyde KD, Jones EBG, Moss ST. Mycelial adhesion to surfaces. In: Moss ST, ed. The biology of marine fungi. Cambridge: Cambridge University Press, 1986a:331–340.
Hyde KD, Jones EBG, Moss ST. How do fungal spores attach to surfaces? In: Barry S, Houghton DR, Llewellyn GC, O'Rear CE, eds. Biodeterioration 6. London: CAB International Mycological Institute and The Biodeterioration Society, 1986b:584–589.
Hyde KD, Moss ST, Jones EBG. Attachment studies in marine fungi. Biofouling 1989;1:287–298.
Johnson RG. Ultrastructure of ascospore appendages of marine ascomycetes. Bot Mar. 1980;23:501–527.

Johnson RG. Ultrastructure and histochemistry of the ontogeny of ascospores, and their appendages in marine ascomycetes. Ph.D. Thesis, Council for National Academic Awards, Portsmouth Polytechnic: Portsmouth, UK, 1982.

Johnson RG, Jones EBG, Moss ST. Taxonomic studies of the Halosphaeriaceae: *Remispora* Linder, *Marinospora* Cavaliere and *Carbosphaerella* Schmidt. Bot Mar. 1984;27:557–566.

Johnson RG, Jones EBG, Moss ST. Taxonomic studies of the Halosphaeriaceae: *Ceriosporopsis, Haligena* and *Appendichordella* gen. nov. Can J Bot. 1987;65:931–942.

Johnson RG, Jones EBG, Moss ST. Histochemical and electron microscope studies of ascospore ontogeny in three marine fungi with cleistothecia: *Amylocarpus encephaloides* Currey, *Dryosphaera navigans* Koch *et* Jones and *Eiona tunicata* Kohlm. Bot Mar. 1991;34:229–239.

Jones EBG. *Haligena spartinae* sp. nov. A Pyrenomycete on *Spartina townsendii*. Trans Br Mycol Soc. 1962;45:246–248.

Jones EBG. Marine fungi – spore dispersal, settlement and colonization. In: Acker RF, Floyd BB, De Palma JR, Inverson P, eds. Proceedings of the 3rd international congress on marine corrosion and fouling. Boston: Northeastern University Press, 1973:640–647.

Jones EBG. Fungal adhesion. Mycol Res. 1994;98:961–981.

Jones EBG. Ultrastructure and taxonomy of the aquatic ascomycetous order Halosphaeriales. Can J Bot. 1995;73:S790–S801.

Jones EBG. Marine fungi: some factors influencing biodiversity. Fungal Divers. 2000;4:53–73.

Jones EBG. Form and function of fungal spore appendages. Mycoscience 2006;47:167–183.

Jones EBG, Moss ST. Ascospore appendages of marine ascomycetes: an evaluation of appendages as taxonomic criteria. Mar Biol. 1978;49:11–26.

Jones EBG, Choeyklin R. Ecology of marine and freshwater basidiomycetes. In: Boddy L, Frankland JC, van West P, eds. Ecology of saprotrophic Basidiomycetes. London: Elsevier, 2008:301–324.

Jones EBG, Johnson RG, Moss ST. Taxonomic studies of the Halosphaeriaceae: *Corollospora* Werdermann. Bot J Linn Soc. 1983a;87:193–212.

Jones EBG, Johnson RG, Moss ST. *Ocostaspora apilongissima* gen. *et* sp. nov: a new marine Pyrenomycete from wood. Bot Mar. 1983b;24:353–360.

Jones EBG, Johnson RG, Moss ST. Taxonomic studies of the Halosphaeriaceae: *Halosphaeria* Linder. Bot Mar. 1984;27:129–143.

Jones EBG, Johnson RG, Moss ST. Taxonomic studies of the Halosphaeriaceae. Philosophy and rationale for the selection of characters in the delineation of genera. In: Moss ST, ed. The biology of marine fungi. Cambridge: Cambridge University Press, 1986:211–230.

Jones EBG, Hyde KD, Read SJ, Moss ST, Alias SA. *Tirisporella* gen. nov., an ascomycete from the mangrove palm *Nypa fruticans*. Can J Bot. 1996;74:1487–1495.

Jones EBG, Sakayaroj J, Suetrong S, Somrithipol S, Pang KL. Classification of marine Ascomycota, anamorphic taxa and Basidiomycota. Fungal Divers. 2009;35:1–187.

Kirk PW Jr. Cytochemistry of marine fungal spores. In: Jones EBG, ed. Recent advances in aquatic mycology. New York: Wiley, 1976:177–192.

Koch J, Jones EBG. The identity of *Crinigera maritima* and three new genera of marine cleistothecial ascomycetes. Can J Bot. 1989;67:1183–1197.

Kohlmeyer J, Kohlmeyer E. On the life history of marine Ascomycetes: *Halosphaeria mediosetigera* and *H. circumvestita*. Nova Hedw. 1966;12:189–202.

Kohlmeyer J, Kohlmeyer E. Marine mycology – the higher fungi. New York: Academic Press, 1979.

Kohlmeyer J, Volkmann-Kohlmeyer B. Reflections on the genus *Corollospora* (Ascomycetes). Trans Br Mycol Soc. 1987;88:181–188.

Kohlmeyer J, Volkmann-Kohlmeyer B. *Corollospora armoricana* sp. nov., an arenicolous ascomycete from Brittany (France). Can J Bot. 1989;67:1281–1284.

Leightly LE, Eaton RA. *Nia vibrissa* – a marine white rot fungus. Trans Br Mycol Soc. 1979;73: 35–40.
Lloyd LS, Wilson IM. Development of the perithecium in *Lulworthia medusa* (Ell. *et* Ev.) Cribb *et* Cribb, a saprophyte on *Spartina townsendii*. Trans Br Mycol Soc. 1962;45:359–372.
Lutley M, Wilson IM. Development and fine structure of ascospores in the marine fungus *Ceriosporopsis halima*. Trans Br Mycol Soc. 1972;58:393–402.
McKeown TA, Alias SA, Moss ST, Jones EBG. Ultrastructural studies of *Trematosphaeria malaysiana* sp. nov. and *Leptosphaeria pelagica*. Mycol Res. 2001;105:615–624.
Mouzouras R. Pattern of timber decay caused by marine fungi. In: Moss ST, ed. The biology of marine fungi. Cambridge: Cambridge University Press, 1986:341–353.
Mouzouras R, Jones EBG, Venkatasamy R, Holt DM. Microbial decay of lignocellulose in the marine environment. In: Thompson MF, Sarojini R, Nagabhushanaim R, eds. Marine biodeterioration. New Delhi: Oxford and J B H Publishing, 1988:329–354.
Nakagiri A. Two new species of *Lulworthia* and evaluation of genera-delimiting characters between *Lulworthia* and *Lindra* (Halosphaeriaceae). Trans Mycol Soc Jpn. 1984;25:377–388.
Nakagiri A. Intertidal mangrove fungi from Iriomote Island. IFO Res Commun. 1993;16:24–62.
Nakagiri A, Ito T. Basidiocarp development of the cyphelloid gasteroid aquatic basidiomycetes *Halocyphina villosa* and *Limnoperdon incarnatum*. Can J Bot. 1991;69:2320–2327.
Nakagiri A, Tokura R. Taxonomic studies of the genus *Corollospora* (Halosphaeriaceae, Ascomycotina) with descriptions of seven new species. Trans Mycol Soc Jpn. 1987;28:413–436.
Nakagiri A, Okane I, Ito T. Zoosporangium development, zoospore release and culture properties of *Halophytophthora mycoparasitica*. Mycoscience 1998;3:223–230.
Palmer JG, Murmanis L, Highley TL. Visualisation of hyphal sheaths in wood-decay Hymenomycetes. I. Brown rotters. Mycologia 1983;75:995–1004.
Pang KL, Jheng JS, Jones EBG. Marine mangrove fungi of Taiwan. Chilung, Taiwan (ROC); National Taiwan Ocean University Press, 2011.
Porter D. The appendaged ascospores of *Trichomaris invadens* (Halosphaeriaceae), a marine ascomycetous parasite of the Tanner crab, *Chionoecetes bairdi*. Mycologia 1982;74:363–375.
Read SJ, Moss ST, Jones EBG. Ultrastructure of asci and ascospores sheath of *Massarina thalassiae* (Loculoascomycetes, Ascomycotina). Bot Mar. 1994;37:547–553.
Read SJ, Jones EBG, Moss ST. Ultrastructural observations of asci, ascospores and appendages of *Massarina armatispora* (Ascomycota). Mycoscience 1997a;38:141–146.
Read SJ, Moss ST, Jones EBG. Ultrastructure of asci, ascospores and appendages of *Massarina ramunculicola* (Loculoascomycetes, Ascomycota). Bot Mar. 1997b;40:465–471.
Rees G. Factors affecting the sedimentation rates of spores. Bot Mar. 1980;23:375–385.
Rees G, Jones EBG. Observations on the attachment of spores of marine fungi. Bot Mar. 1984;27:145–160.
Reeves F. The fine structure of ascospore formation in *Pyronema domesticum*. Mycologia 1967;59:1018–1033.
Sakayaroj J, Pang KL, Jones EBG. Multi-gene phylogeny of the Halosphaeriaceae: its ordinal status, relationships between genera and morphological character evolution. Fungal Divers. 2011;46:87–109.
Schmidt I. *Corollospora intermedia*, nov. spec., *Carbosphaerella leptosphaerioides*, nov. spec. und *Crinigera maritima*, nov. gen., nov. spec., 3 neue marine Pilzarten von der Ostseeküste. Natur und Naturschutz in Mecklenburg. 1969;7:5–14.
Seifert K, Morgan-Jones G, Gams W, Kendrick B. The genera of Hyphomycetes. Utrecht: CBS, 2011.
Shearer CA, Miller M. Fungi of the Chesapeake Bay and its tributaries V. *Aniptodera chesapeakensis* gen. *et* sp. nov. Mycologia 1977;69:887–898.

Suetrong S, Schoch CL, Spatafora JW, Kohlmeyer J, Volkmann-Kohlmeyer B, Sakayaroj J, Phongpaicht S, Tanaka K, Hairayama K, Jones EBG. Molecular systematics of the marine Dothideomycetes. Stud Mycol. 2009;64:155–173.

Szaniszlo PJ, Wirsen C, Mitchell R. Production of a capsular polysaccharide by a marine fungus. J Bacteriol. 1968;96:1474–1483.

Wilson IM. *Ceriosporopsis halima* Linder and *Ceriosporopsis cambrensis* sp. nov.: two Pyrenomycetes on wood. Trans Br Mycol Soc. 1954;37:272–285.

Wilson IM. Some new marine Pyrenomycetes on wood and rope: *Halophiobolus* and *Lindra*. Trans Br Mycol Soc. 1956;39:401–415.

Yanna, Ho WH, Hyde KD. Can ascospores ultrastructure differentiate the genera *Linocarpon* and *Neolinocarpon* and species therein? Mycol Res. 2003;107:1305–1313.

Yusoff M, Jones EBG, Moss ST. A taxonomic reappraisal of the genus *Ceriosporopsis* based on ultrastructure. Can J Bot. 1994a;72:1550–1559.

Yusoff M, Moss ST, Jones EBG. Ascospore ultrastructure of *Pleospora gaudefroyi* Patouillard (Pleosporaceae, Loculoascomycetes, Ascomycotina). Can J Bot. 1994b;72:1–6.

Yusoff M, Jones EBG, Moss ST. Ascospore ultrastructure in the marine genera *Lulworthia* Sutherland and *Lindra* Wilson. Crypt Bot. 1995;5:307–315.

Phylogeny of fungal-like organisms

9 An introduction to fungus-like microorganisms

Sigrid Neuhauser, Sally L. Glockling,
Eduardo M. Leaño, Osu Lilje, Agostina V. Marano
and Frank H. Gleason

The term "fungus-like" is difficult to define, as it is used to describe a diverse set of eukaryotic microorganisms that either share some characteristics with true fungi or were once classified as true fungi. In particular, fungus-like microorganisms and true fungi share certain features, such as heterotrophic nutrition, zoospores, spores with cell walls that contain chitin, hyphae, resistant fungus-like resting structures and fungus-like lifecycles. Prior to the use of DNA-based phylogenetic data, fungus-like microorganisms were placed into groups according to morphological characteristics. With the increased use of DNA-based methods in phylogeny, many taxonomic groups of fungus-like microorganisms are no longer included in the Kingdom Fungi and have been placed elsewhere in the eukaryote tree of life (Adl et al. 2005). In this chapter we list some of the fungus-like groups of microorganisms that can be found in estuarine or marine environments and are still often referred to as fungi, lower fungi or fungus-like organisms in ecological or biodiversity studies. An overview of the current taxonomic affiliation of these fungus-like groups, together with zoosporic true fungi, is given in Table 9.1.

Often fungus-like microorganisms are still placed into ecological categories such as "aquatic phycomycetes" (Sparrow 1960) or "heterotrophic flagellates" or HFs (Sime-Ngando et al. 2011) because they have similar morphological structures, thrive in similar habitats, interact with one another, and have similar ecological roles. Sparrow (1960) included all groups of fungi and fungus-like organisms that produce motile spores propelled by flagella (zoospores) in the aquatic phycomycetes. Using categories based on size, many of these fungal-like organisms are included in the HFs, which include all groups of eukaryotic microorganisms that produce zoospores that are approximately equal to or below 5 µm in diameter and lack chlorophyll (Sime-Ngando et al. 2011). Other ecological categories include nano- and picoeukaryotes or nano- and picoplankton (Moreira and Lopez-Garcia 2002).

These ecological categories are sometimes used as alternative frames of reference to phylogenetic groups in ecological investigations because they are based on unifying ecological features like size or trophic modes. For large-scale ecological studies of complex microbial communities, such morphological or functional ecological categories are useful because they allow a quick and feasible means of processing components of the microbial communities of large samples. However, although it can be hypothesized that morphologically and physiologically similar organisms have similar ecological functions (e.g., in food webs), this is not necessarily the case for all ecological functions and there is still a considerable lack of detailed knowledge about the taxa within these groups. For example, Glücksman et al. (2010) found that phylogenetically-related heterotrophic

Table 9.1 Groups of zoosporic true fungi and fungus-like microorganisms.

Supergroup	Kingdom	Name of group/phylum	Subdivision/order	Habitat	Reference
Opisthokonta (Unikonta)	Basal to the Fungi	Cryptomycota		Marine and freshwater	Lara et al. (2010); Jones et al. (2011)
		Mesomycetozoea	Ichthyophonida	Marine	Cafaro (2005)
			Amoebidiales*	Freshwater	
			Dermocystidia	Marine	Marshall and Berbee (2011)
			Eccrinales	Both	
	Fungi	Chytridiomycota		Marine and freshwater	James et al. (2006)
		Blastocladiomycota*		Freshwater	
		Neocallimastigomycota		Marine and freshwater	
		Olpidium group		Marine and freshwater	
Chromoalveolata (Heterokonta)	Heterotrophic stramenopiles	Oomycota	Saprolegniales*	Freshwater	Kitancharoen et al. (1994); Roza and Hatai (1999); Diggles (2001); Nakagiri et al. (2001); Leaño (2002); Lara and Belbahri (2011); Beakes et al. (2012)
			Leptomitales	Marine	
			Atkinsielliales	Marine	
			Peronosporales	Freshwater and terrestrial	
			Pythiales		
			Rhipidiales		
			Olpidiopsidales		
			Haliphthorales	Marine	
			Eurychrasmatales		
		Hyphochytriomycota		Marine	Honda et al. (1999); Leander et al. (2004)
		Labyrinthulomycota	Labyrinthulids	Marine	
			Aplanochytrids	Marine	
			Thraustochytrids	Marine	
Rhizaria (Heterokonta)		Phytomyxea	Phagomyxida, Plasmodiophorida		Neuhauser et al. (2011)

biflagellates belonging to the protist phylum Cercozoa had different mechanisms of feeding on bacterial communities and therefore differed markedly in their ecological impact. Consequently, a great deal of information on the ecology, physiology and genetics of these fungal-like organisms has yet to be elucidated, especially in marine ecosystems.

This chapter provides an overview of the biology, taxonomy and ecology of fungus-like organisms. In addition, the biodiversity and ecological preferences of fungal-like organisms are discussed along with their potential role in marine food webs and their mode of dispersal and distribution. The higher-level phylogeny, trophic modes and modes of nutrition of fungus-like organisms are discussed in greater detail in Chapters 10, 11, 12 and 13.

Higher level phylogeny

During the past decade, eukaryotic organisms were divided into five to eight supergroups in the tree of life (e.g., Baldauf 2003; Adl et al. 2005; Cavalier-Smith 2009). In this chapter we will follow the subdivision of eukaryotic organisms into the six supergroups as proposed by Adl et al. (2005), with references to newer literature where necessary. Two of the supergroups, the Opisthokonta and the Amoebozoa, are Unikonta and include organisms that have as motile stages either amoebae or zoospores with a single flagellum. The four remaining supergroups, the Chromalveolata, Rhizaria, Excavata and Plantae, are Heterokonta and have as motile stages either amoebae or zoospores with two flagella. Fungus-like organisms are found in four of the supergroups: the Opisthokonta (which includes the "true" fungi, Mesomycetozoea, Cryptomycota), the Amoebozoa (Myxomycetes), the Chromalveolata (Oomycetes, Labyrinthulomycota and Hyphochytridiomycota) and the Rhizaria (Phytomyxea). Since there are no known marine fungus-like species in the Amoebozoa (Myxomycota), this group is not included in this chapter.

Traditionally (i.e., until the 1990s) species were assigned to taxonomic groups using morphological characteristics that were assumed to be of phylogenetic significance. Although these morphological criteria have provided a good basis for taxonomy, detailed phylogenies combining morphology and nucleic acid sequences are now providing a more in-depth assessment of the relatedness of fungi and fungus-like organisms. Combining phylogenies based on genetic information with morphological and ultrastructural features and with biochemical properties is currently the most powerful taxonomic tool. Such studies have resulted in some major changes to the traditional classification of microorganisms, such as the removal of the Oomycetes from the Kingdom Fungi and their inclusion in the Straminipila super kingdom along with several groups of photosynthetic organisms (Dick 2001; Cavalier-Smith 2009). There have also been fundamental reclassifications within major groups including the Oomycetes (Beakes and Sekimoto 2009; Lara and Belbahri 2011) and the Chytridiomycota (James et al. 2006).

The Tree of Life project (Maddison and Schulz 2007) is providing a (regularly updated) synthesis of current phylogenetic studies enabling an overview of the distribution of higher fungi, lower fungi, fungus-like microorganisms and other eukaryotes within the tree of life. As more groups are being analyzed and sequences of individual genes and entire genomes are added to public databases, this analysis is becoming increasingly comprehensive. However, genetic information from many species has yet to be sequenced, and the effort to collect data from marine and estuarine species in particular, needs to be intensified. The lack of genetic data for species in marine and estuarine ecosystems is partly due to the complexity of the marine environment and its unidentified biodiversity.

The commonly parasitic or host-dependent nature of some marine species also means that they are difficult to isolate and grow in culture (Neuhauser et al. 2011). The main reason for the lack of sequence data, however, is that the fungus-like organisms in marine and estuarine environments have in general been poorly studied. The addition of DNA sequences from well-defined, cultured species and type material to public databases will help to resolve many of the currently unassigned environmental sequences. The amount of unassigned sequence data that can be determined through such efforts was recently exemplified in a study with the Mortierellales, a group of zygomycetous fungi (Nagy et al. 2011). Sequencing genes and genomes from unculturable, host-dependent parasites isolated with their hosts is now also increasing our understanding of the less well-known taxa of lower fungi and fungal-like organisms (Beakes et al. 2006; James et al. 2006).

The increasing number of environmental sequences has recently resulted in the discovery of a new clade of fungi, the Cryptomycota (see Chapter 10), which is basal to the fungi and which includes the endobiotic parasite, *Rozella* (Jones et al. 2011). Species in the genus *Rozella* had previously been considered to be part of the earliest branch of the fungal tree (James et al. 2006). The Cryptomycota appear to be abundant in both marine and freshwater environments (Jones et al. 2011). In another group of fungus-like microorganisms, the Mesomycetozoea (see Chapter 10), which is also considered to be close to the point where animals and fungi diverge, Marshall and Berbee (2011) have recently described new species which they isolated from invertebrate hosts. Inclusion of sequences from soil samples in phylogenetic studies of the Chytridiomycota revealed currently unknown clades of species associated with high-elevation soils (Freeman et al. 2009). The taxa of marine Oomycetes (see Chapter 11) that are currently known are almost all parasitic. *Eurychasma*, which is parasitic on brown algae, has been shown to represent one of the most basal clades in oomycete phylogenies (Beakes et al. 2006; Beakes and Sekimoto 2009). A recent phylogeny of the Oomycetes, which included environmental sequences, has revealed several previously unknown basal clades, all derived from the marine environment (Lara and Belbahri 2011). The combination of organism-based observations with modern phylogenetic tools will facilitate the exploration and understanding of many previously poorly known or unknown branches of the tree of life, such as fungus-like organisms.

Trophic modes: saprobes, mutualists and parasites

Recent research on the interactions between organisms has revealed that these interactions are often not exclusively saprophytic, mutualistic or parasitic, but are frequently a mixture of trophic modes depending on environmental factors (McCreadie et al. 2011), see chapter 12. The precise nature of these interactions, and hence the ecological functions of the species involved, can only be understood with intensive metagenomic (e.g., transcriptome profiles, monitoring gene expression levels), physiological (e.g., monitoring of nutrient uptake, metabolic products) and morphological investigations, which have to date rarely been conducted with fungus-like microorganisms. The development and use of metagenomic analysis for ecological investigations has recently been reviewed by Simon and Daniel (2011). Rising numbers of such studies can be expected with the availability of inexpensive sequencing techniques and the evolving bioinformatics solutions for data analysis. However, most heterotrophic microorganisms are currently assigned to the arbitrary categories discussed below, according to their primary modes of nutrition. The primary mode of nutrition of organisms is one determining factor of their ecological placement.

Saprobes (saprotrophs) are species that use organic material derived from dead organisms as a food source. Saprobic organisms commonly cause the decay of organic material and are therefore important in decomposing leaf litter and other dead organic material. Saprobic fungus-like microorganisms are well represented by *Halophytophthora* spp. (Oomycetes) and *Thraustochytrium* spp. (Labyrinthulomycetes), which are part of the normal biota of marine and estuarine environments. Of the *Halophytophthora* spp. identified, all but one (*H. mycoparasitica*) are saprobic and most species were isolated from fallen mangrove leaves during the early to late stages of leaf decay (Nakagiri et al. 2001, see Chapter 12). The abundance of *Halophytophthora* species is attributed to their tolerance of broad environmental parameters (temperature, salinity and pH), the production of abundant zoospores, and efficient mechanisms that enable zoospore attachment to suitable substrata (Leaño et al. 2000). Saprobic thraustochytrids and labyrinthulids are also common in marine and estuarine waters (Naganuma et al. 1998), on littoral algae and seaweeds (Miller and Jones 1983), and fallen mangrove leaves (Bremer 1995; Leaño 2001; Fan et al. 2002a).

Mutualists are organisms that have an intimate and often inter-dependent relationship with another organism, which is beneficial for both parties. For example, rumen fungi derive nourishment from the nutrient-rich environment of the digestive systems of herbivorous mammals and reptiles by the decomposition of plant cell walls and fibre while providing the animal with sugars and short-chain fatty acids (Trinci et al. 1994).

Parasites can roughly be divided into three groups: biotrophic, necrotrophic, and facultative (Gleason et al. 2010). Biotrophic parasites depend on a living host for their growth and propagation, and often cannot grow outside the host. Specialized forms include obligate biotrophic parasites that are incapable of growth without a host. These organisms are therefore host-dependent and, where infectivity is limited to a narrow host range, they are also host-specific. Species of Phytomyxea are good examples of obligate biotrophic parasites (Neuhauser et al. 2011).

Necrotrophic parasites first kill living host cells by releasing toxins, for example, or by stimulating apoptosis of the host cells. Digestive enzymes are then excreted by the parasite to consume the host tissue. Facultative parasites can grow well either as saprotrophs or parasites, and these organisms often have a wide host range. Examples of facultative parasites are the oomycetous pathogens of marine aquatic animals, including species of *Lagenidium, Haliphthoros, Sirolpidium, Atkinsiella* and *Halocrusticida* (Ramaiah 2006; Yasunobu 2010; Leaño 2011). Stephen and Kurtböke (2011) recently described isolates of *Halophytophthora* that may be weak facultative parasites of midge larvae in the intertidal zone. However, the precise role of *Halophytophthora* in the death of midge larvae remains unknown. Some zoosporic true fungi have also been identified as facultative parasites. It is thought that *Rhizophydium littoreum* may be a weak facultative parasite of siphonaceous marine algae in estuaries, but its impact on these hosts is not understood (Amon 1984). These facultative parasites are usually non-pathogenic in the natural environment but can become pathogenic when unfavorable conditions occur for the host (stressful environment) or when the hosts become immunocompromised. One example of this is the presence of concurrent infections in fish and shrimps (Noga 1990). The complete lifecycles and trophic relationships of many fungus-like organisms have not yet, however, been fully determined.

Modes of nutrition: osmotrophic, phagotrophic and mixotrophic

The osmotrophic mode of nutrition depends on the ability of an organism to synthesize and release extracellular digestive enzymes. The osmotroph also needs to manufacture transporters and place them in the cell membrane for the uptake of small molecules. In osmotrophic nutrition, the digestion of macromolecules occurs entirely outside the cell and the resulting small organic molecules are transported inside the cell. Saprobic Labyrinthulomycetes exhibit this mode of nutrition by recycling nutrients in marine and coastal ecosystems and by the chemical alteration of detritus through extracellular enzymes (Raghukumar 2002).

The phagotrophic mode of nutrition depends on the formation of pseudopodia to engulf food particles and on the presence of actin filaments and vacuoles containing digestive enzymes in the cytoplasm. In phagotrophic nutrition, the digestion of macromolecules occurs entirely within a food vacuole inside the cell.

In mixotrophic nutrition, both osmotrophic and phagotrophic processes occur simultaneously.

Feeding experiments with dyed or radioactively labeled substrates are often necessary to demonstrate the mode of nutrition. However, the mode of nutrition of many marine eukaryotic microorganisms is not yet fully understood. More research in this field is needed because the mode of nutrition influences the biological strategies of fungus-like marine organisms (e.g., the time of attachment to a certain substrate). Understanding the resource-seeking strategies of these eukaryotic microorganisms is important not only for understanding their ecology and biology, but also their evolution. A study of the implications of parasitism by zoosporic true fungi on the relative abundance of chemotypes of the freshwater cyanobacterium, *Planktothrix* by Sønstebø and Rohrlack (2011) provides an excellent model for future research in this field.

Transitions between marine and freshwater environments

Most fungus-like microorganisms have been described from either freshwater, brackish water or marine environments. Since the transition from marine to freshwater environments is a process that has had a huge impact on the evolution of most microorganisms, this topic has been of particular interest to evolutionary biologists. Cavalier-Smith (2009); Logares et al. (2009) and Heger et al. (2010) have recently proposed that transitions of eukaryotic microorganisms between marine and freshwater ecosystems are rare on an evolutionary scale. Sharp differentiation into freshwater and marine forms is one of the most distinctive features of biodiversity in most groups of organisms (Logares et al. 2009; Heger et al. 2010). The distribution of euglyphid testate amoebae into distinct marine supra littoral genotypes and freshwater genotypes is a good example of this well-defined differentiation (Heger et al. 2010). However, an exception to the clear-cut distinctions regarding evolution in marine and freshwater environments can be found in the Oomycetes. The data provided by Lara and Belbahri (2011) suggest that transitions between freshwater and marine environments have been common during the evolution of Oomycetes, as some closely-related species, such as those in the Pythiales, have both marine and freshwater representatives. Furthermore, some parasitic species of Oomycetes appear to be able to cross the marine–freshwater barrier (Newell et al. 1977).

Marine and freshwater ecosystems are fundamentally different in many aspects. Salinity is usually considered to be the most important distinguishing abiotic feature, but many other abiotic and biotic factors also need to be considered. The marine–freshwater boundary can have a steep salinity gradient, which is considered to be one of the most difficult barriers to cross because the osmotic potential changes rapidly with salt concentration (Cavalier-Smith 2009; Logares et al. 2009; Heger et al. 2010). Gradients in osmotic pressure and ion concentrations are among the most important factors in preventing marine–freshwater cross colonization. Many microorganisms have specialized mechanisms that allow them to adapt to gradual changes in salinity, but these may not be sufficient for large, sudden, or permanent osmotic changes. Protists without cell walls need contractile vacuoles to pump water out in freshwater environments and sodium ion pumps in the cell membrane to pump sodium out in marine environments (Cavalier-Smith 2009). As these adaptive mechanisms are highly specialized, changes in the adaptation to a certain environment would not be expected to happen frequently.

Besides the need for osmoregulation, competition and predation can also limit plasticity (Logares et al. 2009), causing the composition of species along a salinity gradient to change. According to Logares et al. (2009) well-defined changes in the composition of communities along a salinity gradient are expected. Using molecular techniques, Mohamed and Martiny (2011) recently documented changes in fungal diversity and composition along a salinity gradient in three tidal marshes. If local species are better adapted to an environment and immigrant species are poorly adapted, the immigrant species may not survive. This concept was tested in the laboratory by Amon and Yei (1982), who investigated the effect of competition on the growth of an estuarine ecotype of *R. littoreum* (Chytridiomycota) and a marine ecotype of *Thraustochytrium striatum* (Labyrinthulomycota). It was found that *R. littoreum* grew better at low to intermediate salinities, while *T. striatum* grew better at intermediate to hypersaline conditions in both single-member and two-member cultures. However, these two microbes differed more significantly in their response to salinity when competition was a factor.

Marine food webs

A generalized microbial food web showing the trophic links between organisms in a typical marine and estuarine environment is illustrated in Figure 9.1. The food web includes a pool of zoospores released from true fungi and from fungus-like organisms, which provide a food resource for other organisms (Figure 9.1).

Dense populations of heterotrophic prokaryotic and eukaryotic microorganisms are often found growing on the surface of seaweeds in estuarine and marine ecosystems (Armstrong et al. 2000). These microorganisms benefit from the exudates released by the seaweeds. In particular, saprobic species of Labyrinthulomycota (Raghukumar 2002) and Oomycetes (Leaño 2001) participate in the decomposition of insoluble organic material from seaweeds and mangrove leaves.

In tropical and sub-tropical mangrove habitats, *Halophytophthora* species (Oomycetes) and thraustochytrids (Labyrinthulomycetes) are reported to be among the most abundant initial colonizers of fallen mangrove leaves (Nakagiri 1998; Leaño et al. 2000). Mangroves are highly productive ecosystems, which are not only self-sustaining but also able to produce sufficient nutrients to support large trophic webs (Ong 1995). The

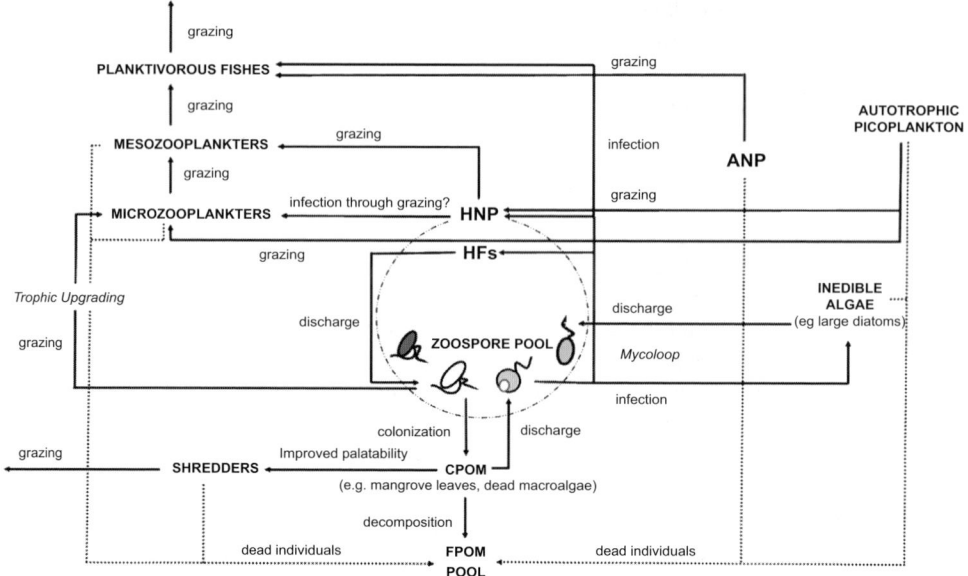

Figure 9.1 Trophic links of a generalized marine and estuarine microbial food web in which zoosporic true fungi and fungal-like organisms are involved. Zoospores from parasitic and saprotrophic zoosporic true fungi and fungal-like organisms are included in the zoospore pool. Key: CPOM: coarse particulate organic matter; FPOM: fine particulate organic matter; HNP: heterotrophic nanoplankton; HF: heterotrophic flagellates; ANP: autotrophic nanoplankton.

bulk of the energy input in mangrove habitats is derived from leaf litter, which is decomposed by fungi and fungus-like microorganisms (Leaño et al. 2000; see Chapters 6 and 22) as well as by other microbes such as bacteria. Oomycetes and thraustochytrids play a major role as detrital microbes in mangrove environments because they alter the chemistry of the detritus (fallen leaves), degrade a wide variety of organic compounds (Raghukumar et al. 1994), and enrich the nutrient contents of the degraded leaves with their own biomass (Nakagiri 1998). This process contributes significantly to the nutrition of many bottom-feeding and filter-feeding organisms, making mangroves excellent nursery grounds for many fish and crustacean species (Leaño 2001).

Many species of eukaryotic microorganisms belonging to the Oomycota, Labyrinthulomycota and Phytomyxea, are parasites of seaweeds and sea grasses (Küpper and Müller 1999; Gachon et al. 2010; Neuhauser et al. 2011). For example, *Pythium porphyrae* (Oomycota) is an important parasite of the red alga, *Porphyra yezoensis* (Rhodophyta), and causes significant losses in the production of *P. yezoensis* along the north coast of China (Ding and Ma 2005). The labyrinthulomycete *Labyrinthula zosterae* is the primary pathogen causing wasting disease in eel grass (*Zostera marina*). This significantly reduces eel grass populations in North America and Europe (Muehlstein et al. 1991; Ralph and Short 2002). Furthermore, many species of Labyrinthulomycota are thought to be capable of decomposing terrestrial organic matter from riverine input (Kimura and Naganuma 2001).

Heterotrophic flagellates (HFs) make up an important fraction of the heterotrophic nanoplankton (HNP). The HNP consist of assemblages of heterotrophic organisms that range from 2 to 20 μm in size (Yoshida et al. 2009). In the past, many lineages of HNP have been overlooked or classified as unidentified HFs because of their lack of distinctive morphological features (Sime-Ngando et al. 2011). However, during the past few years, the development of molecular techniques has allowed the identification of many of these previously unidentified lineages, which include parasitic taxa of zoosporic true fungi and fungal-like organisms (Lefèvre et al. 2008; Jones et al. 2011). These molecular studies have also revealed the importance of these taxa as components of the HNP ecosystem (Sime-Ngando et al. 2011).

Both osmotrophic and phagotrophic nutrition occurs in some groups of HFs. In freshwater ecosystems, all groups of HF, including the zoospores of fungi and fungal-like eukaryotes, are known to provide valuable food resources for grazing and filter-feeding zooplankton and metazoan invertebrates (Sime-Ngando et al. 2011). Similarly, the large numbers of zoospores released by parasites of algae and invertebrate animals and also by saprotrophs would be expected to provide important food resources in coastal marine environments. Saprobes can also reach large population densities when colonizing detritus (Raghukumar 2002; Sime-Ngando et al. 2011).

The addition of saprophytic and parasitic HFs to marine food web models greatly increases the complexity of the food web (see Figure 9.1). No quantitative data on energetics of zoosporic true fungi and fungus-like organisms in marine food webs, however, are yet available. One quantitative study on the population dynamics of dinoflagellates did estimate the amount of carbon transferred from a host to a parasite. This food chain involved energy transfer from *Neoceratium falcatiforme* (a dinoflagellate host) to *Amoebophyra* spp. (a dinoflagellate parasite) and then to tintinnid ciliates (predators) (Salomon et al. 2009). This research provided a model that could be adapted to studies with parasitic zoosporic true fungi and fungus-like organisms because it demonstrated a rapid increase in parasite density and a movement of energy into the cells of the parasite. The study also demonstrated the importance of frequently monitoring such parasites, because their populations can reach high densities within a short time but can decline at the same rapid rate.

Eukaryotic microorganisms are known to synthesize compounds that are essential as nutrients for the growth of many grazing and filter-feeding invertebrates in a process called trophic upgrading. For example, many freshwater and marine eukaryotic microorganisms are known to synthesize polyunsaturated fatty acids and sterols. These essential nutrients have been found in species in the Chytridiomycota (Southall et al. 1977; Weete et al. 1989), Blastocladiomycota (Southall et al. 1977; Weete et al. 1989), Neocallimastigomycota (Koppová et al. 2008), Oomycetes (Gandhi and Weete 1991; Kim et al. 1998), Hyphochytriomycota (Weete et al. 1989), Labyrinthulomycota (Weete et al. 1997; Leaño et al 2003; Raghukumar 2008) and Phytomyxea (Sundelin et al. 2010).

Several species in the Labyrinthulomycota are considered to be potential sources for the commercial production of polyunsaturated fatty acids and sterols (Raghukumar 2008; Nagano et al. 2009). These lipids are considered to be important storage products for these microorganisms, providing energy for the movement of flagella in the zoospores and for the subsequent development of other stages in the lifecycle (Gleason and Lilje 2009).

The process of trophic upgrading also occurs when zoosporic true fungi provide nutrition in the form of zoospores for component zooplankton species in freshwater ecosystems (Kagami et al. 2007; Gleason et al. 2008; Kagami et al. 2011). As many species of Oomycota, Labyrinthulomycota, Phytomyxea, and other fungal-like organisms are abundant in marine ecosystems, the zoospores produced by these species would be expected to transfer significant amounts of energy to secondary consumers (see Figure 9.1).

Dispersal and distribution

A large variety of propagules are produced by zoosporic true fungi and fungus-like organisms. These include zoospores, zoospore cysts, aplanospores, amoeboid cells, endospores, oospores, hyphal fragments and other structures. The conditions that stimulate sporulation and encystment of zoospores are poorly understood. Motile propagules, such as zoospores, may be chemotactic so that they can sense diffusion gradients and swim toward utilizable substrates (active dispersal). Chemotaxis has been studied in marine species of *Rhizophydium* (Muehlstein et al. 1988), *Halophytophthora* (Leaño et al. 1998), and mangrove thraustochytrids including *Thraustochytrium*, *Schizochytrium* and *Ulkenia* (Fan et al. 2002b). Non-motile propagules rely on currents for dispersal and reach their substrates by chance (passive dispersal). Therefore, species that produce chemotactic zoospores may reach utilizable substrates quicker than those that produce non-motile spores. On the other hand, non-motile propagules may be produced in higher numbers or may be more resistant and able to survive longer than short-lived zoospores. In laboratory cultures, fungus-like parasites of invertebrates in the Oomycetes (*Myzocytiopsis* and *Chlamydomyzium*) and the zoosporic true fungi (*Olpidium*) may produce abundant zoospores while healthy host organisms are available. These parasites will also produce thick-walled resting spores or oospores, especially towards the decline of infection when conditions become unfavorable or host populations diminish (Glockling and Beakes 2006).

Both motile and non-motile propagules are frequently swept away by currents and end up as food for secondary consumers or as components of the detritus. Species must therefore produce large numbers of spores in order to increase the probability of successful propagation following location and attachment of infective spores to utilizable substrates and susceptible hosts. Zoospores would be expected to remain motile until they either attach to a substrate or encyst because they have either exhausted their energy reserves or have encountered unfavorable environmental conditions. Amoeboid cells are able to crawl over surfaces and attach to them. Propagules that remain attached to substrates or those that are produced endobiotically inside the host can be carried with the currents together with the host or substrate. Some propagules may remain attached to detritus and settle to the bottom, where oxygen can be limited. Other propagules can be carried into the intertidal zone, where the salinity varies greatly. If propagules can withstand drying, they may survive for long periods of time in a dormant state. Therefore the type of cells formed determines the survival and distribution of fungus-like organisms. Having more than one type of propagule allows fungus-like organisms to combine more than one dissemination strategy during their lifecycle.

It is hypothesized that a species may become widespread and even ubiquitous if its propagules remain viable for long enough to be dispersed by the water currents over long distances. However, for widespread distribution, the propagules also need to find appropriate substrates or hosts and experience favorable environmental conditions. The same hypothesis would indicate that species which do not form resistant structures or have a limited host range might be expected to be endemic rather than cosmopolitan. In contradiction to this hypothesis, however, some species, such as the oomycete nematode parasites in *Haptoglossa* (a genus basal in the oomycete lineage; Beakes et al. 2006), appear to be ubiquitous despite the absence of resistant spores and have even been recorded from littoral marine nematodes (Newell et al. 1977). It is difficult to prove the hypothesis for endemic or cosmopolitan species because the actual distribution of many species of eukaryotic microorganisms is poorly known due to lack of sampling data.

Furthermore, so-called cryptic species confuse the estimation of the global distribution of some taxa. Recent molecular work on protists indicates that some isolates that are morphologically identical have distinct genotypes, suggesting they are in fact different species or varieties; whereas other taxa that are morphologically distinct share the same genotype, suggesting the species has several morphotypes. The geographical distribution patterns of certain species of ciliates have been discussed thoroughly by Foissner (2008). Some data suggest that ciliate species are endemic, while other data suggest that they are cosmopolitan. Nonetheless, the high frequency at which some microbial DNA sequences appear in environmental samples from diverse and often extreme environments (Lefèvre et al. 2008; Stoeck et al. 2010) supports the cosmopolitan hypothesis and suggests a relatively high biomass of zoosporic true fungi and fungus-like microorganisms at the primary and secondary consumer levels in food webs.

Acknowledgements

SN was supported by a Hertha-Firnberg research grant (Austrian Science Fund (FWF) grant T379-B16).

References

Adl SM, Simpson AGB, Farmer MA, Andersen RA, Anderson OR, Barta JR, Bowser SS, Brugerolle G, Fensome RA, Fredericq S, James TY, Karpov S, Kugrens P, Krug J, Lane CE, Lewis LA, Lodge J, Lynn DH, Mann DG, Mccourt RM, Mendoza L, Moestrup O, Mozley-Standridge SE, Nerad TA, Shearer CA, Smirnov AV. The new higher level classification of Eukaryotes with emphasis on the taxonomy of protists. J Eukaryot Microbiol. 2005;52:399–451.

Amon JP. *Rhizophydium littoreum*: a chytrid from siphonaceous marine algae: an ultrastructural examination. Mycologia 1984;76:132–139.

Amon JP, Yei SP. The effect of salinity on the growth of two marine fungi in mixed culture. Mycologia 1982;74:117–122.

Armstrong E, Rogerson A, Leftley JW. The abundance of heterotrophic protists associated with intertidal seaweeds. Estuar Coast Shelf Sci. 2000;50:415–424.

Baldauf SL. The deep roots of Eukaryotes. Science 2003;300:1703–1706.

Beakes GW, Sekimoto S. The evolutionary phylogeny of oomycetes – insights gained from studies of holocarpic parasites of algae and invertebrates. In: Lamour K, Kamoun S, eds. Oomycete genetics and genomics: diversity, interactions, and research tools. London: Wiley-Blackwell, 2009:1–24.

Beakes GW, Glockling SL, James TY. The diversity of oomycete pathogens of nematodes and its implications to our understanding of oomycete phylogeny. In: Meyer W, Pearce C, eds. Proceedings of the 8th International Mycological Congress, Cairn, Australia. 2006:7–12.

Beakes GW, Glockling SL, Sekimoto S. The evolutionary phylogeny of the oomycete "fungi". Protoplasma 2012;249:3–19.

Bremer GB. Lower marine fungi (Labyrinthulomycetes) and decay of mangrove leaf litter. Hydrobiol. 1995;295:89–95.

Cafaro MJ. Eccrinales (Trichomycetes) are not fungi, but a clade of protists at the early divergence of animals and fungi. Mol Phylogenet Evol. 2005;35:21–34.

Cavalier-Smith T. Megaphylogeny, cell body plans, adaptive zones: causes and timing of Eukaryote basal radiations. J Eukaryot Microbiol. 2009;56:26–33.

Dick MW. Straminipilous fungi. Dordrecht: Klewer Academic Publishers, 2001:670.

Diggles BK. A mycosis of juvenile spiny rock lobster, *Jasus edwardsii* (Hutton, 1875) caused by *Haliphthoros* sp., and possible methods of chemical control. J Fish Dis. 2001;24:99–110.

Ding H, Ma J. Simultaneous infection by red rot and chytrid diseases in *Porphyra yezoensis* Ueda. J Appl Phycol. 2005;17:51–56.

Fan KW, Vrijmoed LLB, Jones EBG. Physiological studies of subtropical mangrove thraustochytrids. Bot Mar. 2002a;45:50–57.

Fan KW, Vrijmoed LLB, Jones EBG. Zoosporic chemotaxis of mangrove thraustochytrids from Hong Kong. Mycologia 2002b;94:569–578.

Foissner W. Protist diversity and distribution: some basic considerations. Biodivers Conserv. 2008;17:235–242.

Freeman KR, Martin AP, Karki D, Lynch RC, Mitter MS, Meyer AF, Longcore JE, Simmons DR, Schmidt SK. Evidence that chytrids dominate fungal communities in high-elevation soils. Proc Nat Acad Sci. 2009;106:18315–18320.

Gachon CMM, Sime-Ngando T, Strittmater M, Chambouvet A, Kim GH. Algal diseases: spotlight on a black box. Trends Plant Sci. 2010;15:633–640.

Gandhi S, Weete JD. Production of the polyunsaturated fatty acids arachidonic acid and eicosapentaenoic acid by the fungus *Pythium ultimum*. J Gen Microbiol. 1991;137:1825–1831.

Gleason F, Lilje O. Structure and function of fungal zoospores: ecological implications. Fungal Ecol. 2009;2:53–59.

Gleason FH, Kagami M, Lefèvre E, Sime-Ngando T. The ecology of chytrids in aquatic ecosystems: roles in food web dynamics. Fungal Biol Rev. 2008;2:17–25.

Gleason FH, Marano AM, Johnson P, Martin WW. Blastocladian parasites of invertebrates. Fungal Biol Rev. 2010;24:56–67.

Glockling SL, Beakes GW. An ultrastructural study of development and reproduction in the nematode parasite, *Myzocytiopsis vermicola*. Mycologia 2006;98:7–21.

Glücksman E, Bell T, Griffiths RI, Bass D. Closely related protist strains have different grazing impacts on natural bacterial communities. Environ Microbiol. 2010;12:3105–3113.

Heger TJ, Mitchell EAD, Todorov M, Golemansky V, Lara E, Leander BS, Pawlowski J. Molecular phylogeny of euglyphid testate amoebae (Cercozoa: Euglyphida) suggests transitions between marine supralittoral and freshwater/terrestrial environments are infrequent. Mol Phylogenet Evol. 2010;55:113–122.

Honda D, Yokochi T, Nakahara T, Raghukumar S, Nakagiri A, Schauman K, Higashihara T. Molecular phylogeny of labyrinthulids and thraustochytrids based on the sequencing of 18S ribosomal RNA gene. J Eukaryot Microbiol. 1999;46:637–647.

James TY, Letcher PM, Longcore JE, Mozley-Standridge SE, Porter D, Powell MJ, Griffith GW, Vilgalys R. A molecular phylogeny of the flagellated fungi (Chytridiomycota) and description of a new phylum (Blastocladiomycota). Mycologia 2006;98:860–871.

Jones MDM, Forn I, Gadelha C, Egan MJ, Bass D, Massana R, Richards TA. Discovery of novel intermediate forms redefines the fungal tree of life. Nature 2011;474:200–203.

Kagami M, de Bruin A, Ibelings BW, van Donk E. Parasitic chytrids: their effects on phytoplankton communities and food-web dynamics. Hydrobiol. 2007;578:113–129.

Kagami M, Helmsing NR, van Donk E. Parasitic chytrids could promote copepod survival by mediating material transfer from inedible diatoms. Hydrobiol. 2011;659:49–54.

Kim H, Gandhi SR, Moreau RA, Weete DJ. Lipids of *Haliphthoros philippinensis*: an oomycetous marine microbe. J Amer Oil Chem Soc. 1998;75:1657–1665.

Kimura H, Naganuma T. Thraustochytrids: a neglected agent of the marine microbial food chain. Aquat Ecosys Health Manage. 2001;4:13–18.

Kitancharoen N, Nakamura K, Wada S, Hatai K. *Atkinsiella awabi* sp. nov. isolated from stocked abalone, *Haliotis sieboldii*. Mycoscience 1994;35:265–270.

Koppová I, Novotná Z, Štrosová L, Fliegerová K. Analysis of fatty acid composition of anaerobic rumen fungi. Folia Microbiol. 2008;53:217–220.

Küpper FC, Müller DG. Massive occurrence of the heterokont parasites *Anisolpidium*, *Eurychasma* and *Chytridium* in *Pylaiella littoralis* (Ectocarpales, Phaeophyceae). Nova Hedw. 1999;69: 381–389.

Lara E, Belbahri L. SSU and rRNA reveals major trends in oomycete evolution. Fungal Divers. 2011;49:93–100.

Lara E, Moreira D, Lopez-Garcia P. The environmental clade LKM11 and *Rozella* from the deepest branching clade of fungi. Protist 2010;161:116–121.

Leander CA, Porter D, Leander BS. Comparative morphology and molecular phylogeny of aplanochytrids (Labyinthulomycota). Eur J Protistol. 2004;40:317–328.

Leaño EM. Straminipilous organisms from fallen mangrove leaves from Panay Island, Philippines. Fungal Divers. 2001;6:75–81.

Leaño EM. *Haliphthoros* spp. from spawned eggs of captive mud crab, *Scylla serrata*, broodstocks. Fungal Divers. 2002;9:93–103.

Leaño EM. Fungal diseases. In: Lio-Po GD, Inui Y, eds. Health management in aquaculture, 2nd ed. SEAFDEC Aquaculture Department, Iloilo, Philippines. 2011:39–51.

Leaño EM, Vrijmoed LLB, Jones EBG. Zoospore chemotaxis of two mangrove strains of *Halophytophthora vesicula* from Mai Po, Hong Kong. Mycologia 1998;90:1001–1008.

Leaño EM, Jones EBG, Vrijmoed LLP. Why are *Halophytophthora* species well adapted to mangrove habitats? Fungal Divers. 2000;5:131–151.

Leaño EM, Gapasin RSJ, Polohan B, Vrijmoed LLP. Growth and fatty acid production of thraustochytrids from Panay mangroves, Philippines. Fungal Divers. 2003;12:111–122.

Lefèvre E, Roussel B, Amblard C, Sime-Ngando T. The molecular diversity of freshwater picoeukaryotes reveals high occurrence of putative parasitoids in the plankton. PLoS ONE 2008;3(6):e2324.

Logares R, Bråte J, Bertilsson S, Clasen JL, Shalchian-Tabrizi K, Rengefors K. Infrequent marine-freshwater transitions in the microbial world. Trend Microbiol. 2009;17:414–422.

Maddison DR, Schulz KS (ed). The tree of life web project, 2007. (Accessed April 11, 2012 at http://tolweb.org).

Marshall WL, Berbee ML. Facing unknowns: living cultures (*Pirum gemmata* gen. nov., sp. nov., and *Abeoforma whisleri*, gen. nov., sp. nov.) from invertebrate digestive tracts represent an undescribed clade within the unicellular opisthokont lineage Ichthyosporea (Mesomycetozoea). Protist 2011;162:33–57.

McCreadie JW, Adler PH, Beard CE. Ecology of symbionts of larval black flies (Diptera: Simuliidae): distribution, diversity, and scale. Environ Entomol. 2011;40:289–302.

Miller JD, Jones EBG. Observations on the association of thraustochytrid marine fungi with decaying seaweed. Bot Mar. 1983;26:245–251.

Mohamed DJ, Martiny JBH. Patterns of fungal diversity and composition along a salinity gradient. ISME J. 2011;5:379–388.

Moreira D, Lopez-Garcia P. The molecular ecology of microbial eukaryotes unveils a hidden world. Trends Microbiol. 2002;10:31–38.

Muehlstein LK, Amon JP, Leffler DL. Chemotaxis in the marine fungus *Rhizophydium littoreum*. Appl Environ Microbiol. 1988;54:1668–1672.

Muehlstein LK, Porter D, Short D, Short FT. *Labyrinthula zosterae* sp. nov., the causative agent of wasting disease of eelgrass, *Zostera marina*. Mycologia 1991;83:180–191.

Nagano N, Taoka Y, Honda D, Hayashi M. Optimization of culture conditions for growth and docosahexaenoic acid production by a marine thraustochytrid, *Aurantiochytrium limacinum* mh0186. J Oleo Sci. 2009;58:623–628.

Naganuma T, Takasugi H, Kimura H. Abundance of thraustochytrids in coastal planktons. Mar Ecol Prog Ser. 1998;162:105–110.

Nagy LG, Petkovits T, Kovacs GM, Voigt K, Vagvolgyi C, Papp T. Where is the unseen fungal diversity hidden? A study of *Mortierella* reveals a large contribution of reference collections to the identification of fungal environmental sequences. New Phytol. 2011;191:789–794.

Nakagiri A. Diversity of halophytophthoras in subtropical mangroves and factors affecting their distribution. In: Proceedings of the Asia-Pacific Mycological Conference on Biodiversity and Biotechnology. Hua Hin, Thailand. 1998:109–113.

Nakagiri A, Ito T, Manoch L, Tanticharoen M. A new *Halophytophthora* species, *H. porrigovesica*, from subtropical and tropical mangroves. Mycoscience 2001;42:33–41.

Neuhauser S, Kirchmair M, Gleason FH. Ecological roles of the parasitic phytomyxids (plasmodiophorids) in marine ecosystems – a review. Mar Freshw Res. 2011;62:365–371.

Newell SY, Cafalu R, Fell JW. *Myzocytium*, *Haptoglossa* and *Gonimochaete* (fungi) in littoral marine nematodes. Bull Mar Sci. 1977;27:177–207.

Noga EJ. A synopsis of mycotic diseases of marine fishes and invertebrates. In: Perkins FO, Cheng TC, eds. Pathology in marine science. San Diego, USA: Academic Press, 1990:143–159.

Ong JE. The ecology of mangrove conservation and management. Hydrobiol. 1995;295:343–351.

Raghukumar S. Ecology of the marine protists, the Labyrinthulomycetes (Thraustochytrids and Labyrinthulids). Eur J Protistol. 2002;38:127–145.

Raghukumar S. Thraustochytrid marine protists: production of PUFAs and other emerging technologies. Mar Biotechnol. 2008;10:631–640.

Raghukumar S, Sharma S, Raghukumar C, Sathe-Pathak V, Chandramohan D. Thraustochytrids and fungal components of marine detritus: IV. Laboratory studies on decomposition of leaves of the mangrove *Rhizophora apiculata* Blume. J Exp Mar Biol Ecol. 1994;183:113–131.

Ralph PJ, Short FT. Impact of the wasting disease pathogen, *Labyrinthula zosterae*, on the photobiology of eelgrass *Zostera marina*. Mar Ecol Prog Ser. 2002;226:265–271.

Ramaiah N. A review of fungal diseases of algae, marine fishes, shrimps and corals. Ind J Mar Sci. 2006;35:380–387.

Roza D, Hatai K. *Atkinsiella dubia* infection in the larvae of Japanese mitten crab *Eriocheir japonicus*. Mycoscience 1999;40:235–240.

Salomon PS, Granéli E, Neves MCB, Rodriguez EG. Infection by *Amoebophrya* spp. parasitoids of dinoflagellates in a tropical marine coastal area. Aquat Microbial Ecol. 2009;55:143–153.

Sime-Ngando T, Lefèvre E, Gleason FH. Hidden diversity among aquatic heterotrophic flagellates: ecological potentials of zoosporic fungi. Hydrobiol. 2011;659:5–22.

Simon C, Daniel R. Metagenomic analyses: past and future trends. Appl Environ Microbiol. 2011;77:1153–1161.

Sønstebø JH, Rohrlack T. Possible implications of chytrid parasitism for population subdivision in freshwater cyanobacteria of the genus *Planktothrix*. Appl Environ Microbiol. 2011;77:1344–1351.

Southall MA, Motta JJ, Patterson GW. Identification and phylogenetic implications of fatty acids and sterols in three genera of aquatic phycomycetes. Am J Bot. 1977;64:246–252.

Sparrow FK. Aquatic Phycomycetes, 2nd ed. Ann Arbor: University of Michigan Press, 1960:1137.

Stephen K, Kurtböke DI. Screening of oomycete fungi for their potential role in reducing the biting midge (Diptera: Ceratopogonidae) larval populations in Hervey Bay, Queensland, Australia. Int J Environ Res Public Health. 2011;8:1560–1574.

Stoeck T, Bass D, Nebel M, Christen R, Jones MDM, Breiner HW, Richards TA. Multiple marker parallel tag environmental DNA sequencing reveals a highly complex eukaryotic community in marine anoxic water. Mol Ecol. 2010;19:21–31.

Sundelin T, Christensen CB, Larsen J, Moller K, Lubeck M, Bodker L, Jensen B. In planta quantification of *Plasmodiophora brassicae* using signature fatty acids and real-time PCR. Plant Dis. 2010;94:432–438.

Trinci APJ, Davies DR, Gull K, Lawrence M, Nielsen BB, Rickers A, Theodorou MK. Anaerobic fungi in herbivorous animals. Mycol Res. 1994;98:129–152.

Weete JD, Fuller S, Huang MQ, Gandhi S. Fatty acids and sterols of selected Hyphochytriomycetes and Chytridiomycetes. Exp Mycol. 1989;13:183–195.

Weete JD, Gandhi S, Kim H, Wang HY, Dute RR. Lipids and ultrastructure of *Thraustochytrium* sp., ATCC 26185. Lipids 1997;32:839–845.

Yasunobu H. Prevention of fungal infection in the swimming crab *Portunus trituberculatus* larvae by high pH of rearing water. Bull Hyogo Pref Tech Cent Agr Forest Fish (Fish. Sec.) 2010;41: 1–58 (in Japanese with English abstract).

Yoshida N, Nishimura M, Inoue K, Yoshizawa S, Kamiy E, Taniguchi A, Hamasaki K, Kogure K. Analysis of nanoplankton community structure using flow sorting and molecular techniques. Microbes Environ. 2009;4:297–304.

10 Cryptomycota (Rozellida) and Mesomycetozoea (Ichthyosporea) the Super-group Opisthokonta

Sally L. Glockling, Meredith D.M. Jones and Frank H. Gleason

During the past decade there has been a great deal of interest in the groups of microorganisms that are considered to be basal to the fungi and animals. Although a small number of such species had been known for a long time, they had previously been classified as fungi, protists or algae according to their morphology. Following the introduction of new techniques, such as second-generation sequencing, there have been many changes to our concepts of the phylogenetic relationships among these microorganisms, especially since this topic was reviewed by Cavalier-Smith (1987). In this chapter we consider two groups of fungus-like microorganisms: Cryptomycota (Rozellida) and Mesomycetozoea (Ichthyosporea), both of which are thought to be basal to the fungi and include taxa that have been reported from marine environments.

In basal groups in the super-group Opisthokonta, the propagules can be zoospores with posteriorly-directed whiplash flagella, amoebae or non-motile spores with cell walls. Some species appear to have lost the ability to produce centrioles and flagella, and some species never produce amoebae. The mitochondria in the super-group Opisthokonta have flat plate-like cristae, although some species lack mitochondria (Adl et al. 2005; Ginger et al. 2010). Burger et al. (2003) studied the mitochondrial genomes in *Monosiga brevicollis* (Choanoflagellates) and *Amoebidium parasiticum* (Mesomycetozoea). They found that the mitochondrial DNA was gene-rich and much larger than the DNA strands found in fungi and animals. This indicated that fungi and animals have lost mitochondrial genes. Chitin is found in the cell walls of true fungi and in the exoskeleton of some groups of animals, but is only present in small amounts in the cell walls of the Mesomycetozoea, and is absent within the lifecycle stages of the Cryptomycota subgroups observed by Jones et al. (2011a).

Cryptomycota

Historical background

The Cryptomycota are an early branching clade of organisms comprised largely of environmental sequences; the only known species being those in the genus *Rozella* (Lara et al. 2010; Jones et al. 2011a; James and Berbee 2012). Parasites described in *Rozella* were originally placed in the fungal phylum Chytridiomycota (Barr 2001) because they released uniflagellate zoospores from their thalli and produced resting spores. Parasites in *Rozella* differed from other members of the Chytridiomycota, however, because their thalli grew as naked protoplasts within host cells, only separated from the host protoplasm by the plasma membrane (Held 1981). A multi-gene molecular phylogeny of the Chytridiomycota, which described the new phylum Blastocladiomycota, did not place *Rozella* in either the Blastocladiomycota or the Chytridiomycota (James et al. 2006). Instead, *Rozella* unexpectedly emerged as a separate lineage basal to the Chytridiomycota, which was interpreted at the time as being the most basal lineage within the Fungi kingdom (James et al. 2006).

Prior to this, several phylogenetic analyses of the small subunit ribosomal genes (SSU rDNA) data from environmental surveys had revealed a group of unknown species that did not cluster with any of the main fungal groups or with any known basal clades within the Opisthokonta. The first sequence representative of this group was termed LKM11 (van Hannen et al. 1999). This group, initially entirely comprised of environmental sequences, was later shown to be related to fungi, but its precise phylogenetic position remained unresolved (Berney et al. 2004).

Lara et al. (2010) analyzed the phylogenetic position of the LKM11 clade using sequences from SSU rDNA. This analysis included three new environmental sequences from peat bogs and sequences from the database derived from freshwater and marine sediments. Importantly, this analysis also included two species of *Rozella*. The phylogenetic analysis showed that twenty-six environmental sequences, including those from the peat bog, clustered in the LKM11 clade as did the two *Rozella* species. This was the first time that the LKM11 clade, known only from environmental samples, had been phylogenetically associated with morpholohically described species. The name *Rozellida* was suggested for the LKM11 clade, which was shown to be a robust monophyletic group and interpreted as either the deepest branching clade of the fungi or the nearest sister group to this (Lara et al. 2010). The name Cryptomycota was subsequently suggested as the phylum name (Jones et al. 2011a) and this name was later validated (Jones et al. 2011b).

Recently, Jones et al. (2011a) compiled a comprehensive phylogeny of the LKM11 clade using multiple SSU rRNA genes that incorporated over 40 environmental sequences derived from freshwater and marine environments around the world. Fluorescence *in situ* hybridization-based techniques were used to visualize Cryptomycota cells within freshwater samples, and counterstaining with specific antibodies produced evidence of uniflagellate zoospores within the subgroups tested (Jones et al. 2011a), see Figure 10.1. Non-flagellate putative cycts of Cryptomycota did not bind with wheat germ agglutinin, suggesting the absense of any chitinous cell wall (Jones et al. 2011a). This combined environmental DNA analysis and fluorescent detection using DNA probes led Jones

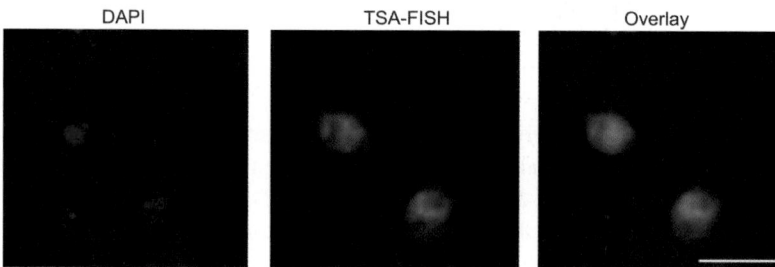

Figure 10.1 Micrographs showing tyramine signal amplication (TSA)-fluorescence *in situ* hybridization detection of Cryptomycota cells within a freshwater sample. Scale bar = 10 μm.

et al. (2011a) to conclude that the Cryptomycota was an intermediate group between fungi and protists. James and Berbee (2012), however, advise caution in interpreting the Cryptomycota as the earliest divergence in the fungal tree of life because the naked thallus of *Rozella* could signify loss of the cell wall as an adaptation to parasitism, and because artifacts and phylogenetic errors can occur in tree topologies with certain problematic species.

Characteristics of the genus *Rozella*

Rozella is a genus of 26 parasitic species (Lara et al. 2010) that infect the thalli and sporangia of true fungi in the Chytridiomycota and Blastocladiomycota and also the thalli and sporangia of some Oomycota (heterotrophic stramenopiles). Its most characteristic feature is the absence of a cell wall during its trophic phase and the unwalled mature sporangium. In common with zoosporic true fungi, all species of *Rozella* produce uniflagellate zoospores with a posteriorly-directed whiplash flagellum. In addition, some species form pigmented spiny resting spores that may later develop into sporangia (Held 1981).

The early research on *Rozella*, together with details of its confusing taxonomic history, was reviewed by Held (1981). Only one species, *R. allomycis*, has been cultivated in dual culture with its host and thoroughly studied in the laboratory; other species described have been listed by Sparrow (1960) and Karling (1977). Held (1981) organized *Rozella*, and species thought to be in related genera, into groups based on their morphology and life histories.

Propagation in *Rozella* is by uniflagellate zoospores that infect the host cells. The zoospores of *Rozella* tend to be smaller than those of its hosts (Held 1981). In *R. allomycis*, a parasite of the blastocladian fungus *Allomyces*, the *Rozella* zoospores were found to exhibit chemotaxis and to attach to receptors on the surface of the host cell (Held 1974). The encysted zoospore penetrates the host cell wall with a narrow germ tube, and a protoplast then grows inside the host cell and develops into a naked, multinucleate, sporangium. Following a transmission electron microscopy (TEM) study of *R. polyphagi*, which showed host mitochondria inside a vacuole in the naked thallus of the parasite, the uptake of nutrients during this trophic phase was thought to be by phagocytosis (Powell 1984). However, it is not known whether all *Rozella* species are phagotrophic or to what extent. The plasma membrane of the developing sporangium becomes closely adpressed to the host cell wall and infected cells were reported to be hypertrophied in some species to up to four times their usual size, although Held (1981) maintained that the dimensions were not outside their normal range. The protoplasm of the sporangium becomes cleaved and fully-formed zoospores swarm before being explosively discharged via an exit papilla. *Rozella* species may form one or several exit papillae and these may be either short or long and branched (Karling 1942). The protoplast may alternatively develop into a resting spore, often with a spiny exterior, which may later become a sporangium and produce zoospores, usually after the host cell has perished. The echinulate wall of the resting spores was thought to be derived from the host cell (Karling 1942). The naked thallus of *Rozella* suggested that this parasite was unable to form chitin (Held 1981). However, recent studies have shown that, while the outer spiny layer of the mature resting spore is composed of non-chitinous material, the wall layer of immature resting spores stained with calcofluor white, indicating the presence of chitin in the inner spore wall (James and Berbee 2012). Furthermore, James and Berbee (2012) have shown that *Rozella* contains a fungal specific chitin synthase gene (although it is not yet known whether this contains the myosin domain common to true fungi) and have mapped the occurrence of chitin synthases to a diagrammatic tree of fungi and other opisthokonts.

In some *Rozella* species the multinucleate protoplast may become segmented by septation and a linear series of sporangia subsequently develop. The septal walls that divide the protoplast belong to the host (Held 1981). Electron microscopy of the *Rozella* sporangium and studies of infections using chemically-induced plasmolysis have confirmed that the *Rozella* sporangium lacks a wall and that the exit papillae are continuous with the cell wall of the host (Held 1980, 1981).

Marine species

Only one species of *Rozella*, *R. marina*, has been recorded from marine ecosystems (Figure 10.2). This species was initially described as a parasite of *Chytridium polysiphoniae* that was growing on the red alga *Polysiphonia* off the coast of North America (Sparrow 1936). The host of *R. marina*, *C. polysiphoniae*, is a zoosporic true fungus and is also one of several common parasites of brown algal species (Küpper and Müller 1999; Müller et al. 1999; Gleason et al. 2011). The morphology and lifecycle of *R. marina* was later documented by Johnson (1966), who isolated it and its chytrid host from the alga *Polysiphonia violaceae*, collected from the sea around Iceland. *Rozella marina* forms a solitary sporangium inside the host sporangium and zoospores are released via one or several exit papillae (Figure 10.2). Neither Sparrow (1936) nor Johnson (1966) observed any resting spores in their isolates. No quantitative data on the incidence of parasitism by *R. marina* on *C. polysiphoniae* are currently available, but this parasite may be an important factor in controlling the population sizes of *C. polysiphoniae*. Unlike the freshwater species *R. allomycis*, the occurrence of chemotaxis in the zoospores of *R. marina* has yet to be tested.

Mesomycetozoea

Historical background

Before the advent of phylogenetic analyses, when classification was based upon morphological features, several fish parasites and other similar microorganisms, which are now known to be in the Mesomycetozoea, were described as fungi, algae or protists. This was because they shared morphological characteristics with these groups, such as spherical walled cells containing spores, hyphae and uniflagellate zoospores. Other similar microorganisms were unclassified. One of these was the fish parasite referred to as the rosette agent. When this microorganism, a parasite of Chinook salmon, was included in an 18S rDNA phylogeny, it was found to be closely related to the choanoflagellates (Kerk et al. 1995).

Subsequent phylogenetic studies showed that several other parasites that were unclassified or assigned to fungi, algae or protist genera, also belonged to this same monophyletic clade that diverged at the base of the fungal and animal lineages (Spangaard et al. 1996; Ragan et al. 1998). This clade, now known as the class Mesomycetozoea, currently

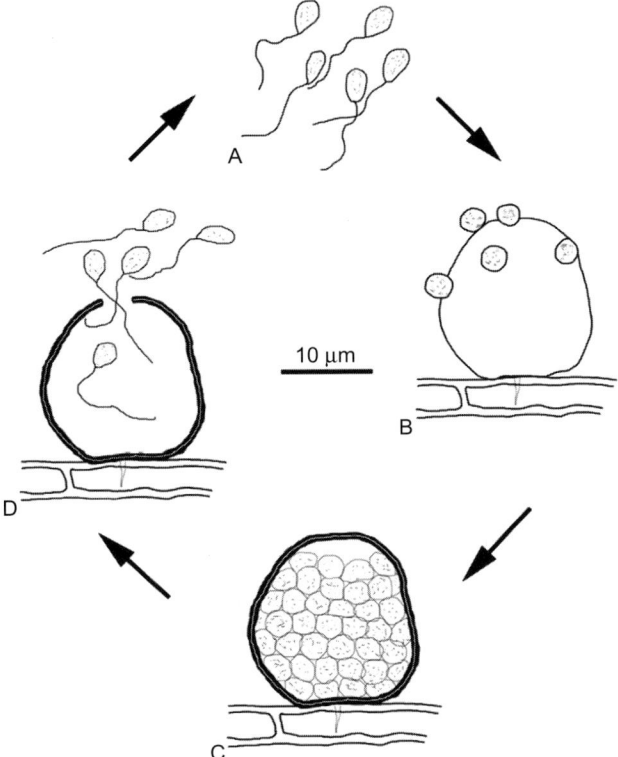

Figure 10.2 Lifecycle of *Rozella marina:* (A) Zoospores. (B) Zoospores of *R. marina* encysted on a sporangium of *Chytridium polysiphoniae* that is growing on an alga. (C) Cleaved spores of *R. marina* inside the sporangium. (D) Release of *R. marina* zoospores from the sporangium. (Adapted from Johnson 1966.)

contains about 30 genera. Many Mesomycetozoean species have now been well studied and documented, but others are known only from environmental sequence data. The Mesomycetozoea was initially referred to as the DRIP clade, which was an acronym for the four taxa of sequenced fish parasites: *Dermocystidium*, rosette agent, *Ichthyophonus* and *Psorospermium* (Ragan et al. 1998). The class Ichthyosporea was initially proposed to accommodate these taxa (Cavalier-Smith 1998). This class name, which pertained to fish parasites, was subsequently renamed Mesomycetozoea as more diverse species were discovered and incorporated into this group (Mendoza et al. 2001). These additions to the Mesomycetozoea included arthropod and amphibian gut parasites and *Rhinosporidium seeberi*, which infects birds and mammals (including humans) (Mendoza et al. 2001).

Taxonomy and phylogeny

The Mesomycetozoea contain two orders: the Dermocystida and the Ichthyophonida (Mendoza et al. 2002), see Figure 10.3. Following phylogenetic sequencing of arthropod gut parasites, members of two former fungal orders – the Amoebidiales and the Eccrinales – were added to the Mesomycetozoea (Cafaro 2005). The Amoebidiales and Eccrinales had been classified as fungi in the Trichomycetes (Zygomycota) along with the Harpellales (Lichwardt 1986). As well as sharing the ecological niche of the arthropod gut with the Harpellales, the Eccrinales also shared some morphological features with this true fungal order. However, a phylogenetic study showed that these two taxa of gut parasites were phylogenetically unrelated (Cafaro 2005). The addition of 17 genera of eccrinid arthropod symbionts to the Icthyophonida more than doubled the number of genera in this order. Seven of these transferred genera inhabit the gut of marine crustaceans.

Two new species of Mesomycetozoea were later described with detailed morphological and ultrastructural studies and placed in the Ichthyophonida by Marshall and Berbee (2011a). These species, *Pirum gemmata* and *Abeoforma whisleri*, were isolated from marine invertebrate digestive tracts, and their phylogenetic sequencing placed them in a Mesomycetozoean clade that otherwise contained species that are only known from environmental sequences (Marshall and Berbee 2011a). The two orders Dermocystida and Ichthyophonida are strongly supported subdivisions within the class (Marshall et al. 2008; Marshall and Berbee 2011a). Some of the significant differences between these two orders are discussed in the following sections.

Phylogenetic sequencing enabled many species previously misclassified as fungi, algae or protists – some for over a century – to be correctly accommodated in the Mesomycetozoea. For example, a species infecting clams – formerly placed in the unrelated alveolate genus, *Perkinsus* – was renamed *Pseudoperkinsus* and placed in the Mesomycetozoea (Figueras et al. 2000). These genera are morphologically indistinguishable but phylogenetically distant (Mendoza et al. 2002).

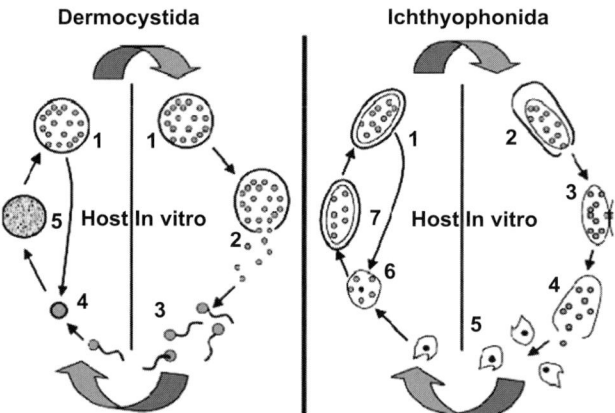

Figure 10.3 Life cycle of members of the order Dermocystida (left) and Icthyophonida (right). Dermocystida: 1) spherical cells containing endospores; 2) release of endospores; 3) uniflagellate zoospores; 4) encysted spore; 5) enlarged spore. Icthyophonida: 1) ovoid cell; 2) release of spore receptacle; 3) receptacle containing spores; 4) release of spores from receptacle; 5) motile amoeboid cells; 6) small receptacle; and 7) walled receptacle containing endospores.

Ecology

The class Mesomycetozoea contains both ecologically and morphologically diverse groups of microbes present in both marine and freshwater ecosystems. Many species are parasitic and most of these are thought to be host-dependent. Some species, such as the fish parasite *Dermocystidium salmonis* (Dermocystida), kill the host; while other species are non-pathogenic. The endospores of *D. salmonis* are released in the fish gills in such quantities that the fish is starved of oxygen (Mendoza et al. 2002). The rosette agent also primarily affects fish gills, although infected salmon also develop anemia and lymphocytosis, as the spherical cells of the rosette agent are concentrated in the macrophages of the kidneys and spleen (Mendoza et al. 2002). The only fish parasite in the Ichthyophonida, *Ichthyophonus hoferi*, is also lethal to its hosts. The thick-walled spherical cells of *I. hoferi* invade the internal organs of the fish including the heart, liver, muscles, and spleen (Mendoza et al. 2002). Other members of the Ichthyophonida, such as the gut parasite *A. parasiticum*, appear to live as non-pathogenic symbionts in the nutrient-rich environment of the digestive tract where they absorb nutrients (osmotrophic nutrition). One species, isolate LKM51 – known only from its environmental sequence – was recovered from phytoplankton, but it is not known whether this reflects only a small part of its lifecycle (Mendoza et al. 2002). One freshwater species, *Caullerya mesnili*, is a parasite of *Daphnia*, a common zooplankter (Lohr et al. 2010).

Within the Dermocystida, all of which are associated with vertebrates, no cultivation of cells outside the host has been achieved. Although many of the species in the Ichthyophonida are host-dependent, some have been axenically cultured (Jostensen et al. 2002; Marshall et al. 2008; Marshall and Berbee 2011a). The Dermocystida are all parasites of fish, except for the mammalian parasite *R. seeberi* (which affects the mucous membranes and causes the disease rhinosporidiosis), and a few species are gut parasites of amphibians. Most Ichthyophonida are associated with the host digestive tract and have a broad host range including both vertebrates and invertebrates, which include fish, shellfish, amphibians, insects and crustaceans from terrestrial, freshwater and marine habitats (Mendoza et al. 2002; Pekkarinen et al. 2003; Cafaro 2005; Marshall and Berbee 2011a). *Ichthyophonus* and *Psorospermium* are the only genera in the Ichthyophonida known to infect the internal tissues of the host (Marshall et al. 2008). Multiple isolations of the marine species *Creolimax fragrantissima* have revealed a broad host range covering four invertebrate phyla, which include the peanut worm, a tunicate species, a burrowing sea cucumber and a chiton (Marshall et al. 2008). Multiple hosts have also been recorded for *A. parasiticum*, *Anurofeca richardsi* and *I. hoferi* (Whistler 1960; Wong et al. 1994; Jones and Dawe 2002). The fish parasite *I. hoferi* has been recorded from fish in both freshwater and marine habitats (Mendoza et al. 2002).

Biochemistry and ultrastructure

The Mesomycetozoea are all osmotrophic unicellular symbionts that have spherical or ovoid cells with thick cell walls. The fish parasite *I. hoferi* was found to contain low levels of chitin in its cell walls (Spangaard et al. 1995) and *R. seeberi* was found to possess chitin synthase genes (Herr et al. 1999).

An ultrastructural study of assimilative cells of *C. fragrantissima* revealed cells containing a large central vacuole surrounded by an area of glycogen (Marshall et al. 2008). Cells also contained areas of branching tubules in the plasma membrane and the nuclei had prominent nucleoles, while some cells also had spindle pole bodies (Marshall et al. 2008). Spindle pole bodies had also been identified with transmission electron microscopy (TEM) in *I. hoferi* (Franco-Sierra and Alvarez-Pellitero 1999) but these differed in size from those of *C. fragrantissima* (Marshall et al. 2008). Centrioles have been observed in the zoospore-forming species *Dermocystidium percae* (Pekkarinen et al. 2003). Studies of cells of *Pseudoperkinsus* species (Marshall 2009) revealed a two-layered cell wall and Golgi, sometimes having up to seven stacked cisternae, which were associated with the nuclei. In other cells, nuclei had up to three associated Golgi bodies but the cisternae in these were not closely stacked (Marshall 2009). As with almost all other members of the Mesomycetozoea, the mitochondria of *C. fragrantissima* had flat plate-like cristae (Marshall et al. 2008). The rosette agent was initially reported to have tubular cristae (Arkush et al. 1998), but this was later considered to be a misinterpretation of the TEM data (Mendoza et al. 2002). *Ichthyophonus hoferi* is the only species reported to have tubular mitochondrial cristae (Spangaard et al. 1996; Franco-Sierra and Alvarez-Pelliteo 1999) and these findings have since been corroborated by Mendoza et al. (2002). Mendoza et al. (2002) consider this exception to indicate that mitochondrial composition may not in fact be the well-conserved trait it has been believed. However, the discovery of different mitochondrial types within a monophyletic group is extraordinary and requires further investigation.

Lifecycle

The lifecycles of the Mesomycetozoea can be very complex, especially in the Ichthyophonida where at least half of the lifecycle is in the host digestive tract (Marshall 2009). It is less complex in the Dermocystida (Figure 10.3). Infection in the Dermocystida may be by a uniflagellate zoospore that develops into a spherical multinucleate walled cell or sporangium. Endospores develop inside the sporangium and, following release, these may give rise to zoospores (Figure 10.3). However, many descriptions of Dermocystida species are incomplete as studies have only been possible during the parasitic stages (Mendoza et al. 2002). The uniflagellate zoospores of the Dermocystida, which have a single posterior whiplash flagellum, can be very small; those of the rosette agent are reported to be only 1–2 microns in diameter (Mendoza et al. 2002). In the Ichthyophonida the host may be infected by motile amoeboid cells that become multinucleate once inside the host and develop into sporangia. The sporangia become surrounded by a thick wall. *In vitro* studies have shown that the sporangia emerge from the walled enclosure and then release endospores that develop into amoebae (Vogt and Rug 1999; Mendoza et al. 2002), see Figure 10.3. Many species of Mesomycetozoea have no known motile stage but nearly all have sporangiospores or endospores. These are always spherical with the exception of *A. parasiticum*, which has narrow crescent-shapes spores (Mendoza et al. 2002).

Cultivation of some species of Mesomycetozoea has greatly aided our understanding of their lifecycles. However, of the five genera of Dermocystida none have been cultured except for *Sphaerothecum destruens*, which was successfully cultured along with its salmonid host cell line (Marshall 2009). In the Ichthyophonida, five species have now been grown in pure culture (Marshall 2009).

No amoeboid phase has ever been seen in the Dermocystida and zoospores are not present in the lifecycle of the Ichthyophonida (Figure 10.3). There was a report of uniflagellate zoospores in the Icthyophonida species, *Pseudoperkinsus tapetis* (Ordas and Figueras 1998), but it is thought that this could have been a contaminant of *Perkinsus* that is morphologically similar and shares the same niche (Mendoza et al. 2002). Cultivation of *I. hoferi* showed that amoebae production can be induced by raising the pH (Okamoto et al. 1985). In addition to amoebae, the Ichthyophonida may form plasmodia.

Mesomycetozoean species may release large quantities of endospores or sporangiospores from the mature sporangia. These sporangia generally measure about 70 microns in diameter in members of the Ichthyophonida but in the *Dermocystidium* fish parasites and in *Rhinosporidium* the sporangia reach several hundred microns in diameter and release thousands of endospores (Mendoza et al. 2002). Marshall et al. (2008) and Marshall and Berbee (2011a) studied the mechanisms of release of endospores from a number of species of Ichthyophonida. In *C. fragrantissima* endospores and amoebae can be released through a pore in the cell wall (Marshall et al. 2008). However, in *A. whisleri* and *Pirum gemmata*, endospores and amoebae are released through tears in the cell wall (Marshall and Berbee 2011a).

Since DNA sequences from Ichthyophonida species have been found in water and sediment samples, viable propagules probably persist outside the host. Cultured cells of *C. fragrantissima* remained viable for six months in sea water (Marshall et al. 2008). Eventually these endospores might be ingested passively by the hosts and resume growth as parasites in a nutrient-rich digestive tract. The cells of Mesomycetozoean species do not produce attachment structures, such as rhizoids or holdfasts, but they can be sticky at certain stages of their lifecycle and are therefore able to attach to solid surfaces, including cells in the digestive tract of the host (Marshall et al. 2008).

Members of the Mesomycetozoea reproduce asexually and until recently their ploidy was unknown and there had been no reported sexual cycle. Nuclear fusion was observed in *Enteropogon sexuale* but no subsequent growth or division followed (Hibbits 1978). Marshall and Berbee (2011b) searched for heterozygosity in cultures of *Pseudoperkinsus* by analysis of polymorphic populations. However, they found no evidence of heterozygosity and concluded that *Pseudoperkinsus* and probably all the Mesomycetozoeans are haploid. In the same study, population level analyses indicated that *Pseudoperkinsus* has a history of recombination but that sex occurred infrequently (Marshall and Berbee 2011b).

Conclusion

Following the inclusion of the Mesomycetozoea and Cryptomycota, there are now a total of eight unicellular lineages in the super-group Opisthokonta in the tree of life (Mendoza et al. 2002; Baldauf 2003; Adl et al. 2005; Marshall and Berbee 2011a). Recent molecular data suggest that the Cryptomycota, Mesomycetozoea, Choanoflagellates and Nucleariida all cluster around the point in the super-group Opisthokonta where animals and fungi diverge (Lara et al. 2010; Jones et al. 2011a). These diverse and distantly-related groups of microorganisms are considered basal to both true fungi and animals, and for that reason they are of great interest to evolutionary biologists.

There are contradictory statements in the literature concerning which group – the Cryptomycota or the Mesomycetozoea – is the deepest branching clade of fungus-like microorganisms. Some phylogenies have indicated that the Cryptomycota is the most basal branch (Jones et al. 2011a), while others claim it is the Mesomycetozoea (Ruiz-Trillo et al. 2006; Shalchian-Tabrizi et al. 2008). However, support for the relatedness of the Cryptomycota and Mesomycetozoea to each other, and to other groups in the Opisthokonta super-group, is poor (Mendoza et al. 2002; Lara et al. 2010; Jones et al. 2011b). The precise relationships of these early-evolved groups to one another cannot be satisfactorily elucidated until more species are discovered, more genes are sequenced and more comprehensive phylogenies can be compiled. We must therefore await further genomic investigation. Although the Mesomycetozoea, Cryptomycota, Choanoflagellates and Nucleariida are indeed related (Adl et al. 2005; White et al. 2006; Lara et al. 2010), these groups do not represent a monophyletic cluster and they cannot be put into the same phylum.

The presence of a large number of DNA sequences obtained from environmental surveys, some of which are from extreme environments, suggests that the microorganisms in the Cryptomycota and Mesomycetozoea may be numerous and ubiquitous (Held 1981; Marshall et al. 2008; Lara et al. 2010; Jones et al. 2011a; Marshall and Berbee 2011a). Poor sampling, particularly of marine environments, has previously limited the number of taxa discovered. Many of the host species for *Rozella* and for the Mesomycetozoeans are common in poorly-sampled freshwater and marine ecosystems, therefore we surmise that the number of species in these groups of organisms in aquatic environments is considerably larger than is presently recognized. Furthermore, the impact of these parasites on the physiology and ecology of their hosts may be substantial.

Acknowledgements

Figure 10.3 is reproduced from Mendoza et al., Ann Rev Microbiol 2002;56:315–344, with Leonel Mendoza's permission.

References

Adl SM, Simpson ACB, Farmer MA, Andersen RA, Anderson OR, Barta JR, Bowser SS, Brugerolle G, Fensome RA, Fredericq S, James TY, Karpov S, Kugrens P, Krug J, Lane CE, Lewis LA, Lodge J, Lynn DH, Mann DG, Mccourt RM, Mendoza L, Moestrup O, Mozley-Standridge SE, Nerad TA, Shearer CA, Smirnov AV. The new higher level classification of Eukaryotes with emphasis on the taxonomy of protists. J Eukaryot Microbiol. 2005;52:399–451.

Arkush KD, Frasca S Jr, Hedrick RP. Pathology associated with the rosette agent, a systemic protist infecting salmonid fishes. J Aquat Animal Health. 1998;10:1–11.

Baldauf SL. The deep roots of Eukaryotes. Science 2003;300:1703–1706.

Barr DJS. Chytridiomycota. In: McLaughlin DL, McLaughlin EG, Lemke PA, eds. The Mycota. vol. VII, Part A. New York: Springer-Verlag, 2001:93–112.

Berney CJ, Fahni J, Pawlowski J. How many novel eukaryotic kingdoms? Pitfalls and limitations of environmental DNA surveys. BMC Biol. 2004;2:1–13.

Burger GL, Forget Y, Zhu M, Gray W, Lang BF. Unique mitochondrial genome architecture in unicellular relatives of animals. Proc Nat Acad Sci. 2003;100:892–897.

Cafaro MJ. Eccrinales (Trichomycetes) are not fungi, but a clade of protists at the early divergence of animals and fungi. Mol Phyl Evol. 2005;35:21–34.

Cavalier-Smith T. The origin of fungi and pseudofungi. In: Rayner A, Brasier C, Moore D, eds. Evolutionary biology of fungi. Cambridge: Cambridge University Press, 1987:339–353.

Cavalier-Smith T. A revised six kingdom system of life. Biol Rev. 1998;73:203–266.

Figueras A, Lorenzo G, Ordas MC, Novoa B. Sequence of the small subunit ribosomal RNA gene of *Perkinsus atlanticus*-like isolated from carpet shell clam in Galicia, Spain. Mar Biotechnol. 2000;2:419–428.

Franco-Sierra A, Alvarez-Pellitero P. The morphology of *Icthyophonus* sp. in their mugulid hosts (Pisces: Teleostei) and following cultivation *in vitro*. A light and electron microscopy study. Parasitol Res. 1999;85:562–575.

Ginger ML, Fritz-Laylin LK, Fulton C, Cande WZ, Dawson SC. Intermediary metabolism in protists: a sequence-based view of facultative anaerobic metabolism in evolutionarily diverse eukaryotes. Protist 2010;161:642–671.

Gleason FH, Küpper FC, Amon JP, Picard K, Gachon CMM, Marano AV, Sime-Ngando T, Lilje O. Zoosporic true fungi in marine environments: a review. Mar Freshw Res. 2011;62:383–393.

Held AA. Attraction and attachment of zoospores of the parasitic chytrid *Rozella allomycis* in response to host-dependent factors. Arch Microbiol. 1974;95:97–114.

Held AA. Development of *Rozella* in *Allomyces*: a single zoospore produces numerous zoosporangia and resistant sporangia. Can J Bot. 1980;58:959–979.

Held AA. *Rozella* and *Rozellopsis*: naked endoparasitic fungi which dress-up as their hosts. Bot Rev. 1981;47:451–515.

Herr AR, Ajello L, Mendoza L. Chitin synthase class 2 (CHS2) gene from the human and animal pathogen *Rhinosporidium seeberi*. Proc. 99th Gen Meet Am Soc Microbiol. 1999:296 (Abstract).

Hibbits J. Marine Eccrinales (Trichomycetes) found in crustaceans of the San Juan Archipelago, Washington. Syesis 1978;11:213–261.

James TY, Berbee ML. No jacket required – new fungal lineage defies dress code. Bioessays 2012;94–102.

James TY, Letcher PM, Longcore JE, Mozley-Standridge SE, Porter D, Powell MJ, Griffith GW, Vilgalys R. A molecular phylogeny of the flagellated fungi (Chytridiomycota) and description of a new phylum (Blastocladiomycota). Mycologia 2006;98:860–871.

Johnson TW Jr. *Rozella marina* in *Chytridium polysiphoniae* from Icelandic waters. Mycologia 1966;58:490–494.

Jones MDM, Forn I, Gadelha C, Egan MJ, Bass D, Massana R, Richards TA. Discovery of novel intermediate forms redefines the fungal tree of life. Nature 2011a;474:200–203.

Jones MDM, Richards TA, Hawsworth DJ, Bass D. Validation and justification of the phylum name *Cryptomycota* phyl. nov. with notes on its recognition. IMA Fungus 2011b;2:173–175.

Jones SRM, Dawe SC. *Icthyophonus hoferi* Plehn & Mulsow in British Columbia stocks of Pacific herring, *Clupea pallasi* Valenciennes, and its infectivity to Chinook salmon, *Onchorhynchus tshawytscha* (Walbaum). J Fish Dis. 2002;25:415–421.

Jostensen JP, Sperstad S, Johansen S, Landfeld B. Molecular, phylogenetic, structural and biochemical features of a cold adapted, marine icthyosporean near the animal-fungi divergence, described from in vitro cultures. Eur J Parasitol. 2002;38:93–104.

Karling JS. Parasitism among the chytrids. Am J Bot. 1942;29:24–34.

Karling JS. Chytridiomycetarum Iconographia. Vaduz: J Cramer, 1977.

Kerk D, Gee A, Standish M, Wainright PO, Drum AS. The rosette agent of Chinook salmon (*Oncorhyncus tshawytscha*) is closely related to choanoflagellates, as determined by phylogenetic analyses of its small ribosomal subunit RNA. Mar Biol. 1995;122:187–192.

Küpper FC, Müller DG. Massive occurrence of the heterokont parasites *Anisolpidium*, *Eurychasma* and *Chytridium* in *Pylaiella littoralis* (Ectocarpales, Phaeophyceae). Nova Hedw. 1999;69:381–389.

Lara E, Moreira D, Lopez-Garcia P. The environmental clade LKM11 and *Rozella* from the deepest branching clade of fungi. Protist 2010;161:116–121.

Litchwardt RW. The Trichomycetes. Fungal associates of arthropods. New York: Springer-Verlag, 1986.

Lohr JN, Laforsch C, Koerner H, Wolinska J. A *Daphnia* parasite (*Caullerya mesnili*) constitutes a new member of the Ichthyosporea, a group of protists near the animal-fungi divergence. J Eukaryot Microbiol. 2010;57:328–336.

Marshall WL. New Ichthyosporean protists and their biology as inferred from molecular genetic and light and electron microscopic study. PhD Thesis, Department of Botany, University of British Columbia, Vancouver, BC, Canada, 2009.

Marshall WL, Berbee ML. Facing unknowns: living cultures (*Pirum gemmata* gen. nov., sp. nov., and *Abeoforma whisleri*, gen. nov., sp. nov.) from invertebrate digestive tracts represent an undescribed clade within the unicellular Opisthokont lineage Ichthyosporea (Mesomycetozoea). Protist 2011a;162:33–57.

Marshall WL, Berbee ML. Population-level analyses indirectly reveal cryptic sex and life history traits of *Pseudoperkinsus tapetis* (Ichthyosporea, Opisthokonta): a unicellular relative of animals. Mol Biol Evol. 2011b;27:2014–2020.

Marshall WL, Celio G, McLaughlin DJ, Berbee ML. Multiple isolations of a culturable, motile Ichthyosporean (Mesomycetozoa, Opisthokonta), *Creolimax fragrantissima* n. gen., n. sp., from marine invertebrate digestive tracts. Protist 2008;159:415–433.

Mendoza L, Ajello L, Taylor JW. The taxonomic status of *Lacazia loboi* and *Rhinosporidium seeberi* has been finally resolved with the use of molecular tools. Rev Iberoam Micol. 2001;18:95–98.

Mendoza L, Taylor JW, Ajello L. The class Mesomycetozoea: a heterogeneous group of microorganisms at the animal – fungal boundary. Ann Rev Microbiol. 2002;56:315–344.

Müller DG, Küpper FC, Küpper H. Infection experiments reveal broad host ranges of *Eurychasma dicksonii* (Oomycota) and *Chytridium polysiphoniae* (Chytridiomycota), two eukaryotic parasites in marine brown algae (Phaeophyceae). Phycol Res. 1999;47:217–223.

Okamoto N, Nakase K, Suzuki H, Nakai H, Fujii Y, Sano T. Life history and morphology of *Icthophonus hoferi* in vitro. Fish Pathol. 1985;20:273–285.

Ordas MC, Figueras A. *In vitro* cultivation of *Perkinsus atlanticus*, a parasite of the carpet shell clam *Ruditapes decussates*. Dis Aquat Org. 1998;33:129–136.

Pekkarinen M, Lom J, Murphy CA, Ragan MA, Dyková D. Phylogenetic position and ultrastructure of two *Dermocystidium* species (Ichthyosporea) from the common perch (*Perca fluviatilis*). Acta Protozool. 2003;42:287–307.

Powell MJ. Fine structure of the unwalled thallus of *Rozella polyphagi* in its host *Polyphagus euglenae*. Mycologia 1984;76:1039–1048.

Ragan MA, Goggins CL, Cawthorn RJ, Cerenius AVC, Jamienson L. A novel clade of protistan parasites near the animal-fungal divergence. Proc Nat Acad Sci USA. 1998;93:11907–11912.

Ruiz-Trillo I, Lane CE, Archibald JM, Roger AJ. Insights into the evolutionary origin and genome architecture of the unicellular opisthokonts *Capsaspora owczarzaki* and *Sphaeroforma arctica*. J Eukaryot Microbiol. 2006;53:1–6.

Shalchian-Tabrizi K, Minge MA, Espelund M, Orr R, Ruden T, Jakobsen KS, Cavalier-Smith T. Multigene phylogeny of choanozoa and the origin of animals. PLoS ONE 2008;3:e2098

Spangaard B, Huss HH, Brescianni J. Morphology of *Icthyophonus hoferi*, assessed by light and scanning electron microscopy. J Fish Dis. 1995;18:567–577.

Spangaard B, Skouboe P, Rossen L, Taylor JW. Phylogenetic relationships of the intracellular fish pathogen *Icthyophonus hoferi*, and fungi, choanoflagellates and the rosette agent. Mar Biol. 1996;126:109–115.

Sparrow FK Jr. Biological observations on the marine fungi of Woods Hole Waters. Biol Bull. 1936;70:236–263.

Sparrow FK Jr. Aquatic Phycomycetes. 2nd ed. Ann Arbor, MI: University of Michigan Press, 1960.

van Hannen EJ, Moji W, van Agterveld MP, Gons HJ, Laanbroek P. Detritus-dependent development of the microbial community in an experimental system: quantitative analysis by denaturing gradient gel electrophoresis. Appl Environ Microbiol. 1999;65:2478–2484.

Vogt G, Rug M. Life stages and tentative life cycle of *Psorospermium haeckeli*, a species of novel DRIPs clade from the animal-fungal dichotomy. J Exp Zool. 1999;283:31–42.

Whistler HC. Pure culture of the trichomycete *Amoebidium parasiticum*. Nature 1960;186: 732–733.

White MM, James TY, O'Donnell K, Cafaro MS, Tanabe Y, Sugiyama J. Phylogeny of the Zygomycota based on nuclear ribosomal sequence data. Mycologia 2006;96:872–884.

Wong ALC, Berbee TJC, Griffiths RA. Factors affecting the distribution and abundance of an unpigmented heterotrophic alga *Prototheca richardsi*. Freshw Biol. 1994;32:33–38.

11 Hyphochytriomycota, Oomycota and Perkinsozoa (Super-group Chromalveolata)

Agostina V. Marano, Carmen L.A. Pires-Zottarelli, José I. de Souza, Sally L. Glockling, Eduardo Leaño, Claire M.M. Gachon, Martina Strittmatter and Frank H. Gleason

Cavalier-Smith (2002) established the name Chromalveolata by hypothesizing the presence of a common plastid in the dinoflagellates and heterokonts, which is thought to have arisen from a single event of secondary endosymbiosis with an ancestral red alga (Cavalier-Smith 1999; Adl et al. 2005). The name specifically refers to the bag-like sacs called alveoli that underlie the cell membrane (*alveus*, Lat. = small cavity) and to the fact that many members in this group are photosynthetic (*chroma*, Lat. = color).

There are six lineages within the Chromalveolata: apicomplexans, ciliates, cryptomonads, dinoflagellates, haptophytes and stramenopiles. The apicomplexans, ciliates, dinoflagellates and stramenopiles form a monophyletic assemblage, while the cryptomonads and haptophytes form a weakly-supported group (Harper et al. 2005; Hackett et al. 2007). Cryptomonads, haptophytes, and stramenopiles are grouped in the Chromista based on the shared features of their plastids (i.e., the presence of four bounding membranes and confluence of the outermost plastid membrane with the nuclear envelope). The Alveolata contain the apicomplexans, ciliates and dinoflagellates and, in contrast to the Chromistan groups, their outermost plastid membrane is not connected to the nuclear membrane (Schnepf and Elbrächter 1999).

Recently, Strittmatter et al. (2009) reviewed the current knowledge of basal lineages of marine Oomycetes. Most groups of marine fungal-like organisms have been poorly explored, however, mainly due to methodological limitations, the limited number of researchers working on marine fungal-like taxa, and the biotrophic lifecycle of most parasitic forms.

In this chapter we will briefly review the information available on marine and estuarine Chromalveolata, *sensu* Adl et al. (2005), which includes the Alveolata, the Stramenopiles (or Chromista), the Haptophyta, and the Cryptophyceae. In particular, we will focus on the heterotrophic members of the super-group, i.e., the phyla Oomycota and Hyphochytriomycota within the subgroup Chromista (commonly referred to as "heterotrophic stramenopiles" or "heterokonts") and the Perkinsozoa ("perkinsids") within the subgroup Alveolata. The fungal-like Labyrinthulomycota and the Phytomyxea will be addressed in the following two chapters of this book (Chapters 12 and 13). Species names used in this contribution are taken from the current scientific names in Index Fungorum (www.indexfungorum.org), from MycoBank (www.mycobank.org) for fungal-like organisms and Algaebase (www.algaebase.org) for algae.

Biodiversity

The Chromalveolata represent a large fraction of the eukaryotic diversity in both marine and estuarine ecosystems, ranging from microscopic parasites to macroscopic algae (Harper et al. 2005). According to Cavalier-Smith and Chao (2004), this super-group accounts for approximately half the recognized species of protists and algae.

As shown in Table 11.1, the large number of known marine Oomycota species include pathogens of algae, dinoflagellates, marine invertebrates (e.g., crustaceans, nematodes and snails), and also saprobes (e.g., decomposers of crustacean exoskeletons, dead algae and fallen leaves) (Dick 2001; Sekimoto et al. 2007; Beakes and Sekimoto 2009a). The small numbers of known marine Hyphochytriomycota species are parasites of algae (Dick 2001; Sekimoto et al. 2007; Beakes and Sekimoto 2009a).

Many species in the 24 genera of Hyphochytriomycota and Oomycota have been recorded in marine and estuarine environments (Dick 2001; Strittmatter et al. 2009). Approximately half of these genera, i.e., *Ectrogella, Eurychasmidium, Eurychasma, Haliphthoros, Halodaphnea, Halophytophthora, Lagenisma, Pontisma, Salilagenidium, Salisapilia,* and *Sirolpidium,* are exclusively marine and/or estuarine. Most of these genera are parasitic on algae and small invertebrates in marine ecosystems, while species of *Halophytophthora, Salisapilia* and *Pythium* are decomposers of fallen leaves in both marine and estuarine habitats (Table 11.1).

All known Perkinsozoa are exclusively intracellular parasites of microalgae (especially dinoflagellates) and molluscs such as oysters, clams, and abalones (Mangot et al. 2011). Since they have never been assimilated into the "marine fungi", they are not listed in Table 11.1. To date, this phylum consists of only two genera, *Perkinsus* and *Parvulucifera* (Hoppenrath and Leander 2009).

The three *Parvilucifera* species described are intracellular parasites of planktonic, mostly toxic, dinoflagellates (Norén et al. 2000; Park et al. 2004; Figueroa et al. 2008). *Parvilucifera sinerae,* for example, was first isolated from a bloom of the toxic dinoflagellate *Alexandrium minutum* in the Mediterranean Sea in Spain (Figueroa et al. 2008); *P. prorocentri* infects the marine benthic and non-toxic dinoflagellate, *Prorocentrum fukuyoi* (Leander and Hoppenrath 2008) and *Parvilucifera infectans* infects many dinoflagellate species and does not appear to be host-specific (Norén et al. 2000). Experimental infection of various dinoflagellate species with *P. infectans* and *P. sinerae* revealed that several species can be infected but intra-species differences occur (Norén et al. 2000; Figueroa et al. 2008).

In contrast, species of *Perkinsus* are parasites of molluscs. Most of the six currently described species of *Perkinsus* seem to be specific to one type of bivalve (oysters or clams). *Perkinsus mediterraneus,* for example, only infects the European flat oyster *Ostrea edulis* (Casas et al. 2004), while *Pe. chesapeaki* and *Pe. beihaiensis* only infect clams from the genera *Crassostrea, Mya* and *Venerupis* (McLaughlin et al. 2000; Dungan and Reece 2006; Moss et al. 2008). Other species, such as *Pe. marinus*, infect both oysters and clams. Likewise, *Pe. olseni* infects both clams (e.g., *Ruditapes decussatus*) and abalones (e.g., *Haliotis ruber*) (Lester and Davis 1981; Azevedo 1989).

Biogeography

Most species of estuarine and marine fungal-like organisms occur along the coasts of more than one continent or in more than one ocean basin. Not surprisingly, the distribution

Table 11.1 Fungal-like organisms (Hyphochytriomycota and Oomycota) from marine and estuarine environments.

Taxa	Habitat	Type of host/Substrate	Distribution	Host species	Reference
STRAMINIPILA					
HYPHOCHYTRIOMYCOTA					
ANISOLPIDIALES					
Anisolpidium ectocarpi Karling	M	A	USA	*Ectocarpus siliculosus, Hincksia michelliae* (as *Ectocarpus michellae*)	Karling (1943); Wilson (1960); Dick (2001)
A. joklianum M. W. Dick	M	A	Adriatic Sea (Mediterranean)	*Hincksia granulosa*	Sparrow (1960); Dick (2001)
A. minutum (H. E. Petersen) M. W. Dick	M	A	Denmark	*Chorda filum*	Sparrow (1960); Dick (2001)
A. olpidium (H. E. Petersen) M. W. Dick	M	A	Denmark	*Ectocarpus siliculosus* (as *Ectocarpus confervoides*)	Sparrow (1960); Dick (2001)
A. rosenvingei (H. E. Petersen) Karling	M	A	Denmark, Ireland, Shetland, Sweden, UK	*Pylaiella littoralis*	Küpper and Müller (1999); Küpper (2001)
A. sphacelatum (Kny) Karling	M	A	Denmark, France, Ireland, Japan, Sweden, UK	*Chaetopteris plumosa, Cladostephus spongiosus, C. spongiosorus* f. *verticillatus* (as *C. verticillatus*), *Sphacelaria apicalis, S. cirrosa, S. subfusca, S. tribuloides*	Karling (1943); Johnson and Sparrow (1961); Dick (2001)
HYPHOCHYTRIALES					
Hyphochytrium peniliae N. J. Artemczuk et Zelez	M	Cl	Russia	*Penilia avirostris*	Dick (2001)

Table continued on next page.

Table 11.1 (Continued)

Taxa	Habitat	Type of host/Substrate	Distribution	Host species	Reference
OOMYCOTA					
HAPTOGLOSSALES					
Haptoglossa heterospora Drechsler	M	N	USA	*Rhabditis marina*	Newell et al. (1977); Karling (1981)
LAGENISMATALES					
Lagenisma coscinodisci Drebes	M, E	A	Canada, Germany, Jamaica, North Sea	*Coscinodiscus centralis, C. concinnus, C. granii, C. perforatus, Palmeria hardmaniana*	Gotelli (1971); Grahame (1976); Karling (1981); López and MacCarthy (1985); Raghukumar (1996); Dick (2001)
MYZOCYTIOPSIDALES					
Gonimochaete latitubus S.Y. Newell, Cefalon et Fell	M	N	USA	*Rhabditis marina*	Newell et al. (1977); Dick (2001)
Myzocytiopsis vermicola (Zopf) M. W. Dick	M	N	USA	*Rhabditis marina*	Newell et al. (1977); Karling (1981); Dick (2001)
OLPIDIOPSIDALES					
Olpidiopsis bostrychiae S. Sekimoto, T. A. Klochkova, J. A. West, G. W. Beakes *et* D. Honda	M	A	Korea, Madagascar	*Bostrychia moritziana, B. intricata* (as *Stictosiphonia intricata*), *B. radicans, B. radicosa, B. tenella, Heterosiphonia japonica, Bostrychia* spp., *Porphyra* spp.	West et al. (2006); Sekimoto et al. (2009a)

Species			Host	Location	References
O. magnusii (J. et G. Feldman) M. W. Dick	M	A	*Ceramium flabelligerum*	France	Karling (1981); Sekimoto et al. (2009a)
O. porphyrae Sekimoto, Yokoo, Y. Kawam. et D. Honda	M	A	*Bangia fuscopurpurea, B. gloiopeltidicola, Porphyra kuniedae, P. seriata, P. tenera, P. tenuipedalis, P. yezoensis, Bangia* spp., *Porphyra* spp.	China, Japan	Ding and Ma (2005); Ma et al. (2007); Sekimoto et al. (2008, 2009a)
PERONOSPORALES **PERONOSPORACEAE**					
Halophytophthora avicenniae (Gerr.-Corn. et J. A. Simpson) H. H. Ho et S. C. Jong*	E	ML	*Aegiceras corniculatum, Avicennia lanata, A. marina* var. *australasica, A. germanis, Rhizophora mangle*	Australia, Bahamas, Philippines, USA	Fell and Master (1975); Newell et al. (1987); Gerrettson-Cornell and Simpson (1984); Ho and Jong (1990); Newell (1992); Leaño (2001)
H. batemanensis (Gerr.-Corn. *et* J. A. Simpson) H. H. Ho *et* S. C. Jong*	E	ML	*Avicennia* sp.	Japan	Ho and Jong (1990)
H. elongata H. H. Ho *et* H. S. Chang*	M	ML	*Avicennia marina, Kandelia obovata*	Taiwan	Ho et al. (2003); Leaño and Pang (2010)
H. epistomia (Fell *et* Master) H. H. Ho *et* S. C. Jong*	E	ML	*Bruguiera gymnorrhiza, Hymenocallis latifolia, Kandelia candel, Rhizophora apiculata, R. mangle, R. stylosa, Sonneratia* sp., *Xylocarpus granatum*	Costa Rica, Japan, Philippines, USA	Fell and Master (1975); Ho et al. (1990); Nakagiri (2000); Leaño (2001)
H. exoprolifera H. H. Ho, Nakagiri *et* S. Y. Newell*	E	ML	*Bruguiera gymnorrhiza, Rhizophora apiculata, R. mangle, R. mucronata, R. stylosa*	Bahamas, Okinawa, Singapore	Ho et al. (1992); Tan and Pek (1997)

Table continued on next page.

Table 11.1 (Continued)

Taxa	Habitat	Type of host/Substrate	Distribution	Host species	Reference
H. kandeliae H. H. Ho, H. S. Chang *et* S. Y. Hsieh*	E, M	ML	Japan, Philippines, Singapore, Taiwan, USA	*Aegiceras corniculatum, Avicennia lanata, A. alba, A. germanis, A. rumphiana, Bruguiera gymnorrhiza, Kandelia candel, Rhizophora apiculata, R. stylosa, Sonneratia sp., Xylocarpus granatum*	Ho et al. (1991); Newell (1992); Tan and Pek (1997); Nakagiri (2000); Leaño (2001)
H. masteri Nakagiri *et* S. Y. Newell*	E	ML	Bahamas	*Avicennia germinans*	Nakagiri et al. (1994)
H. mycoparasitica (Fell *et* Master) H. H. Ho *et* S.C. Jong	E	ML	Malaysia	*Bruguiera gymnorrhyza, Rhizophora* sp., *R. mangle, R. stylosa*	Fell and Master (1975); Ho and Jong (1990); Nakagiri et al. (1998); Leaño (2000); Nakagiri (2000)
	E	F	Japan	*Penicilium* sp., *Pestalotia* spp., *Pestalotiopsis* on *Rhizophora stylosa*	Fell and Master (1975); Nakagiri (2000)
H. operculata (Pegg *et* Alcorn) H. H. Ho *et* S. C. Jong*	E	ML	Australia, Singapore	*Avicennia alba, A. marina, Rhizophora apiculata*	Pegg and Alcorn (1982); Ho and Jong (1990); Tan and Pek (1997); Nakagiri (2000); Leaño (2001)
H. polymorphica (Gerr.-Corn. *et* J. A. Simpson) H. H. Ho *et* S.C. Jong*	E	DL	USA	*Avicennia marina, Bruguiera gymnorrhiza, Eucalyptus* sp., *Rhizophora stylosa*	Ho and Jong (1990)
H. porrigovesica Nakagiri, Tad. Ito, Manoch *et* Tantich.*	E	ML	Japan, Thailand	*Avicennia alba, Bruguiera gymnorrhiza, Sonneratia alba*	Nakagiri et al. (2001); Jones et al. (2006)

Species			Location	Host/Substrate	References
H. spinosa var. *lobata* (Fell *et* Master) H. H. Ho *et* S.C. Jong*	E	ML	Singapore, Taiwan, Thailand, USA	*Bruguiera gymnorrhyza, Kandelia obovata; Rhizophora apiculata, R. mangle, R. mucronata, R. stylosa, Xylocarpus moluccensis*	Ho and Jong (1990); Tan and Pek (1997); Nakagiri (2000); Leaño (2001); Jones *et al.* (2006); Leaño and Pang (2010)
H. spinosa var. *spinosa* (Fell *et* Master) H. H. Ho *et* S.C. Jong*	E	ML	Japan, Philippines, Thailand, USA	*Rhizophora mangle*	Fell and Master (1975); Jones *et al.* (2006)
H. vesicula (Anastasiou *et* Churchland) H. H. Ho *et* S.C. Jong*	E	ML	Australia, Bahamas, Bonaire, Cayman Island, Costa Rica, Haiti, Hawaii, Hong Kong, Jamaica, Japan, Malaysia, Philippines, Seychelles, Singapore, Taiwan, Trinidad, Venezuela, USA, Vietnam	*Acanthus ilicifolius, Avicennia lanata, A. officinalis, A. germanis, A. marina, Bruguiera gymnorrhyza, Ceriops decandra, Conocarpus erecta, Kandelia candel, Laguncularia racemosa, Rhizophora apiculata, R. mangle, R. mucronata, R. stylosa, Sonneratia alba, Xylocarpus granatum, X. moluccensis*	Newell *et al.* (1987); Raghukumar *et al.* (1995); Leaño *et al.* (1998a); Leaño (2001); Leaño and Pang (2010)
SALISAPILIACEAE					
Salisapilia nakagirii Hulvey, Nigrelli, Telle, Lamour *et* Thines*	Bra	DL	USA	*Spartina alterniflora*	Hulvey *et al.* (2010)
Sa. tartarea (Nakagiri *et* S. Y. Newell) Hulvey, Nigrelli, Telle, Lamour *et* Thines*	Bra	DL	USA	*Spartina alterniflora*	Hulvey *et al.* (2010)
Sa. sapeloensis Hulvey, Nigrelli, Telle, Lamour *et* Thines*	Bra	DL	USA	*Spartina alterniflora*	Hulvey *et al.* (2010)

Table continued on next page.

Table 11.1 (Continued)

Taxa	Habitat	Type of host/Substrate	Distribution	Host species	Reference
PYTHIALES					
Pythium chondricola de Cock	E	A/DL	Netherlands	*Chondrus crispus, Ulva lactuca, Zostera marina*	de Cock (1986)
P. grandisporangium Fell *et* Master	Bra	ML	Japan, Thailand, USA	*Rhizophora mangle, Phragmites australis*	Fell and Master (1975); Newell (1992); Jones et al. (2006); Kurowaka and Tojo (2010)
	E	A/DL	Canada, Netherlands	*Chondrus crispus, Ulva lactuca, Zostera marina*	de Cock (1986)
P. marinum Sparrow	M	A	Denmark, Japan, USA	*Ceramium virgatum* (as *C. rubrum*), *Porphyra miniata, P. perforata*	Höhnk (1936); Kazama and Fuller (1970); Kazama (1979)
P. maritimum Höhnk	M	A	Germany	*Ceramium* sp.	van der Plaats-Niterink (1981); Dick (2001)
P. porphyrae M. Takah. et M. Sasaki	M	A	China, Japan, Korea, USA	*Porphyra cuneiformes, P. haitanensis, P. nereocystis, P. lanceolata, P. perforata, P. pulchra, P. schizophylla, P. yezoensis*	Höhnk and Vallin (1953); Fuller et al. (1966); Andrews (1976); Kazama (1979); Park et al. (2001a, 2001b); Ding and Ma (2005); Park (2006)
P. thalassium D. Atkins	M	C	UK	*Macropodia* sp., *Pinnotheres pisum, Liocarcinus depurator* (as *Portunus depurator*)	Atkins (1955)
P. salinum Höhnk	Bra	Sa	Germany	N/A	Johnson and Sparrow (1961); van der Plaats-Niterink (1981); Dick (2001)

SALILAGENIDIALES

Species			Location	Host	References
Salilagenidium callinectes (J. N. Couch) M. W. Dick	M	A	USA		Fuller et al. (1964)
	M	B	USA	*Ceramium* spp., *Chordaria* spp., *Cladophora* spp., *Ectocarpus* spp.	Sindermann (1977); Karling (1981)
	M	C	Indonesia, Japan, USA	*Chelonibia patula*	
				Callinectes sapidus, *Cancer magister*, *Neopanope texana*, *Pinnotheres ostreum*, *Portunus pelagicus*, *P. trituberculatus*, *Scylla serrata*	Armstrong et al. (1976); Roza and Hatai (1999a); Hatai et al. (2000)
	M	S	N/A	*Artemia salina*	Karling (1981); Fisher (1983)
	M	L	USA	*Homarus americanus*	Nilson et al. (1976)
S. chthamalophilum (T. W. Johnson) M. W. Dick	M	B	USA	*Chthamalus fragilis denticulata*	Johnson (1958); Fuller et al. (1964); Karling (1981); Dick (2001)
S. marinum (D. Atkins) M. W. Dick	M	Bi	UK	*Barnea candida*, *Acanthocardia echinata* (as *Cardium echinatum*), *Mytilus edulis*	Atkins (1929, 1954b); Johnson and Sparrow (1961); Dick (2001)
S. myophilum (Hatai et Lawhav.) M. W. Dick	M	S	Japan	*Pandalus boreales*	Hatai and Lawhavinit (1988); Nakamura et al. (1994)
S. scyllae (Bian, Hatai, Po et Egusa) M. W. Dick	M	C	Philippines	*Scylla serrata*	Bian et al. (1979); Dick (2001)
S. thermophilum (K. Nakam., Miho Nakam., Hatai et Zafran) M. W. Dick	E	C	Bali, Indonesia, Thailand	*Penaeus monodon*, *Scylla serrata*	Nakamura et al. (1995); Muraosa et al. (2006)
	M	P	Indonesia	*Crangon vulgaris*, *Palaemon serratus* (as *Leander serratus*)	Nakamura et al. (1995)

Table continued on next page.

Table 11.1 (Continued)

Taxa	Habitat	Type of host/Substrate	Distribution	Host species	Reference
SAPROLEGNIALES					
Aphanomyces invadans Willoughby, R.J. Roberts et Chinabut	E	Fi	Australia, Japan, USA	*Brevoortia tyrannus*, *Mugil cephalus*	Blazer et al. (2002); Fraser et al. (1992)
Leptolegnia baltica Höhnk et Vallin	M	Co	Sweden (Gulf of Bothnia)	*Eurytemora hirulzdoides*	Höhnk and Vallin (1953); Johnson et al. (2002)
Saprolegnia parasitica	E	C	India	*Penaeus monodon*	Gopalan et al. (1980)
Coker	M	Fi	Japan	*Oncorhynchus kisutch*	Hatai and Hoshiai (1994)
FAMILY *incertae sedis*					
ECTROGELLACEAE					
Ectrogella eurychasmoides Feldmann et Feldm.-Maz.	M	A	France	*Licmophora lyngbyei*	Feldmann and Feldmann (1955)
E. licmophorae Scherffel	M	A	Italy	*Licmophora* spp.	Johnson (1966)
E. perforans H. E. Petersen	M	A	Denmark, France, Germany, North Sea, Sweden, USA	*Fragilaria islandica*, *Licmophora gracilis*, *L. lyngbyei* (syn. *L. abbreviata*), *L. flabellata*, *Podocystis adriatica*, *Striatella unipunctata*, *Synedra ulna*, *Tabularia fasciculata* (as *Synedra tabulata*), *Thalassionema nitzschioides*, *Punctaria* sp.	Sparrow (1969); Raghukumar (1978, 1996); Karling (1981)

FAMILY *incertae sedis*
EURYCHASMATACEAE

Eurychasma dicksonii (E.P. Wright) Magnus	M	A	Algeria, Argentina, Australia, Canada, Denmark, Faroe Islands, Finland, France, Germany, Greenland, India, Ireland, Italy, Japan, Kerguelen, Norway, Scotland, Shetland islands, Siberia, Spitsbergen, Sweden, UK, USA	*Acinetospora crinita* (as *Ectocarpus crinitus* and *E. pusillus*), *Ectocarpus constanciae*, *E. siliculosus* (syn. *Ectocarpus confervoides*), *Pylaiella littoralis*, *Striaria attenuata*, *Hincksia granulosa* (as *Ectocarpus granulosus*), *H. sandriana* (as *E. sandrianus*), *Stictyosiphon soriferus*, *S. tortillis*, *Feldmannia* spp., *Punctaria* spp.	Karling (1981); Aleem (1955); Müller et al. (1999); Dick (2001); Sekimoto et al. (2008); Gachon et al. (2009)
Eurychasmidium joyceae (Sparrow) M. W. Dick	M	A	USA	*Polysiphonia* sp., *Pterosiphonia* sp.	Dick (2001)
E. sacculum (H. E. Petersen) M. W. Dick	M, Bra	A	Greenland, North Sea	*Devaleraea ramentacea* (as *Halosaccion ramentaceum*), *Palmaria palmata* (as *Rhodymenia palmata*)	Karling (1981); Dick (2001)
E. tumefaciens (Magnus) Sparrow	M	A	Denmark, France, Ireland, Italy, Scotland, USA	*Ceramium diaphanum*, *C. shuttleworthianum* (as *C. acanthonotum*), *C. spiniferum*, *C. virgatum* (as *C. flabelliferum* and *C. rubrum*)	Karling (1981)

Table continued on next page.

Table 11.1 (Continued)

Taxa	Habitat	Type of host/ Substrate	Distribution	Host species	Reference
FAMILY *incertae sedis* **HALIPHTHORACEAE**					
Haliphthoros milfordensis Vishniac	M	Ab	Japan	*Haliotis gigantea* (as *Haliotis sieboldii*)	Hatai (1982)
	M	A	USA	*Enteromorpha* sp.	Fuller et al. (1964)
	M, E	C	Indonesia, Philippines	*Callinectes sapidus*, *Pinnotheres* spp., *Portunus pelagicus*, *P. pisum*, *P. trituberculatus*, *Scylla serrata*	Visniac (1958); Tharp and Bland (1977); Nakamura and Hatai (1995); Roza and Hatai (1999a); Hatai et al. (2000); Leaño (2002)
	M	Lo	N/A	*Homarus americanus*, *H. gammarus*	Fisher et al. (1975); Abrahams and Brown (1977)
	M	P	Vietnam	*Farfantepenaeus duorarum* (as *Penaeus duorarum*), *Penaeus japonicus*, *P. monodon*, *Litopenaeus setiferus* (as *P. setiferus*)	Fuller et al. (1964); Tharp and Bland (1977); Hatai et al. (1992); Chukanhom et al. (2003)
	E	S	N/A	*Artemia salina*	Overton and Bland (1981)
	M	Sn	USA	*Urosalpinx cinerea*	Visniac (1958)
H. philippinensis Hatai, Bian, Batic. *et* Egusa	M	C	Philippines	*Scylla serrata*	Leaño (2002)
H. zoopthorum (Vishniac) M.W. Dick	M	P	Philippines	*Penaeus monodon*	Hatai et al. (1980)
	M	Bi	USA	*Crassostrea virginica*, *Mercenaria mercenaria* (as *Venus mercenaria*)	Dick (2001)

FAMILY *incertae sedis*
PONTISMATACEAE

Species			Location	Host	Reference
Petersenia lobata (H. E. Petersen) Sparrow	M	A	Denmark, France, Ireland, Sweden, USA	*Aglaothamnion* sp., *A. hookeri* (as *Callithamnion hookeri*), *Callithamnion corymbosum*, *Ceramium pedicellatum*, *C. tenuicornis* (as *C. C. corticullatum*), *C. virgatum* (as *C. rubrum*), *Gymnothamnion elegans*, *Herposiphonia secunda* f. *tenella* (as *H. tenella*), *Polysiphonia stricta* (as *P. urceolata*), *Pylaiella littoralis*, *Seirospora interrupta* (syn *S. griffithsiana*), *Spermothamnion repens* (syn *S. turneri*)	Johnson and Sparrow (1961); Dick (2001)
Pe. palmariae van der Meer *et* Pueschel	M	A	Canada	*Palmaria mollis*	van der Meer and Peuschel (1985)
Pontisma antithamnionis (Whittick *et* South) M. W. Dick	M	A	Canada	*Antithamnionella floccosa* (as *Antithamnion floccosum*)	Porter (1986); Dick (2001)
Po. dangeardii (Feldmann *et* Feldm.-Maz.) M. W. Dick	M	A	France	*Radicilingua reptans*	Dick (2001)
Po. feldmannii (Aleem) M. W. Dick	M	A	France	*Falkenbergia rufolanosa*, *Trailliella* spp.	Aleem (1952)
Po. inhabile (H. E. Petersen) M. W. Dick	M	A	Denmark	*Polysiphonia fibrilosa*	Dick (2001)

Table continued on next page.

Table 11.1 (Continued)

Taxa	Habitat	Type of host/Substrate	Distribution	Host species	Reference
Po. lagenidiosis H. E. Petersen	M	A	Denmark, France, Germany, Japan, India, North Sea, Sweden, USA	*Ceramium diaphanum, C. fructiculosum, C. pedicellatum, C. tenuicornis* (as *C. corticullatum* and *C. strictum*), *C. virgatum* (as *C. rubrum*), *Chaetomorpha antennina* (syn. *C. media*), *Cladophora japonica, Valoniopsis* spp.	Sparrow (1960); Raghukumar (1987, 1996); Raghukumar and Raghukumar (1994)
Po. magnusii (Feldmann et Feldm.-Maz.) M. W. Dick	M	A	France	*Ceramium virgatum* (as *C. flabelliferum*)	Dick (2001)
FAMILY *incertae sedis* **SIROLPIDIACEAE**					
Sirolpidium andreei (Lagerh.) M. W. Dick	M	A	Adriatic Sea, Denmark, France, Greenland, King Charles land, North Sea, USA	*Acrosiphonia incurva, Ceramium diaphanum, Ectocarpus siliculosus, Feldmannia simplex* (as *Ectocarpus simplex*), *Hincksia granulosa* (as *Ectocarpus granulosus*), *Spongomorpha hystrix* (as *Acrosiphonia hystrix*), *Spongomorpha* sp., *Striaria attenuata*	Karling (1981)

Species			Host	Location	Reference
S. bryopsidis (de Bruyne) H. E. Petersen	M	A	*Bryopsis plumosa, Cladophora japonica, Cladophora liebetruthii* (as *C. frascatii*), *C. flexuosa* (as *C. gracilis*), *Pseudobryopsis myura, Rhizoclonium* spp.	Denmark, France, India, Italy, Japan, North Sea, Sweden, USA	Karling (1981); Raghukumar (1986, 1987); Dick (2001)
S. glenodinianum (P. A. Dang.) M. W. Dick	M	Di	*Glenodinium* sp.	France	Sparrow (1960); Dick (2001)
S. globosum (Anastasiou et Churchl.) M. W. Dick	M	F	*Papulaspora halima* on *Arbutus* debris	Canada	Dick (2001)
S. marinum (P. A. Dang.) M. W. Dick	M	A	*Prasinocladus lubricus* f. *subsalsa* (as *Chlorodendron subsalsum*)	France	Dangeard (1912); Dick (2001)
S. paradoxum (H. E. Petersen) M. W. Dick	M	F	*Rhizophydium distinctum* on *Spongomorpha vernalis*	Norway	Dick (2001)
S. salinum (P. A. Dang.) M. W. Dick	M	A	*Cladophora laetevirens, C. flavescens, C. sericea*	France	Karling (1981); Dick (2001)
GENERA *incertae sedis*					
Atkinsiella dubia (D. Atkins) Vishniac	M	Ab	*Haliotis gigantea* (as *Haliotis sieboldii*)	Japan	Nakamura and Hatai (1995)
	M	A	*Chordaria* sp., *Cladospora* sp.	USA	Fuller et al. (1964)
	M	C	*Eriocheir japonicus, Gonoplax angulata, G. rhomboides, Hyas* sp., *Macropodia* sp., *Oregonia* sp., *Pinnotheres pisum, Portunus trituberculatus, P. depuratus, Typton spongicola*	Japan, UK	Atkins (1954a); Fuller et al. (1964); Karling (1981); Roza and Hatai (1999b); Sparrow (1973)

Table continued on next page.

Table 11.1 (Continued)

Taxa	Habitat	Type of host/ Substrate	Distribution	Host species	Reference
Halodaphnea awabi (Kitancharoen et al.) Nakamura *et Hatai*	M	Ab	Japan	*Haliotis gigantea* (as *Haliotis sieboldii*)	Kitancharoen et al. (1994)
H. baliensis Hatai, Roza *et* T. Nakay.	E	C	Indonesia	*Scylla serrata*	Hatai et al. (2000)
H. hainanensis Bian *et* Egusa	M	C	Japan	*Scylla serrata*	Bian and Egusa (1980)
H. okinawensis (Nakamura *et* Hatai) Nakamura *et* Hatai	M	Lo	Japan	*Portunus pelagicus*	Dick (2001)
H. panulirata (Kitanch. *et* Hatai) K. Nakam. *et* Hatai	E	C	Indonesia	*Eriocheir japonicus*, *Scylla serrata*	Roza et al. (1999a, 1999b)
	M	L	Japan	*Panulirus japonicus*	Kitancharoen and Hatai (1995); Dick (2001)
Ha. parasitica (K. Nakam. *et* Hatai) K. Nakam. *et* Hatai	M	C	Japan	*Portunus trituberculatus*	Nakamura and Hatai (1994); Dick (2001)
	M	R	Japan	*Brachionus plicatilis*	Nakamura and Hatai (1994); Dick (2001)
Halioticida noduliformans Muraosa *et* Hatai	M	Ab	Japan	*Haliotis midae, Haliotis rufescens, Haliotis gigantea* (as *Haliotis sieboldii*)	Muraosa et al. (2009)

Key: A: algae, Ab: abalones, B: barnacles, Bi: bivalves, Bra: brackish, C: crabs, Cl: cladocerans, Co: copepods, D: dinoflagellates, DL: decaying leaves, E: estuarine, Fi: fishes, F: fungi, Lo: lobsters, M: marine, ML: mangle leaves, N: nematodes, P: prawns, R: rotifers, S: shrimps, Sa: saprobe, Sn: snails.
*Indicates exclusively saprotrophic forms.

of fungal-like organisms in marine habitats appears to depend on (Schmidt and Shearer 2003):

(i) the distribution of their hosts;
(ii) the organic substances resulting from the decay of dead algae and animals;
(iii) input of fallen leaves from the riverine vegetation, particularly in estuarine coastal habitats; and
(iv) physicochemical factors, such as salinity, water temperature and pH.

Species of Oomycetes involved in leaf decay are almost entirely intertidal and most are found in only one ocean basin. The greatest diversity of mangrove fungi and fungal-like organisms, are found in tropical and subtropical regions (Schmidt and Shearer 2003). The distribution of *Halophytophthora* species is dependent upon the distribution of mangroves (Leaño et al. 2000; see Chapter 12).

Species of *Perkinsus* and *Parvilucifera* are widely distributed along the coasts of the world. For instance, *Parvilucifera infectans* has been reported along the Swedish west coast, the North Sea, the Atlantic coast, the Mediterranean Sea, the Indian Ocean, eastern North America, Australia, and the southern coasts of Korea (Norén et al. 2000; Park et al. 2004; Salomon 2004). This distribution appears to reflect the distribution of their hosts (Mangot et al. 2011). *Perkinsus* species were once prevalent in endemic areas such as Chesapeake Bay in the United States, parasitizing both natural and farmed populations of the oyster *Crassostrea virginica* (Burreson and Ragone Calvo 1996). In these conditions, the oyster fishery was not significantly affected by oyster mortality; however, when oysters were moved to other areas due to consecutive periods of droughts and warm winters, parasitic infections spread along with their hosts and the populations of the oyster consequently drastically declined (Burreson and Ragone Calvo 1996).

The greatest diversity of marine fungal-like organisms has been recorded in Europe, followed by Asia and America. The highest number of records came from the United States, followed by Japan (Table 11.2). Surveys of zoosporic fungal-like organisms have traditionally been conducted in the temperate regions of Europe and North America, so any conclusions on their distribution could be inaccurate because most groups have been poorly sampled in other climates (e.g., Africa and Oceania). These distribution patterns only reflect the regions that have been intensively sampled by researchers.

Phylogeny

The monophyly of both the Alveolata and Chromista is currently widely accepted (Patterson 1999; Yoon et al. 2002). The Chromista are a nutritionally diverse group of eukaryonts, second only to the Animalia (Cavalier-Smith 2009), and originally included only predominantly algal groups: Heterokonta, Haptophyta and Cryptomonada (Beakes and Sekimoto 2009a). Currently, as defined by Cavalier-Smith (2009), this group includes organisms possessing one or both of the following characters:

(i) chlorophyll c-containing plastids lying within the periplastid membrane inside the lumen of the rough endoplasmic reticulum; and/or
(ii) tripartite or bipartite rigid tubular hairs on one or both flagella.

Table 11.2 Number of species recorded per country in each continent. In each continent, the countries are listed in descending order according to the number of species.

Europe		Asia		America		Africa		Oceania	
France	16	Japan	29	USA	35	Madagascar	1	Australia	4
Denmark	14	Philippines	9	Canada	8	Seychelles	1	Hawaii	1
Sweden	7	Indonesia	7	Bahamas	4	Argelia	1		
UK	6	Singapore	5	Greenland	3	Kerguelen	1		
North Sea	6	Thailand	5	Jamaica	2				
Germany	6	Taiwan	4	Costa Rica	2				
Ireland	5	India	4	Bonaire	1				
Italy	4	Korea	2	Cayman Island	1				
Scotland	2	China	2	Haiti	1				
Shetland	2	Malaysia	2	Trinidad	1				
Netherlands	2	Vietnam	2	Venezuela	1				
Adriatic Sea	2	Russia	1	Argentina	1				
Faroe Islands	1	Okinawa	1						
Finland	1	Hong Kong	1						
Norway	1	Siberia	1						
Spitsbergen	1								
King Charles Land	1								
Total number of species	77		75		60		4		5

Both characters appear to have evolved simultaneously in the ancestral chromist and some of the lineages have lost at least one character. For example, it has now been proven that the loss of chloroplasts has occurred many times within the lineage, including at least twice among heterokonts (Cavalier-Smith and Chao 2006; Tsui et al. 2009). As a result, the Chromista include three heterotrophic lineages: free-living bacteriotrophic flagellates (e.g., bicoecids), absorptive gut commensals/parasites (opalanids, proteromonads, and *Blastocystis*), and fungal-like osmotrophs (labyrinthulids, hyphochytrids and Oomycetes). The Rhizaria (phyla Cercozoa and Retaria) and Heliozoa have been found to be phylogenetically related and are included within Chromista, as shown by multigene trees (Burki et al. 2007, 2008; Hackett et al. 2007).

The Alveolata consist of three extremely diverse and morphologically distinctive subgroups: ciliates, dinoflagellates, and apicomplexans (e.g., Cavalier-Smith 1993; Fensome et al. 1993; Ellis et al. 1998; Katz 2001; Leander and Keeling 2003, 2004). Photosynthetic dinoflagellates share the presence of chlorophyll c in their chloroplasts with the Chromista, and on the basis of having the same red algal ancestor and the acquisition of plastids by a unique secondary endosymbiotic event (Cavalier-Smith 1999); the Alveolata are grouped together with the Chromista in the Chromalveolata (Leander and Keeling 2003). The delimitation of the Chromalveolata as a monophyletic group is, however, controversial (Adl et al. 2005; Parfrey et al. 2006), mainly because one putative chromalveolate group, the haptophytes, robustly branches within the Archaeplastida (Hampl et al. 2009). Two possible origins for the Chromalveolata have been proposed (Hampl et al. 2009):

(i) polyphyletic: the lineage of haptophytes and cryptophytes might have acquired the secondary plastid through an endosymbiosis event that was independent to that which occurred in the Chromista and the Alveolata; or
(ii) monophyletic: in this case, the Rhizaria should be included as a basal clade in the "SAR" (Straminipila + Alveolata + Rhizaria) group.

In this section, we will comment on the origin and phylogenetic relationships within some of the Chromista and Alveolata lineages.

Chromista

Oomycota

The "crown lineages" of Oomycetes are mostly terrestrial (peronosporalean lineage) or freshwater organisms (saprolegnialean lineage), although there are some predominantly marine lineages, such as the "basal Oomycetes". Dick (2001) considered that marine stramenopiles were descendants of a terrestrial or freshwater ancestor, but Beakes and Sekimoto (2009a) have speculated that the earliest diverging oomycete genera were predominantly marine. As recently highlighted by Lara and Belbahri (2011), a larger number of sequences should be analyzed before the origin of the Oomycetes can be firmly established, particularly as molecular markers are still unavailable for some major clades.

"Basal Oomycetes": Haptoglossales, Lagenismatales, Eurychasmataceae and Ectrogellaceae

The large majority of the Oomycetes fall into two main galaxies, which were recognized before the development of modern-day molecular genetics (Sparrow 1976; Dick 1990). These two galaxies were the peronosporalean orders, which included many plant-pathogenic species, and the saprolegnialean orders, which included water molds and fish pathogens. The two galaxies were initially distinguished by morphological features and biochemical differences, but their separation into two major clades has since been strongly supported by molecular data (Riethmüller et al. 1999; Petersen and Rosendahl 2000; Lara and Belbahri 2011). Dick (1998, 2001) had assigned subclass status to the Peronosporales and Saprolegniales but, due to the high phylogenetic support for these major clades, Beakes and Thines (2012) have assigned them informal class status (the Saprolegniomycetes and Peronosporomycetes). Much of the formal classification of the Oomycetes, largely based on morphology in straminipilous fungi, was published by Dick (2001). Molecular phylogenies have shown that several morphological traits are not reliable indicators of phylogenetic relationships, however, so the classification of many oomycete genera needs to be reassessed.

The early phylogenies of Oomycetes mainly included gene-sequence data from higher taxa of economically-important terrestrial and freshwater species that were readily accessible in culture (Dick et al. 1999; Riethmüller et al. 1999; Förster et al. 2000; Hudspeth et al. 2000; Leclerc et al. 2000; Petersen and Rosendahl 2000). However, many oomycete taxa are holocarpic parasites and many are from marine and estuarine habitats (Sparrow 1943, 1960; Karling 1981; Glockling and Beakes 2000a; Dick 2001; Nakagiri 2002; Sekimoto et al. 2008b). Later phylogenies, which included some of these parasitic taxa such as the holocarpic parasites of nematodes and marine parasites of seaweeds and crustaceans (Cook et al. 2001; Nakagiri 2002; Beakes et al. 2006; Küpper et al. 2006; Hakariya et al. 2007; Sekimoto et al. 2007, 2008a, 2008b; Beakes et al. 2012), showed that oomycete phylogeny was much more complicated than originally thought. Some of these holocarpic parasitic taxa were shown to be basal within the peronosporalean and saprolegnialean clades, while others fell into deeply branching clades that diverged in the oomycete lineage before the saprolegnialean and peronosporalean divide. Further basal clades of morphologically unknown taxa revealed as environmental sequences were incorporated into oomycete phylogenies (Massana 2002; Lara and Belbahri 2011). Most of the early diverging genera in the Oomycetes are marine or estuarine parasites and many are holocarpic with very simple morphology, lacking gametangia and oospores and, in very early taxa, also lacking a cell wall in the early stages of development.

Resolving the phylogenetic positions of the marine and holocarpic taxa that lie at the cusp of divergence of the saprolegnialean and peronosporalean lineages has proven to be very problematic, largely due to the poor sampling and under-representation of taxa (Beakes and Thines 2012). This is true for species in the Rhipidiales and Leptolegniaceae at the base of the Peronosporalean lineage and for the Leptomitales and the marine parasites of crustaceans in the genus *Atkinsiella* at the base of the saprolegnialean lineage (Beakes and Sekimoto 2009a; Beakes and Thines 2012).

Phylogenies based on small subunit (SSU) rRNA genes and mitochondrial $cox2$ genes (Beakes et al. 2006; Küpper et al. 2006; Hakariya et al. 2007; Sekimoto et al. 2008) have revealed that the two most basal genera in the Oomycetes fall into a clade comprising the marine seaweed parasite *Eurychasma* and the terrestrial nematode parasite *Haptoglossa*. The genus *Eurychasma* has a wide distribution and contains only two or three species, the best known being *E. dicksonii* (a marine parasite of filamentous brown algae; Küpper and Müller 1999). This species has been shown to have a very broad host range (Müller et al. 1999; Gachon et al. 2009). *Eurychasma* develops holocarpic thalli inside the host cells, which later develop into sporangia with a network of walled spores around the periphery (Figure 11.1).

Haptoglossa encompasses 12 host-specific obligate parasitic species of mostly terrestrial nematodes (Glockling and Beakes 2000b, 2001a, 2001b), although the type species, *H. heterospora*, has been reported from littoral marine nematodes (Newell et al. 1977). *Haptoglossa* species are either aplanosporic or zoosporic, the spore cysts later producing structurally-complex gun cells that infect healthy nematodes by a microinjection process. Aplanosporic species have two or more morphologically different cell types, only one of which is capable of infecting nematodes. The alternative spore type, which could target an unidentified host, may be bi-nucleate with an elaborate internal tubular system and a fibrillar pad that may be adhesive (Glockling and Beakes 2000b, 2001a). One *Haptoglossa* species, *H. beakesii*, has vesiculate spore discharge, which is considered to be

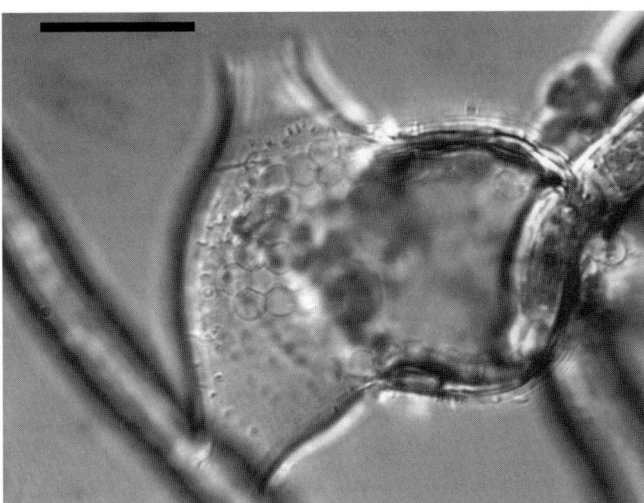

Figure 11.1 Dehiscent sporangium of *Eurychasma dicksonii* (strain CCAP 4018/1) grown in the brown algal host *Ectocarpus siliculosus* (strain CCAP 1310/299). The honeycomb structure is made of the remnants of encysted primary spore cell walls, and is a defining feature of the *Eurychasma* genus. Photograph taken with a Zeiss Axioscop 2 microscope coupled to an Axiocam camera. Bar: 40 μm.

a pythiacious trait (Glockling and Serpell 2010). There was some doubt as to whether *Haptoglossa*, which was originally placed in the Saprolegniaceae (Drechsler 1940), was indeed an oomycete after Beakes and Glockling (1998) had depicted a whole mount zoospore of *H. dickii* showing two whiplash flagella. This led to speculation that *Haptoglossa* could be more closely related to the plasmodiophorids (Beakes and Glockling 1998; Dick 2001). However, phylogenetic analysis confirmed that *Haptoglossa* is indeed a basal oomycete (Beakes et al. 2006; Hakariya et al. 2007) and species of *Haptoglossa* with tripartite tubular hairs along one flagellum and tripartite tubular hair packets in the spore initials have since been discovered (S.L. Glockling unpublished). *Haptoglossa* zoospores have an intracellular arrangement of a central nucleus closely surrounded by mitochondria, with dense body vesicles around the spore periphery. The dense body vesicles lack the fingerprint striations found in the zoospores of later-evolved nematode parasites in *Myzocytiopsis* (Glockling and Beakes 2000a).

It has been informally proposed that *Haptoglossa* and *Eurychasma* are accommodated together in a new class, the Eurychasmomycetes, because they form a single clade, albeit separated by long branches (Beakes and Thines 2012). Our phylogenetic tree by maximum likelihood based on 18S rDNA sequences from GenBank also shows *Haptoglossa* and *Eurychasma* as a monophyletic group (Figure 11.2). *Eurychasma* lacks a cell wall during the early assimilative stages of thallus formation but thalli of *Haptoglossa* are always walled; the infective sporidium having a fibrillar coat outside a wall layer immediately following firing of the gun cell into the nematode host (Glockling and Beakes 2001a). Despite the walled condition of *Haptoglossa*, the lack of a cell wall in the assimilative stages of growth generally appears to be a primitive feature in the Oomycetes.

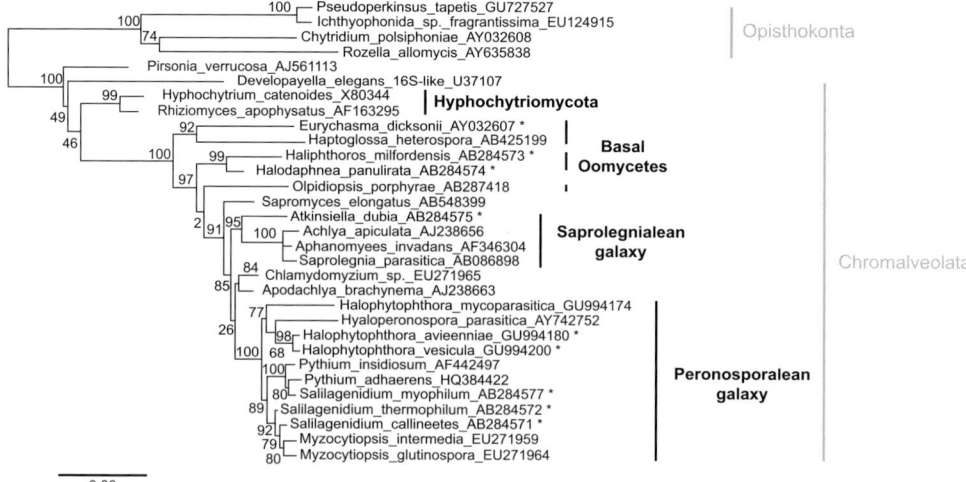

Figure 11.2 Phylogenetic tree inferred by maximum likelihood analysis of Chromalveolata 18S sequence data (only sequences from representatives of Chromista were included in the analysis). Analysis run in the www.Phylogeny.fr platform (Dereeper et al. 2008) on "Advanced Mode" with a curation step (Gblocks) set for a less stringent selection allowing smaller final blocks, gap positions within the final blocks and less strict flanking positions. Substitution model HKY85, aLRT test = SH-like, number of substitution rate categories = 4, gamma parameter: estimated, proportion of invariable sites: estimated, transition/transversion ratio = 4. *Pseudoperkinsus tapetis, Ichthyophonida* sp. *fragrantissima, Chytridium polysiphoniae* and *Rozella allomycis* (Opisthokonta) were included in the analysis as outgroups. References: numbers above or below the branches indicate approximate likelihood-ratio test (aLRT) branch support (%); bar: number of substitutions per site, * indicates sequences of exclusively marine or estuarine species.

Two genera parasitic in diatoms, *Ectrogella* and the monotypic genus *Lagenisma*, also have naked thalli during early thallus development. Therefore, although the rDNA and *cox*2 genes of these genera have not yet been sequenced, it is likely that they fall into the *Eurychasma/Haptoglossa* clade (Beakes and Thines 2012). The parasites of red algae in the genus *Olpidiopsis* are also unwalled in early thallus development, and this genus has also been shown to form an early diverging clade (Sekimoto et al. 2008b). *Ectrogella* (Raghukumar 1980a, 1980b; Hanic et al. 2009) and *Olpidiopsis*, like some other higher oomycete genera, contain both marine and non-marine (i.e., terrestrial and freshwater) species. *Lagenisma coscinodisci* (Schnepf and Drebes 1977) is a marine endoparasite infecting the centric diatoms *Coscinodiscus* and *Palmeria* (Table 11.1). This parasite has an asexual reproductive cycle, producing zoospores that are diplanetic, and also a sexual cycle whereby haploid zoospore cysts fuse to form a diploid oospore following karyogamy (Schnepf et al. 1978a, 1978b).

Recent phylogenetic studies have indicated that other parasitic and mainly marine species, including *Halodaphnea, Haliphthoros* and *Olpidiopsis*, also diverge before the division of the major oomycete clades (Sekimoto et al. 2008b; Beakes et al. 2012). This is also observed in the 18S rDNA phylogenetic tree in Figure 11.2.

Halodaphnea and *Haliphthoros* are the only basal genera that can be easily grown in artificial media (Sekimoto et al. 2007). Both *Halodaphnea* (syn. *Halocrusticida*; Dick 1998) and *Haliphthoros* are marine parasites of crustaceans (Cook et al. 2001;

Sekimoto et al. 2007); while *Olpidiopsis* species infect red algae (West et al. 2006; Sekimoto et al. 2008b, 2009) but also have representatives in freshwater ecosystems (Beakes and Thines 2012). The freshwater parasite *Olpidiopsis varians* is unusual as it is the only basal oomycete known to form oospores as a result of fusion between oogonial and antheridial thalli (Martin and Miller 1986a). Also, during mitosis, *O. varians* formed polar fenestrae – gaps in the nuclear envelope (Martin and Miller 1986b) – but these were absent during mitosis in *Lagenisma coscinodisci* (Schnepf and Drebes 1977). *Halodaphnea* and *Haliphthoros* group together in SSU rDNA phylogenies while *Olpidiopsis* falls into a separate clade (Beakes and Sekimoto 2009a), as is also shown in Figure 11.2. In *cox*2 phylogenies, however, these clades are not statistically separable (Sekimoto et al. 2008a, 2008b; Beakes and Thines 2012).

Other marine parasites of seaweeds, such as *Petersenia*, *Pontisma* and *Sirolpidium*, for which there is currently no available sequence data, might be expected to fall close to these basal genera. Dick (2001) included these genera in the order Olpidiopsidales and Beakes and Thines (2012) propose combining the Olpidiopsidales with the Haliphthorales (*Haliphthoros* and *Halodaphnea*) in a new class, the Olpidiopsidiomycetes, as modern phylogenies show strong support for the relatedness of these groups (Beakes and Sekimoto 2006).

Several environmental sequences that do not cluster with any known clade also fall between the basal-most *Eurychasma* and *Haptoglossa* clade and the two main oomycete galaxies (Massana et al. 2004, 2006; Lara and Belbahri 2011).

The Rhipidiales are freshwater saprotrophs but there is presently only sequence data for a single species, *Sapromyces elongatus*. However, *S. elongatus* has been reported to be the basal clade of the saprolegnialean lineage in LSU rDNA phylogenies (Petersen and Rosendahl 2000) and the basal clade to the peronosporalean lineage in *cox*2 phylogenies (Hudspeth et al. 2000). It is therefore a particularly difficult species to place. Based on rRNA phylogenies, Lara and Belbahri (2011) stated that *S. elongatus*, along with an unidentified abalone parasite (AB178865), could not be affiliated with any certainty with either the peronosporalean or the saprolegnialean lineages, as also observed in Figure 11.2. In SSU rDNA trees, *S. elongatus* falls closest to *Chlamydomyzium* parasite of nematodes (Beakes and Sekimoto 2009a).

Chlamydomyzium is a genus that has also proved difficult, as it too falls in different places on the tree depending on the gene sequenced and the type of analysis used. *Chlamydomyzium*, *Apodachlya brachynema* and *Atkinsiella dubia* clustered with reasonable support in the saprolegnialean galaxy in the SSU rRNA phylogeny of Lara and Belbahri (2011). In our 18S rDNA phylogenetic tree, only *A. dubia* clustered with high support in the saprolegnialean clade, while *Chlamydomyzium* and *A. brachynema* appeared as the sister group of the peronosporalean galaxy (Figure 11.2).

The genus *Chlamydomyzium* had been included in the family Myzocytiopsidaceae along with *Myzocytiopsis*, another genus of nematode parasites, and *Syzygangia*, an obligate endoparasite of plants (Dick 1997). Although *Chlamydomyzium* and *Myzocytiopsis* have many morphological features in common and have similar ecological traits, the family Myzocytiopsidaceae is not, however, supported by molecular genetics. *Chlamydomyzium* diverges earlier than members of *Myzocytiopsis*, which mostly fall in the peronosporalean galaxy close to *Lagenidium* at the base of the Pythiales (Beakes and Sekimoto 2006), as also observed in Figure 11.2. Ultrastructural examination of *Chlamydomyzium oviparasiticum*

(Glockling and Beakes 2006) showed that this species had ultrastructural features pertaining to both the saprolegnialean and peronosporalean galaxies. Members of the Leptolegniales are also considered to belong to the Rhipidiales clade (Beakes and Sekimoto 2009).

The crustacean parasite *Atkinsiella dubia* is another taxon that falls as a monophyletic clade basal to the saprolegnialean galaxy in SSU rDNA and *cox*2 phylogenies (Cook et al. 2001; Sekimoto et al. 2007; Beakes and Sekimoto 2009a; Lara and Belbahri 2011) and in Figure 11.2. *Apodachlya*, in the Leptomitales, also forms an early diverging clade in the saprolegnialean galaxy (Petersen and Rosendahl 2000; Beakes and Sekimoto 2009a; Lara and Belbahri 2011). In our 18S rDNA phylogenetic tree, however, *Apodachlya* appeared to be more closely related to the peronosporalean than to the saprolegnialean clade (Figure 11.2). The Leptomitales comprise four families and include the sewage fungi as well as holocarpic keratinaceous species. *Cornumyces*, isolated from keratin baits, was sequenced and found to be closely related to *Apodachlya* and *Leptomitus* but formed a separate clade (Inaba and Hariyama 2006).

It has been postulated that the Oomycetes could have migrated from sea to land with their nematode hosts or by switching hosts to terrestrial species (Beakes and Sekimoto 2009; Beakes et al. 2012). This hypothesis is based on the dominance of marine parasites among the early diverging clades of oomycete taxa, and the occurrence of marine and estuarine taxa throughout some predominantly terrestrial and freshwater genera, such as *Pythium* and *Aphanomyces*. However, Lara and Belbahri (2011) maintain that a larger sampling effort is required before assumptions can be made as to oomycete origins, especially as *Haptoglossa* is largely terrestrial and an environmental sequence derived from a peat bog was also found to cluster in this basal-most position. Some marine environmental sequences fell into clades that were basal to *Olpidiopsis* and had no morphologically-known representatives (Lara and Belbahri 2011). The presence of these unknown basal oomycete clades, and of other marine environmental sequences falling within the peronosporalean lineage, suggest that there is a diversity of undiscovered oomycete clades in the marine environment that, with further sampling, should aid future understanding of the evolution of this group of organisms.

Saprolegniales

Species in the order Saprolegniales occur worldwide and can be frequently found in freshwater and nonsaline soils (Hulvey et al. 2007). All genera in this order form a well-defined clade (Dick et al. 1999; Riethmüller et al. 1999; Leclerc et al. 2000; Petersen and Rosendahl 2000; Inaba and Tokumasu 2002; Spencer et al. 2002) that has been considered ancestral due to the similarity between *Saprolegnia* and the alga *Vaucheria* (Beakes 1987). As previously mentioned, more recent results from the analysis of LSU rDNA data showed two major lineages: one formed by the Leptomitales, Rhipidiales and Saprolegniales (Saprolegniomycetidae); and the other by the Peronosporales and Pythiales (Peronosporomycetidae).

Most species in this order are decomposers of vegetable and animal debris, but some of them are reported to be pathogens of economically-important plants (e.g., *Aphanomyces euteiches*), fresh water fishes (e.g., *Saprolegnia parasitica*) and crustaceans (e.g., *Aphanomyces astaci*) (Dieguez-Uribeondo et al. 1997; Anderson 2001; van West 2006; Dieguez-Uribeondo et al. 2007; Gaulin et al. 2007). The first phylogenetic analysis, based on data from the ITS region, attempted to map the zoospore release, a character that has traditionally been used to define the genera in the family Saprolegniaceae (Daugherty et al. 1998). However, it is now accepted that this character, together with other morphological characters, is not informative for classifying the genera in the Saprolegniaceae (Dick et al. 1999; Riethmüller et al. 1999; Inaba and Tokumasu 2002; Spencer et al. 2002). Phylogenetic studies have shown *Achlya* and *Saprolegnia* to be non-monophyletic genera, having representatives scattered in several different "genus-level" clades (Leclerc et al. 2000; Inaba and Tokumasu 2002; Spencer et al. 2002). Other studies also showed the polyphyly of the morphologically-defined genera of Saprolegniaceae. Rietmüller et al. (2002) demonstrated the polyphyly of centric and excentric lineages of *Achlya*, while Hulvey et al. (2007) showed the polyphyly of *Saprolegnia* based on trees from ITS data. Based on molecular data (particularly sequences from the 28S rDNA), Beakes and Sekimoto (2008) showed that the genus *Aphanomyces*, together with *Plectospira*, *Pachymetra* and *Verrucalvus*, form a well-supported clade that is separated from other members of the Saprolegniaceae. It is evident that further studies are required to accurately elucidate the phylogenetic relationships between the major clades of this order.

Peronosporales and Pythiales

The genera *Phytophthora* and *Pythium* have traditionally been placed together based on morphological similarities and characteristics of sexual reproduction (Waterhouse 1973; Dick 1995). More recently however, Petersen and Rosendahl (2000) – based on data from partial sequences of LSU rDNA and several ecological and morphological differences – separated these two genera, placing *Phytophthora* in the Peronosporales and *Pythium* in the Pythiales.

Most species in the Peronosporales are found in terrestrial and freshwater ecosystems (Sparrow 1960). The genus *Halophytophthora* was originally erected to accommodate species found as decomposers of leaf litter in marine environments, which were formerly described in *Phytophthora* (Ho and Jong 1990). Members of the genus exhibit zoospore release with or without the presence of a vesicle or with a semi-persistent vesicle (Hulvey et al. 2010). These characters, together with other asexual characters, were initially employed to segregate the species of *Halophytophthora* from *Phytophthora*. Subsequently, additional species of *Halophytophthora* have been described (Ho et al. 1991, 1992, 2003) but no comprehensive phylogenetic analysis of this genus has so far been conducted. Only a single *Halophytophthora* species (*H. spinosa*) had been included in multigene phylogenetic studies of the Peronosporales (Göker et al. 2007), until very recently when Lara and Belbahri (2011) included more species. These authors showed that some *Halophytophthora* species (i.e., *H. elongata, H. polymorphica* and *H. batemanensis*) do not appear to be monophyletic but paraphyletic and are located at the base of the Peronosporales. In addition, one *Halophytophthora* species (*H. tartarea*) was transferred to the new genus *Salisapilia* (Hulvey et al. 2010). Therefore, *Halophytophora* seems to be composed of a heterogeneous assemblage of organisms that have a combination of characteristics between *Pythium* and *Phytophthora* (Cooke et al. 2000; Nakagiri 2002) and which are only defined by their ecological preference. Lara and Belbahri (2011), however, attributed the results to misidentification of the strains deposited in culture collections.

The genus *Salisapilia* appears to be morphologically similar to some isolates of *Halophytophthora*, but diverges on the basis of a combined sequence analysis of two nuclear loci and by some morphological characters, i.e., the absence of a vesicle during spore discharge, and the presence of a protruding plug of material at the discharge tube apex (Hulvey et al. 2010). It has been proposed that *Salisapiliaceae* is an ancient lineage of the Peronosporales, which is the sister group to the monophyletic group containing both the Pythiaceae and the Peronosporaceae.

Two possible origins for *Halophytophthora* have been hypothesized (Nakagiri 2002):

(i) a terrestrial origin and a secondary migration and acclimatization to estuarine/marine habitats; or
(ii) a marine origin – this genus representing a vestigial marine lineage of the Peronosporales.

If the organisms in this genus have their origin in the open sea, they made the transition to terrestrial environments by inhabiting the benthos of estuaries (Beakes and Sekimoto 2008, 2009). As in the case of Rhipidiales and Albuginales (Thines et al. 2009), a reversal adaptation to marine environments also appeared to be the most parsimonious hypothesis for the origin of both *Halophytophthora* and *Salisapilia*.

Pythium is a diverse genus of approximately 150 recognized species (Kirk et al. 2008), which are saprobic or parasitic on plants of economic importance (Frezzi 1956) and animals including humans (de Cock et al. 1987; Mendoza et al. 1987). Only a few species are found in marine and estuarine habitats, mostly as saprotrophs on decaying leaves or as parasites of red algae (Table 11.1). Species in this genus have been subjected to several phylogenetic analyses in the past few years and many new species have been described (e.g., Lévesque and de Cock 2004; Villa et al. 2006; Moralejo et al. 2008; Bala et al. 2010). Results from Lévesque and de Cock (2004) first showed two major clades corresponding to *Pythium* species with filamentous and globose sporangia and a small clade between these two, represented by the species with ovoid-papillate sporangia. More recently, results obtained by Uzuhashi et al. (2010) showed that *Pythium* is polyphyletic and is composed of five monophyletic clades, each characterized by a particular type of sporangium. Consequently, these authors split the genus into four new genera: *Ovatisporangium, Globisporangium, Elongisporangium*, and *Pilasporangium*, and emended *Pythium*. In addition, in the same year a new genus named *Phytopythium* was introduced by Bala et al. (2010) to include a *Pythium* species from clade K in Lévesque and de Cock (2004) that was already included in the genus *Ovatisporangium*, and which is morphologically and phylogenetically between *Pythium* and *Phytophthora*. *Phytopythium* includes exclusively terrestrial plant pathogens and appears as the sister group of *Phytophthora* and *Halophytophthora* (Bala et al. 2010).

Hyphochytriomycota

The microorganisms in the phylum Hyphochytriomycota are closely related to the Oomycota (Adl et al. 2005). All members of the Hyphochytriomycota produce zoospores with a single anteriorly-directed flagellum with mastigonemes. Sparrow (1960) recognized three families: Anisolpidiaceae, Rhizidiomycetaceae, and Hyphochytriaceae. The Anisolpidiaceae is primarily marine; while the Rhizidiomycetaceae and Hyphochytriaceae are found in freshwater and soil ecosystems (Sparrow 1960).

Several species in the family Anisolpidiaceae are common parasites of a number of genera of brown algae (Phaeophyta), such as *Ectocarpus, Cladostephus, Sphacelaria* and *Pylaiella* in marine environments (Table 11.1). For example, *Anisolpidium rosenvingei* is a monocentric parasite of the cosmopolitan filamentous alga *Pylaiella littoralis*. It infects only unilocular and pleurilocular sporangia of the algae and can cause epidemics in populations of this host (Küpper and Müller 1999). These authors studied the abundance of *A. rosenvingei* in tissues of *P. littoralis* along the European coast, and at one site up to 70% of the population of *P. littoralis* was infected by the end of the growing season. This parasite probably disrupts the reproductive processes of the host. Field data collected over three consecutive years suggested a correlation between the prevalence of the parasite and the decline in host populations (Küpper and Müller 1999).

Another *Anisolpidium* species, *A. ectocarpii*, is a parasite of *Ectocarpus siliculosus* populations on the east coast of North America (Table 11.1). Both vegetative cells and pleurilocular sporangia in the host are infected by this parasite (Johnson 1957). Johnson (1957) described the lifecycle of this parasite in hosts collected from the intertidal zone at Taylor Creek in North Carolina. Hosts were experimentally infected in the laboratory. *Anisolpidium ectocarpii* produces thick-walled zygotes during sexual reproduction. These zygotes become resting spores that can possibly survive a wide range of temperatures. Other species in the Hyphochytriomycota can produce resting spores (i.e., thick-walled structures), but only by asexual reproduction. Both *A. ectocarpii* and *A. rosenvingei* are found on hosts in the intertidal zone where temperatures and salinities vary considerably (Johnson 1957; Küpper and Müller 1999). Temperature and salinity are important abiotic factors controlling parasite and host populations in marine ecosystems. Resistance to environmental extremes is likely to be a characteristic of thick-walled resting spores (Gleason et al. 2008a, 2010).

Species of *Hyphochytrium* and *Rhizidiomyces* are frequently found in soil (Sparrow 1960; Barr 1970; Letcher et al. 2004) and are sometimes found in extreme environments (Barr 1970; Longcore 2005). For example, *Hyphochytrium* has been isolated from hot deserts and extremely cold climates in the Arctic (Barr 1970) and *Rhizidiomyces* has been isolated from canopy detritus collected in tropical rain forests in the south Pacific (Longcore 2005). Only one *Hyphochytrium* species has been reported to be a parasite of cladocerans in marine environments (Table 11.1).

In general, Hyphochytriomycota species have been poorly studied. Ribosomal DNA sequences have been recorded in public databases for only three isolates: two in the genus *Hyphochytrium* (van der Auwera et al. 1995; Gleason et al. 2008b); and one in the genus *Rhizidiomyces* (Hausner et al. 2000). No sequences are currently available for species in the Anisolpidiaceae.

Alveolata: Perkinsozoa

The Perkinsozoa is currently considered the sister group to free-living dinoflagellates within the Alveolata (Leander and Keeling 2003; Mangot et al. 2011). Originally the Perkinsozoa was considered to be the sister group of the Apicomplexa, but their phylogenetic position has been controversial for many years (Norén et al. 2000). *Perkinsus marinus*, for example, was originally described as a protozoan (*Dermocystidium marinus*) and subsequently transferred to the Labyrinthomorpha (*Labyrinthomyxa marina*) and

Apicomplexa (Mackin and Ray 1966) due to morphological characteristics in the zoospore stage (Levine 1978). Molecular phylogenies have supported the exclusion of *Pe. marinus* from the Apicomplexa (Siddall et al. 1997). The taxonomic affinities of *Pe. marinus* and *Parvilucifera infectans* led to the establishment of a new phylum, the Perkinsozoa (Norén et al. 2000). Recently, molecular phylogenies based on gene sequences from proteins (e.g., tubulins and actins) and the SSU of the rRNA, suggested that the Perkinsozoa are the earliest divergent group from the dinoflagellate lineage (Leander and Kelling 2004) and have retained several characteristics inferred to be ancestral for the dinoflagellates and apicomplexans (Kuvardina et al. 2002; Leander and Keeling 2003, 2004; Cavalier-Smith and Chao 2004; Moore et al. 2008). Until now only two genera, *Parvilucifera* and *Perkinsus*, have been described in the Perkinsozoa and the validity of most species is still considered doubtful (Burreson et al. 2005).

Ecological role and importance

Fungal-like organisms are adapted to aquatic habitats by the production of zoospores that range from $\cong 2$ to 12 µm (Sparrow 1960). They are characterized as obligate marine microorganisms when they grow and sporulate exclusively in marine or estuarine habitats under permanently or intermittently submerged conditions (Kohlmeyer et al. 2004). Oomycetes and Hyphochytriomycetes play a major role in aquatic ecosystems as saprobes or parasites (Sparrow 1960). The Perkinsozoa are known exclusively as intracellular parasites in marine ecosystems, although environmental sequences have also been obtained from freshwater bodies (Mangot et al. 2011).

The occurrence and distribution of fungal-like taxa in marine and estuarine habitats is poorly documented worldwide, probably due to the limited number of mycologists studying these groups (Jones 2011). There is compelling evidence that these organisms, together with the asexual (anamorphic) fungi and the ascomycetes, are the primary decomposers of plant materials in coastal marine ecosystems and estuaries (Nakagiri 2000; Raghukumar 2004; Bärlocher 2005; Gulis et al. 2006; see Chapter 22). In marine pelagic ecosystems, however, their importance and impact have not been extensively quantified (Shearer et al. 2007) and they are often regarded as being of minor importance in food-web dynamics (Kurtzman and Fell 2004). However, with the advent of molecular techniques, fungal-like parasites increasingly appear to be quantitatively important in pelagic food webs (Lefèvre et al. 2008; Jobard et al. 2010; Rasconi et al. 2012). Some studies have revealed a strong correlation between phytoplankton cell counts and parasite infection (e.g., Canter and Lund 1948; Hohlfeld 1998). *Parvilucifera* species (Perkinsozoa), for example, were proven to have a regulatory effect on their hosts, controlling dinoflagellate blooms (Gisselson et al. 2002; Park et al. 2004). Modeling strongly suggested that *Amoebophrya* parasites played a bigger role than grazers in controlling the population of toxic *Alexandrium* populations (Montagnes et al. 2008). Parasitic fungal-like taxa thus not only control phytoplankton and zooplankton populations but also feed "depauperate" pelagic communities with autochthonous organic matter from dead hosts.

Environmental rDNA surveys showed that small heterotrophic flagellates (HFs) are a heterogeneous assemblage mainly composed of zoospores of parasitic and saprotrophic zoosporic fungi, alveolates (e.g., perkinsids) and stramenopiles (Gleason et al. 2008b; Sime-Ngando et al. 2011). Thus, zoospores significantly contribute to the pool of

HFs in the marine plankton. These zoospores are grazed by zooplankters (Buck et al. 2011) and are considered to be excellent food sources for these organisms in terms of size, shape and nutrient content (Kagami et al. 2004, 2007), see Chapter 9. By grazing zoospores, zooplankters obtain important supplementary nutrients, such as polyunsaturated fatty acids and cholesterol from HFs. In this way, nutrients are transferred to higher trophic levels, a process known as trophic upgrading (Kagami et al. 2007). Additionally, grazing might affect the density of infective zoospores in the plankton and thus help to regulate the parasites. This type of regulation has been observed for the chytrid parasite, *Batrachochytrium dendrobatidis*, which is the causal agent of "chytridiomycosis" in amphibians in freshwater habitats (Buck et al. 2011). In marine ecosystems, mature trophozoites of *Parvilucifera* species are liberated from the host tissues in order to disseminate the parasite (Perkins 1976). These trophozoites enlarge and become prezoosporangia, which release large numbers of zoospores into the water column (Perkins 1996). Some of these can be grazed by zooplankters while others infect the host tissues by filtration from bivalves (Andrews 1996).

The following example illustrates the complexity of the trophic relationships in which zoosporic parasites are involved. In aquatic habitats, large or colonial planktonic algae (e.g., diatoms) are inedible due to their size and, in some marine species, due to the production of substances that are deleterious for the reproduction of potential grazers (Ban et al. 1997). These algae are therefore usually not grazed but are left to sink (Lefèvre et al. 2008). These inedible algae appeared to be more frequently parasitized by zoosporic fungi than edible algae (Kagami et al. 2007), however, and act as trophic intermediaries through the grazing of fungal zoospores by zooplankters (Lefèvre et al. 2008). This concept, known as "Mycoloop" (Kagami et al. 2007), demonstrates that zoosporic parasites have a complex role in food-webs, impacting several trophic levels through different cascades. It is not yet known whether grazing can represent a source of infection for grazers (Jobard et al. 2010) or whether parasitic stramenopiles can also act as trophic intermediaries between inedible hosts and grazers. These top-down and bottom-up mechanisms of regulation significantly add to the complexity of linkages between trophic levels in marine food webs.

Saprotrophs

Role as decomposers of fallen leaves

There has been a great deal of interest in mangrove-inhabiting fungi and fungal-like organisms and, as a consequence, a number of papers that summarize their diversity have been written in the past ten years (e.g., Leaño et al. 2000; Schmidt and Shearer 2003; Vittal and Sarma 2006; Alias and Jones 2009; Pang et al. 2011). Marine and estuarine saprotrophic Oomycetes seem to be present on almost all dead mangrove leaves and do not appear to exhibit substrate specificity (Newell et al. 1987; Schmidt and Shearer 2003) but rather substrate preference, i.e., some species are more common on certain substrates than on others (Jones and Alias 1997; Alias and Jones 2009). Nonetheless, they remain important microbiota in many marine and estuarine habitats all over the world.

Zoospores of fungal-like organisms appear to be well-adapted to constantly changing conditions of salinity, pH and temperature that occur daily in estuarine ecosystems due to tidal fluctuations (Leaño et al. 1998a). Species of *Pythium*, *Halophytophthora* and *Salisapilia* are among the few Oomycetes reported from marine leaf litter worldwide (de Cock 1986; Newell et al. 1987; Newell 1992; Leaño et al. 2000; Nakagiri et al. 2001; Kohlmeyer et al. 2004; Kis-Papo 2005; Hulvey et al. 2010). In particular, the presence of *Halophytophthora* on mangrove leaf litter is well-documented worldwide (e.g., Newell et al. 1987; Raghukumar et al. 1995; Nakagiri 2000; Leaño 2001; Nakagiri et al. 2001), Table 11.1. *Halophytophthora* species are considered to be the most frequent colonizers of fallen mangrove leaves (Newell et al. 1987; Nakagiri et al. 1989; Newell and Fell 1992; Tan and Pek 1997) and play a major role as decomposers during the early to late stages of leaf decay in these ecosystems (Nakagiri 1993). The recently-erected genus *Salisapilia* includes three species that are important decomposers of leaves of *Spartina alterniflora* in salt marshes (Hulvey et al. 2010).

Preferences for particular types of substrate and conditions of salinity and temperature have been documented as important factors influencing the local, seasonal and geographic distribution of *Halophytophthora* species (Nakagiri 2000). Their abundance in many mangrove habitats has been attributed to their efficient production of zoospores (Nakagiri et al. 1989), their ability to compete against higher fungi and other microbes in colonizing fallen mangrove leaves (Newell and Fell 1997), and their wide growth tolerance to varying levels of pH, temperature and salinity (Nakagiri 1993; Nakagiri et al. 1994; Leaño et al. 1998a). Moreover, their success in the colonization of fallen mangrove leaves depends on the homing ability of zoospores, their positive chemotactic response to some chemical components and nutrients coming from decomposing leaves (Leaño et al. 1998b), and on their ability to securely attach to mangrove leaves for rapid germination and colonization (Leaño et al. 2000). Leaño (2001) also pointed out that the decomposition of fallen mangrove leaves by marine Stramenopiles (i.e., *Halophytophthora* and thraustochytrids) contributes to the enrichment of mangrove detritus for consumption by organisms at higher trophic levels, making mangroves excellent nurseries for many species of fishes and crustaceans.

Role as members of the marine plankton

One of the most important fractions of plankton in marine ecosystems is the nanoplankton, which consists of assemblages of organisms from 2 to 20 µm in size with autotrophic and heterotrophic nanoplankton (HNP) components (Yoshida et al. 2009). The HFs, particularly the heterotrophic stramenopiles, constitute an important fraction of the HNP, which are components of the microbial loop and play essential roles in marine food webs by transferring carbon from bacterioplankton up to metazooplankton (Carrias et al. 1996). Nevertheless, the quantification of HNP has been scarcely studied due to methodological limitations, i.e., difficulty in measuring population sizes and identifying this fraction using traditional techniques (Rasconi et al. 2009). In addition, many lineages in the assemblages of HNP have been overlooked, classified as unidentified HFs or misidentified as bacterivorous flagellates (Lefèvre et al. 2007). During the past few years, the advent of molecular techniques has allowed the identification of many previously-unidentified lineages and shown their relative importance as components of the HNP. Recently, numerous studies have characterized marine nanoplanktonic assemblages from different locations worldwide using amplification and sequencing of the 18S rRNA genes. Data on 18S rRNA sequences have increasingly demonstrated the broad diversity of sequences affiliated to numerous phylogenetic groups of eukaryotes as components of the HNP in oceanic systems.

An unexpected diversity of eukaryotic (SSU) rDNA sequences has been recovered from extreme marine habitats such as the ocean surface (Díez et al. 2001; Moon-van der Staay et al. 2001) and from anoxic marine sediments near fumaroles (Takishita et al. 2005). Some of these sequences have appeared to represent novel lineages within the alveolates and stramenopiles. In addition to lineages of existing straminipile clades (e.g., hyphochytrids and Oomycetes), a great diversity of unidentified lineages has been found (Massana et al. 2004, 2006).

Environmental SSU rRNA sequences with previously unknown affiliations obtained from freshwater, soil and marine samples worldwide were recently shown to form a new clade that branched at the base of the Fungi. This new clade was tentatively named Rozellida, since the only formally-described taxon within this clade was *Rozella* (Lara et al. 2010; James and Berbee 2012). Jones et al. (2011) further analyzed the environmental SSU rRNA sequences used by Lara et al. (2010) and added a large sample of environmental DNA sequences obtained from GenBank, and described a well-supported clade named Cryptomycota (Jones et al. 2011), see Chapter 10. These results demonstrate the number of sequences still with uncertain affinity and the necessity for further analysis of marine assemblages, particularly from extreme habitats. In addition, databases still need to be expanded with further information about eukaryotic HNP.

Parasites

Parasitism appears to have evolved early in the Oomycetes (Beakes and Sekimoto 2009) and this could explain why the majority of parasitic lineages are among the basal oomycete clades. Parasitic interactions increase the complexity of food webs through many direct (e.g., mortality of hosts and zoospores as food resources for grazers) and indirect effects (e.g., competition between zoospores while locating a suitable host). This complexity is increased by extending the network, increasing the connectance (percentage of possible links), nestedness (degree of structure and order), and the efficiency of the transfers (Lafferty et al. 2006; Mangot et al. 2011). Thus, an understanding of host–parasite interactions on all trophic levels is required to properly describe the energy fluxes in marine food webs.

Oomycete parasites range in morphology from holocarpic to mycelial and eucarpic forms (Porter 1986). The majority of marine parasites belong to the "lower Oomycetes", a group of intracellular and holocarpic taxa that are predominantly, but not exclusively, parasites of several marine organisms. Species of *Ectrogella, Eurychasma, Haliphthoros, Halodaphnea, Lagenisma, Pontisma* and *Sirolpidium* are exclusively marine pathogens; while species of *Haptoglossa, Olpidiopsis* and *Petersenia* are also found in freshwater and terrestrial habitats (Strittmatter et al. 2009). The majority of species are parasites of marine algae and crustaceans, but they are also parasites of fishes, molluscs, nematodes, and rotifers (Table 11.1).

Diseases of algae due to fungal-like infection are quite common in the marine habitat, both in nature and in mariculture (Raghukumar 1996; Hyde et al. 1998; Murray and

Peeler 2005; Das et al. 2006; Gachon et al. 2010). Many of the taxa of Oomycetes listed in Table 11.1 are parasites of algae (reviewed in Strittmatter et al. 2009). Diseases caused by these organisms appear to be host-specific and show characteristic symptoms, such as changes in color, rot lesions and abnormal growth in the host (Li et al. 2010). *Olpidiopsis* is a holocarpic genus that parasitizes several species of marine green, brown and red algae (Dick 2001), as listed in Table 11.1. Some species of marine fungal-like organisms are parasites of economically-important seaweeds, such as *Porphyra* species. This alga is parasitized by *Pythium porphyrae* (van der Plaats-Niterink 1981), and by *Olpidiopsis porphyrae* (Sekimoto et al. 2008b). The infection of *Porphyra* (commonly called nori) by *Olpidiopsis porphyrae* and *O. bostrychiae* has been shown to cause significant decreases in nori production (Fujita 1990; Gachon et al. 2010). *Eurychasma dicksonii* is currently considered the most common and widespread species of marine fungal-like organisms (Sparrow 1960; Strittmatter et al. 2009), see Table 11.1. This parasite has mainly been reported as affecting populations of the filamentous brown alga *Pylaiella littoralis*, causing so-called brown tides (Wilce et al. 1982), but has also shown a broad host range, infecting other brown algae such as *Ectocarpus* sp. and the gametophytic stage of kelps (Müller et al. 1999).

Besides the algae, various economically-important crustaceans (lobsters, crabs, prawns, and shrimps) are infected by fungal-like taxa, such as *Atkinsiella, Haliphthoros, Halodaphnea, Leptolegnia, Pythium* and *Salilagenidium*. The genus *Salilagenidium* is represented by six species that are of great importance in mariculture, since they all parasitize algae, crustaceans and molluscs in marine and estuarine habitats (Table 11.1). *Salilagenidium callinectes* is the most commonly observed species due to the high variety of hosts that it can parasitize (see Table 11.1 for references). Among the main crab parasites are *Atkinsiella dubia*, which produced 100% mortality of larvae of the Japanese mitten crab, *Eriocheir japonicus* (Roza and Hatai 1999a), and *Haliphthoros milfordensis*, which is commonly observed as a parasite of economically-important marine crustaceans (Nakamura and Hatai 1995; Hatai et al. 2000; Diggles 2001; Chukanhom et al. 2003; Sekimoto et al. 2007). *Haliphthoros* species are known to parasitize of a wide range of marine animals (Table 11.1).

Three species of Oomycetes – *Gonimochaete latitubus, Haptoglossa heterospora* and *Myzocytiopsis vermicola* – have been isolated from the littoral marine nematode *Rhabditis marina* (Newell et al. 1977). These nematode parasites normally occur in terrestrial habitats that are rich in organic materials and include high densities of bacteria and nematodes. They produce holocarpic thalli within the host body and many species infect with zoospores, but *Gonimochaete latitubus* and *Haptoglossa heterospora* only produce aplanospores (Glocking and Beakes 2000b). Both *H. heterospora* and *M. vermicola* have been recorded several times from terrestrial species of *Rhabditis*, while *G. latitubus* is the only nematode parasite known from a single marine isolate. *Halioticida* is a recently-erected and monotypic genus, with *H. noduliformis* isolated from white nodules found on the mantle of three species of abalones (Muraosa et al. 2009).

Within the Saprolegniales, three species have been cited as important pathogens: *Leptolegnia baltica, Aphanomyces invadans* and *Saprolegnia parasitica* (Table 11.1). *Leptolegnia baltica* is a poorly known taxon that has been responsible for mass mortality of the copepod *Eurytemora hirundoides* in the Gulf of Bothnia (Bothnia and Herzegovina; Johnson et al. 2002). *Aphanomyces invadans* and *S. parasitica* are known as parasites of fishes in freshwater, estuarine and marine habitats and have been responsible for devastating infections in aquaculture, fish farms and home fish tanks (Willoughby 2003; van West 2006; Cerenius et al. 2009; Robertson et al. 2009).

Aphanomyces invadans has frequently been reported as a parasite of fishes in estuaries. This species causes epizootic ulcerative syndrome, primarily in the economically-important fish *Brevoortia tyrannus*, in low-salinity estuaries (Fraser et al. 1992; Blazer et al. 2002; Vandersea et al. 2006). It also infects more than 30 freshwater fishes worldwide (Lilley et al. 1998). In the United States, since the early 1980s *B. tyrannus* has been reported to have a high prevalence of skin ulcers, often located near the anus and deeply penetrating, with peripheral hemorrhage, extensive necrosis, and tissue loss (Noga 1993; Kiryu et al. 2002).

Saprolegnia parasitica has been cited as an important pathogen of many fish species, commonly causing significant economic losses in the fish-farming industry, mainly of freshwater fishes (Willoughby 2003; Diéguez-Uribeondo et al. 2007; Robertson et al. 2009). Due to overfishing, fish production has become dependent on fish farming for an adequate supply, giving aquaculture a social and economic importance worldwide. Saprolegniosis, the disease caused by *Saprolegnia* species, commonly affects the epidermis of the fishes, but internal infection has also been cited (Gieseker et al. 2006; Robertson et al. 2009). This disease appeared to be an important factor in economic loss in aquaculture mainly after the use of the toxic organic dye malachite green, which was banned due to its carcinogenic effects on both fish and consumers (Almeida et al. 2009).

A few marine ecotypes belonging to the Hyphochytriomycota are important parasites of algae and crustacea (Table 11.1). All species of *Anisolpidium* listed in Table 11.1 are obligate endoparasites of marine algae, particularly of the brown algae *Ectocarpus, Hincksia* and *Sphacelaria* (Dick 2001; Kirk et al. 2008). The genus *Hyphochytrium* is represented by a single marine species, *H. peniliae*, which is parasitic on *Penilia avirostris*, a planktonic cladoceran. *Perkinsus* species are known as parasites of many economically-important molluscs, such as oysters, clams and abalones. The loss of farmed and wild molluscs due to *Perkinsus* infection contributes to a significant economic loss worldwide, and therefore research efforts focus on the underlying molecular mechanisms of infection and disease control (e.g., Park and Choi 2001; Bushek et al. 2004; Joseph et al. 2010). The infection of dinoflagellates with *Parvilucifera* species has been attributed to the control and dissipation of dinoflagellate populations (e.g., Park et al. 2004; Mangot et al. 2011). Recently it has been shown that the presence of *Parvilucifera infectans* stimulates the adaptation of life history strategy in the host *Alexandrium minutum* (Figueroa et al. 2008).

Concluding remarks

Data on zoosporic fungal-like organisms from marine ecosystems have been scattered throughout the literature and have never been extensively reviewed. The information presented in this chapter highlights two significant issues.

First, these organisms have traditionally received little attention from mycologists, except for studies that have primarily focused on parasites of economically-important algae, bivalves, crustaceans and fishes. Several studies, however, have demonstrated that both saprotrophs and parasites impact on the growth and survival of many marine microorganisms, directly or indirectly influencing their lifecycles. Zoosporic fungal-like organisms are involved in many trophic links and play important roles in maintaining the equilibrium of food webs in marine habitats.

Second, molecular data are increasingly indicating that heterotrophic fungal-like organisms are surprisingly diverse and represent an important fraction of the marine nanoplankton. With the advent of molecular techniques, some of the hidden diversity of uncultured microorganisms in marine assemblages is now becoming evident (e.g., recent discoveries of new chromalveolate lineages). In addition, new tools such as fluorescence *in situ* hybridization (FISH) can target these organisms in environmental samples and putatively identify previously unknown clades. Despite these advances, a great part of this diversity will remain hidden from our observations because of methodological limitations (e.g., species that occur at densities below our limit of detection).

Extensive research and sampling of marine fungal-like organisms are much needed in order to extend our knowledge on the diversity of zoosporic organisms in marine ecosystems and to further elucidate the identity of environmental sequences. New potential hosts should be examined in order to reveal the additional diversity of zoosporic fungal-like organisms, articularly in previously unexplored geographical regions and extreme marine habitats. Efforts to increase the number of researchers and to stimulate collaboration between existing groups are essential in order to advance our knowledge and understanding of the roles of zoosporic fungal-like organisms in marine ecosystems.

References

Abrahams D, Brown WD. Evaluation of fungicides for *Haliphthoros milfordensis* and their toxicity to juvenile European lobsters. Aquaculture 1977;12:31–40.

Adl SM, Simpson AGB, Farmer MA, Andersen RA, Anderson OR, Barta JR, Bowser SS, Brugerolle. G, Fensome RA, Fredericq S, James TY, Karpov S, Kugrens P, Krug J, Lane CE, Lewis LA, Lodge J, Lynn DH, Mann DG, Mccourt RM, Mendoza L, Moestrup Ø, Mozley-Standridge SE, Nerad TA, Shearer CA, Smirnov AV. The new higher level classification of Eukaryotes with emphasis on the taxonomy of protists. J Eukaryot Microbiol. 2005;52:399–451.

Aleem AA. *Olpidiopsis feldmanni* sp. nov. champignon marin parasite d'Algues de la famille des Bonnemaisoniacees. Comp Rend Hebd Acad Sci Paris. 1952;235:1250–1252.

Aleem AA. Marine fungi from the West-Coast of Sweden. Arkiv fur Botanik 1955;3:1–33.

Alias SA, Jones EBG. Marine fungi from mangroves of Malaysia. Kuala Lumpur: University of Malaya, 2009.

Almeida A, Cunha A, Gomes NCM, Alves E, Costa L, Faustino MAF. Phage therapy and photodynamic therapy: low environmental impact approaches to inactivate microorganisms in fish farming plants. Mar Drugs. 2009;7:268–313.

Andersson MG. Differentiation and pathogenicity within the *Saprolegniaceae*. Studies on physiology and gene expression patterns in *Saprolegnia parasitica* and *Aphanomyces astaci*. Acta Universitatis Upsaliensis. Sweden: Comprehensive Summaries of Uppsala Dissertations from the Faculty of Science and Technology, 2001.

Andrews JD. History of *Perkinsus marinus* a pathogen of oysters in Chesapeake Bay 1950–1984. J Shellfish Res. 1996;15:13–16.

Andrews JH. The pathology of marine algae. Biol Rev. 1976;51:211–253.

Armstrong DA, Buchanan DV, Caldwell RS. A mycosis caused by *Lagenidium* sp. in laboratory-reared larvae of the Dungeness crab *Cancer magister* and possible chemical treatments. J Inv Pathol. 1976;28:329–336.

Atkins D. On a fungus allied to Saprolegniaceae found in the pea-crab, *Pinnotheres*. J Mar Biol Ass UK. 1929;16:203–219.

Atkins D. A marine fungus *Plectospira dubia* n. sp. (Saprolegniaceae) infecting crustaceans eggs and small Crustacea. J Mar Biol Assoc UK. 1954a;33:721–732.

Atkins D. Further notes on a marine member of the Saprolegniaceae *Leptolegnia marina* n. sp. infecting certain invertebrates. J Mar Biol Assoc UK. 1954b;33:613–625.

Atkins D. *Pythium thalassium* sp. nov. infecting the egg-mass of the pea-crab *Pinnotheres pisum*. Trans Br Mycol Soc. 1955;38:31–46.

Azevedo C. Fine structure of *Perkinsus atlanticus* n. sp. (Apicomplexa Perkinsea) parasite of the clam *Ruditapes decussatus* from Portugal. J Parasitol. 1989;75:627–635.

Bala K, Robideau GP, Lévesque CA, de Cock AWAM, Abad GZ, Lodhi AM, Shahzad S, Ghaffar A, Coffey MD. *Phytopythium sindhum* Lodhi Shahzad and Lévesque sp. nov. Persoonia 2010;24:136–137.

Ban S, Burns C, Castel J, Christou E, Escribano R, Fonda Umani S, Gasparini S, Guerrero Ruiz F, Hoffmeyer H, Ianora A, Kang H, Laabir M, Lacoste A, Miralto A, Poulet S, Ning X, Rodriguez V, Runge J, Shi J, Starr M, Uye S, Wang Y. The paradox of diatom-copepod interactions. Mar Ecol Prog Ser. 1997;157:287–293.

Bärlocher F. Freshwater fungal communities. In: Dighton J, White JF, Oudemans P, eds. The fungal community: its organization and role in the ecosystem. Boca Raton: CRC Press, 2005:39–59.

Barr DJS. *Hyphochytrium catenoides*: a morphological and physiological study of North American isolates. Mycologia 1970;62:492–503.

Beakes GW. Oomycete phylogeny: ultrastructural perspectives. In: Rayner ADM, Brasier CM, Moore D, eds. Evolutionary biology of the fungi. Cambridge: Cambridge University Press, 1987: 405–421.

Beakes GW, Glockling SL. An ultrastructural analysis of organelle arrangement during gun (infection) cell differentiation in the nematode parasite *Haptoglossa dickii*. Mycol Res. 1998;104: 1258–1269.

Beakes GW, Sekimoto S. The evolutionary phylogeny of Oomycetes-Insights gained from studies of holocarpic parasites of algae and invertebrates. In: Lamour K, Kamoun S, eds. Oomycete genetics and genomics: diversity, interactions, and research tools. London: Wiley-Blackwell, 2009:1–24.

Beakes GW, Thines M. Phylum Oomycota. In: Margulis L, eds. Handbook of protoctista. Sudbury, Massachusetts: Jones and Bartlett Publishers, Inc, 2012 [in press].

Beakes GW, Glockling SL, James T. The diversity of oomycete pathogens of nematodes and its implications to our understanding of oomycete phylogeny. In: Meyer W, Pearce C, eds. Proc Eigth Intern Mycol Congress, Cairns, Australian Mycology Association. 2006:7–12.

Beakes GW, Glockling SL, Sekimoto S. The evolutionary phylogeny of oomycete 'fungi'. Protoplasma 2012;249:3–19.

Bian BZ, Egusa S. *Atkinsiella hamanaensis* sp. nov. isolated from cultivated ova of the mangrove crab *Scylla serrata* (Forsskal). J Fish Dis. 1980;3:373–385.

Bian Z, Hatai K, Po GL, Egusa S. Studies on the fungal diseases in Crustaceans. I. *Lagenidium sycttae* sp. nov. isolated from cultivated ova and larvae of the mangrove crab (*Scylla serrata*). Trans Mycol Soc Jpn. 1979;20:115–124.

Blazer VS, Lilley JH, Schill WB, Kiryu Y, Densmore CL, Panyawachira V, Chinabut S. *Aphanomyces invadans* in Atlantic menhaden along the east coast of the United States. J Aquat Anim Health. 2002;14:1–10.

Buck JC, Truong L, Blaustein AR. Predation by zooplankton on *Batrachochytrium dendrobatidis*: biological control of the deadly amphibian chytrid fungus? Biodivers Conserv. 2011;20: 3549–3553.

Burki F, Shalchian-Tabrizi K, Minge M, Skjaeveland A, Nikolaev SI, Jakobsen KS, Pawlowski J. Phylogenomics reshuffles the eukaryotic supergroups. PLoS ONE 2007;2:e790.

Burki F, Shalchian-Tabrizi K, Pawlowski J. Phylogenomics reveals a new 'megagroup' including most photosynthetic eukaryotes. Biol Lett. 2008;4:366–369.

Burreson EM, Ragone Calvo LM. Epizootiology of *Perkinsus marinus* disease of oysters in Chesapeake Bay with emphasis on data since 1985. J Shellfish Res. 1996;15:17–34.

Burreson EM, Reece KS, Dungan CF. Molecular morphological and experimental evidence support the synonymy of *Perkinsus chesapeaki* and *Perkinsus andrewsi*. J Eukaryot Microbiol. 2005;52:258–270.

Bushek D, Richardson D, Bobo MY, Coan LD. Quarantine of oyster shell cultch reduces the abundance of *Perkinsus marinus*. J Shellfish Res. 2004;23:369–373.

Canter HM, Lund JWG. Studies on plankton parasites. I. Fluctuations in the numbers of *Asterionella formosa* Hass in relation to fungal epidemics. New Phytol. 1948;47:238–261.

Carrias JF, Amblard C, Bourdier G. Protistan bacterivory in an oligomesotrophic lake: importance of attached ciliates and flagellates. Microb Ecol. 1996;31:249–268.

Casas SM, Grau A, Reece KS, Apakupakul K, Azevedo C, Villalba A. *Perkinsus mediterraneus* n. sp. a protistan parasite of the European flat oyster *Ostrea edulis* (L.) from the coast of Balearic Islands Mediterranean Sea. DAO. 2004;58:231–244.

Cavalier-Smith T. Kingdom Protozoa and its 18 phyla. Microbiol Rev. 1993;57:953–994.

Cavalier-Smith T. Principles of protein and lipid targeting in secondary symbiogenesis: euglenoid dinoflagellate and sporozoan plastid origins and the eukaryote family tree. J Eukaryot Microbiol. 1999;46:347–366.

Cavalier-Smith T. The phagotrophic origin of eukaryotes and phylogenetic classification of Protozoa. IJSEM. 2002;52:297–354.

Cavalier-Smith T. Megaphylogeny cell body plans adaptive zones: causes and timing of eukaryote basal radiations. J Eukaryot Microbiol. 2009;56:26–33.

Cavalier-Smith T, Chao EEY. Protalveolate phylogeny and systematics and the origins of Sporozoa and dinoflagellates (phylum Myzozoa nom. nov.). Eur J Protistol. 2004;40:185–212.

Cavalier-Smith T, Chao EEY. Phylogeny and megasystematics of phagotrophic heterokonts (Kingdom Chromista). J Mol Evol. 2006;62:388–420.

Cerenius L, Gunnar Anderson G, Soderhall K. *Aphanomyces astaci* and crustaceans. In: Lamour K, Kamoun S, eds. Oomycete genetics and genomics: diversity interactions and research tools. London: Wiley-Blackwell, 2009:425–433.

Chukanhom K, Borisutpeth P, Khoa LV, Hatai K. *Haliphthoros milfordensis* isolated from black tiger prawn larvae (*Penaeus monodom*) in Vietnam. Mycoscience 2003;44:23–127.

Cook KL, Hudspeth DSS, Hudspeth MES. A *cox*2 phylogeny of representative marine peronosporomycetes (Oomycetes). Nova Hedw. 2001;122:231–243.

Cooke DEL, Drenth A, Duncan JM, Wayek G, Brasier CM. A molecular phylogeny of *Phytophthora* and related Oomycetes. Fungal Genet Biol. 2000;30:17–32.

Dangeard PA. Recherches sur quelques algues nouvelles ou peu connues. Le Botaniste 1912; 12:1–26.

Das S, Lyla PS, Khan SA. Marine microbial diversity and ecology: importance and future perspectives. Curr Sci. 2006;90:1325–1335.

Daugherty J, Evans TM, Skillom T, Watson LE, Money NP. Evolution of spore release mechanisms in the Saprolegniaceae (Oomycetes): evidence from a phylogenetic analysis of internal transcribed spacer sequences. Fungal Genet Biol. 1998;24:354–363.

de Cock AWAM. Marine Pythiaceae from decaying seaweeds in the Netherlands. Mycotaxon 1986;25:101–110.

de Cock AWAW, Mendoza L, Padhye AA, Ajello L, Kauffman L. *Pythium insidiosum* sp. nov. the etiological agent of pythiosis. J Clin Microbiol. 1987;25:344–349.

Dick MW. Phylum Oomycota. In: Margulis L, Corliss JO, Melkonian M, Chapman D, eds. Handbook of Protoctista. Boston: Jones and Bartlett, 1990:661–685.

Dick MW. Sexual reproduction in the Peronosporomycetes (chromistan fungi). Can J Bot. 1995;73:5712–5724.
Dick MW. The Myzocytiopsidaceae. Mycol Res. 1997;101:878–882.
Dick MW. The species and systematic position of *Crypticola* in the Peronosporomycetes and new names for the genus *Halocrusticida* and species therein. Mycol Res. 1998;102:1062–1066.
Dick MW. Straminipilous Fungi. Systematics of the Peronosporomycetes including accounts of the marine straminipilous protists the plasmodiophorids and similar organisms. Dordrecht: Kluwer Academic Publishers, 2001.
Dick MW, Vick MC, Gibbings JG, Hedderson TA, Lopez Lastra CC. 18S rDNA for species of *Leptolegnia* and other Peronospormycetes: justification of the subclass taxa Saprolegniomycetidae and Peronosporomycetidae and division of the Saprolegniaceae *sensu lato* into the Leptolegniaceae and Saprolegniaceae. Mycol Res. 1999;103:1119–1125.
Diéguez-Uribeondo J, Temiño C, Mùzquiz. JL. The crayfish plague fungus *(Aphanomyces astaci)* in Spain. Bull Fr Pêche Piscic. 1997;347:753–763.
Diéguez-Uribeondo J, Fregeneda-Grandes JM, Cerenius L, Perez-Iniesta M, Aller-Gancedo JM, Tellerıa MT, Soderhall K, Martın MP. Re-evaluation of the enigmatic species complex *Saprolegnia diclina-Saprolegnia parasitica* based on morphological physiological and molecular data. Fungal Genet Biol. 2007;44:585–601.
Diggles BK. A mycosis of juvenile spiny lobster *Jasus edwardsii* (Hutton 1875) caused by *Haliphthoros* sp. and possible methods of chemical control. J Fish Dis. 2001;24:99–110.
Ding H, Ma J. Simultaneous infection by red rot and chytrid diseases in *Porphyra yezoensis* Ueda. J Appl Phycol. 2005;17:51–56.
Drechsler C. Three fungi destructive to free-living terricolous nematodes. J Wash Acad Sci. 1940;30:240–254.
Dungan CF, Reece KS. *In vitro* propagation of two *Perkinsus* spp. Parasites from Japanese Manila clams *Venerupis philippinarum* and description of *Perkinsus honshuensis* n. sp. J Eukaryotic Microbiol. 2006;53:316–326.
Ellis JT, Morrison DA, Jeffries AC. The phylum Apicomplexa: an update on the molecular phylogeny. In: Coombs G, Vickerman K, Sleigh M, Warren A, eds. Evolutionary relationships among Protozoa. Boston: Kluwer Academic Publishers, 1998:255–274.
Feldmann J, Feldmann G. Observations sur quelques Phycomycetes marins nouveaux ou peu connus. Rev Mycol. 1955;20:231–251.
Fell JW, Master IM. Phycomycetes (*Phytophthora* spp. nov. and *Pythium* sp. nov.) associated with degrading mangrove (*Rhizophora mangle*) leaves. Can J Bot. 1975;53:2908–2922.
Fensome RA, Taylor FJR, Norris G, Sarjeant WAS, Wharton DI, Williams GL. A classification of living and fossil dinoflagellates. Micropaleontology Special Publication 7. Pennsylvania: Sheridan Press, 1993.
Figueroa RI, Garces E, Massana R, Camp J. Description host-specificity and strain selectivity of the dinoflagellate parasite *Parvilucifera sinerae* sp nov (Perkinsozoa). Protist 2008;159:563–578.
Fisher WS. Eggs of *Palaemon macrodactylus*: III. Infection by the fungus *Lagenidium callinectes*. Biol Bull. 1983;164:214–226.
Fisher WS, Nilson EH, Shleser RA. Effect of the *Haliphthoros milfordensis* on the juvenile stage of the American lobster *Homarus americanus*. J Invert Pathol. 1975;26:41–45.
Förster H, Cummings MP, Coffey MD. Phylogenetic relationships of *Phytophthora* species based on ribosomal ITS I DNA sequence analysis with emphasis on waterhouse groups V and VI. Mycol Res. 2000;104:1055–1061.
Fraser GC, Callinan RB, Calder LM. *Aphanomyces* species associated with red spot disease: an ulcerative disease of estuarine fish from eastern Australia. J Fish Dis. 1992;15:173–181.
Frezzi MJ. Especies de *Pythium* fitopatógenas identificadas en la República Argentina. Rev Invest Agríc. 1956;10:1–241.

Fujita Y. Disease of cultivated *Porphyra* in Japan. In: Akatsuka I, ed. Introduction to Applied Phycology. The Hague: SPB Academic Publishing, 1990:177–190.

Fuller MS, Fowles BE, McLaughlin DJ. Isolation and pure culture study of marine Phycomycetes. Mycologia 1964;56:745–756.

Fuller MS, Lewis B, Cook P. Occurrence of *Pythium* sp. on the marine alga *Porphyra*. Mycologia 1966;58:313–318.

Gachon CMM, Strittmatter M, Müller DG, Kleinteich J, Küpper FC. Detection of differential host susceptibility to the marine oomycete pathogen *Eurychasma dicksonii* by Real-Time PCR: not all algae are equal. Appl Environ Microbiol. 2009;75:322–328.

Gachon CMM, Sime-Ngando T, Strittmatter M, Chambouvet A, Kim GH. Algal diseases: spotlight on a black box. Trends Plant Sci. 2010;15:633–640.

Gaulin E, Jacquet C, Bottin A, Dumas B. Root rot disease of legumes caused by *Aphanomyces euteiches*. Molec Plant Pathol. 2007;8:539–548.

Gerrettson-Cornell L, Simpson J. Three new marine *Phytophthora* species from New South Wales. Mycotaxon 1984;19:453–470.

Gieseker CM, Serfling SG, Reimschuessel R. Formalin treatment to reduce mortality associated with *Saprolegnia parasitica* in rainbow trout *Oncorhynchus mykiss*. Aquaculture 2006;253: 120–129.

Gisselson L, Carlsson P, Granéli E. *Dinophysis* blooms in the deep euphotic zone of the Baltic Sea: do they grow in the dark? Harmful Algae 2002;1:401–418.

Gleason FH, Letcher PM, McGee PA. Freeze tolerance of soil chytrids from temperate climates in Australia. Mycol Res. 2008a;112:976–982.

Gleason FH, Letcher PM, Evershed N, McGee PA. Recovery of growth of *Hyphochytrium catenoides* after exposure to environmental stress. J Eukaryot Microbiol. 2008b;55:351–354.

Gleason FH, Schmith SK, Marano AV. Can zoosporic true fungi grow or survive in extreme environments? Extremophiles 2010;14:417–425.

Glockling SL, Beakes GW. A review of the biology and infection strategies of biflagellate zoosporic parasites of nematodes. Fungal Divers. 2000a;4:1–20.

Glockling SL, Beakes GW. The dimorphic infection cells of *Haptoglossa heteromorpha* illustrate the developmental plasticity of infection structures in a nematode parasite. Can J Bot. 2000b;78:1095–1107.

Glockling SL, Beakes GW. An ultrastructural study of sporidium formation during infection of rhabditid nematodes by large gun cells of *Haptoglossa heteromorpha*. J Invert Path. 2001a;76: 208–215.

Glockling SL, Beakes GW. Two species of *Haptoglossa* from N. E. England *H. northumbrica* and *H. polymorpha*. J Linn Soc Bot. 2001b;136:329–336.

Glockling SL, Beakes GW. Structural and developmental studies of *Chlamydomyzium oviparasiticum* from *Rhabditis* nematodes in culture. Mycol Res. 2006;110:1119–1126.

Glockling SL, Serpell LC. A new species of aplanosporic *Haptoglossa, H. beakesii*, with vesiculate spore release. Botany 2010;88:93–101.

Göker M, Voglmayr H, Riethmüller A, Oberwinkler F. How do obligate parasites evolve? A multigene phylogenetic analysis of downy mildews. Fungal Genet Biol. 2007;44:105–122.

Gopalan UK, Meenakshikunjamma PP, Purushan KS. Fungal infection in the tiger prawn (*Penaeus monodon*) and in other crustaceans from the cochin backwaters. Mahasagar Bull Nat Inst Ocean. 1980;13:359–365.

Gotelli D. *Lagenisma coscinodisci*, a parasite of the marine diatom *Coscinodiscus*, occurring in the Puget Sound, Washington. Mycologia 1971;63:171–174.

Grahame ES. The occurrence of *Lagenisma coscinodisci* in *Palmeria hardmaniana* from Kingston Harbour, Jamaica. Eur J Phycol. 1976;11:57–61.

Gulis V, Kuehn K, Suberkropp K. The role of fungi in carbon and nitrogen cycles in freshwater ecosystems. In: Gadd GM, ed. Fungi in biogeochemical cycles. Cambridge: Cambridge University Press, 2006:404–435.

Hackett JD, Yoon HS, Li S, Reyes-Prieto A, Rummele SE, Bhattacharya D. Phylogenomic analysis supports the monoplyly of Cryptophytes and Haptophytes and the association of Rhizaria with Chromalveolates. Mol Biol Evol. 2007;24:1702–1713.

Hakariya MD, Hirose D, Tokumasu S. A molecular phylogeny of *Haptoglossa* species, terrestrial peronosporomycetes (Oomycetes) endoparasitic on nematodes. Mycoscience 2007;48:169–175.

Hampl V, Hug L, Leigh JW, Dacks JB, Franz LB, Simpson AGB, Roger AJ. Phylogenomic analyses support the monophyly of Excavata and resolve relationships among eukaryotic "supergroups". Proc Natl Acad Sci USA. 2009;106:3859–3864.

Hanic LA, Sekimoto S, Bates SS. Oomycete and chytrids infections of the marine diatom *Pseudonitzschia pungens* (Bacillariophyceae) from Prince Edward island, Canada. Botany 2009;87:1096–1105.

Harper JT, Waanders E, Keeling PJ. On the monophyly of chromalveolates using a sixprotein phylogeny of eukaryotes. Int J Syst Evol Microbiol. 2005;55:487–496.

Hatai K. On the fungus *Haliphthoros milfordensis* isolated from temporarily held abalone *Haliotis sieboldii*. Fish Pathol. 1982;17:199–204.

Hatai K, Hoshiai GI. Pathogenicity of *Saprolegnia parasitica* Coker. In: Mueller GJ, ed. Salmon Saprolegniasis. Oregon: Department of Energy, Bonneville Power Administration, 1994:87–98.

Hatai K, Lawhavinit O. *Lagenidium myophilum* sp. nov., a new parasite on adult northern shrimp (*Pandalus borealis* Kryer). Trans Mycol Soc Jpn. 1988;29:175–184.

Hatai K, Bian BZ, Batigados MCL, Egusa S. Studies on the fungal diseases in Crustaceans. II. *Haliphthoros philippinensis* sp. nov. isolated from cultivated larvae of the jumbo tiger prawn (*Penaeus monodon*). Trans Mycol Soc Jpn. 1980;21:47–55.

Hatai K, Rhoobunjongde W, Wada S. *Haliphthoros milfordensis* isolated from gills of juvenile kuruna prawn (*Penaeus japonicus*) with black gill disease. Trans Mycol Soc Jpn. 1992;33:185–192.

Hatai K, Roza D, Nakayama T. Identification of lower fungi isolated from larvae of mangrove crab, *Scylla serrata*, in Indonesia. Mycoscience 2000;41:565–572.

Hausner G, Belkhiri A, Klassen GR. Phylogenetic analysis of the small subunit ribosomal RNA gene of the hyphochytrid *Rhizidiomyces apophysatus*. Can J Bot. 2000;78:124–128.

Ho HH, Jong SC. *Halophytophthora* gen. nov., a new member of the family Pythiaceae. Mycotaxon 1990;36:377–382.

Ho HH, Chang HS, Hsieh SY. *Halophytophthora kandeliae*, a new marine fungus from Taiwan. Mycologia 1991;83:419–424.

Ho HH, Nakagiri A, Newell SY. A new species of *Halophytophthora* from Atlantic and Pacific subtropical islands. Mycologia 1992;84:548–554.

Ho HH, Chang HS, Huang SH. *Halophytophthora elongata*, a new marine species from Taiwan. Mycotaxon 2003;85:417–422.

Höhnk W. On three pyhtiaceous Oomycetes. Beih Bot Centralbl. 1936;55:89–99.

Höhnk W, Vallin S. Epidemisches Absterben von Eurytemora im Bottnischen Meerbusen, verursacht durch *Leptolegnia baltica* nov. spec. Veröff Inst Meeresf Bremerhaven. 1953;2:215–223.

Holfeld H. Fungal infections of the phytoplankton: seasonality, minimal host density, and specificity in a mesotrophic lake. New Phytol. 1998;138:507–517.

Hoppenrath M, Leander BS. Molecular phylogeny of *Parvilucifera prorocentri* (Alveolata, Myzozoa): Insights into perkinsid character evolution. J Eukaryot Microbiol. 2009;56:251–256.

Hudspeth DSS, Nadler SA, Hudspeth MES. A Cox II molecular phylogeny of the Peronosporomycetes. Mycologia 2000;99:674–684.

Hulvey J, Telle S, Nigrelli L, Lamour K, Thines M. Salisapiliaceae – a new family of Oomycetes from marsh grass litter of Southeastern North America. Persoonia 2010;25:109–116.

Hulvey JP, Padgett DE, Bailey JC. Species boundaries within *Saprolegnia* (Saprolegniales, Oomycota) based on morphological and DNA sequence data. Mycologia 2007;99:421–429.

Hyde KD, Jones EBG, Leãno E, Pointing SB, Poonyth AD, Vrijmoed LLP. Role of marine fungi in marine ecosystems. Biodivers Conserv. 1998;7:1147–1161.

Inaba S, Hariyama S. The phylogenetic studies on the genus *Cornumyces* (Oomycetes) based on the nucleotide sequences of the nuclear large subunit ribosomal RNA and the mitochondrially-encoded cox2 genes. Proceedings of the 8th International Mycological Congress Handbook and Abstracts, Cairns IMC, 2006.

Inaba S, Tokumasu S. The genus *Brevilegnia* (Saprolegniales, Oomycetes) in Japan. Mycoscience 2002;43:59–66.

James TY, Berbee ML. No jacket required – new fungal lineage defies dress code. Bioassays 2012;34:94–102.

Jobard M, Rasconi S, Sime-Ngando T. Diversity and functions of microscopic fungi: a missing component in pelagic food webs. Aquat Sci. 2010;72:255–268.

Johnson Jr TW. Resting spore development in the marine Phycomycete *Anisolpidium ectocarpii*. Am J Bot. 1957;44:875–878.

Johnson Jr TW. Fungi in planktonic *Synedra* from brackish waters. Mycologia 1966;58:373–382.

Johnson Jr TW, Seymour RL, Padgett DE. Biology and systematics of the Saprolegniaceae, 2002. (Accessed April 11, 2012 at: http://dl.uncw.edu/digilib/biology/fungi/taxonomy%20and%20systematics/padgett%20book/SYSTEMATIC/CHAPTER_39/LEPTOLEGNIA.pdf

Johnson TW. A fungus parasite in ova of the barnacle *Chthamalus fragilis denticulata*. Biol Bull. 1958;114:205–214.

Johnson TW Jr, Sparrow FK Jr. Fungi in oceans and estuaries. Weinheim: Cramer, 1961.

Jones EBG. Fifty years of marine mycology. Fungal Divers. 2011;50:73–112.

Jones EBG, Alias SA. Biodiversity of mangrove fungi. In: Hyde KD, ed. Biodiversity of tropical microfungi. Hong Kong: Hong Kong University Press, 1997:71–92.

Jones EBG, Palantanapak A, Chatmala I, Sakayaroj J, Phongpaicit S, Choeyklin R. Thai marine fungal diversity. Songklanakarin J Sci Technol. 2006;28:687–708.

Jones MDM, Richards TA, Hawksworth DL, Bass D. Validation and justification of the phylum name *Cryptomycota* phyl. nov. IMA Fungus. 2011;2:173–175.

Joseph SJ, Fernandez-Robledo JA, Gardner MJ, El-Sayed NM, Kuo CH, Schott EJ, Wang H, Kissinger JC, Vasta GR. The alveolate *Perkinsus marinus*: Biological insights from EST gene discovery. BMC Genomics. 2010;11:228.

Kagami M, van Donk E, de Bruin A, Rijkeboer M, Ibelings BW. *Daphnia* can protect diatoms from fungal parasitism. Limnol Ocean. 2004;49:680–685.

Kagami M, de Bruin A, Ibelings BW, van Donk E. Parasitic chytrids: their effects on phytoplankton communities and food-web dynamics. Hydrobiol. 2007;578:113–129.

Karling JS. The life history of *Anisolpidium ectocarpii* gen. nov. *et* sp. nov., and a synopsis and classification of other fungi with anteriorly uniflagellate zoospores. Am J Bot. 1943;30:637–648.

Karling JS. Predominantly holocarpic and eucarpic simple biflagellate Phycomycetes, Cramer: Vaduz, 1981.

Katz LA. Evolution of nuclear dualism in ciliates: a reanalysis in light of recent molecular data. Int J Syst Evol Microbiol. 2001;51:1587–1592.

Kazama FY. *Pythium* 'red rot disease' of *Porphyra*. Experientia 1979;35:443–444.

Kazama FY, Fuller MS. Ultrastructure of *Porphyra perforata* infected with *Pythium marinum*, a marine fungus. Can J Bot. 1970;48:2103–2107.

Kirk PM, Cannon PF, Minter DW, Stalpers JA. Dictionary of the fungi. Wallingford: CABI Bioscience Publishing, 2008.
Kiryu Y, Shields JD, Vogelbein WK, Zwerner DE, Kator H, Blazer VS. Induction of skin ulcers in Atlantic menhaden by injection and aqueous exposure to the zoospores of *Aphanomyces invadans*. J Aquat Animal Health. 2002;14:11–24.
Kis-Papo T. Marine fungal communities. In: Dighton J, White JF, Oudemans P, eds. The fungal community Its organization and role in the ecosystem. Boca Raton: CRC Press, 2005:61–92.
Kitancharoen N, Hatai K. A marine oomycete *Atkinsiella panulirata* sp. nov. from philozoma of spiny lobster, *Panulirus japonicus*. Mycoscience 1995;38:97–104.
Kitancharoen N, Nakamura K, Wada S, Hatai K. *Atkinsiella awabi* sp. nov. isolated from stocked abalone, *Haliotis sieboldii*. Mycoscience 1994;35:265–270.
Kohlmeyer J, Volkmann-Kohlmeyer B, Newell SY. Marine and estuarine mycelia Eumycota and Oomycota. In: Müeller GM, Bills GF, Foster MS, eds. Biodiversity of fungi: Inventory and monitoring methods. San Diego: Elsevier Academic Press, 2004:533–545.
Küpper FC. New record of *Anisolpidium rosenvingei* (H.E. Petersen) Karling in Ireland. Ir Nat J. 2001;26:470–471.
Küpper FC, Müller DG. Massive occurrence of the heterokont and fungal parasites *Anisolpidium*, *Eurychasma* and *Chytridium* in *Pylaiella littoralis* (Ectocarpales, Phaeophyceae). Nova Hedw. 1999;69:381–389.
Küpper FC, Maier I, Müller DG, Loiseaux-de Goer S, Guillou L. Phylogenetic affinities of two eukaryotic pathogens of marine microalgae, *Eurychasma dicksonii* (Wright) Magnus and *Chytridium polysiphoniae* Cohn. Cryptogamie Algol. 2006;27:165–184.
Kurtzman CP, Fell, JW. Yeast. In: Müeller GM, Bills GF, Foster MS, eds. Biodiversity of fungi: inventory and monitoring methods. San Diego: Elsevier Academic Press, 2004:337–342.
Kuvardina ON, Leander BS, Aleshin VV, Myl'nikov AP, Keeling PJ, Simdyanov TG. The phylogeny of colpodellids (Eukaryota, Alveolata) using small subunit rRNA genes suggests they are the free-living ancestors of apicomplexans. J Eukaryot Microbiol. 2002;49:498–504.
Lafferty KD, Dobson AP, Kuris AM. Parasites dominate food web links. PNAS. 2006;103: 11211–11216.
Lara E, Belbahri L. SSU rRNA reveals major trends in oomycete evolution. Fungal Divers. 2011;49:93–100.
Lara E, Moreira D, Lopez-Garcia P. Environmental clade LKM11 and *Rozella* form the deepest branching clade of Fungi. Protist 2010;161:116–121.
Leander BS, Hoppenrath M. Ultrastructure of a novel tube-forming, intracellular parasite of dinoflagellates: *Parvilucifera prorocentri* sp. nov. (Alveolata, Myzozoa). Eur J Protistol. 2008;44:55–70.
Leander BS, Keeling PJ. Morphostasis in alveolate evolution. Trends Ecol Evol. 2003;18:395–402.
Leander BS, Keeling PJ. Early evolutionary history of dinoflagellates and apicomplexans (Alveolata) inferred from HSP90 and actin phylogenies. J Phycol. 2004;40:341–250.
Leaño EM. Straminipilous organisms from fallen mangrove leaves from Panay Island, Philippines. Fungal Divers. 2001;6:75–81.
Leaño EM. *Haliphthoros* spp. from spawned eggs of captive mud crab, *Scylla serrata*, broodstocks. Fungal Divers. 2002;9:93–103.
Leaño EM, Pang KL. Effect of copper(II), lead(II), and zinc(II) on growth and sporulation of *Halophytophthora* from Taiwan mangroves. Water Air Soil Pollut. 2010;213:85–93.
Leaño EM, Vrijmoed LLP, Jones EBG. Physiological studies on *Halophytophthora vesicula* (straminipilous fungi) isolated from fallen mangrove leaves from Mai Po, Hong Kong. Bot Mar. 1998a;41:411–419.
Leaño EM, Vrijmoed LLP, Jones EBG. Zoospore chemotaxis of two mangrove strains of *Halohytophthora vesicula* from Mai Po, Hong Kong. Mycologia 1998b;90:1001–1008.

Leaño EM, Jones EBG, Vrijmoed LLP. Why are *Halophytophthora* species well adapted to mangrove habitats? Fungal Divers. 2000;5:131–151.

Leclerc MC, Guillot J, Deville M. Taxonomic and phylogenetic analysis of saprolegniaceae (Oomycetes) inferred from LSU rDNA and ITS sequence comparisons. Antonie Leeuwenhoek 2000;77:369–377.

Lefèvre E, Bardot C, Noel C, Carrias J-F, Viscogliosi E, Amblard C, Sime-Ngando T. Unveiling fungal zooflagellates as members of freshwater picoeukaryotes: evidence from a molecular diversity study in a deep meromictic lake. Environ Microbiol. 2007;9:61–71.

Lefèvre E, Roussel B, Amblard C, Sime-Ngando T. The molecular diversity of freshwater picoeukaryotes reveals high occurrence of putative parasitoids in the plankton. PloS One 2008;3:e2324.

Lester RJG, Davis GHG. A new *Perkinsus* species (Apicomplexa, Perkinsea) from abalone *Haliotis ruber*. J Invert Pathol. 1981;37:181–187.

Letcher PM, McGee PA, Powell MJ. Distribution and diversity of zoosporic fungi from soils of four vegetation types in New South Wales, Australia. Can J Bot. 2004;82:1490–1500.

Lévesque CA, de Cock AW. Molecular plylogeny and taxonomy of the genus *Pythium*. Mycol Res. 2004;108:1363–1383.

Levine ND. *Perkinsus* gen. n. and other new taxa in the protozoan phylum Apicomplexa. J Parasitol. 1978;64:54.

Li W, Zhang T, Tang X, Wang B. Oomycetes and fungi: important parasites on marine algae. Acta Oceanol Sin. 2010;29:74–81.

Lilley JH, Callinan RG, Chinabut S, Kanchanakhan S, Macrae IH, Phillips MJ. Epizootic ucerative syndrome (EUS) technical handbook. Bangkok, Thailand: Aquatic Animal Health Research Institute, 1998.

Longcore JE. Zoosporic fungi from Australia and New Zealand tree-canopy detritus. Austr J Bot. 2005;53:259–272.

López SE, MacCarthy S. Presencia de "ficomicetes" parásitos en hongos y algas en la Argentina. Darwiniana 1985;26:61–70.

Ma J, Lin Q, Jian M. Preliminary study on the *Olpidiopsis* disease of *Porphyra yezoensis*. J Fish China. 2007;31:860–864.

Mackin JG, Ray SM. The taxonomic relationships of *Dermocystidium marinum* Mackin, Owen and Collier. J Invert Path. 1966;8:544–545.

Mangot J-F, Debroas D, Domaizon I. Perkinsozoa, a well-known marine protozoan flagellate parasite group, newly identified in lacustrine systems: a review. Hydrobiol. 2011;659:37–48.

Martin RW, Miller CE. Ultrastructure of zoosporogenesis in the endoparasite *Olpidiopsis varians*. Mycologia 1986a;78:230–241.

Martin RW, Miller CE. Ultrastructure of sexual reproduction in *Olpidiopsis varians*. Mycologia 1986b;78:359–370.

Massana R, Guillou L, Diex B, Pedros-Alio C. Unveiling the organisms behind novel eukaryotic ribosomal DNA sequences from the ocean. Appl Environ Microbiol. 2002;68:4554–4558.

Massana R, Castresana J, Balagué V, Guillou L, Romari K, Groisillier A, Valentine K, Pedró-Alió C. Phylogenetic and ecological analysis of novel marine stramenopiles. Appl Environ Microbiol. 2004;70:3528–3534.

Massana R, Terrado R, Forn I, Lovejoy C, Pedró-Alió C. Distribution and abundance of uncultured heterotrophic flagellates in the world oceans. Environ Microbiol. 2006;8:1515–1522.

McLaughlin SM, Tall BD, Shaheen A, Elsayed E, Faisal M. Zoosporulation of a new *Perkinsus* species isolated from the gills of the softshell clam *Mya arenaria*. Parasite 2000;7:115–122.

Mendoza L, Kaufman L, Standard P. Antigenic relationship between the animal and human pathogen *Pythium insidiosum* and nonpathogenic *Pythium* species. J Clin Microbiol. 1987;25:2159–2162.

Montagnes DJS, Chambouvet A, Guillou L, Fenton A. Responsibility of microzooplankton and parasite pressure for the demise of toxic dinoflagellate blooms. Aquat Microb Ecol. 2008;53:211–225.

Moon-van der Staay SY, De Wachter R, Vaulot D. Oceanic 18S rDNA sequences from picoplankton reveal unsuspected eukaryotic diversity. Nature 2001;409:607–610.

Moore RB, Obornik M, Janouskovec J, Chrudimsky T, Vancova M, Green DH, Wright SW, Davies NW, Bolch CJS, Heimann K, Slapeta J, Hoegh-Guldberg O, Logsdon JM, Carter DA. A photosynthetic alveolate closely related to apicomplexan parasites. Nature 2008;451:959–963.

Moralejo E, Clemente A, Descals E, Belbahri L, Calmin G, Lefort F, Spies CFJ, McLeod A. *Pythium recalcitrans* sp. nov. revealed by multigene phylogenetic analysis. Mycologia 2008;100:310–319.

Moss JA, Xiao J, Dungan CF, Reece KS. Description of *Perkinsus beihaiensis* n. sp., a new *Perkinsus* sp. parasite in oysters of southern China. J Eukaryot Microbiol. 2008;55:117–130.

Müller DG, Küpper FC, Küpper H. Infection experiments reveal broad host ranges of *Eurychasma dicksonii* (Oomycota) and *Chytridium polysiphoniae* (Chytridiomycota), two eukaryotic parasites in marine brown algae (Phaeophyceae). Phycol Res. 1999;47:217–223.

Muraosa Y, Lawhavinit O, Hatai K. *Lagenidium thermophilum* isolated from eggs and larvae of black tiger shrimp *Penaeus monodon* in Thailand. Fish Pathol. 2006;41:35–40.

Muraosa Y, Morimoto K, Sano A, Nishimura K, Hatai K. A new peronosporomycete, *Halioticida noduliformans* gen. et sp. nov., isolated from white nodules in the abalone *Haliotis* spp. from Japan. Mycoscience 2009;50:106–115.

Murray AG, Peeler EJ. A framework for understanding the potential for emerging diseases in aquaculture. Prevent Vet Med. 2005;67:223–235.

Nakagiri A. Growth and reproduction of *Halophytophthora* species. Trans Mycol Soc Jpn. 1993;34:87–99.

Nakagiri A. Ecology and diversity of *Halophytophthora* species. In: Hyde KD, Ho WH, Pointing SB, eds. Aquatic mycology across the millennium. Fungal Divers. 2000;5:153–164.

Nakagiri A. Diversity and phylogeny of *Halophytophthora* (Oomycetes). Abstracts 7th International Mycological Congress, Oslo, 2002:19.

Nakagiri A, Tokumasu S, Araki H, Koreeda S, Tubaki K. Succession of fungi in decomposing mangrove leaves in Japan. In: Hattori T, Ishida Y, Maruyama Y, Morita R, Uchida A, eds. Recent advances in microbial ecology. Tokyo: Japan Sci Soc Press, 1989:297–301.

Nakagiri A, Newell SY, Ito T. Two new *Halophytophthora* species, *H. tartarea* and *H. masteri*, from intertidal decomposing leaves in saltmarsh and mangrove regions. Mycoscience 1994;35:223–232.

Nakagiri A, Okane I, Ito T. Zoosporangium development, zoospore release and culture properties of *Halophytophthora mycoparasitica*. Mycoscience 1998;39:223–230.

Nakagiri A, Ito T, Manoch L, Tantichareon M. A new *Halophytophthora* species, *H. porrigovesica*, from subtropical and tropical mangroves. Mycoscience 2001;42:33–41.

Nakamura K, Hatai K. *Atkinsiella parasitica* sp. nov. isolated from rotifer, *Brachionus plicatilis*. Mycoscience 1994;35:383–389.

Nakamura K, Hatai K. Three species of Lagenidiales isolated from the eggs and zoeae of the marine crab *Portunus pelagicus*. Mycoscience 1995;36:87–95.

Nakamura K, Wada S, Hatai K, Sugimoto T. *Lagenidium myophilum* infection in the coonstripe shrimp, *Pandalus hypsinotus*. Mycoscience 1994;35:99–104.

Nakamura K, Nakamura M, Hatai K, Zafran. *Lagenidium* infection in eggs and larvae of mangrove crab (*Scylla serrata*) produced in Indonesia. Mycoscience 1995;36:399–404.

Newell SY. Autumn distribution of marine Pythiaceae across a mangrove-saltmarsh boundary. Can J Bot. 1992;70:1912–1916.

Newell SY, Fell JW. Distribution and experimental responses to substrate of marine oomycetes (*Halophytophthora* spp.) in mangrove ecosystems. Mycol Res. 1992;96:851–856.

Newell SY, Fell JW. Competition among mangrove oomycetes, and between oomycetes and other microbes. Aquat Microb Ecol. 1997;12:21–28.

Newell SY, Cefalu R, Fell JW. *Myzocytium, Haptoglossa* and *Gonimochaete* (fungi) in littoral marine nematodes. Bull Mar Sci. 1977;27:177–207.

Newell SY, Miller JD, Fell JW. Rapid and pervasive occupation of fallen mangrove leaves by a marine zoosporic fungus. Appl Environ Microbiol. 1987;53:2464–2469.

Nilson EH, Fisher WS, Shleser RA. A new mycosis of larval lobster (*Homarus americanus*). J Invert Pathol. 1976;27:177–183.

Noga EJ. Fungal diseases of marine and estuarine fish. In: Couch JA, Fournie JV, eds. Pathobiology of marine and estuarine organisms. Boca Raton: CRC Press, 1993:85–110.

Norén F, Moestrup Ø, Rehnstam-Holm A-S, Larsen J. Worldwide occurrence and host specificity of *Parvilucifera infectans*: a parasitic flagellate capable of killing toxic dinoflagellates. In: Hallegraeff GM, Blackburn SI, Bolch CJ, Lewis RJ, eds. Harmful algal blooms 2000. Intergovernmental Oceanographic Commission of UNESCO 2001, 2000:481–483.

Overton SV, Bland CE. Infection of *Artemia salina* by *Haliphthoros milfordensis*: a scanning and transmission electron microscope study. J Invert Pathol. 1981;37:249–257.

Pang KL, Jheng JS, Jones EBG. Marine mangrove fungi of Taiwan. Keelung: National Taiwan Ocean University Press, 2011.

Parfrey LW, Barbero E, Lasser E, Dunthorn M, Bhattacharya D, Patterson DJ, Katz LA. Evaluating support for the current classification of eukaryotic diversity. PLoS Genet. 2006;2:e220.

Park CS. Rapid detection of *Pythium porphyrae* in commercial samples of dried *Porphyra yezoensis* sheets by polymerase chain reaction. J Appl Phycol. 2006;18:203–207.

Park CS, Kakinuma M, Amano H. Detection of the red rot disease fungi *Pythium* spp. by polymerase chain reaction. Fisheries Sci. 2001a;67:197–199.

Park CS, Kakinuma M, Amano H. Detection and quantitative analysis of zoospores of *Pythium porphyrae*, causative organism of red rot disease in *Porphyra*, by competitive PCR. J Appl Phycol. 2001b;13:433–441.

Park KI, Choi KS. Spatial distribution of the protozoan parasite *Perkinsus* sp. found in Manila clams, *Ruditapes philippinarum*, in Korea. Aquaculture 2001;203:9–22.

Park MG, Yih W, Coats DW. Parasites and phytoplankton, with special emphasis on dinoflagellate infections. J Eukaryot Microbiol. 2004;51:145–155.

Patterson DJ. The diversity of eukaryotes. Am Nat. 1999;154:96–124.

Pegg KG, Alcorn JL. *Phytophthora operculata* sp. nov., a new marine fungus. Mycotaxon 1982;16:99–102.

Perkins FO. Zoospores of the oyster pathogen *Dermocystidium marinum*. I. Fine structure of the conoid and other sporozoan-like organelles. J Parasitol. 1976;62:959–974.

Perkins FO. The structure of *Perkinsus marinus* (Mackin, Owen and Collier, 1950) Levine, 1978 with comments on taxonomy and phylogeny of *Perkinsus* spp. J Shellfish Res. 1996;15:67–87.

Petersen AB, Rosendahl S. Phylogeny of the Peronosporomycetes (Oomycota) based on partial sequences of the large ribosomal subunit (LSU rDNA). Mycol Res. 2000;104:1295–1303.

Porter D. Mycoses of marine organisms: an overview of pathogenic fungi. In: Moss ST, ed. The biology of marine fungi. Cambridge: Cambridge University Press, 1986:141–154.

Raghukumar C. Physiology of infection of the marine diatom *Licmorphora* sp. by the fungus *Ectrogella perforans*. Veroff Inst Meeresforsch Bremerhaven. 1978;17:1–14.

Raghukumar C. An ultrastructural study of the marine diatom *Licmorphora hyaline* and its parasite *Ectrogella perforans*. I. Infectin of host cells. Can J Bot. 1980a;58:1280–1290.

Raghukumar C. An ultrastructural study of the marine diatom *Licmorphora hyaline* and its parasite *Ectrogella perforans*. II. Development of the fungus in its host. Can J Bot. 1980b;58:2557–2574.

Raghukumar C. Thraustochytrid fungi associated with marine algae. Ind J Mar Sci. 1986;15:121–122.

Raghukumar C. Fungal parasites of marine algae from Mandapam (South India). Dis Aquat Org. 1987;3:137–145.
Raghukumar C. Zoosporic fungal parasites of marine biota. In: Dayal R, ed. Advances in zoosporic fungi. New Delhi: MD Publications Pvt Ltd, 1996:61–83.
Raghukumar S. The role of fungi in marine detrital processes. In: Ramaiah N, ed. Marine microbiology: Facets and opportunities. Goa: Nat Inst Oceano, 2004:125–140.
Raghukumar S, Sathe-Pathak V, Sharma S, Raghukumar C. Thraustochytrid and fungal component of marine detritus. III. Field studies on decomposition of leaves of the mangrove *Rhizophora apiculata*. Aquat Microb Ecol. 1995;9:117–125.
Rasconi S, Jobard M, Jouve L, Sime-Ngando T. Use of calcofluor white for detection, identification, and quantification of phytoplanktonic fungal parasites. Appl Environ Microbiol. 2009;75:2545–2553.
Rasconi S, Niquil N, Sime-Ngando T. Phytoplankton chytridiomycosis: community structure and infectivity of fungal parasites in aquatic ecosystems. Environ Microbiol. 2012, doi: 10.1111/j.1462-2920.2011.02690.x.
Riethmüller A, Weiß M, Oberwinkler F. Phylogenetic studies of Saprolegniomycetidae and related groups based on nuclear large subunit ribosomal DNA sequences. Can J Bot. 1999;77:1790–1800.
Riethmüller A, Volgmayr H, Göker M, Weiß M, Oberwinkler F. Phylogenetic relationships of the downy mildews (Peronosporales) and related groups based on nuclear large subunit ribosomal DNA sequences. Mycologia 2002;94:834–849.
Robertson E, Anderson VL, Phillips AJ, Secombers CJ, Diéguez-Uribeondo J, van West P. *Saprolegnia*-fish interactions. In: Lamour K, Kamoun S, eds. Oomycete genetics and genomics: diversity, interactions and research tools. Hoboken: John Wiley and Sons Inc: 2009:407–424.
Roza D, Hatai K. *Atkinsiella dubia* infection in the larvae of Japanese mitten crab, *Eriocheir japonicus*. Mycoscience 1999a;40:235–240.
Roza D, Hatai K. Pathogenicity of fungi isolated from the larvae of the mangrove crab, *Scylla serrata*, in Indonesia. Mycoscience 1999b;40:427–431.
Salomon PS. Importance of pathogens on the ecology of marine phytoplankton populations. PhD dissertation, Kalmar University, Kalmar, 2004.
Schmidt JP, Shearer CA. A checklist of mangrove-associated fungi, their geographical distribution and known host plants. Mycotaxon 2003;85:423–477.
Schnepf E, Drebes G. Uber die Entwicklung des marinen parasitischen Phycomyceten *Lagenisma coscinodisci* (Lagenidiales). Helgol wiss Meersunters. 1977;29:291–301.
Schnepf E, Elbrächter M. Dinophyte chloroplasts and phylogeny – a review. Grana 1999;38:81–97.
Schnepf E, Deichgräber G, Drebes G. Development and ultrastructure of the marine parasitic oomycete, *Lagenisma coscinodisci* (Lagenidiales): the infection. Archiv für Mikrobiol. 1978a;116:133–139.
Schnepf E, Deichgräber G, Drebes G. Development and ultrastructure of the marine parasitic oomycete, *Lagenisma coscinodisci* (Lagenidiales): sexual reproduction. Can J Bot. 1978b;56:1315–1325.
Sekimoto S, Hatai K, Honda D. Molecular phylogeny of an unidentified *Haliphthoros*-like marine oomycete and *Haliphthoros milfordensis* inferred from nuclear-encoded small- and large-subunit rRNA genes and mitochondrial-encoded *cox*2 gene. Mycoscience 2007;48:212–221.
Sekimoto S, Beakes GW, Gachon CMM, Müller DG, Küpper FC, Honda D. The development, ultrastructural cytology, and molecular phylogeny of the basal oomycete *Eurychasma dicksonii*, infecting the filamentous Phaeophyte algae *Ectocarpus siliculosus* and *Pylaiella littoralis*. Protist 2008a;159:299–318.
Sekimoto S, Yokoo K, Kawamura Y, Honda D. Taxonomy, molecular phylogeny, and ultrastructural morphology of *Olpidiopsis porphyrae* sp. nov. (Oomycetes, Stramenopiles), a unicellular obligate endoparasite of *Porphyra* spp. (Bangiales, Rhodophyta). Mycol Res. 2008b; 112:361–374.

Sekimoto S, Klochkova TA, West JA, Beakes GW, Honda D. *Olpidiopsis bostrychiae*: a new species of endoparasitic oomycete that infects *Bostrychia* and other red algae. Phycologia 2009a;48:460–472.

Shearer CA, Descals E, Kohlmeyer B, Kohlmeyer J, Marvanová L, Padgett D, Porter D, Raja HA, Schmit JP, Thorton HA, Voglymayr H. Fungal biodiversity in aquatic habitats. Biodivers Conserv. 2007;16:49–67.

Siddall ME, Reece KS, Graves JE, Burreson EM. "Total evidence" refutes the inclusion of *Perkinsus* species in the phylum Apicomplexa. Parasitology 1997;115:165–176.

Sime-Ngando T, Lefèvre E, Gleason FH. Hidden diversity among aquatic heterotrophic flagellates: ecological potentials of zoosporic fungi. Hydrobiol. 2011;659:5–22.

Sindermann CJ. Disease diagnosis and control in North American marine aquaculture. New York: Elsevier, 1977.

Sparrow FK. The aquatic Phycomycetes, exclusive of the Saprolegniaceae and *Pythium*. Ann Arbor: University of Michigan Press, 1943.

Sparrow FK. Aquatic Phycomycetes. Ann Arbor: University of Michigan Press, 1960.

Sparrow FK. The peculiar marine phycomycete *Atkinsiella dubia* from crab eggs. Arch Mikrobiol. 1973;93:137–144.

Sparrow FK. The present status of classification in biflagellate fungi. In: Jones EBG, ed. Recent advances in aquatic mycology. London: Elek Science, 1976:213–222.

Spencer MA, Vick MC, Dick MW. Revision of *Aplanopsis, Pythiopsis,* and 'subcentric' *Achlya* species (Saprolegniaceae) using 18S rDNA and morphological data. Mycol Res. 2002;106:549–560.

Strittmatter M, Gachon CMM, Küpper FC. Ecology of lower Oomycetes. In: Lamour K, Kamoun S, eds. Oomycete genetics and genomics: diversity, interactions, and research tools. Hoboken: John Wiley and Sons Inc, 2009:25–46.

Takishita K, Miyake H, Kawato M, Maruyama T. Genetic diversity of microbial eukaryotes in anoxic sediment around fumaroles on a submarine caldera floor based on the small-subunit rDNA phylogeny. Extremophiles 2005;9:185–196.

Tan TK, Pek CL. Tropical mangrove leaf litter fungi in Singapore with an emphasis on *Halophytophthora.* Mycol Res. 1997;101:165–168.

Tharp TP, Bland CE. Biology and host range of *Haliphthoros milfordensis.* Can J Bot. 1977;55:2936–2944.

Thines M, Choi YJ, Kemen E, Ploch S, Holub EB, Shin H-D, Jones JDG. A new species of *Albugo* parasitic to *Arabidopsis thaliana* reveals new evolutionary patterns in white blister rusts (Albuginaceae). Persoonia 2009;22:123–128.

Tsui CKM, Marshall W, Yokoyama R, Honda D, Craven KL, Peterson PD, Lippmeier JC, Berbee ML. Labyrinthulomycetes phylogeny and its implication for the evolutionary loss of chloroplasts and gain of ectoplasmic gliding. Mol Phylogenet Evol. 2009;50:129–140.

Uzuhashi S, Tojo M, Kakishima M. Phylogeny of the genus *Pythium* and description of new genera. Mycoscience 2010;51:337–365.

van der Auwera G, de Baere R, van de Peer Y, de Rijk P, van den Broeck I, de Wachter R. The phylogeny of the Hyphochytriomycota as deduced from ribosomal RNA sequences of *Hyphochytrium catenoides.* Mol Biol Evol. 1995;12:671–678.

van der Meer JP, Pueschel CM. *Petersenia palmariae* n. sp. (Oomycetes): a pathogenic parasite of the red alga *Palmaria mollis* (Rhodophyceae). Can J Bot. 1985;63:404–408.

van der Plaats-Niterink AJ. Monograph of genus *Pythium*. Baarn: Centraalbureau voor Schimmelcultures Stud Mycol, 1981.

van West P. *Saprolegnia parasitica*, an oomycete pathogen with a fishy appetite: new challenges for an old problem. Mycologist 2006;20:99–104.

Vandersea MW, Wayne Litaker R, Yonnish B, Sosa E, Landsberg JH, Pullinger C, Moon-Butzin P, Green J, Morris JA, Kator H, Noga EJ, Tester PA. Molecular assays for detecting *Aphanomyces invadans* in ulcerative mycotic fish lesions. Appl Environ Microbiol. 2006;72:1551–1557.

Villa NO, Kageyama K, Asano T, Suga H. Phylogenetic relationships of *Pythium* and *Phytophthora* species based on ITS rDNA, cytochrome oxidase II and beta-tubulin gene sequences. Mycologia 2006;98:410–422.

Vishniac HS. A new marine Phycomycete. Mycologia 1958;50:66–79.

Vittal BPR, Sarma V. Diversity and ecology of fungi on mangroves of Bay of Begal region – An overview. Indian J Mar Sci. 2006;35:308–317.

Waterhouse GM. Peronosporales. In: Ainsworth GC, Sparrow FK, Sussman AS, eds. The fungi: an advanced treatise. Vol. IV. New York: Academic Press, 1973:165–183.

West JA, Klochkova TA, Kim GH, Loiseaux-de Goër S. *Olpidiopsis* sp., an oomycete from Madagascar that infects *Bostrychia* and other red algae: host species susceptibility. Phycol Res. 2006;54:72–85.

Wilce RT, Schneider CW, Quinlan AV, Bosch KV. The life history and morphology of free-living *Pylaiella littoralis* (L) Kjellm (Ectocarpaceae, Ectocarpales) in Nahant Bay, Massachusetts. Phycologia 1982;21:336–354.

Willoughby LG. Diseases of freshwater fishes. In: Tsui CKM, Hyde KD, eds. Freshwater Mycology. Hong Kong: Fungal Diversity Press, 2003:111–126.

Wilson IM. Marine fungi: a review of the present position. Proc Linn Soc London. 1960;171:53–70.

Yoon HS, Hackett JD, Pinto G, Bhattacharya D. The single, ancient origin of chromist plastids. Proc Natl Acad Sci USA. 2002;99:15507–15512.

Yoshida N, Masahiko N, Inoue K, Yoshizawa S, Kamiya E, Taniguchi A, Hamasaki K, Kogure K. Analysis of nanoplankton community structure using flow sorting and molecular techniques. Microb Environ. 2009;24:297–304.

12 Labyrinthulomycota

Eduardo M. Leaño and Varada Damare

Labyrinthulomycetes are a group of fungus-like organisms that produce motile zoospores and are classified under microbial eukaryotes (or protists). Microbial eukaryotes are defined loosely as eukaryotic organisms that are not plants, animals or fungi (McGrath and Katz 2004). Labyrinthulomycetes are a relatively poorly studied group of heterotrophic unicellular eukaryotes (Scharer et al. 2007) and due to their unique characteristics of being non-photosynthetic (like fungi) and producing motile zoospores (like algae), they have been transferred from one kingdom to another (Figure 12.1). Originally classified in the Kingdom Fungi, they were transferred to Kingdom Protista (Whittaker 1969) – Protoctista (Margulis and Schwartz 1988), Stramenopila (Alexopoulos et al. 1996), Chromista (Cavalier-Smith 1998) – and are currently under the super-group Chromalveolata (Adl et al. 2005; Parfrey et al. 2006). Stramenopiles (Chromists) in general are characterized by having tripartite, tubular, flagellar hairs and mitochondria with tubular cristae (Dick 2001).

Labyrinthulomycetes are primarily marine osmoheterotrophic, straminipilan protists that have been isolated from a wide variety of habitats around the world (Raghukumar 2002). They are considered to be important components of marine microbial communities (Scharer et al. 2007). Some species seem to extend their existence into brackish water as well as terrestrial environments (Raghukumar 2002). Analyzing genetic libraries of the 18S rDNA for novel stramenopile sequences indicates that they are found in ecologically diverse systems, including oceanic, coastal, and anoxic habitats (Massana et al. 2004). Their distinguishing feature, in general, is the production of an ectoplasmic net, which is an extension of the plasma membrane secreted by a specialized organelle (sagenogenetosome) that attaches the cell to its substrate and secretes digestive enzymes for absorptive nutrition (Moss 1986; Tsui et al. 2009).

Labyrinthulomycota is comprised of three distinct groups (lineages) of marine heterotrophic stramenopiles that are readily distinguished on the basis of gross morphological characters and molecular phylogenetic data (Leander et al. 2004):

(i) Labyrinthulids (slime nets; slime molds);
(ii) Thraustochytrids; and
(iii) Aplanochytrids.

At present, they are still being referred to as lower (zoosporic) fungi, slime molds (Labyrinthulids) and, in the industrial sector, as microalgae (Thraustochytrids).

Labyrinthulids are a family of mainly marine unicellular protists that live, for the most part, on sea grasses and marine algae where they exist either as parasites, commensals or mutualists. Labyrinthulids grow colonially wherein numerous spindle-shaped cells are linked to each other by means of an ectoplasmic net (Raghukumar 2002). The ectoplasmic membrane is secreted by an organelle called a bothrosome, which in turn produces a network of filaments along which cells move and absorb nutrients (Moss 1986). These filaments are reminiscent of a net, and give the labyrinthulids their common name of "slime nets" (Figure 12.2).

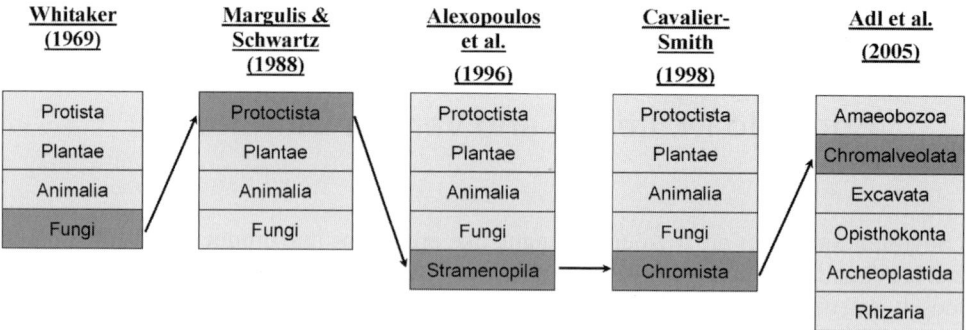

Figure 12.1 Kingdom classification of Labyrinthulomycetes from the Kingdom Fungi to the recently-established super-group Chromalveolata.

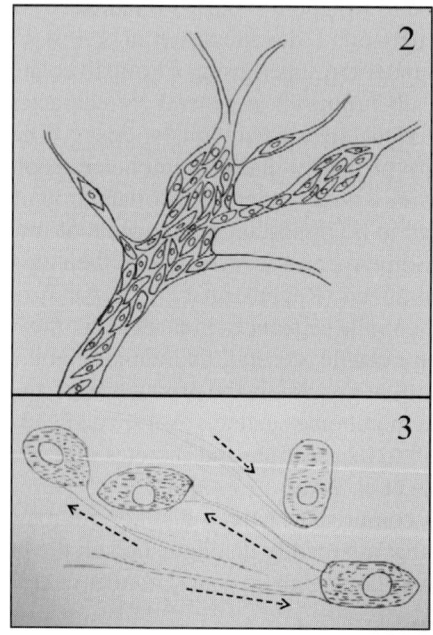

Figures 12.2–12.3 Diagrammatic illustrations of labyrinthulids (2) aplanochytrids (3). Dashed arrows in (3) indicate movement direction of aplanochytrid cells. Illustration by C.C.L. Leaño.

Members of this group of fungal-like microbes were originally placed in the slime mold category, but their phylogeny suggests that they are related to the stramenopiles.

Thraustochytrids are a group of non-photosynthetic marine fungal-like microbes characterized by the presence of a sagenogenetosome, an ectoplasmic net, a cell wall composed of non-cellulosic scales (Cavalier-Smith et al. 1994), and the production of biflagellate heterokont zoospores in many of the described genera (Moss 1991). Vegetative cells of thraustochytrids consist of single cells that are globose to sub-globose (Figures 12.4–12.6). They measure 4–20 μm in diameter and usually grow epibiontically on various substrata

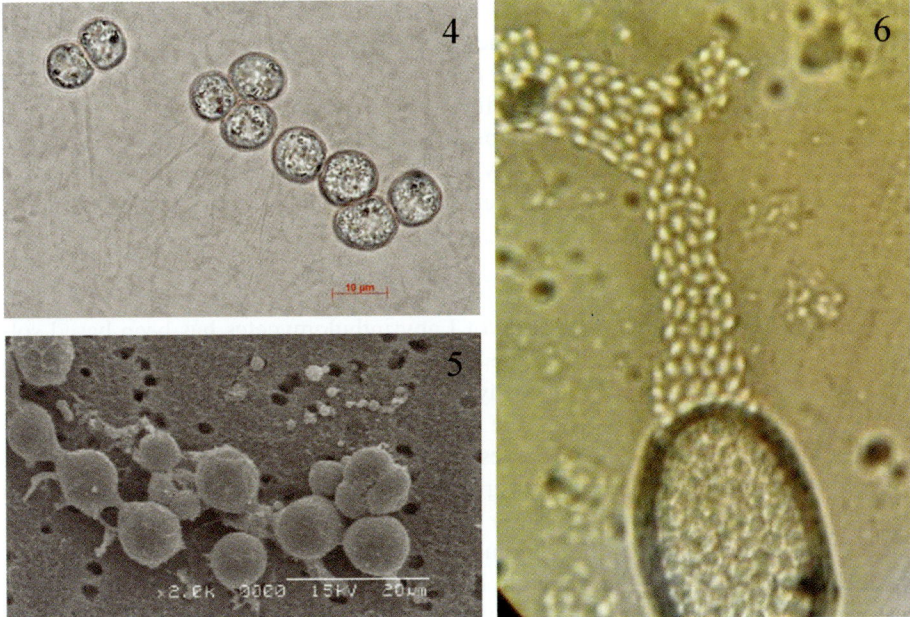

Figures 12.4–12.6 Cells of thraustochytrids: light and scanning electron micrographs of *Schizochytrium* sp. (4) and (5); *Thraustochytrium* sp. releasing zoospores (6). Photo courtesy of V. Damare (4), Y.C. Lin (5) and K.H.V. Arafiles (6).

(Raghukumar 2002). They are common marine microheterotrophs that are mostly saprobes and have a wide geographic distribution, including tropical (Leaño 2001; Arafiles et al. 2011), sub-tropical (Nakahara et al. 1996; Honda et al. 1998; Fan et al. 2002a), temperate and even Arctic/Antarctic regions (Bahnweg and Sparrow 1974a, 1974b; Moro et al. 2003; Stoeck et al. 2007). They are commonly isolated from dead organic materials such as decaying seaweeds and mangrove leaves (Raghukumar et al. 1994; Leaño 2001; Fan et al. 2002a). Populations of thraustochytrids increase after the senescence and decay of macroalgae but not during their healthy phase (Raghukumar et al. 1992), clearly indicating the saprobic nature of most thraustochytrid species. Other species are also reported to be associated with living marine organisms, however, such as the surface mucus of hermatypic coral (Harel et al. 2008), pathogens of marine animals (Polglase 1980), and marine zooplankton (Damare and Raghukumar 2006). Other species are parasitic, such as *Thraustochytrium caudivorum*, which parasitizes *Macrostomum lignano*, marine free-living flatworms (Scharer et al. 2007).

Aplanochytrids is a group of fungal-like microbes formerly classified as Labyrinthuloides, but Leander and Porter (2001) suggested that this genus comprises a distinct lineage among the Labyrinthulomycetes. Aplanochytrids also produce globose sporangia and a multi-laminar scaly wall, as in the thraustochytrids. They multiply by means of aplanospores, however, which move in a gliding (crawling) manner using ectoplasmic net elements (Leander and Porter 2001) rather than the highly motile biflagellated zoospores seen in other members of Labyrinthulomycota (labyrinthulids and thraustochytrids;

Leander et al. 2004). The ectoplasmic net of aplanochytrids does not completely enrobe the cells, a characteristic that distinguishes them from the labyrinthulids (Figure 12.3). They are often associated with dead and decaying plant material, but some species are also known to be pathogens of marine animals (Bower 1987).

In this chapter, the three distinct groups of Labyrinthulomycetes will be discussed in detail in terms of their taxonomy and phylogeny, ecology, their important roles as parasites or saprobes, and their use as producers of commercially important bio-products.

Taxonomy and phylogeny

For a long time taxonomic classification of Labyrinthulomycetes has been based solely on morphological characteristics and the mode of zoospore release (except for aplanochytrids). As an example, for thraustochytrids generic identification was based on the morphology of vegetative cells (e.g., the presence or absence of tetrads and octads, or the formation of amoeboid cells or possession of proliferous bodies) and the process of zoospore release. For labyrinthulids and aplanochytrids, which comprised only one genus each, species differentiation depended on the shape of vegetative cells (globose, oblong, spindle, etc.) and production or non-production of biflagellate spores. Labyrinthulids and thraustochytrids were once considered to be unrelated organisms, but ultrastructural and physiological examination later suggested that they were closely related (Honda et al. 1999). A detailed identification scheme for the families and genera of Labyrinthulomycetes, based morphological characters (cell shape, presence or absence of ectoplasmic nets, presence of amoeboid cells), zoospore formation, and production of pigments (mainly β-carotene), was partially developed by Yokoyama et al. (2007).

Nowadays, the use of modern molecular tools for species identification has resulted in further delineation of different genera and species. Although a young field, marine protistan genomics has already greatly expanded the understanding of eukaryotes (Worden and Allen 2010). The genomes that are now available are invaluable tools for exploring evolutionary relationships, gene function, metabolism and the development of complex traits among different groups of organisms (Worden and Allen 2010), and will continue to uncover features that are either not present in the representative species or are exaggerated in microbial groups (McGrath and Katz 2004). Elucidating the scope of genome dynamics is essential for understanding the relationship between genomes and phenotypes (Parfrey et al. 2008). Moreover, interpreting the evolution of eukaryotic genomes requires knowledge of the evolutionary relationships among eukaryotes, particularly among the microbial lineages (McGrath and Katz 2004). In general, marine protists are highly diverse, which is the result of multiple different ancient endosymbiotic events after the mitochondrion was initially established. For this reason, their genomes reveal many new insights into eukaryotes overall (Worden and Allen 2010).

Labyrinthulids

Labyrinthulids (Family Labyrinthulaceae) is comprised of only one genus, *Labyrinthula*. *Labyrinthula* was first described by Cienkowski (1867), who established the genus and reported two marine species: *L. vitellina* and *L. macrocystis*. These were isolated from marine algae and referred to as a marine slime or net-slime molds. Labyrinthulas produce motile cells that move in a net-like extracellular matrix. In general, *Labyrinthula* species inhabit various plant materials in marine environments by digesting microorganisms (bacteria, yeasts, fungi and microalgae) on plant surfaces (Nakagiri 2001). Most *Labyrinthula* species are indistinguishable when observed at the vegetative stage, but one morphological character used to distinguish species is cell size (Watson and Raper 1957), which is still used in the identification and differentiation of new species (Bigelow et al. 2005). At present, there are at least 16 described species of *Labyrinthula*: *L. algariensis; L. apis; L. cienkowski; L. chattoni; L. coenocystis; L. jeremarina; L. macrosystis; L. magnifica; L. pohlia; L. roscoffensis L. terrestris; L. thais; L. valkanovi; L. vittelina L. zopfii;* and *L. zosterae*.

Only a few reports are available on the use of modern molecular tools for the phylogenetic classification of *Labyrinthula*, including the study by Honda et al. (1999) and that by Tsui et al. (2009). This might be due to the lone genus and non-discovery of new species since the last described species (*L. terrestris*) by Bigelow et al. (2005), which is the causative agent for rapid blight disease of turf grass irrigated with saline water or effluents. Honda et al. (1999) analyzed the phylogenetic relationship of *Labyrinthula* sp. and thraustochytrids and found that the two groups form a monophyletic clade within the Stramenopila. Through 18S rDNA molecular phylogeny, *Labyrinthula* spp. are grouped with *Aplanochytrium kerguelense, Aplanochytrium minuta* and *Schizochytrium minutum*, and this clade was separated from the other thraustochytrid species used in their study. Tsui et al. (2009), however, investigated the phylogenetic position of Labyrinthulomycetes in relation to the non-photosynthetic bicoeceans and oomycetes and the photosynthetic ochrophytes based on sequences of actin, beta-tubulin, *elongation factor 1-alpha* gene fragments, and ribosomal small subunit (SSU) genes. The multilocus phylogenies generated from the different protein-coding sequences revealed that Labyrinthulomycetes and bicoecean are sister groups. This phylogeny suggested that the microorganisms that glide via an ectoplasmic net (e.g., *Labyrinthula* and *Aplanochytrium*) have evolved from species that have mainly used the ectoplasmic net for anchorage and assimilation rather than motility (e.g., thraustochytrids).

Thraustochytrids

Of the three lineages under Labyrinthulomycota, thraustochytrids appears to be the most studied group as it is composed of several genera and has been recognized as a potential source of essential fatty acids for commercial production. Previously composed of seven genera (*Althornia, Diplophrys, Elina, Japonochytrium, Thraustochytrium, Schizochytrium, and Ulkenia*), additional genera were created after taxonomic rearrangement of *Ulkenia* and *Schizochytrium* based on morphology, chemotaxonomical characteristics and 18S rRNA gene phylogeny (Yokoyama and Honda 2007; Yokoyama et al. 2007). As such, thraustochytrids are now composed of 12 genera: *Althornia, Aurantiochytrium, Botryochytrium, Diplophrys, Elina, Japonochytrium, Oblongichytrium, Parietichytrium, Schizochytrium, Sicyoidochytrium, Thraustochytrium,* and *Ukenia*.

Thraustochytrids are broadly characterized by single cells that are not interconnected, the presence of ectoplasmic net elements (except *Althornia*), and reproduction by zoospores (Bongiorni et al. 2005a). The type genus of Thraustochytriaceae is *Thraustochytrium* (Sparrow 1936), which is characterized by globose sporangia with or without a proliferous body; and partial dissolution of the sporangial cell wall for zoospore release.

From this initial description of thraustochytrids, other genera have been proposed using the morphological features of *Thraustochytrium* as a basis. However, the delineation of genera has been problematical as the characteristic features overlap among some species of the type genus *Thraustochytrium* (Yokoyama and Honda 2007). An example of this is the formation of amoeboid cells by *Ulkenia*, which was originally unique but was later observed in *Schizochytrium* and some species of *Thraustochytrium* (under some culture conditions).

The other genera of thraustochytrids are briefly described and differentiated from the type species as follows:

(i) *Althornia*: characterized by the absence of an ectoplasmic net (Jones and Alderman 1971).

(ii) *Aurantiochytrium*: vegetative cells do not form large colonies and are mostly dispersed as single cells; ectoplasmic nets are poorly developed; and zoospores are ovoid in shape (Yokoyama and Honda 2007).

(iii) *Botryochytrium*: colonies are comparatively large with a well-developed ectoplasmic net; at maturity, the cell wall deliquesces completely leaving the protoplast fully naked, becomes botryose by centripetal division and forms a star shape before zoospore formation (Yokoyama et al. 2007).

(iv) *Diplophrys*: vegetative cells are ovoid, uninucleate and covered with scales; cells display irregular gliding motility due to the use of a fine filamentous ectoplasmic net extending from the polar extremities; refractive lipid body. Repeated cell divisions produce flat circular patches of cells in culture (Dykstra and Porter 1984).

(v) *Elina*: thallus globose; the zoospores cleave in the sporangia and are released by deliquescence of the whole zoosporangial wall; the ectoplasmic net is present (Artemtchuk 1972; Dick 2001).

(vi) *Japonochytrium*: possesses an apophysis-like structure not found in other genera (Sparrow 1960).

(vii) *Oblongichytrium*: the thallus is thin-walled, globose and pale yellow; vegetative cells form large colonies with a well-developed ectoplasmic net; the zoospores are a narrow elliptical to oblong in shape and are only released when the sporangia are transferred from agar (solid) to liquid media (Yokoyama and Honda 2007).

(viii) *Parietichytrium*: the colonies are comparatively large; the ectoplasmic nets comparatively well-developed; the cell wall is persistent after the release of the protoplast, with the protoplast becoming botryose by a centripetal division then becoming star-shaped before zoospore formation (Yokoyama et al. 2007).

(ix) *Schizochytrium*: the thallus is thin-walled, globose and pale yellow; vegetative colonies are large and are created by continuous binary cell division; the ectoplasmic nets are well developed; zoospores are reniform to ovoid in shape (Yokoyama and Honda 2007).

(x) *Sicyoidochytrium*: the vegetative cells and colonies are somewhat small; the ectoplasmic nets are not well-developed; at maturity, the cell wall deliquesces completely and the protoplast ends up totally naked; the protoplast divides several times, and at the final division zoospores are formed by pinching and pulling of the divided protoplast (Yokoyama et al. 2007).

(xi) *Ulkenia*: thallus epi-, endo- and interbiotic, thin-walled during active growth, globose, sub-globose or pear-shaped, very variable in diameter; the ectoplasmic net is endo- or interbiotic and is not well-developed; at maturity, the cell wall disappears totally and the protoplast ends up fully naked or creeps like an amoeba out of the sporangium wall through a small opening at the apical part of the sporangium; after settlement, the protoplast becomes zoosporangium and heterokont zoospores are released (Yokoyama et al. 2007).

The erection of the five new genera of thraustochytrids by Yokoyama and Honda (2007) and Yokoyama et al. (2007) resulted in the re-classification of some *Schizochytrium*, *Thraustochytrium* and *Ulkenia* species. *Oblongichytrium* was established for *Schizochytrium minutum*, while *Ulkenia radiata* was moved to a new genus *Botryochytrium*. Other reclassified species of thraustochytrids are listed in Table 12.1.

Aplanochytrids

Like labyrinthulids, aplanochytrids are classified within a single genus, *Aplanochytrium*. There are a total of eight recognized species, but species identification is difficult due to the plasticity and ambiguity of the fundamental morphological features (Leander et al. 2004).

Table 12.1 Reclassified thraustochytrid species upon erection of five new genera after taxonomic rearrangement of *Schizochytrium* (Yokoyama and Honda 2007) and *Ulkenia* (Yokoyama et al. 2007).

Genus	Species	Previous classification
Oblongichytrium	*Oblongichytrium minutum** (A. Gaertn.) R. Yokoyama *et* D. Honda	*Schizochytrium minutum* A. Gaert.
	Oblongichytrium multirudimentale (S. Goldst.) R. Yokoyama *et* D. Honda	*Thraustochytrium multirudimentale* S. Goldst.
	Oblongichytrium octosporum (Raghuk.) R. Yokoyama *et* D. Honda	*Schizochytrium octosporum* Raghuk.
Aurantiochytrium	*Aurantiochytrium limacinum** (D. Honda *et* Yokochi) R. Yokoyama *et* D. Honda	*Schizochytrium limacinum* D. Honda *et* Yokochi
	Aurantiochytrium mangrovei (Raghuk) R. Yokoyama *et* D. Honda	*Schizochytrium mangrovei* Raghuk.
Botryochytrium	*Botryochytrium radiatum** (A. Gaert.) R. Yokoyama, B. Salleh *et* D. Honda	*Ulkenia radiata* A. Gaertn.
Parietichytrium	*Parietichytrium sarkarianum** (A. Gaertn.) R. Yokoyama, B. Salleh *et* D. Honda	*Ulkenia sarkariana* A. Gaertn.
Sicyoidochytrium	*Sicyoidochytrium minutum** (Raghuk.) R. Yokoyama, B. Salleh *et* D. Honda	*Ulkenia minuta* Raghuk.

*Type species.

The type species, *Aplanochytrium kerguelensis*, is characterized by the formation of spores that crawl along the substrate. This species was originally isolated from sub-Antarctic waters. All the *Aplanochytrium* species are listed in Table 12.2 and are based on the classification by Leander and Porter (2000), and a new species described by Moro et al. (2003).

Leander et al. (2004) studied the molecular phylogeny of the genus in detail based on the small subunit ribosomal gene sequences (SSU rDNA) and morphological characters derived from light, scanning and electron microscopy. Putative synapomorphies for aplanocytrids include:

(i) an ectoplasmic network within or over which cells crawl (Leander and Porter 2000);
(ii) a cell wall comprised mainly of fucose (Honda et al. 1999); and
(iii) polygonal cell surface patterns (Leander et al. 2004).

Table 12.2 Morphological characteristics of different species of *Aplanochytrium* (Bahnweg and Sparrow 1972; Leander and Porter 2000; Moro et al. 2003).

Species	Important morphological features	Isolate source
A. kerguelensis	Cells are globose; production of crawling spores	Sub-Antarctic waters
A. yorkensis	Cells are globose; production of biflagellate spores in addition to crawling spores	Oyster mantle, water samples, sediments and detritus
A. minuta	Cells are oblong and divide into tetrads; vegetative cells are motile for the entire lifecycle	Green alga (*Ulva* sp.); other algae (chlorophytes and rhodophytes)
A. saliens	Cells with pointed posterior and rounded/inflated anterior; sporangia are spherical and appear rough due to compartmentalization	Marine grass *Halophila englemannii*
A. schizochytrops	Vegetative cells are spherical to ovoid, and become enlarge to produce sporangia	Seagrass *Halodule wrightii*
A. haliotidis	Sporangia and vegetative cells are spherical; crawling spores not formed; biflagellate spores are produced	Pathogen of abalone
A. thaisii	Sporangia-rich colonies alternating with vegetative-rich colonies (both occur in mono-layers); produce tetrads, biflagellate spores and motile plasmodia	Marine gastropod *Thais haemastoma floridana*
A. stocchinoi	Spherical vegetative cells attached to substrate by ectoplasmic net; sporangia and vegetative cells surrounded by multi-layered wall; crawling spores released by complete cleavage	Antarctic chlorophyte *Urospora* sp.

Among these, the polygonal array of ectoplasmic filaments on the surface of aplanochytrids (Leander et al. 2004) seems to be the most unique feature of this group. Polygonal ectoplasmic filament patterns have not been observed among thraustochytrids (Harrison and Jones 1974) and are very unlikely to be present among labyrinthulids as their spindle-shaped cells are completely enrobed by an ectoplasmic net.

Distribution and ecology

Labyrinthulomycetes are exclusively marine in nature and cosmopolitan in distribution. They are found almost everywhere in marine and brackish waters. One exception, *Labyrinthula terrestris*, has been reported from the terrestrial environment, infecting turf grass irrigated with water of an elevated salinity (Olsen 2007). The first thraustochytrid, *Thraustochytrium proliferum*, was described by Sparrow (1936) from marine algae in the vicinity of Woods Hole (Atlantic coastal waters). Many species of *Thraustochytrium, Japonochytrium, Schizochytrium* and *Ulkenia* were subsequently described from coastal regions (see Raghukumar 2002). While the majority of the species belonging to *Aplanochytrium* were isolated from coastal environments, *Aplanochytrium kerguelensis* was isolated from oceanic waters of Antarctica (Bahnweg and Sparrow 1972) and equatorial Indian Ocean (Damare and Raghukumar 2010). Bahnweg and Sparrow (1972) also reported several *Thraustochytrium* species from Antarctic region.

Occurrence in the water column

The occurrence of Labyrinthulomycetes in the water column was detected by means of culture methods. Gaertner (1968) put forward the most probable number technique based on a pine pollen baiting method for the detection and enumeration of these organisms from the water column and sediments. Later, Raghukumar and Schaumann (1993) designed an epifluorescence technique for the direct detection of these organisms from the environment (Figure 12.7) and this technique has been widely used for detecting Labyrinthulomycetes from water columns, e.g., from the coastal regions of India, Japan

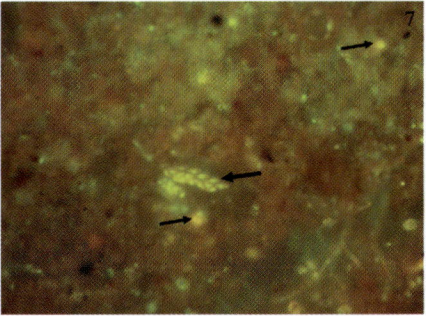

Figure 12.7 Epifluorescence technique for the direct detection of thraustochytrid cells: photomicrograph of aggregates stained with acriflavine under blue light. The rightward arrows are thraustochytrid cells while the leftward arrow points to *Labyrinthula*-like cells (note the spindle-shaped cells). Scale bar: 10 µm. Photo by V. Damare.

and Italy (Naganuma et al. 1998; Kimura et al. 1999; Raghukumar et al. 2001; Bongiorni and Dini 2002) and in some oceanic environments (Naganuma et al. 2006; Damare and Raghukumar 2008).

Naganuma et al. (2006) enumerated Labyrinthulomycetes in Arctic and sub-Arctic surface waters, distant from land masses. Damare and Raghukumar (2008) detected the presence of Labyrinthulomycetes in the oceanic environment of equatorial Indian Ocean waters. Their studies suggested that Labyrinthulomycetes are common inhabitants of the water column of oceanic waters, as they are in coastal waters. At their maxima, their biomass in the water column can occasionally be equivalent to that of bacteria (Raghukumar et al. 2001). Although cells detected through the epifluorescence method have been termed as thraustochytrids, it is also quite likely that many of these belong to the aplanochytrids and labyrinthulids, since the latter two stain in the same manner as the thraustochytrids.

Relation of Labyrinthulomycetes with particulate matter in the water column

Labyrinthulomycetes occur in patches in the water column (Damare and Raghukumar 2008). They are generally found at high densities at different depths, usually appearing as two peaks in the water column (Raghukumar et al. 2001, Damare and Raghukumar 2008). It is likely that these organisms are associated with the transparent exopolymeric particles (TEPs) present there. By virtue of their ectoplasmic net elements, Labyrinthulomycetes can attach to different particles and draw nutrients (see Raghukumar 2002). Kimura et al. (2001) and Raghukumar et al. (2001) found that they are positively related to particulate organic carbon during certain seasons. The patchy distribution of these organisms in the water column and their positive relation with particulate organic carbon suggest their probable association with the TEPs and microbial aggregates present there (Damare and Raghukumar 2008).

A plume of organic matter trails behind a sinking aggregate as a result of diffusion of dissolved organic matter from the aggregate into the surrounding waters due to microbial action (Alldredge and Youngbluth 1985; Simon et al. 1990; Smith et al. 1992; Kiørboe 2001). The aggregate, as well as the plume, are the hotspots of microbial activity and therefore might harbor Labyrinthulomycetes within it, leading to its patchy distribution. This is supported by high abundance of bacteria and Labyrinthulomycetes nearly 20 m above peak TEPs concentrations at some places in the water column, reflecting their active growth in such a plume (Damare and Raghukumar 2008). Lyons et al. (2005) reported the dense presence of the thraustochytrid pathogen QPX (Quahog Parasite Unknown) in marine aggregates and suggested that these might provide a means for survival and transport of the pathogen. As Labyrinthulomycetes can survive in elevated hydrostatic pressure similar to that at the depths of 1000 m (Raghukumar and Raghukumar 1999), they can travel to greater depths in the water column by hitchhiking on the sinking aggregates.

Relation of Labyrinthulomycetes with phytoplankton from the water column

Labyrinthulomycetes are found in low numbers during the season of high primary productivity (Raghukumar et al. 2001). Labyrinthulomycetes are mostly associated with non-phytoplanktonic particles rather than particles bearing phytoplankton (Kimura et al. 2001). This was evident in Greenland and the Norwegian seas where the abundance of Labyrinthulomycetes did not correlate with particulate adenosine 5´-triphosphate (ATP) (Naganuma et al. 2006). Phytoplankton, heterotrophic flagellates and ciliates contribute greatly to the *in situ* ATP pool there rather than the planktonic Labyrinthulomycetes and bacteria (Naganuma et al. 2006). Hence they were found to be associated with non-ATP-related rather than ATP-related organic materials. The non-ATP-related or the non-phytoplanktonic particles include the post-bloom phytodetritus and terrigenous matter. During an attempt to examine the growth and survival of Labyrinthulomycetes on laboratory-generated aggregates, these organisms were capable of multiplying in conditions similar to aggregate formation in the presence of microzooplankton and picoplankton in nature (Damare 2009).

Interestingly, in yet another experiment Labyrinthulomycetes grew well in the presence of diatom polysaccharides and absence of phytoplankton cells (Damare 2009). Therefore, while phytoplankton polysaccharides and their TEPs are probably utilized by Labyrinthulomycetes for growth and multiplication, the presence of live phytoplankton might be inhibitory. Raghukumar et al. (1992) also observed that, with respect to macroalgae, their populations only increase after their senescence and death.

Relation of Labyrinthulomycetes to bacteria in the water column

Labyrinthulomycetes and bacteria coexist in the marine environment (Vishniac 1956). Both being osmoheterotrophic in their mode of nutrition and capable of producing various degradative enzymes, they are likely to share an ecological niche by either surviving on different nutrient sources or competing for the same nutrient source. Damare (2009) observed that in presence of diatom polysaccharides, both Labyrinthulomycetes and bacteria grew well, whereas in the presence of bacterial polysaccharides only bacteria grew well. Thus it could be that in the presence of an adequate supply of nutrients, both groups share them explicitly; whereas when nutrients are limiting, bacteria outcompete Labyrinthulomycetes in spite of bacterivory observed in some of their species (Raghukumar 1992; Damare 2009). The substrates that the bacteria colonize support the good growth of Labyrinthulomycetes once the bacteria reach their stationary phase, and the Labyrinthulomycetes thrive on the residual nutrients left over by the bacteria (Raghukumar and Damare 2011).

Occurrence in mangrove habitats

Mangroves are unique intertidal wetlands that are largely confined to coastal regions, and are reported to be the most selected sites for the isolation of thraustochytrids (Fan and Chen 2006). Labyrinthulomycetes (e.g., thraustochytrids) are common in mangrove habitats as one of the initial and major colonizers and they play an important role in the degradation of mangrove detritus (Leaño 2001; Fan et al. 2002a). Thraustochytrids are actively involved in the carbon cycle and represent valuable sources of nutrients (e.g., polyunsaturated fatty acids) in the microbial loop of marine sediments (Fan et al. 2007). Their biomass also contributes to the organic detritus (Nakagiri 1998). As such, the degradation and enrichment of fallen mangrove leaves by these microbes contribute significantly to the nutrition of bottom- and filter-feeding organisms, making mangroves excellent nursery grounds for many fish and crustacean species (Leaño 2001).

Occurrence in other habitats

Aside from mangroves and the open ocean, Labyrinthulomycetes can be found on algal surfaces (Booth and Miller 1968; Miller and Jones 1983), in estuarine habitats (Ulken 1981), in saline soils (Booth 1971), some extreme environments, such as, hyper-saline lakes, the deep sea (Amon 1978; Raghukumar et al. 2001), Arctic habitats (Naganuma et al. 2006; Stoeck et al. 2007) and shallow water hydrothermal vents (Colaco et al. 2006). They play profound roles in the global carbon cycle by processing of organic carbon in the marine environment (Raghukumar 2004).

Due to their ubiquitous occurrence, they are being isolated from various substrates all over the world. Sea grasses, algae and sediments are the most widely-known substrates next to mangrove leaves and detritus. Zooplankton could also serve as a potent substrate, as Labyrinthulomycetes were also isolated epibiontically from their bodies (Damare and Raghukumar 2006). Table 12.3 lists details of different substrates and places where they have been isolated into culture. Thraustochytrids have been found to be important in primary film formation on freshly immersed surfaces in seawater. Their cell surface hydrophobicity induces the settlement of barnacle larvae on surfaces (Raghukumar et al. 2000). Thraustochytrids can also be useful indicators of fish farm biodeposition in sediments, as

Table 12.3 Some examples of studies aiming to culture Labyrinthulomycetes from various substrates all over the world.

Substrates	Area of isolation	References
Mangrove leaves and detritus	India, Japan, China, Hong Kong, Indonesia, Italy, Philippines, Taiwan, Thailand	Raghukumar (1988b); Bowles et al. (1999); Leaño (2001); Yokochi et al. (2001); Fan et al. (2002a, 2002b, 2007); Kamlangdee and Fan (2003); Bongiorni et al. (2005c); Unagul et al. (2005); Wong et al. (2005); Kumon et al. (2006); Perveen et al. (2006); Juntaban et al. (2007); Lin et al. (2010)
Sea grasses	Florida (USA), Puerto Rico (USA); Nova Scotia (Canada), Newfoundland (Canada), Labrador (Canada); west coast of Europe; Japan; Argentina	Short et al. (1987); Muehlstein et al. (1988); Short et al. (1993); Ralph and Short (2002); Leander et al. (2004); Bigelow et al. (2005); Burja et al. (2006); Rosa et al. (2006)
Algae	Maine (USA); Hampshire (UK); India	Booth and Miller (1968); Miller and Jones (1983); Raghukumar (1988a); Raghukumar et al. (1992); Sathe-Pathak et al. (1993)
Sediments	59–61°N, 50–51°N; Fladen Ground area of the North Sea (UK); Canada; Portugal; Indonesia, Italy	Raghukumar (1980); Bowles et al. (1999); Bongiorni et al. (2005b); Burja et al. (2006); Jakobsen et al. (2007)

they are highly abundant in such environments due to the higher availability of food resources than in control sites that were located 0.7–1.5 miles away from fish farms (Bongiorni et al. 2005b).

Ecological roles

Raghukumar (2002) carried out a comprehensive review of the ecology of Labyrinthulomycetes and found that thraustochytrids are usually associated with dead autochthonous and allochthonous plant materials, such as macroalgae and submerged mangrove leaves. They play an important role as saprobes by virtue of extracellular enzyme production and the chemical alteration of detritus. Thraustochytrids are rarely observed on living marine plant materials, unlike the labyrinthulids and aplanochytrids which are regularly found on living algae and sea grasses (Muehlstein et al. 1988; Porter 1990; Steele et al. 2005) either as parasites (pathogens) or commensals. The roles of Labyrinthulomycetes as saprobes, parasites and commensals are discussed further in the following sections.

Saprobes

The decomposition of dead organic material by Labyrinthulomycetes and their proliferous growth on detritus clearly show their saprophytic nature in marine ecosystem. These organisms are able to degrade a range of substrates present in the environment (Bahnweg 1979; Bremer and Talbot 1995; Bongiorni et al. 2005c). They may degrade recalcitrant compounds that are left over after initial degradation by other microorganisms like bacteria (Bongiorni et al. 2005b).

During field studies on the decomposition of the brown alga *Sargassum cinereum*, thraustochytrids as well as bacteria increased in numbers with the age of the detritus (Sathe-Pathak et al. 1993). While *Aplanochytrium minutum* was initially present in the healthy alga as well as in the ageing detritus, *Ulkenia visurgensis* arose later in the ageing detritus. *Aplanochytrium minutum* was frequently isolated from the brown algae *Padina tetrastromatica* and *Sargassum cinereum*, but did not show any disease-like symptoms in the algae (Raghukumar et al. 1992; Sathe-Pathak et al. 1993). Other saprobic associations of Labyrinthulomycetes on marine macroalgae are summarized in Table 12.4.

During field studies on decomposition of leaves of the mangrove *Rhizophora apiculata*, thraustochytrids colonized the detritus seven days after senescence (Raghukumar et al. 1995). Laboratory studies on the same substrate manifested their capability to elaborate various degradative enzymes in the detritus (Raghukumar et al. 1994). This and several other studies show the involvement of Labyrinthulomycetes, as one of the dominant colonizers, in the degradation of fallen mangrove leaves (Table 12.5). This indicates their importance in the detrital processes as saprobes.

Thraustochytrids colonize the fecal pellets of various invertebrates, like the salp (*Pegea confoederata*; Raghukumar and Raghukumar 1999) and sea urchin (*Lytechinus variegatus*; Wagner-Merner et al. 1980). These organisms might passively inhabit the guts of these animals which feed upon them or on Labyrinthulomycetes-infested aggregates in the water column, and hence they occur in their faeces. Alternatively, they might colonize fresh faeces in order to degrade the organic matter present therein. Thraustochytrids were found to be endolithic in mollusk shell fragments, especially mussel and clam

Table 12.4 Saprobic associations of Labyrinthulomycetes with marine macroalgae.

Substrate	Organism	Location	Reference
Cladophora sp., *Rhizosolenia* spp. (diatom), *Bryopsis plumose, Ceramium diaphanum*	*Labyrinthula, Thraustochytrium proliferum*	Vicinity of Woods Hole	Sparrow (1936)
Algal fragments	*Thraustochytrium roseum*	Maine	Booth and Miller (1968)
Fucus serratus	Thraustochytrid	Southsea Castle, Hampshire	Miller and Jones (1983)
Cladophora sp.	*Ulkenia visurgensis*	Beach, Goa	Raghukumar (1988)
Centroceras clavulatum, Sargassum cinereum, Padina tetrastromatica	Thraustochytrids	Goa	Raghukumar et al. (1992)
Sargassum cinereum	*Aplanochytrium minutum, Ulkenia visurgensis*	Dona Paula Bay and Zuari river, Goa	Sathe-Pathak et al. (1993)
Small pieces of various macroalgae	Labyrinthulids	Kagoshima Bay, Japan	Sakata et al. (2000)
Padina arborescens, Sargassum sp.	Labyrinthulids	Iriomote and Ishigaki Islands, Okinawa Prefecture, Japan	Yokochi et al. (2001)
Dictyota cervicornis, Chaetomorpha sp., *Cladophora* sp.	*Aplanochytrium*	San Juan (Puerto Rico), Miami Harbor (Florida)	Leander et al. (2004)
Ulva rigida	Thraustochytrids	Palombina Beach, (Italy)	Bongiorni et al. (2005b)

shells (Porter and Lingle 1992). Bongiorni et al. (2005c) also found and isolated thraustochytrids growing on a dead crab exoskeleton. The dying and dead eggs of the blue crab *Callinectes sapidus* harbor large masses of thraustochytrids (Shields and Overstreet 2003). The subsistence of Labyrinthulomycetes in coral mucus and the seawater surrounding the coral reef area also suggests a definite role for saprobic Labyrinthulomycetes in association with these invertebrates (Raghukumar and Balasubramanian 1991; Nakahara et al. 1996; Harel et al. 2008).

Parasites

Among Labyrinthulomycetes, the labyrinthulids are well-known parasites of sea grasses. The wasting disease of eelgrass *Zostera marina* that has led to a severe loss of populations of eelgrass in different parts of the world for decades is caused by *Labyrinthula zosterae* (Armiger 1964; Short et al. 1987; Muehlstein et al. 1988; Hily et al. 2002).

Table 12.5 Saprobic associations of Labyrinthulomycetes with fallen mangrove leaves.

Substrate	Organism	Location	Reference
Fallen decaying leaves	*Schizochytrium mangrovei*, *Labyrinthula*	Panay Island (Philippines), Orda (Goa), Okinawa Prefecture (Japan), Kaohsiung, Tainan and Chunan (Taiwan)	Raghukumar (1988); Leaño (2001); Perveen et al. (2006); Lin et al. (2010)
Kandelia candel	*Schizochytrium mangrovei*, *Thraustochytrium striatum*, *Ulkenia* sp.	Futial National Nature Reserve (Shenzhen, China), 3 mangrove areas of Hong Kong	Fan et al. (2002a, 2002b); Kamlangdee and Fan (2003); Wong et al. (2005); Fan et al. (2007)
Rhizophora sp.	*Thraustochytrium*, *Schizochytrium*, *Labyrinthula*, *Aplanochytrium*	Panay Island (Philippines), Iriomote & Ishigaki Islands and Okinawa Prefecture (Japan), Sweetings Cay (Bahamas)	Leaño (2001); Yokochi et al. (2001); Leander et al. (2004)
Bruguiera sp.	*Labyrinthula*, thraustochytrids	Panay Island (Philippines), Iriomote & Ishigaki Islands and Okinawa Prefecture (Japan), Bang Khun Thean District, Bangkok (Thailand)	Leaño (2001); Yokochi et al. (2001); Bongiorni et al. (2005a); Juntaban et al. (2007)

In addition, a *Labyrinthula* spp. also infects turtle grass *Thalassia testudinum* (Steele et al. 2005). *Labyrinthula terrestris* infects turf grass, causing the rapid blight disease found on golf courses (Bigelow et al. 2005; Craven et al. 2005). *Labyrinthula* spp. also parasitize several algae, such as *Chaetomorpha*, *Lyngbya*, *Cladophora* and *Rhizoclonium* (Raghukumar 1987a, 1987b). The hyperparasitic nature of *Labyrinthula* was demonstrated upon its growth on the fungal sporangia of *Pontisma lagenidioides*, which were found to infect algae collected from the intertidal region of Mandapam in India (Raghukumar 1987a).

Among thraustochytrids, the QPX is a well-known pathogen of the clam *Mercenaria mercenaria* (Whyte et al. 1994; Smolowitz and Leavitt 1997; Anderson et al. 2003). QPX is thought to be an opportunistic pathogen occurring naturally in clam-growing waters and causing sudden disease outbreak under stressful conditions. The infected clams show poor growth and a weakened response to predators, leading to increased mortality in nature and a reduced shelf life.

Diatoms including *Coscinodiscus* sp., *Navicula* sp., *Nitzschia* sp., *Grammatophora* sp. and *Melosira* sp. have been shown to be infected by the thraustochytrid *Ulkenia amoeboidea* (Raghukumar 2006). Other parasitic associations of thraustochytrids are listed in Table 12.6.

Table 12.6 List of parasitic associations of Labyrinthulomycetes.

Disease	Causal organism	Reference
Skin ulcers in octopus (*Eledone cirrhosa*)	Thraustochytrids and labyrinthulids	Polglase (1980)
Yellow spot disease in the skin of the nudibranch *Tritonia diomedea*	Thraustochytrids	McLean and Porter (1982)
Infected gills of squid (*Illex illecebrosus lesueur*)	*Schizochytrium*	Jones and O'Dor (1983)
Ulcerative dermal lesions in caged rainbow trout (*Salmo gairdneri*)	Thraustochytrids	Polglase et al. (1986)
Infection in abalone *Haliotis kamtschatkana*	*Aplanochytrium haliotidis*	Bower (1987); Bower et al. (1989)
Ulcerative lesions of fishes *Lates calcarifer* and *Etroplus suratensis*	*Schizochytrium*	Loganathan et al. (1985)
Lesions and dissolution of posterior part of flatworm *Macrostomum lignano*	*Thraustochytrium caudivorum*	Scharer et al. (2007)

Commensals

Labyrinthulomycetes also exist in non-parasitic associations with invertebrates. Kramarsky-Winter et al. (2006) reported Labyrinthulomycetes-like organisms on the surface and in the tissue of the solitary coral *Fungia granulosa*. Molecular investigation identified them as *Labyrinthula* and *Oblongichytrium multirudimentale*. These species were also found in the coral gastrodermal layer in areas devoid of zooxanthellae. Raghukumar and Balasubramanian (1991) isolated thraustochytrids and aplanochytrids from the polyps of the *Acropora* and *Pocillopora* corals from Lakshadweep Islands. When present in the corals, Labyrinthulomycetes provide them with nutritional sources that aid survival during bleaching events (Harel et al. 2008).

Aplanochytrid cells were found within the guts, as well as surrounding the gut region of mesozooplankton, when treated with *in situ* hybridization using the probes designed against aplanochytrids isolated from zooplankton by Damare and Raghukumar (2010). They were also consistently present within different parts of the body in the chaetognaths, suggesting that aplanochytrids may occur as a regular symbiotic component in their bodies. These organisms may therefore serve as a source of docosahexaenoic acid (DHA) to the zooplankton, which needs them for growth and reproduction.

Similarly, the thraustochytrid *Ulkenia visurgensis* was detected in the coelenteron and hydranth of a healthy hydroid by immunofluorescence (Raghukumar 1988a). Thraustochytrids were also present in the gut contents of the sea urchin *Lytechinus variegates* (Wagner-Merner et al. 1980). Although present within oyster tissues, these organisms did not appear to be pathogenic (Perkins 1973; Alderman 1976). *Schizochytrium*-like cells were present within the shells of mollusks (Porter and Lingle 1992). Though heterotrophic nutrition seems arduous in the calcareous shells, the shell matrix proteins, such as conchiolin, appear to provide thraustochytrids with sufficient carbon and nitrogen.

Commercial/industrial importance

Aside from the ecological roles of Labyrinthulomycetes as saprobes, pathogens or commensals in many marine ecosystems, some members of this group of fungal-like microorganisms have been reported on due to their commercial importance in the production of novel compounds. This is in particular reference to thraustochytrids, which were reported to produce polyunsaturated fatty acids (PUFAs), carotenoids and several degrading enzymes.

PUFA production

PUFAs are necessary constituents of cell membranes and of many cell signaling systems. Deficiencies in PUFA can be associated with defects in cellular functions that may lead to disease (Lewis et al. 1999). PUFAs have been shown to be essential dietary components for humans (Takahata et al. 1998) and also for marine aquaculture operations, especially in larval production (Sorgeloos and Leger 1992).

The biotechnological potential of thraustochytrids was first recognized by the discovery of significantly high PUFA contents of their cells (Lewis et al. 1999), particularly DHA. With the increased interest in the nutritional importance of PUFAs in the past two decades, these microbes have been tapped as an alternative source of PUFAs, which are currently mainly sourced from fish oils. Compared to fish oils, PUFAs produced by thraustochytrids have a less fishy smell, are highly purified, and contain relatively low concentrations of structurally-related PUFAs that simplify the downstream processing of DHA (Singh et al. 1996). Therefore, thraustochytrid-derived oils can decrease the expense currently involved in producing high-purity microbial oils, if these microbes can be mass produced at an appropriate cost (Lewis et al. 1999).

Heterotrophic cultivation of thraustochytrids offers several challenges, since biomass, lipid content, and fatty acid profiles are dependent on growth conditions (Shene et al. 2010). As such, several studies were undertaken to assess different thraustochytrid isolates for their production of DHA and other PUFAs, and to optimize cultural conditions for maximum production of these fatty acids. Early studies on DHA production by *Schizochytrium* (Nakahara et al. 1996; Yaguchi et al. 1997; Vazhappilly and Chen 1998; Yokochi et al. 1998) and *Thraustochytrium* (Bajpai et al. 1991a, 1991b; Li and Ward 1994; Singh and Ward 1996; Singh et al. 1996) detailed various culture conditions for the mass production of PUFAs (using flasks or fermenters). The DHA yields of the different strains tested ranged from as low as 0.4 mg/liter for *Scizochytrium aggregatum* (Vazhappilly and Chen 1998) to a high of 13,300 mg/liter for *Schizochytrium* sp. SR21 (Yaguchi et al. 1997).

More recent studies, especially on tropical and sub-tropical strains of thraustochytrids isolated from fallen mangrove leaves, also showed that *Schizochytrium* (*Aurantiochytrium*) species can produce high yields of DHA compared to *Thraustochytrium* spp. (Yokochi et al. 1998; Bowles et al. 1999; Fan et al. 2001; Kamlangdee and Fan 2003; Leaño et al. 2003; Unagul et al. 2005; Yamasaki et al. 2006; Chi et al. 2007; Shene et al. 2010; Arafiles et al. 2011). Table 12.7 summarizes the DHA production of some species of tropical and sub-topical thraustochytrids.

Due to the high DHA contents of thraustochytrid cells, their commercial utilization has been investigated and applied in several industries, including aquaculture and livestock.

Table 12.7 Docosahexaenoic acid (DHA) production by selected thraustochytrids.

Species	DHA (mg/liter)	Culture method	Reference
Aurantiochytrium mangrovei	2,762	Shake flask	Fan et al. (2001)
	1,400	Shake flask	Leaño et al. (2003)
Aurantiochytrium limacinum	13,300	Bioreactor	Yaguchi et al. (1997)
	4,700	–	Nakahara et al. (1996)
	4,200	Shake flask	Yokochi et al. (1998)
Thraustochytrium aureum ATCC 34304	270	Shake flask	Bajpaj et al. (1991a)
	510	Shake flask	Li and Ward (1994)
Thraustochytrium roseum ATCC 28210	841	Shake flask	Li and Ward (1994)
	2,000	Shake flask	Singh and Ward (1996)
Thraustochytrium sp. ATCC 20892	707	Shake flask	Singh et al. (1996)
Thraustochytrium sp. G13	1,956	Bioreactor	Bowles et al. (1999)
Ulkenia sp. SAM 2179	5,500	Bioreactor	Tanaka et al. (2003)

Freeze-dried *Schizochytrium* spp. (Barclay and Zeller 1996; Jaritkhuan 2002; Jaritkhuan and Jones 2007) and *Schizochytrium mangrovei* (Estudillo-del Castillo et al. 2009) cells were used to enrich *Brachionus* and *Artemia,* which resulted in the high DHA-content of these important live food organisms for larval fish and crustaceans. *Schizochytrium* was also used to partially replace fish oil or was used as a feed supplement for Atlantic salmon (Carter et al. 2003), channel catfish (Li et al. 2009), sea bream (Ganuza et al. 2008) and marine mussels (Langdon and Onal 1999).

For livestock, *Schizochytrium* sp. was also used to enhance the DHA contents of poultry eggs (Sims et al. 2000) and omega-3 fatty acid composition of poultry breast meat (Mooney et al. 1998). This procedure of enriching poultry meat yields a product that may serve as a substitute for fish-based PUFAs in the human diet. Franklin et al. (1999) reported that fat in the milk from cows fed *Schizochytrium* contained higher concentrations of conjugated linoleic acid, DHA and trans-vaccenic acid. Similar results were obtained with the milk fat content of dairy ewes, where a significant increase in DHA, eicosapentaenoic acid and docosapentaenoic acid contents were observed (Papadopoulos et al. 2002).

By and large, the commercial utilization of thraustochytrids in both aquaculture and livestock industries offers a great potential in replacing conventional sources of fatty acids, such as from fish oil that is presently in short supply and commands a high price. Moreover, the ability of thraustochytrids to produce novel products other than

PUFAs also shows their potential for other industries, such as biodiesel production and pharmaceuticals. In all cases, optimization of culture conditions for the maximum production of biomass and of the different bioactive products is necessary.

Aside from DHA, some strains of thraustochytrids also produce other PUFAs, such as arachidonic acid, docosapentaenoic acid, eicosapentaenoic acid and docosatetraenoic acid (Barclay and Zeller 1996). Moreover, they are also found to produce saturated fatty acid (palmitic acid) and monounsaturated fatty acid (oleic acid; Li and Ward 1994; Ashford et al. 2000; Leaño et al. 2003). The high production of palmitic and oleic acids by thraustochytrids is important in the biofuel industry (Fisher et al. 2008), and may provide an inexpensive alternative to common sources of biofuels, such as plants (Arafiles et al. 2011).

Production of other novel compounds

The numerous studies documenting the production of PUFAs, especially DHA, by thraustochytrids has led to the evaluation of some species for the production of other bioactive compounds including carotenoids, squalene and extracellular enzymes. Shimidzu et al. (1996) confirmed the presence of carotenoids (astaxanthin and phoenicoxanthin) in a *Thraustochytrium* sp., which was found to be positively associated with total biomass concentration. The level of astaxanthin in the total carotenoid production of this species can be as high as 50%. Another thraustochytrid strain (KH105; belonging to the genus *Schizochytrium*), isolated from seawater as a DHA-producer, was also found to accumulate significant levels of β-carotene and xanthophylls including canthaxanthin and astaxanthin (Aki et al. 2003). The isolate is characterized by muddy orange or yellow ochre growth when grown in liquid medium, suggesting the accumulation of endopigments. After four days of cultivation, it produced up to 6.1 mg/liter astaxanthin (cultured in a medium with 0.3% nitrogen source) and 10 mg/liter canthaxanthin (with 6% nitrogen source). Astaxanthin is a superior antioxidant that is also known as "super Vitamin E". It is responsible for the orange-pigmentation in animals (e.g., crustaceans) and is used as a growth enhancer and immune-stimulant for fish and shrimps (Christiansen et al. 1995; Chien et al. 2003; Pan et al. 2003).

Squalene, a strong antioxidant mainly derived from the liver oils of deep-sea sharks, was first reported in a *Thraustochytrium* sp. by Weete et al. (1997), who observed high levels of squalene production, representing 63% of the total non-saponifiable lipids. Production of squalene by a thraustochytrid strain ACEM 6063, rich in omega-3 PUFAs, was significantly affected by both physical and chemical culture conditions (e.g., temperature, dissolved oxygen level, and culture time; Lewis et al. 2001). The squalene content of three strains of *Aurantiochytrium mangrovei* isolated from decaying mangrove leaves in Hong Kong was analyzed by Jiang et al. (2004). They found that the highest squalene production was 0.162 mg/g cell dry weight and that it was significantly affected by culture time. Enhanced production of squalene in a strain of *Aurantiochytrium mangrovei* (FB3) was reported through the medium optimization and treatment of terbinafine, which resulted in increased biomass production at a higher glucose concentration, and enhanced squalene production after treatment with concentrations of 10 and 100 mg/liter terbinafine (Fan et al. 2010).

As Labyrinthulomycetes are abundant in many marine ecosystems, they are reported to probably overcome competition with other microbes, such as bacteria and other higher fungi, by producing unique degrading enzymes that could also have potential industrial applications (Raghukumar 2008). Thraustochytrid strains exhibit a wide spectrum of enzymes including lipases, proteases and carbohydrate-degrading enzymes, which are involved in the hydrolysis of all classes of organic compounds, suggesting that thraustochytrids are capable of degrading a large variety of substrates (Bongiorni et al. 2005c). Strains of *Thraustochytrium, Schizochytrium* and *Aurantiochytrium* were also found to produce protease, lipase, phosphatase, urease and alpha-glucosidase, while amylase was only detected in *Thraustochytrium* (Taoka et al. 2009). Damare and Raghukumar (2006) reported the production of protease, amylase, lipase and chitinase by aplanochytrids.

Besides degrading enzymes, thraustochytrids also produce extracellular polysaccharides (EPSs), which are being explored for various biotechnological applications: anti-tumor agents, anticoagulants; and wound dressing (Jain et al. 2005). In their study, the EPS production of two thraustochytrid strains increased with the age of the culture, reaching a peak at stationary phase. The EPSs produced contained 39–53% sugars, besides proteins, lipids, uronic acids and sulfates.

Conclusion

Labyrinthulomycetes are abundant in many marine habitats all over the world, playing important ecological roles in the marine aquatic ecosystems. They are probably the most dominant group of obligate osmoheterotrophs among marine eukaryotes (Raghukumar 2008). From being important colonizers and degraders of organic detritus (e.g., decaying mangrove leaves and marine macroalgae) to being parasites of marine plants and animals, they have been extensively studied with regard to their isolation, abundance, taxonomy, physiology and pathogenicity. The unique characters of this group of fungus-like microorganisms, sharing some features of fungi and microalgae, means that their classification has been somewhat problematic, having been transferred from one kingdom to another over the years. With the introduction of modern molecular techniques for phylogenetic classification, they are now firmly classified under one super-group (Chromalveolata), together with other stramenopiles, including the oomycetes and hyphochytrids. Phylogenetic reevaluation of some established genera has resulted in the erection of new genera and the reclassification of several species. This trend will continue when more taxa in this super group are subjected to further molecular phylogenetic analyses and reclassifications.

The discovery of PUFA production in many thraustochytrid species has led to countless investigations assessing their DHA production potential and, later, production of other novel compounds. Recent studies have shown that some thraustochytrid strains can be cultured to produce a high biomass, containing substantial amounts of lipid rich in PUFAs, through the manipulation of physical and chemical parameters of the culture (Fan et al. 2001; Leaño et al. 2003; Unagul et al. 2005; Shene et al. 2010; Arafiles et al. 2011). These include the optimization of culture conditions (salinity, temperature) and provision of important nutrients needed for growth, including nitrogen and carbon sources, and even some vitamins and minerals (e.g., vitamin B, phosphates) (Raghukumar 2008). As such, several strains are now being utilized for commercial preparation and the production of valuable products rich in PUFAs, particularly DHA.

With the ecological and industrial importance of Labyrinthulomycetes, they will continue to be one of the most interesting fungus-like microorganisms studied and exploited in the near future. This is because of their abundance in marine and estuarine habitats, their ability to adapt to fluctuating environments (especially sub-tropical and tropical species), their ability to effectively degrade organic detritus and, for some (parasites and commensals), to colonize and infect living marine organisms (plants and animals), and their high potential in the production of important novel compounds. It was reported that the biomass of Labyrinthulomycetes in the water column may often match or even exceed that of bacteria, although their abundance is often seasonal (Raghukumar and Damare 2011). They are often eliminated by competition (mostly by bacteria), however, but can easily survive by resorting to "left-over scavenging" of nutrient substrates. This might explain their persistence in many marine and estuarine ecosystems.

References

Adl SM, Simpson AGB, Farmer MA, Andersen RA, Anderson OR, Barta JR, Bowser SS, Brugerolle G, Fensome RA, Fredericq S, James TY, Karpov S, Kugrens P, Krug J, Lane CE, Lewis LA, Lodge J, Lynn DH, Mann DG, Mccourt RM, Mendoza L, Moestrup Ø, Mozley-Standridge SE, Nerad TA, Shearer CA, Smirnov AV. The new higher level classification of Eukaryotes with emphasis on the taxonomy of protists. J Eukaryot Microbiol. 2005;52:399–451.

Aki T, Hachida K, Yoshinaga M, Katai Y, Yamasaki T, Kawamoto S, Kakizono T, Maoka T, Shigeta S, Suzuki O, Ono K. Thraustochytrid as potential source of carotenoids. J Am Oil Chem Soc. 2003;80:789–794.

Alderman DJ. Fungal diseases of marine animals. In: Jones EBG, ed. Recent advances in aquatic mycology. London: Elek Science, 1976:223–260.

Alexopoulos CJ, Mims CW, Blackwell M. Introductory mycology. 4th ed. John Wiley & Sons: New York, USA, 1996.

Alldredge AL, Youngbluth MJ. The significance of macroscopic aggregates (marine snow) as sites for heterotrophic bacterial production in the mesopelagic zone of the subtropical Atlantic. Deep-Sea Res. 1985;32:1445–1456.

Amon JP. Thraustochytrids and labyrinthulids of terrestrial, aquatic and hypersaline environments of the Great Salt Lake, USA. Mycologia 1978;70:1299–1301.

Anderson RS, Kraus BS, McGladdery SE, Reecec KS, Stoke NA. A thraustochytrid protist isolated from *Mercenaria mercenaria*: molecular characterization and host defense responses. Fish Shellfish Immunol. 2003;15:183–194.

Arafiles KHV, Alcantara JCO, Cordero PRF, Batoon JAL, Galura FS, Leaño EM, Dedeles GR. Cultural optimization of thraustochytrids for biomass and fatty acid production. Mycosphere 2011;2:521–531.

Armiger LC. An occurrence of *Labyrinthula* in New Zealand *Zostera*. NZ J Bot. 1964;2:3–9.

Artemtchuk NJ. The fungi of the White Sea: III. New phycomycete discovered in the Great Salma Strait of the Kandalakshial Bay. Veroff Inst Meeresforch Bremerh. 1972;13:231–237.

Ashford A, Barclay WR, Weaver CA, Giddings TH, Zeller T. Electron microscopy may reveal structure of docosahexaenoic acid-rich oil within *Schizochytrium* sp. Lipids 2000;35:1377–1387.

Bahnweg G. Studies on the physiology of thraustochytriales II. Carbon nutrition of *Thraustochytrium* spp., *Schizochytrium* sp., *Japonochytrium* sp., *Ulkenia* spp. and *Labyrinthuloides* spp. Veroff Inst Meeresforch Bremerh. 1979;17:269–273.

Bahnweg G, Sparrow FK. *Aplanochytrium kerguelensis* gen. nov. sp. nov., a new phycomycete from subantarctic marine waters. Arch Mikrobiol. 1972;81:45–49.

Bahnweg G, Sparrow FK. Four new species of *Thraustochytrium* from Antarctic regions, with notes on the distribution of zoosporic fungi in the Antarctic marine ecosystem. Am J Bot. 1974a;61:754–766.

Bahnweg G, Sparrow FK. Occurrence, distribution and kinds of zoosporic fungi in subantarctic and Antarctic waters. Veroff Inst Meeresforch Bremerh Suppl. 1974b;5:149–157.

Bajpai PK, Bajpai P, Ward OP. Optimization of production of docosahexaenoic acid (DHA) by *Thraustochytrium aureum* ATCC 34304. J Am Oil Chem Soc. 1991a;68:509–514.

Bajpai PP, Bajpai P, Ward OP. Production of docosahexaenoic acid by *Thraustochytrium aureum*. Appl Micobiol Biotechnol. 1991b;35:706–710.

Barclay WB, Zeller B. Nutritional enhancement of ω-3 and ω-6 fatty acids in rotifers and *Artemia* nauplii by feeding spray-dried *Schizochytrium* sp. J World Aquacult Soc. 1996;27:314–322.

Bigelow DM, Olsen MW, Gilbertson RL. *Labyrinthula terrestris* sp. nov., a new pathogen of turf grass. Mycologia 2005;97:185–190.

Bongiorni L, Dini F. Distribution and abundance of thraustochytrids in different Mediterranean coastal habitats. Aquat Microb Ecol. 2002;30:49–56.

Bongiorni L, Jain R, Raghukumar S, Aggarwai RM. *Thraustochytrium gaertnerium* sp. nov.: a new thraustochytrid stramenopilan protist from mangroves of Goa, India. Protist 2005a;156:303–315.

Bongiorni L, Mirto S, Pusceddu A, Danovaro R. Response of benthic protozoa and thraustochytrid protists to fish farm impact in seagrass (*Posidonia oceanica*) and soft-bottom sediments. Microb Ecol. 2005b;50:268–276.

Bongiorni L, Pusceddu A, Danovaro R. Enzymatic of epiphytic and benthic thraustochytrids involved in organic matter degradation. Aquat Microb Ecol. 2005c;41:299–305.

Booth T. Occurrence and distribution of some zoosporic fungi from soils of Hibben and Moresby Islands, Queen Charlotte Islands. Can J Bot. 1971;49:951–965.

Booth T, Miller CE. Comparative morphologic and taxonomic studies in the genus *Thraustochytrium*. Mycologia 1968;60:480–495.

Bower SM. *Labyrinthuloides haliotidis* n. sp. (Protozoa: Labyrinthomorpha), a pathogenic parasite of small juvenile abalone in British Columbia mariculture facility. Can J Bot. 1987;65:1996–2007.

Bower SM, McLean N, Whitaker DJ. Mechanism of infection by *Labyrinthuloides haliotidis* (Protozoa: Labyrinthomorpha), a parasite of abalone (*Haliotis kamtschatkana*) (Mollusca: Gastropoda). J Invertebr Pathol. 1989;53:401–409.

Bowles RD, Hunt AE, Bremer GB, Duchars MG, Eaton RA. Long-chain n-3 polyunsaturated fatty acid production by members of the marine protistan group the thraustochytrids: screening of isolates and optimisation of docosahexaenoic acid production. J Biotechnol. 1999;70:193–202.

Bremer GB, Talbot G. Cellulolytic enzyme activity in the marine protist *Schizochytrium aggregatum*. Bot Mar. 1995;38:37–41.

Burja AM, Radianingtyas H, Windust A, Barrow CJ. Isolation and characterization of polyunsaturated fatty acid producing *Thraustochytrium* species: screening of strains and optimization of omega-3 production. Appl Microbiol Biotechnol. 2006;72:1161–1169.

Carter CG, Bransden MP, Lewis TE, Nichols PD. Potential of thraustochytrids to partially replace fish oil in Atlantic salmon feeds. Mar Biotechnol. 2003;5:480–492.

Cavalier-Smith T. A revised six-kingdom system of life. Biol Rev Cambridge Philosophic Soc. 1998;73:203–266.

Cavalier-Smith T, Allsopp MTEP, Chao EE. Thraustochytrids are chromists, not fungi: 18S rRNA signatures of Heterokonta. Phil Trans R Soc Lond B. 1994;346:387–397.

Chi Z, Pyle D, Wen Z, Frear C, Chen S. A laboratory study of producing docosahexaenoic acid from biodiesel-waste glycerol by microalgal fermentation. Process Biochem. 2007;42:1537–1545.

Chien YH, Pan CH, Hunter B. The resistance to physical stresses by *Penaeus monodon* juveniles fed diets supplemented with astaxanthin. Aquaculture 2003;216:177–191.

Christiansen R, Glette J, Lie O, Torrissen OJ, Waagbo R. Antioxidant status and immunity in Atlantic salmon, *Salmo salar* L., fed semi-purified diets with and without astaxanthin supplementation. J Fish Dis. 1995;18:317–328.

Cienkowski L. Ueber den Ban u. Entwickelung der Labyrinth uleen. Arch f mikr Anat Bd iii, 1867.
Colaco A, Raghukumar C, Mohandass C, Cardigos F, Santos RS. Effect of shallow-water venting in Azores on a few marine biota. Cah Biol Mar. 2006;47:359–364.
Craven KD, Peterson PD, Windham DE, Mitchell TK, Martin SB. Molecular identification of the turf grass rapid blight pathogen. Mycologia 2005;97:160–166.
Damare V. Ecological studies and molecular characterization of thraustochytrids and aplanochytrids from oceanic water column. Ph.D. Thesis, 2009.
Damare V, Raghukumar S. Morphology and physiology of the marine straminipilan fungi, the aplanochytrids isolated from the equatorial Indian Ocean. Indian J Mar Sci. 2006;35:326–340.
Damare V, Raghukumar S. Abundance of thraustochytrids and bacteria in the equatorial Indian Ocean, in relation to transparent exopolymeric particles (TEPs). FEMS Microbiol Ecol. 2008;65:40–49.
Damare V, Raghukumar S. Association of the stramenopilan protists, the aplanochytrids, with zooplankton of the equatorial Indian Ocean. Mar Ecol Prog Ser. 2010;399:53–68.
Dick MW. Straminipilous fungi: systematic of the peronosporomycetes, including cccounts of the marine protists, the plasmodiophorids and similar organisms. Kluwer Academic Publishers: The Netherlands, 2001.
Dykstra MJ, Porter D. *Diplophrys marina*, a new scale-forming marine protist with labyrinthulid affinities. Mycologia 1984;76:626–632.
Estudillo-del Castillo C, Gapasin RSJ, Leaño EM. Enrichment potential of HUFA-rich thraustochytrid *Schizochytrium mangrovei* for the rotifer *Brachionus plicatilis*. Aquaculture 2009;293:57–61.
Fan KW, Chen F. Production of high-value products by marine microalgae thraustochytrids. In: Yang ST, ed. Bioprocessing for value-added products from renewable resources. Elsevier BV, 2006:293–323.
Fan KW, Chen F, Jones EBG, Vrijmoed LLP. Eicosapentaenoic and docosahexaenoic acids production by and okara-utilizing potential of thraustochytrids. J Indust Microbiol Biotechnol. 2001;227:199–202.
Fan KW, Vrijmoed LLP, Jones EBG. Physiological studies of subtropical mangrove thraustochytrids. Bot Mar. 2002a;45:50–57.
Fan KW, Vrijmoed LLP, Jones EBG. Zoospore chemotaxis of mangrove thraustochytrids from Hong Kong. Mycologia 2002b;94:569–578.
Fan KW, Jiang Y, Faan YW, Chen F. Lipid characterization of mangrove thraustochytrid-*Schizochytrium mangrovei*. J Agric Food Chem. 2007;55:2906–2910.
Fan KW, Aki T, Chen F, Jiang Y. Enhanced production of squalene in the thraustochytrid *Aurantiochytrium mangrovei* by medium optimization and treatment with terbinafine. World J Microbiol Biotechnol. 2010;26:1303–1309.
Fisher L, Nicholls D, Sanderson K. Production of biodiesel. World Intellectual Property Organization 067605, 2008.
Franklin ST, Martin KR, Baer RJ, Schingoethe DJ, Hippen AR. Dietary marine algae (*Schizochytrium* sp.) increases concentrations of conjugated linoleic, docosahexaenoic and transvaccenic acids in milk of dairy cows. J Nutr. 1999;129:2048–2054.
Gaertner A. Eine method des qantitativen Nachweises niederer mit pollen koderbarer pilze in meerwasser und im sediment. Veroff Inst Meeresforch Bremerh Suppl. 1968;3:75–92.
Ganuza E, Benitez-Santana T, Atalah E, Vega-Orellana O, Ganga R, Izquierdo MS. *Crytheconidinium cohnii* and *Schizochytrium* sp. as potential substitutes to fisheries-derived oils from seabream (*Sparus aurata*) microdiets. Aquaculture 2008;277:109–116.
Harel M, Ben-Dov E, Rasoulouniriana D, Siboni N, Kramarsky-Winter E, Loya Y, Barak Z, Wiesman Z, Kushmaro A. A new thraustochytrid, strain Fng1, isolated from the surface mucus of the hermatypic coral *Fungia granulosa*. FEMS Microbiol Ecol. 2008;64:378–387.

Harrison JL, Jones EBG. Zoospore discharge in *Thraustochytrium striatum*. Trans Br Mycol Soc. 1974;62:283–288.

Hily C, Raffin C, Brun A, den Hartog C. Spatio-temporal variability of wasting disease symptoms in eelgrass meadows of Brittany (France). Aquat Bot. 2002;72:37–53.

Honda D, Yokochi T, Nakhara T, Erata M, Higashihara T. *Schizochytrium limacinum* sp. nov., a new thraustochytrid from a mangrove area in the West Pacific Ocean. Mycol Res. 1998;102:439–448.

Honda D, Yokochi T, Nakahara T, Raghukumar S, Nakagiri A, Schaumann K, Higashira T. Molecular phylogeny of labyrinthulids and thraustochytrids base on the sequencing of 18S ribosomal RNA gene. J Eukaryot Microbiol. 1999;46:637–647.

Jain R, Raghukumar S, Tharanathan R, Bhosle NB. Extraclellualr polysaccharide production by thraustochytrid protists. Mar Biotechnol. 2005;7:184–192.

Jakobsen AN, Aasen IM, Strøm AR. Endogenously synthesized (-)-proto-quercitol and glycine betaine are principal compatible solutes of *Schizochytrium* sp. Strain S8 (ATCC 20889) and three new isolates of phylogenetically related thraustochytrids. Appl Environ Microbiol. 2007;73:5848–5856.

Jaritkhuan S. Thraustochytrids: a new alternative source of fatty acids for aquaculture. In: Hyde KD, ed. Fungi in marine environments. Hong Kong: Fungal Diversity Press, 2002:345–357.

Jaritkhuan S, Jones EBG. Marine microbe (*Schizochytrium* sp.) from mangrove forest: fatty acid source to *Artemia* enrichment. In: Mangrove forest conference on sufficiency economy based for coastal aquaculture. 12–14 September 2007, Holiday Inn Resort, Cha-am, Petchaburi, Thailand, 2007:534–545.

Jiang Y, Fan KW, Wong RTY, Chen F. Fatty acid composition and squalene content of the marine microalga *Schizochytrium mangrovei*. J Agric Food Chem. 2004;52:1196–1200.

Jones EBG, Alderman DJ. *Althornia crouchii* gen. *et* sp. nov. A marine biflagellate fungus. Nova Hedw. 1971;21:381–399.

Jones GM, O'Dor RK. Ultrastructural observations on a thraustochytrid fungus parasitic in the gills of squid (*Ilex illecebrosus* LeSueur). J Parasitol. 1983;69:903–911.

Juntaban J, Jaritkhuan S, Suanjit S. Diversity of thraustochytrids isolated from mangrove leaves at Bang Khun Thean, Bangkok. In: 8th National graduate research conference "Thai graduate studies under His Royal Beneficence", Mahidol University, Thailand, 2007;1–10.

Kamlangdee N, Fan KW. Polyunsaturated fatty acids production by *Schizochytrium* sp. isolated from mangrove. J Sci Technol. 2003;25:643–650.

Kimura H, Fukuba T, Naganuma T. Biomass of thraustochytrid protoctists in coastal water. Mar Ecol Prog Ser. 1999;189:27–33.

Kimura H, Sato M, Sugiyama C, Naganuma T. Coupling of thraustochytrids and POM, and of bacterio- and phytoplankton in a semi-enclosed coastal area: implication for different substrate preference by the planktonic decomposers. Aquat Microb Ecol. 2001;25:293–300.

Kiørboe T. Formation and fate of marine snow: small-scale processes with large-scale implications. Mar Sci. 2001;65(Suppl. 2):57–71.

Kramarsky-Winter E, Harel M, Siboni N, Ben Dov E, Brickner I. Identification of a protist-coral association and its possible ecological role. Mar Ecol Prog Ser. 2006;317:67–73.

Kumon Y, Yokoyama R, Haque Z, Yokochi T, Honda D, Nakahara T. A new Labyrinthulid isolate that produces only docosahexaenoic acid. Mar Biotechnol. 2006;8:170–177.

Langdon C, Onal E. Replacement of living microalgae with spray-dried diets for the marine mussel *Mytilus galloprovicialis*. Aquaculture 1999;180:283–294.

Leander CA, Porter D. Redefining genus *Aplanochytrium* (Phylum Labyrinthulomycota). Mycotaxon 2000;56:439–444.

Leander CA, Porter D. The labyrinthulomycota is comprised of three distinct lineages. Mycologia 2001;93:159–164.

Leander CA, Porter D, Leander BS. Comparative morphology and molecular phylogeny of aplanochytrids (Labyrinthulomycota). Europ J Protistol. 2004;40:317–328.

Leaño EM. Straminipilous organisms from fallen mangrove leaves from Panay Island, Philippines. Fungal Divers. 2001;6:75–81.

Leaño EM, Gapasin RSJ, Polohan B, Vrijmoed LLP. Growth ad fatty acid production of thraustochytrids from Panay mangroves, Philippines. Fungal Divers. 2003;12:111–122.

Lewis TE, Nichols PD, McMeekin TA. The biotechnological potential of thraustochytrids. Mar Biotechnol. 1999;1:580–587.

Lewis TE, Nichols PD, McMeekin TA. Sterol and squalene content of a docosahexaenoic acid-producing thraustochytrids: influence of culture age, temperature and dissolved oxygen. Mar Biotechnol. 2001;3:439–447.

Li MH, Robinson EH, Tucker CS, Manning BB, Khoo L. Effects of dried algae *Schizochytrium* sp., a rich source of docosahexaenoic acid, on growth, fatty acid composition, and sensory quality of channel catfish *Ictalurus punctatus*. Aquaculture 2009;293:57–61.

Li ZY, Ward OP. Production of docosahexaenoic acid by *Thrasutochytrium roseum*. J Indust Microbiol. 1994;13:238–241.

Lin YC, Leaño EM, Pang KL. Effects of Cu(II) and Zn(II) on growth and cell morphology of thraustochytrids isolated from fallen mangrove leaves in Taiwan. Bot Mar. 2010;53:581–586.

Loganathan B, Venugopalan VK, Raghukumar S. *Schizochytrium* sp. associated with external lesions of some estuarine fishes. J Singapore Natl Acad Sci. 1985;14:132–134.

Lyons MM, Ward JE, Smolowitz R, Uhlinger KR, Gast RJ. Lethal marine snow: pathogen of bivalve mollusc concealed in marine aggregates. Limnol Oceanogr. 2005;50:1983–1988.

Margulis L, Schwartz KV. Five kingdoms: an illustrated guide to the phyla of life on earth. 2nd ed. New York: W.H. Freeman, 1988.

Massana R, Castresana J, Balague V, Guillou L, Romari K, Grosillier A, Valentin K, Pedros-Alio C. Phylogenetic and ecological analysis of novel marine stramenopiles. Appl Environ Microbiol. 2004;70:3528–3534.

McGrath CL, Katz LA. Genome diversity in microbial eukaryotes. Trends Ecol Evol. 2004;19:32–38.

McLean IN, Porter D. The yellow-spot disease of *Tritonia diomedea* Bergh, 1894 (Mollusca: Gastropoda: Nudibranchia): encapsulation of the thraustochytriaceous parasite by host amoebocytes. J Parasitol. 1982;68:243–252.

Miller JD, Jones EBG. Observations on the association of thraustochytrid marine fungi with decaying seaweed. Bot Mar. 1983;26:345–351.

Mooney JW, Hirschler EM, Kennedy AK, Sams AR, Van Elswyk ME. Lipid and flavour quality of stored breast meat from broilers fed marine algae. J Sci Food Agric. 1998;78:134–140.

Moro I, Negrisolo E, Callegaro A, Andreoli C. *Aplanochytrium stocchinoi*: a new Labyrinthulomycota from the Southern Ocean (Ross Sea, Antartica). Protist 2003;154:331–340.

Moss ST. Biology and phylogeny of the Labyrinthulaes and Thraustochytriales. In: Moss ST, ed. The biology of marine fungi. Cambridge: Cambridge University Press, 1986:105–129.

Moss ST. Thraustochytrids and other zoosporic marine fungi. In: Patterson DJ, Larsen J, eds. The biology of freeliving heterotrophic flagellates. Oxford: Clarendon Press, 1991:415–425.

Muehlstein LK, Porter D, Short FT. *Labyrinthula* sp., a marine slime mold producing the symptoms of wasting disease in eelgrass, *Zostera marina*. Mar Biol. 1988;99:465–472.

Naganuma T, Takasugi H, Kimura H. Abundance of thraustochytrids in coastal plankton. Mar Ecol Prog Ser. 1998;162:105–110.

Naganuma T, Kimura H, Karimoto R, Pimenov NV. Abundance of planktonic thraustochytrids and bacteria and the concentration of particulate ATP in the Greenland and Norwegian Seas. Polar Biosci. 2006;20:37–45.

Nakagiri A. Diversity of halophytophthoras in subtropical mangroves and factors affecting their distribution. Proceedings of the Asia-Pacific Mycological Conference on Biodiversity and Biotechnology. Hua Hin, Prachuapkhirikhan: Thailand, 1998:109–113.

Nakagiri A. Ecology of *Labyrinthula* and note on modified methods for cultivation and preservation of the isolates. Aquabiology 2001;23:32–38.

Nakahara T, Yokochi T, Higashihara T, Tanaka S, Yaguchi T, Honda D. Production of docosahexaenoic and docosapentaenoic acids by *Schizochytrium* sp. isolated from Yap islands. J Am Oil Chem Soc. 1996;73:1421–1426.

Olsen MW. *Labyrinthula terrestris*: a new pathogen of cool-season turfgrasses. Mol Plant Path. 2007;8:817–820.

Pan CH, Chien YH, Hunter B. The resistance to ammonia stress of *Penaeus monodon* Fabricius juvenile fed diets supplemented with astaxanthin. J Exp Mar Biol Ecol. 2003;297:107–118.

Papadopoulos G, Goulas C, Apostolaki E, Abril R. Effects of dietary supplements of algae, containing polyunsaturated fatty acids, on milk yield and the composition of milk products in dairy ewe. J Dairy Res. 2002;69:357–365.

Parfrey LW, Barbero E, Lasser E, Dunthorn M, Bhattacharya D, Patterson DJ, Katz LA. Evaluating support for the current classification of eukaryotic diversity. PloS 2006 Dec;2(12):e220. Epub 2006 Nov 13.

Parfrey LW, Lahr DJG, Katz LA. The dynamic nature of eukaryotic genomes. Mol Biol Evol. 2008;25:787–794.

Perkins FO. A new species of marine labyrinthulid *Labyrinthuloides yorkensis* gen. nov. spec. nov. – cytology and fine structure. Arch Mikrobiol. 1973;90:1–17.

Perveen Z, Ando H, Ueno A, Ito Y, Yamamoto Y, Yamada Y, Takagi T, Kaneko T, Kogame K, Okuyama H. Isolation and characterization of a novel thraustochytrid-like microorganism that efficiently produces docosahexaenoic acid. Biotechnol Lett. 2006;28:197–202.

Polglase JL. A preliminary report on the thraustochytrid(s) and labyrinthulid(s) associated with a pathological condition in the lesser octopus *Eledone cirrhosa*. Bot Mar. 1980;23:699–706.

Polglase JL, Alderman DJ, Richards RH. Aspects of the progress of mycotic infections in marine animals. In: Moss ST, ed. The biology of marine fungi. Cambridge: Cambridge University Press, 1986:155–164.

Porter D. Labyrinthulomycota. In: Margulis L, Corliss JO, Melkonian M, Chapman D, eds. Handbook of protoctista. Boston, MA: Jones and Bartlett, 1990:388–398.

Porter D, Lingle WL. Endolithic thraustochytrid marine fungi from planted shell fragments. Mycologia 1992;84:289–299.

Raghukumar C. Fungal parasites of marine algae from Mandapam (South India). Dis Aquat Organ. 1987a;3:137–145.

Raghukumar C. Fungal parasites of the marine green algae, *Cladophora* and *Rhizoclonium*. Bot Mar. 1987b;29:289–297.

Raghukumar C. Algal-fungal interactions in the marine ecosystem: Symbiosis to parasitism. In: Tewari A, ed. Recent advances on applied aspects of Indian marine algae with reference to global scenario. Vol. 1, Central Salt & Marine Chemicals Research Institute, 2006:366–385.

Raghukumar S. *Thraustochytrium benthicola* sp. nov.: a new marine fungus from the North Sea. Trans Br Mycol Soc. 1980;74:607–614.

Raghukumar S. Detection of the thraustochytrid protist *Ulkenia visurgensis* in hydroid, using immunofluorescence. Mar Biol. 1988a;97:253–258.

Raghukumar S. *Schizochytrium mangrovei* sp. nov.: a thraustochytrid from mangroves in India. Trans Br Mycol Soc. 1988b;90:627–631.

Raghukumar S. Bacterivory: a novel dual role for thrausochytrids in the sea. Mar Biol. 1992;113:165–169.

Raghukumar S. Ecology of the marine protists, the Labyrinthulomycetes (Thraustochytrids and Labyrinthulids). Europ J Protistol. 2002;38:127–145.

Raghukumar S. The role of fungi in marine detrital processes. In: Ramaiah N, ed. Marine microbiology: facets and opportunities. Dona Paula, Goa: NIO, 2004:125–140.

Raghukumar S. Thraustochytrid marine protists: production of PUFAs and other emerging technologies. Mar Biotechnol. 2008;10:631–640.

Raghukumar S, Balasubramanian R. Occurrence of thraustochytrid fungi in corals and coral mucus. Indian J Mar Sci. 1991;20:176–181.

Raghukumar S, Damare VS. Increasing evidence for the important role of Labyrinthulomycetes in marine ecosystems. Bot Mar. 2011;54:3–11.

Raghukumar S, Raghukumar C. Thraustochytrid fungoid protists in faecal pellets of the tunicate *Pegea confoederata*, their tolerance to deep-sea conditions and implication in degradation processes. Mar Ecol Prog Ser. 1999;190:133–140.

Raghukumar S, Schaumann K. An epifluorescence microscopy method for direct detection and enumeration of the fungi-like marine protists, the thrasutochytrids. Limnol Oceanogr. 1993;38:182–187.

Raghukumar C, Nagarkar S, Raghukumar S. Association of thraustochytrids and fungi with living marine algae. Mycol Res. 1992;96:542–546.

Raghukumar S, Sharma S, Raghukumar C, Sathe-Pathak V. Thraustochytrid and fungal component of marine detritus. IV. Laboratory studies on decomposition of leaves of *Rhizophora apiculata* Blume. J Exp Mar Biol Ecol. 1994;183:113–131.

Raghukumar S, Sathe-Pathak V, Sharma S, Raghukumar C. Thraustochytrid and fungal component of marine detritus. III. Field studies on decomposition of leaves of the mangrove *Rhizophora apiculata* Blume. Aquat Microb Ecol. 1995;9:117–125.

Raghukumar S, Anil AC, Khandeparker L, Patil JS. Thraustochytrid protists as a component of marine microbial films. Mar Biol. 2000;136:603–609.

Raghukumar S, Ramaiah N, Raghukumar C. Dynamics of thraustochytrid protists in the water column of the Arabian Sea. Aquat Microb Ecol. 2001;24:175–186.

Ralph PJ, Short FT. Impact of the wasting disease pathogen, *Labyrinthula zosterae*, on the photobiology of eelgrass *Zostera marina*. Mar Ecol Prog Ser. 2002;226:265–271.

Rosa SM, Galvagno MA, Vélez CG. Primeros aislamientos de Thraustochytriales (Labyrinthulomycetes, Heterokonta) de ambientes estuariales y salinos de la Argentina. Darwiniana 2006;44:81–88.

Sakata T, Fujisawa T, Yoshikawa T. Colony formation and fatty acid composition of marine labyrinthulid isolates grown on agar media. Fish Sci. 2000;66:84–90.

Sathe-Pathak V, Raghukumar S, Raghukumar C, Sharma S. Thraustochytrid and fungal component of marine detritus. I. Field studies on decomposition of the brown alga *Sargassum cinereum*. J Ag Indian J Mar Sci. 1993;22:159–167.

Scharer L, Knoflach D, Vizoso DB, Rieger G, Peintner U. Thraustochytrids as novel parasitic protists of marine free-living flatworms: *Thraustochytrium caudivorum* sp. nov. parasitizes *Macrostomum lignano*. Mar Biol. 2007;152:1095–1104.

Shene C, Leyton A, Esparza Y, Flores L, Quilodran B, Hinzpeter I, Rubilar M. Microbial oils and fatty acids: effect of carbon source on docosahexaenoic acid (C22:26 N-3, DHA) production by thraustochytrid strains. J Soil Sci Plant Nutr. 2010;10:207–216.

Shields JD, Overstreet RM. Diseases, parasites and other symbionts. In: Kennedy VS, Cronin LE, eds. The blue crab, *Callinectes sapidus*. Maryland, USA: Maryland Sea Grant College, University of Maryland. 2003:223–339.

Shimidzu N, Goto M, Wataku W. Carotenoids as singlet oxygen quenchers in marine organisms. Fish Sci. 1996;62:134–137.

Short FT, Muehlstein LK, Porter D. Eelgrass wasting disease: Cause and recurrence of a marine epidemic. Biol Bull. 1987;173:557–562.

Short FT, Porter D, Iizumi H, Aioi K. Occurrence of the eelgrass pathogen *Labyrinthula zosterae* in Japan. Dis Aquat Org. 1993;16:73–77.

Simon M, Alldredge AL, Azam F. Bacterial carbon dynamics on marine snow. Mar Ecol Prog Ser. 1990;65:205–211.

Sims JS, Nakai S, Guenter W, eds. Egg nutrition and biotechnology. Wallington, UK: CAB International, 2000.

Singh A, Ward OP. Production of high yields of docosahexaenoic acid by *Thraustochytrium aureum* ATCC 28210. J Indust Microbiol. 1996;16:370–373.

Singh A, Wilson S, Ward OP. Docosahexaenoic acid (DHA) production by *Thrausthochytrium* sp. ATCC 20892. World J Microbiol Biotechnol. 1996;12:76–81.

Smith DC, Simon M, Alldredge AL, Azam F. Intense hydrolytic enzyme activity on marine aggregates and implications for rapid particle dissolution. Nature 1992;359:139–142.

Smolowitz R, Leavitt D. Quahog Parasite Unknown (QPX): an emerging disease of hard clams. J Shellfish Res. 1997;16:335–336.

Sorgeloos P, Leger P. Improved larviculture outputs of marine fish, shrimp and prawn. J World Aquacult Soc. 1992;23:251–264.

Sparrow FK. Biological observation on the marine fungi of Woods Hole waters. Biol Bull Mar Biol Lab Woods Hole. 1936;70:236–273.

Sparrow FK. The aquatic phycomycetes. Ann Arbor, MI, USA: Univ of Michigan Press, 1960.

Steele L, Caldwell M, Boettcher A, Arnold T. Seagrass–pathogen interactions: 'pseudo-induction' of turtlegrass phenolics near wasting disease lesions. Mar Ecol Prog Ser. 2005;303:123–131.

Stoeck T, Kasper J, Bunge J, Leslin C, Ilyin V, Epstein S. Protistan diversity in the Arctic: a case of paleoclimate shaping modern biodiversity. PLoS ONE 2007;8:e728.

Takahata K, Monobe K, Tada M, Weber P. The benefits and risks of n-3 polyunsaturated fatty acids. Biosci Biotechnol Biochem. 1998;62:2079–2085.

Tanaka S, Yaguchi T, Shimidzu S, Sogo T, Fujikawa S. Process for preparing docosahexaenoic and deocsapentaenoic acids with *Ulkenia*. US Patent No. 6509178B1, 2003.

Taoka Y, Nagano N, Okita Y, Izumida H, Sugimoto S, Hayashi M. Extracellular enzymes produced by marine eukaryotes, thraustochytrids. Biosci Biotechnol Biochem. 2009;73:180–182.

Tsui CKM, Marshall W, Yokoyama R, Honda D, Lippmeier JC, Craven KD, Peterson PD, Berbee ML. Labyrinthulomycetes phylogeny and its implications for the evolutionary loss of chloroplasts and gain ectoplasmic gliding. Mol Phylogenet Evol. 2009;50:129–140.

Ulken A. On the role of phycomycetes in the food web of different mangrove swamps with brackish waters and waters of high salinity. Kieler Meeresforschungen, Sonderhal. 1981;5:425–428.

Unagul P, Assantachai C, Phadungruengluij S, Suphantharika M, Verduyn C. Properties of the docosahexaenoic acid-producer *Schizochytrium mangrovei* Sk-02: effects of glucose, temperature and salinity and their interaction. Bot Mar. 2005;48:387–394.

Vazhappilly R, Chen F. Eicosapentaenoic acid and docosahexaenoic acid production potential of microalgae and their heterotrophic growth. J Am Oil Chem Soc. 1998;75:393–397.

Vishniac HS. On the ecology of the lower marine fungi. Biol Bull. 1956;111:410–414.

Wagner-Merner BT, Duncan WR, Lawrence JM. Preliminary comparison of Thraustochytriaceae in the guts of a regular and irregular echinoid. Bot Mar. 1980;23:95–97.

Watson SW, Raper KB. *Labyrinthula minuta* sp. nov. J Gen Microbiol. 1957;17:368–377.

Weete JD, Kim H, Gandhi SR, Wang Y, Dute R. Lipids and ultrastructure of *Thraustochytrium* sp. ATCC 26185. Lipids 1997;32:839–845.

Whittaker RH. New concepts of kingdoms or organisms: Evolutionary relations are better represented by new classifications than by the traditional two kingdoms. Science 1969;163:150–160.

Whyte SK, Cawthorn RJ, McGladdery SE. QPX (Quahaug Parasite X), a pathogen of northern quahaug *Mercenaria mercenaria* from the Gulf of St. Lawrence, Canada. Dis Aquat Org. 1994;19:129–136.

Wong MK, Vrijmoed LLP, Au DWT. Abundance of thraustochytrids on fallen decaying leaves of *Kandelia candel* and mangrove sediments in Futian National Nature Reserve. China. Bot Mar. 2005;48:374–378.

Worden AZ, Allen AE. The voyage of the microbial eukaryote. Curr Opinion Microbiol. 2010;13:652–660.

Yaguchi T, Tanaka S, Yokochi T, Nakahara T, Higashihara T. Production of high yields of docosahexaenoic acid by *Schizochytrium* sp. J Am Oil Chem Soc. 1997;74:1431–1434.

Yamasaki T, Aki T, Shinozaki M, Taguchi M, Kawamoto S, Ono K. Utilization of shochu distillery wastewater for production of polyunsaturated fatty acids and xanthophylls using thraustochytrid. J Biosci Bioeng. 2006;102:323–327.

Yokochi T, Honda D, Higashihara T, Nakahara T. Optimization of docosahexaenoic acid production by *Schizochytrium limacinum* SR21. Appl Microbiol Biotechnol. 1998;49:72–76.

Yokochi T, Nakahara T, Higashihara T, Yamaoka M, Kurane R. A new isolation method for Labyrinthulids using a bacterium, *Psychrobacter phenylpyruvicus*. Mar Biotechnol. 2001;3:68–73.

Yokoyama R, Honda D. Taxonomic rearrangement of the genus *Schizochytrium sensu lato* based on morphology, chemotaxonomical characteristics, and 18S rRNA gene phylogeny (Thraustochytriaceae, Labyrinthulomycetes): emendation for *Schizochytrium* and erection of *Aurantiochytrium* and *Oblongichytrium* gen. nov. Mycoscience 2007;48:199–211.

Yokoyama R, Salleh B, Honda D. Taxonomic rearrangement of the genus *Ulkenia* sensu lato based on morphology, chemotaxonomical characteristics, and 18S rRNA gene phylogeny (Thraustochytriaceae, Labyrinthulomycetes): emendation for *Ulkenia* and erection of *Botrochytrium, Parietichytrium*, and *Sicyoidochytrium* gen. nov. Mycoscience 2007;48:329–341.

13 Phytomyxea (Super-group Rhizaria)

Sigrid Neuhauser, Frank H. Gleason and Martin Kirchmair

The Phytomyxea (also known as Plasmodiophorids) is a group of obligate biotrophic parasites of vascular plants, brown algae, diatoms and oomycetes. They depend on a living host to complete their complex lifecycle. The major part of their lifecycle takes place inside the cells of the host, where primary or secondary plasmodia are formed upon infection. These multinucleate plasmodia develop into resting spores (secondary plasmodia) or zoosporangia (primary plasmodia) from which the biflagellate zoospores are released. Both types of zoospores have one long posteriorly-directed whiplash flagellum and one short anteriorly-directed whiplash flagellum. The two different kinds of zoospores are the only stages of the lifecycle outside the host (Karling 1968; Neuhauser et al. 2010, 2011).

The lifecycle begins with the germination of the resting spores. Only one primary zoospore is released from each resting spore. The zoospores encyst and infect a suitable host by forming a tubular cavity ("Rohr") through which a projectile-like structure ("Stachel") is pressed. This structure punctures the cell wall of the host. Through this opening, the parasite enters the host cell as an amoeba. The nucleus of the amoeba divides and a multinucleate plasmodium is formed from which zoosporangia develop. These zoosporangia are usually thin-walled and release one or more zoospores that again infect suitable hosts, form multinucleate plasmodia and finally develop into resting spores. The complete lifecycle is not known in all species (Neuhauser et al. 2011) and differs slightly between species. The general characteristics of the phytomyxean lifecycles were recently summarized by Neuhauser et al. (2011) for marine species, and by Kageyama and Asano (2009) for *Plasmodiophora brassicae*.

Phytomyxean parasites of higher plants have been known since the end of the 19th century. Some economically-important parasites of crop plants, such as *Plasmodiophora brassicae, Polymyxa graminis* or *Polymyxa betae*, have been extensively studied (Dixon 2009; Neuhauser et al. 2010). However, only limited data are available for the abundance, distribution and biodiversity of Phytomyxea in marine environments. The few marine species that have been described infect numerous ecologically-important organisms in marine environments, such as brown algae, diatoms and seagrasses (Neuhauser et al. 2011). The primary reason for this lack of data is that phytomyxean parasites are difficult to identify and recognize, and are therefore often "overlooked" by researchers working with the hosts. In some species, symptoms of infection are easy to spot – for example the formation of prominent galls in the host. Symptoms are not easily visible in non-galling species, however, and in such cases phytomyxean parasites can only be identified by time-consuming microscopic studies (Neuhauser et al. 2011).

This chapter presents a short overview of the taxonomy, biology and identification of marine Phytomyxea, while their ecological roles in marine environments have recently been reviewed by Neuhauser et al. (2011). More research on and increased awareness about this group is clearly needed, and therefore we also point out some future directions for research in the field.

Taxonomy

Before the development of molecular methods for use in phylogenetic studies, the Phytomyxea had been moved from one phylum to another on many occasions (reviewed by Neuhauser et al. 2010). This uncertain position in the tree of life was also the reason that the informal term "Plasmodiophorids" was used without taxonomic rank for the group over the past few decades (Braselton 2001). Although the taxonomy at the lower taxonomic levels might still be subject to future changes with the increasing number of informative DNA sequences, the inclusion of the Phytomyxea into the eukaryote supergroup Rhizaria and their close affiliation to the Endomyxa and the Cercozoa *sensu lato* has been robustly supported in numerous taxonomic studies by both morphological and molecular arguments (Cavalier-Smith 1993; Bulman et al. 2001; Bass et al. 2009; Burki et al. 2010).

Generally, the Phytomyxea are divided into two groups: the Plasmodiophorida – parasites of vascular plants; and the Phagomyxida – parasites of brown algae and diatoms (Bulman et al. 2001; Bass et al. 2009). The taxonomic structure of the phytomyxean parasites of seagrasses and oomycetes is still unclear because DNA sequence data are lacking for most marine species.

Until recently, the concept proposed by Karling (1968) was used and more or less complied with phylogenies based on molecular data. According to Karling's concept, "Plasmodiophorida" were characterized by a full lifecycle including the formation of resting spores and the "Phagomyxida" were characterized by an incomplete lifecycle without resting spores. Recently a second zoosporic state of the brown algal parasite *Maullinia ectocarpi* has, been described (Parodi et al. 2010). The genus *Maullinia* is placed in the Phagomyxida, which makes new taxonomic concepts necessary, especially when more comprehensive data become available. In this chapter we will therefore use ecological categories defined by the hosts for discussing marine phytomyxids (Table 13.1).

Identification and detection

With our current knowledge, it is difficult to give general guidelines for the detection of marine Phytomyxea because of the lack of information about these groups. However, some general characteristics can be used to detect and identify marine members of the Phytomyxea.

Macroscopic symptoms of infection

All marine species that have been described – with the exception of the parasites of diatoms – cause more or less marked hypertrophies ("galls") in their host. Such hypertrophies are especially useful symptoms of infection in seagrasses, where these galls can grow up to a size of more than 10 mm in diameter (Karling 1968). Besides this,

Table 13.1 Phytomyxean parasites found in marine environments including their primary host species. The host range of some of the species is probably wider.

Species name	Host species	Reference
Parasites of Seagrasses		
Plasmodiophora halophilae Ferd. *et* Winge	*Halophila* spp.	Karling (1968)
Pl. bicaudata Feldmann	*Zostera* spp.	den Hartog (1989)
Pl. diplantherae (Ferd. *et* Winge) Ivimery Cook	*Diplanthera* spp.	Karling (1968)
Pl. maritime Feldm.-Maz.	*Triglochin* spp.	Karling (1968)
Tetramyxa parasitica K I Goebel	*Zanichiella* spp. *Ruppia* spp.	Karling (1968)
Parasites of Diatoms		
Phagomyxa bellerochaea Schnepf	*Bellerochea malleus* (Brightwell) Van Heurck	Schnepf et al. (2000)
Ph. odontellae Kühn, Schnepf *et* Bulman	*Odontella sinensis* (Greville) Grunow	Schnepf et al. (2000)
Parasites of Brown algae		
Ph. algarum Karling	*Bachelotia antillarum* (Grunow) Gerloff *Hincksia mitchelliae* (Harvey) Silva	Karling (1944)
Maullinia ectocarpi I Maier et al.	*Ectocarpus siliculosus* (Dillwyn) Lyngbye	Maier et al. (2000)

the overall growth of infected seagrasses generally appears to be reduced, particularly root growth. This causes seagrass plants to be easily uprooted (den Hartog 1989) and subsequently removed from the sites where infection occurs.

Hypertrophies were also reported from studies on brown algae, although these were not as distinct as those of seagrasses (Maier et al. 2000). However hypertrophies are not a necessary or sufficient character for the presence of phytomyxean parasites, because other parasitic organisms may be able to induce the formation of hypertrophies. It is therefore necessary to confirm the presence of a phytomyxid by microscopic observations.

Microscopic characteristics

When distinct galls at the internodes of the stems of seagrasses are formed, these galls should be examined for the relatively thick-walled resting spores. These resting spores are approximately 4–6 μm in diameter, thick-walled and are not connected in *Plasmodiophora* species or are arranged into so-called cytosori forming typical tetrads as in *Tetramyxa* species. Phytomyxid resting spores are uniform in size and shape and usually occur in masses within the infected cells – a fact that allows them to be differentiated from other endophytic organisms or accumulations of plant components.

The life history and characteristics of the two species that are parasitic on brown algae differ markedly from each other (Neuhauser et al. 2011). *Maullinia ectocarpi* induces the formation of spherical, pigment-free appendices or swellings of its host. The thin-walled zoosporangia formed in these hypertrophies release the pigment-free biflagellate zoospores of the parasite. Zoospores of another type, also biflagellate but distinguished by the arrangement of the basal bodies of the flagella, are released from relatively thick-walled zoosporangia (Maier et al. 2000; Parodi et al. 2010). *Phagomyxa algarum* is characterized by a rapid lifecycle during which the host cell becomes depleted of all nutrients and the pigment-free, biflagellate zoospores are released (Karling 1944; Johnson and Sparrow 1961).

Even less is known about the parasites of diatoms. Inside the host, the multinucleate plasmodium digests the contents of the host cell and then the zoosporangia develop. The rest of the lifecycle is either missing or has not been elucidated to date (Schnepf et al. 2000).

Despite the lack of obvious morphological features that can be seen under the light microscope, all phytomyxid species have a special form of nuclear division that is called "cruciform nuclear division". Within the multinucleate plasmodia, the chromatin of the nucleus divides in a cross-like structure (Braselton and Short 1985; Braselton 2001). Before molecular data became available, this cruciform nuclear division was the defining ultrastructural feature used to place organisms in the Phytomyxea.

Future perspectives for research on marine Phytomyxea

At this time we are far from having a clear understanding of the biodiversity of marine Phytomyxea, and therefore more studies are needed. It is very likely that many more species of Phytomyxea that are parasitic on brown algae, diatoms, and marine vascular plants will be discovered in the future. Furthermore, no phytomyxean parasites of marine oomycetes have been described, but it is very likely that this form of parasitism has also evolved in marine ecosystems. Especially with the increased research on marine oomycetes (see Chapter 12) it is likely that phytomyxean parasites of these organisms will be found.

The increasing number of metagenomic, high-throughput sequencing data from different marine ecosystems will increase our knowledge about marine Phytomyxea. To identify phytomyxid sequences from the massive datasets, however, comparative genomic data from marine phytomyxids are needed to allow an annotation of unknown sequences. An increased number of DNA sequences identified and recorded in public databases will allow molecular samplings to be more targeted and will allow identification of novel species when combined with morphological investigation.

It is clear that some species of Phytomyxea are important parasites in some species of seagrasses, large marine algae and small planktonic algae. We do not currently understand the full impact of these parasites on the growth and development or on the population sizes of their hosts. Some species of Phytomyxea in terrestrial environments are known to transmit viruses causing diseases in cultivated plants (e.g. Kanyuka et al. 2003). The possibility that marine species also transmit viruses needs to be investigated, especially as the majority of the described viruses found in marine environments belong to groups that could possibly be transmitted by Phytomyxea (Suttle 2007; Neuhauser et al. 2011). In conclusion, without first understanding the biodiversity, distribution and abundance of marine Phytomyxea, we will not be able to understand their ecological impacts on marine ecosystems.

Acknowledgements

SN was supported by a Hertha-Firnberg research grant (Austrian Science Fund grant T379-B16).

References

Bass D, Chao EEY, Nikolaev S, Yabuki A, Ishida KI, Berney C, Pakzad U, Wylezich C, Cavalier-Smith T. Phylogeny of novel naked Filose and Reticulose Cercozoa: Granofilosea cl. n. and Proteomyxidea. Protist 2009;160:75–109.

Braselton JP. Plasmodiophoromycota. In: McLaughlin DJ, McLaughlin EG, Lemke PA, eds. The Mycota VII Part A. Systematics and evolution. Berlin: Springer-Verlag, 2001:81–91.

Braselton JP, Short FT. Karyotypic analysis of *Plasmodiophora diplantherae*. Mycologia 1985;77: 940–945.

Bulman SR, Kuhn SF, Marshall JW, Schnepf E. A phylogenetic analysis of the SSU rRNA from members of the Plasmodiophorida and Phagomyxida. Protist 2001;152:43–51.

Burki F, Kudryavtsev A, Matz MV, Aglyamova GV, Bulman S, Fiers M, Keeling PJ, Pawlowski J. Evolution of Rhizaria: new insights from phylogenomic analysis of uncultivated protists. BMC Evol Biol 2010;10:377.

Cavalier-Smith T. Kingdom Protozoa and its 18 phyla. Microbiol Rev. 1993;57:953–994.

den Hartog C. Distribution of *Plasmodiophora bicaudata* a parasitic fungus on small *Zostera* species. Dis Aquat Organ. 1989;6:227–230.

Dixon GR. *Plasmodiophora brassicae* in its environment. J Plant Growth Regul. 2009;28: 212–228.

Johnson TW, Sparrow FK. Fungi in oceans and estuaries. Weinheim, Germany: J. Cramer, Pub., 1961.

Kageyama K, Asano T. Life cycle of *Plasmodiophora brassicae*. J Plant Growth Regul. 2009; 28:204–211.

Kanyuka K, Ward E, Adams MJ. *Polymyxa graminis* and the cereal viruses it transmits: a research challenge. Mol Plant Pathol. 2003;4:393–406.

Karling JS. *Phygomyxa algarum* n. gen., n. sp., an unusual parasite with Plasmodiophoralean and Protomyxean characteristics. Am J Bot. 1944;31:38–52.

Karling JS. The Plasmodiophorales, 2nd completely revised edition. New York: Hafner Publishing Company, 1968.

Maier I, Parodi E, Westermeier R, Müller DG. *Maullinia ectocarpi* gen. et sp. nov. (Plasmodiophorea), an intracellular parasite in *Ectocarpus siliculosus* (Ectocarpales, Phaeophyceae) and other filamentous brown algae. Protist 2000;151:225–238.

Neuhauser S, Bulman S, Kirchmair M. Plasmodiophorids: the challenge to understand soil-borne, obligate biotrophs with a multiphasic life cycle. In: Gherbawy Y, Voigt K, eds. Current advances in molecular identification of fungi. Heidelberg: Springer, 2010:51–78.

Neuhauser S, Kirchmair M, Gleason FH. Ecological roles of the parasitic Phytomyxea (plasmodiophorids) in marine ecosystems – a review. Mar Freshw Res. 2011;62:365–371.

Parodi ER, Caceres EJ, Westermeier R, Müller DG. Secondary zoospores in the algal endoparasite *Maullinia ectocarpi* (Plasmodiophoromycota). Biocell 2010;34:45–52.

Schnepf E, Kühn SF, Bulman S. *Phagomyxa bellerocheae* sp. nov. and *Phagomyxa odontellae* sp. nov., Plasmodiophoromycetes feeding on marine diatoms. Helgoland Mar Res. 2000; 54:237–242.

Suttle CA. Marine viruses – major players in the global ecosystem. Nat Rev Microbiol. 2007; 5:801–812.

Biodiversity of marine fungi

14 Mangrove fungi

Kandikere R. Sridhar, S. Aisyah Alias and Ka-Lai Pang

Substrates of marine fungi include woody tissues, leaves, fruits, sea grasses, algae and seaweeds, while marine fungi can also be isolated from animal exoskeletons, keratinaceous substrates, sediments, sea foam and seawater (Vrijmoed 2000). While marine fungi on *Nypa* (Chapter 15), seaweeds (Chapter 17) and salt marsh plants (Chapter 18) are reviewed in other chapters in this volume, this chapter summarizes information on the marine fungi colonizing mangrove substrata, including leaves and drift/attached woody stems and twigs. Over the past 20 years, a great deal of progress has been made in surveying tropical and subtropical habitats for marine fungi, especially mangroves. Kohlmeyer and Kohlmeyer (1979) listed 42 mangrove fungi, while Schmit and Shearer (2003) listed 625 taxa, but this figure included those growing on the terrestrial parts of mangrove trees. Currently some 287 species can be regarded as growing on submerged mangrove substrata (Alias et al. 2010). Illustrated monographs of mangrove fungi have been published for India (80 species: Raveendran and Manimohan 2007), Malaysia (140: Alias and Jones 2009) and Taiwan (69: Pang et al. 2011). Hyde and Jones (1988) recognized that mangrove fungi constituted the second largest ecological group of marine fungi that are widely distributed in old and new world mangroves (Atlantic, Pacific and Indian Oceans).

Mangrove profile

Mangroves are the wetland forests of tropical and subtropical latitudes that exist at the interfaces of fresh- and salt-waters (e.g. estuaries, backwaters, deltas, creeks and lagoons). In standing crop productivity and sustained tertiary yield among the marine habitats, mangroves constitute the second most important ecosystem after coral reefs (Qasim and Wafar 1990; Alongi 2002). Mangrove forests are similar to that of tropical rainforests, as the equatorial mangroves attain a biomass up to 300–700 t/ha (Clough 1992; Alongi 2002). About 25% of the world's circumtropical coastlines in 112 countries are dominated by mangroves, occupying up to 150,000–181,000 km^2 (Spalding et al. 1997; Wilkie and Fortuna 2003).

Distribution of mangrove plant species is strongly influenced by temperature, moisture and large scale currents (Duke 1992; Saenger and Snedaker 1993; DeLonge and DeLonge 1994; Wilkie and Fortuna 2003). The mangrove flora encompass three major categories:

(i) true mangroves with ~80 tree and shrub species restricted to high water levels of neap spring tides, e.g. *Avicennia, Kandelia, Rhizophora* and *Sonneratia*;
(ii) minor vegetation (plant species rarely forming pure stands, e.g. *Aegiceras* and *Excoecaria*); and
(iii) mangrove associates, salt-tolerant plant species restricted to landwards and seawards, e.g. *Acanthus ilicifolius, Dalbergia amerimnion* and *Derris trifloiata* (Tomlinson 1986; Field 1995).

Over 100 mangrove tree species have been listed by Chapman (1976), and 50–60 out of the 80 true mangrove tree and shrub species significantly contribute to the structure of mangrove forests (Tomlinson 1986; Field 1995). However, Mepham and Mepham (1985) use a broader concept of mangroves and list 900 plant species (including epiphytes), but do not include macro-seaweeds.

Productivity of mangroves

Mangrove forests are of great traditional, ecological, economic and social values (Bandaranayake 1995, 1998; Kathiresan and Bingham 2001; Bandaranayake 2002). Their major contribution to the global carbon cycle is by attaining a standing biomass up to 700 t/ha, which is equivalent to 8.7 gigatons (gt) of dry mass (= 4 gt of carbon; Clough 1992; Twilley et al. 1992). Thus, the productivity of mangrove waters is mainly dependent on the supply of carbon, nitrogen and phosphorus from the mangrove stands (Wafar et al. 1997; Kathiresan and Bingham 2001). The average rate of net primary production in mangroves amounts to 64 tons of dry mass/ha/year, which decreases with increasing latitude (Hossain and Hoque 2008; Alongi 2009).

Mangroves are production-based (living mangrove tissues) and detritus-based (fine and coarse particulate matter) forests that generate a large quantity of leaf, twig, branch, bark, wood, inflorescence, seed and other detritus (seedlings, proproots and pneumatophores; Wafar et al. 1997; Ellison 2008). The litter fall in mangrove forests is dependent on the climatic conditions and seasons of different geographic locations (e.g. short dry seasons, long dry winter and post-monsoons; Othman 1989; Duke 1990; Ghosh et al. 1990; Khafaji et al. 1991; Bunt 1995; Day et al. 1996; Mackey and Smail 1996; Mfilinge et al. 2005; Hossain and Hoque 2008; Imgraben and Dittmann 2008; Chen et al. 2009; Bernini and Rszende 2010; Kumar 2011). Besides routine seasonal impacts, deviations occur in the pattern of litter fall, especially by the status of stress in mangroves (e.g. aridity and nutrient-poor soils; Saenger and Snedaker 1993; Imbert and Ménard 1997). Litter fall in mangroves differs significantly from habitat to habitat and is dependent on the composition of plant species and their productivity (Kathiresan and Bingham 2001). The global litter accumulation (as standing stock) on the floor of mangrove forests ranges between 130 and 1870 g/m^2 (Clough 1992), which differs between geographic locations and plant species (e.g. Queensland's *Ceriops* forest: 6 g/m^2; compared to Queensland's *Avicennia* forest: 84 g/m^2; Robertson et al. 1992). Similarly, the annual mangrove litter production was 0.011 t/ha in Kenya (*Ceriops tagal*), 10.2 t/ha in India (*Avicennia officinalis*) and 23.7 t/ha in Australia (*Rhizophora stylosa*) (Bunt 1995; Slim et al. 1996; Wafar et al. 1997).

Mangrove fungi and their niches

Mangrove fungi include terrestrial, phylloplane, parasites, saprobes and endophytes, the latter constituting the second largest group after intertidal woody litter-associated fungi (Hyde 1990b). They grow on a wide variety of substrates, such as wood, leaf, seagrasses, sediments, soil, sand, algae, corals and calcareous material in mangroves (Kohlmeyer and Kohlmeyer 1979). Schmitt and Shearer (2003) listed 625 fungi of terrestrial, freshwater, and marine origin from 72 mangrove plant species of the world (true fungi only: 278 ascomycetes, 277 anamorphic fungi, 30 basidiomycetes, three Chytridiomycota, and 12 Zygomycota). Only 287 species can be regarded as growing on submerged mangrove substrata (Alias et al. 2010). The core marine mangrove fungi are listed in Table 14.1. Schmit and Shearer (2003) and Shearer et al. (2007) reviewed

Table 14.1 Core marine fungi of mangrove wood.

Basidiomycota
Calathella mangrovei E.B.G. Jones *et* Agerer
Halocyphina villosa Kohlm. *et* E. Kohlm.

Ascomycota
Aigialus mangrovei Borse
Aigialus parvus S. Schatz *et* Kohlm.
Aniptodera chesapeakensis Shearer *et* M.A. Mill.
Aniptodera mangrovei K.D. Hyde
Dactylospora haliotrepha (Kohlm. *et* E. Kohlm.) Hafellner
Eutypa bathurstensis K.D. Hyde *et* Rappaz
Halorosellinia oceanica (S. Schatz) Whalley, E.B.G. Jones, K.D. Hyde *et* Lassøe
Halosarpheia marina (Cribb *et* J.W. Cribb) Kohlm.
Halosarpheia fibrosa Kohlm. *et* E. Kohlm.
Kallichroma tethys (Kohlm. *et* E. Kohlm.) Kohlm. *et* Volkm.-Kohlm.
Leptosphaeria australiensis (Cribb *et* J.W. Cribb) G.C. Hughes
Lignincola laevis Höhnk
Lophiostoma acrostichi (K.D. Hyde) Aptroot *et* K.D. Hyde
Lulworthia grandispora Meyers
Marinosphaera mangrovei K.D. Hyde
Nemania maritima Y.M. Ju *et* J.D. Rogers
Neptunella longirostris (Cribb *et* J.W. Cribb) K.L. Pang *et* E.B.G. Jones
Oceanitis cincinnatula (Shearer *et* J.L. Crane) J. Dupont *et* E.B.G. Jones
Passeriniella mangrovei Maria *et* K.R. Sridhar
Patellaria atrata (Hedw.) Fr.
Saagaromyces abonnis (Kohlm.) K.L. Pang *et* E.B.G. Jones
Saagaromyces glitra (J.L. Crane *et* Shearer) K.L. Pang *et* E.B.G. Jones
Savoryella lignicola E.B.G. Jones *et* R.A. Eaton
Savoryella paucispora (Cribb *et* J.W. Cribb) Jørg. Koch
Tirispora unicaudata E.B.G. Jones *et* Vrijmoed
Torpedospora radiata Meyers
Trichocladium alopallonella (Meyers *et* R.T. Moore) Kohlm. *et* Volkm.-Kohlm.
Verruculina enalia (Kohlm.) Kohlm. *et* Volkm.-Kohlm.

Asexual (anamorphic) fungi
Zalerion maritima (Linder) Anastasiou
Bactrodesmium linderi (J.L. Crane *et* Shearer) M.E. Palm *et* E.L. Stewart

Table continued on next page.

Table 14.1 (Continued)

Cirrenalia pseudomacrocephala Kohlm.
Clavatospora bulbosa (Anastasiou) Nakagiri *et* Tubaki
Halenospora varia (Anastasiou) E.B.G. Jones
Hydea pygmea (Kohlm.) K.L. Pang *et* E.B.G. Jones
Matsusporium tropicale (Kohlm.) E.B.G. Jones *et* K.L. Pang
Periconia prolifica Anastasiou
Rhabdospora avicenniae Kohlm. *et* E. Kohlm.
Trichocladium asperum (Meyers *et* R.T. Moore) Dixon

the global distribution of higher mangrove fungi (meiosporic ascomycetes/anamorphic ascomycetes/basidiomycetes) and other fungi in regions including:

(i) north temperate (55/51/11 species respectively);
(ii) tropics (79/60/13);
(iii) Asian temperate (25/17/8);
(iv) tropical Africa (21/14/3);
(v) Madagascar (0/1/1);
(vi) temperate Africa (29/11/1);
(vii) middle east (20/9/0);
(viii) tropical Asia (225/190/33);
(ix) Australasia (67/11/11); and
(x) Pacific Islands (47/56/18).

Among these geographic locations, tropical Asia clearly showed the highest number of fungi, possibly due to high diversity of mangrove plant species, and the extensive surveys undertaken. The ecological habitats supporting fungi are briefly considered below.

Colonization of wood

Fungi on woody litter in mangrove habitats exhibit a distinct pattern of colonization, richness, diversity, succession and interactions. Such attributes are mainly dependent on the wood quality, water chemistry and geographical location. Colonization of fungi in mangroves is influenced by tidal amplitude, the availability of oxygen, temperature regimes, light, pH, osmotic effects, hydrostatic pressure, abundance of propagules, nature of the substrate and efficiency of spore attachment to substrata (Hyde and Lee 1995; Jones and Vrijmoed 1997; Jones 2000; Alias and Jones 2000a; Sarma and Hyde 2001). Newell (1996) predicted that the lower fungi follow the substrate-capture strategy, while higher fungi adapt a mass-accumulation strategy to colonize and exploit the substrata in marine habitats. Newell and Fell (1997) opined that the intermittent wet and dry conditions of litter in mangrove habitats elevate fungal activity, with filamentous fungi tending to compete with the "lower fungi" (Newell and Fell 1992).

Mangrove fungi are known for their ability to colonize natural and artificial substrates under turbulent conditions due to their spore adherence capacity. Three phases in fungal spore adhesion to substrata have been recognized by Jones (1994):

(i) initial passive attachment (impaction, chemotaxis and entrapment, spore appendage mediated attachment);
(ii) active attachment resulting from a thigmotropic response and leading to the production and release of extracellular adhesive; and
(iii) the development of appressoria and penetration of substratum (Hyde et al. 1986).

Many ultrastructural studies have clearly demonstrated the nature and functions of appendages of marine fungal spores in the colonization of substrata (e.g. Jones 1994, 1995; Read et al. 1995). The following sections provide a review of fungal colonization and diversity on random collections of woody litter and the exposure of selected substrata at different geographical locations.

Fungal diversity on driftwood

Fungal diversity and colonization of woody litter in mangroves is influenced by several factors, such as the type of wood, chemical composition, presence or absence of bark, period of submersion, fungal competition, salinity and geographic location (Hyde et al. 1990, 1993; Nakagiri 1993; Hyde and Lee 1995; Kohlmeyer et al. 1995; Jones 2000; Alias and Jones 2000a; Maria and Sridhar 2004; Raveendran and Manimohan 2007). Surveys of fungal diversity of naturally deposited woody litter provide a clear picture on the assemblage, structure and diversity of fungal flora of mangroves (Miller and Whitney 1981; Poonyth et al. 1999; Maria and Sridhar 2002, 2004; Sridhar and Maria 2006). Such studies have shown that random sampling of driftwood supports a greater diversity of fungi than immersed specific woody material (Hyde and Jones 1988; Alias and Jones 2000a; Maria and Sridhar 2002, 2003b; Ananda and Sridhar 2004; Maria and Sridhar 2004; Sakayaroj et al. 2004; Abdel-Wahab 2005; Jones and Abdel-Wahab 2005; Raveendran and Manimohan 2007; Besitulo et al. 2010). For example, Tan et al. (1989) and Leong et al. (1991) compared fungal diversity on driftwood and immersed twigs of the four mangrove trees at Mandai mangrove, Singapore, with 41 species on 188 randomly collected samples and 21 species on *Avicennia alba* (68 samples) and *A. lanata* (58 samples), and 24 on *Bruguiera cylindrica* (60 twigs) and *Rhizophora apiculata* (70 twigs).

Surveys of mangrove fungi in the Atlantic, Pacific and Indian Oceans have identified a core group of mangrove fungi (Table 14.1) with the majority being unique to mangrove habitats. However, variation in fungal diversity occurs from site to site and between geographical locations, and this can be accounted for by different sampling strategies, sample dimensions and substrates, period of time in the mangrove, host specificity, wet *versus* dry seasons, stage of decomposition of the wood and salinity. The factors affecting the distribution of mangrove fungi have been discussed by Nakagiri (1993), Jones (2000), Sarma and Hyde (2001), Hyde and Sarma (2006), and Alias and Jones (2009). A few examples are provided below.

Surveys in the wet and dry seasons in India

Maria and Sridhar (2003b) sampled the woody litter of five mangrove shrubs/trees (*Acanthus ilicifolius, Avicennia officinalis, Bruguiera gymnorrhiza, Rhizophora mucronata* and *Sonneratia caseolaris*) during the summer (dry) and monsoon (wet) seasons and incubated for up to six months. The fungal richness and diversity were highest in the monsoon period, with terrestrial fungi being dominant; while marine fungi were dominant in the summer samples. Fungal diversity was highest on *R. mucronata* samples. *Lignincola laevis, Savoryella lignicola* and *Bactrodesmium linderi* were recovered in both seasons on all substrates. It has been predicted that monsoon season in the mangroves of the west coast of India provide ideal conditions for the colonization and growth of typical terrestrial, freshwater and aero-aquatic fungi due to freshwater intrusion, while marine fungi dominate in the summer season due to saltwater intrusion.

Common fungi on different substrata

The occurrence of fungi on the mangrove palm and seagrasses is documented in Chapters 15 and 16 of this volume, with many of the species reported on the brackish water palm being host-specific (Besitulo et al. 2010). *Acanthus ilicifolius* is a widely distributed shrubby herbaceous plant found in mangroves and a species surveyed for fungi by Sadaba et al. (1995). Forty-four fungi were found on the decaying standing parts of *A. ilicifolius* in the intertidal zone at Mai Po Marshes, Hong Kong, with the aerial portions dominated by asexual fungi. The grass *Phragmites australis* and sedge *Schoenoplectus litoralis* supported 61 (17 ascomycetes, 44 anamorphic taxa) and 31 (six ascomycetes, 25 anamorphic taxa) species, respectively, in Mai Po Marshes, Hong Kong (Poon and Hyde 1998a, 1998b; Wong et al. 1998). For both plants, the fungi present on the submerged regions differed from aerial regions, with the latter supporting the greater diversity. Of the taxa documented, most were typically terrestrial-like, such as, *Cephalosporiopsis* sp., *Septoria*-like sp., *Phomopsis* sp. and *Colletotrichum* sp.

Sridhar et al. (2010) followed fungal colonization, mass loss and biochemical changes during decomposition of the mangrove sedge *Cyperus malaccensis* in the Nethravathi river delta, Katrnataka, India. Nineteen taxa were found (eight ascomycetes, ten anamorphic species and one zygomycete). Initially the terrestrial fungi were dominant, but this was followed by typical mangrove and marine fungi (*Acrocordiopsis patilii, Cumulospora* sp., *Okeanomyces cucullata, Leptosphaeria australiensis, Lignincola laevis, Lulworthia* sp. and *Periconia prolifica*). Mass loss of the different parts of the sedge occurred over four weeks: bract 79%, basal stems 88% and top stems 51%. Enzyme production (cellulase, xylanase and pectinase) also peaked within the first four weeks of exposure.

Intermittently submerged rachis of the mangrove fern (*Acrostichum speciosum*) harbored 22 species of fungi in Brunei, with *Aniptodera chesapeakensis, Cirrenalia pseudomacrocephala* and *Lophiostoma acrostichi* being dominant (Hyde 1989a).

Incubation of mangrove substrata

Ananda and Sridhar (2004) examined the effect of incubation of randomly-collected mangrove substrata and demonstrated a definite succession in the sporulation of terrestrial, marine and arenicolous (sand-inhabiting) fungi. Terrestrial fungi, marine fungi and arenicolous fungi sporulated on wood within 8, 16 and 32 weeks, respectively. Similar to arenicolous fungi, *Halographis runica* (an endolithic lichenoid fungus that commonly occurs in corals and snails) sporulated on woody mangrove litter after 32 weeks of incubation.

Fungal biodiversity on islands

Few studies have been reported of fungi on samples collected on different islands (Kohlmeyer and Kohlmeyer 1979). Volkmann-Kohlmeyer and Kohlmeyer (1993) reported on the mangrove fungi collected on the islands of Moorea and Oahu, with greater diversity occurring on the latter. They attributed this to the greater tree diversity in Oahu. Ananda and Sridhar (2003) investigated the diversity of fungi on the woody litter of two Indian island mangroves after incubation for 32 weeks. Thirty-four fungi were documented; *Savoryella lignicola* and *Tirispora unicaudata* were dominant in the Coconut Island samples, and *Clavatospora bulbosa, Corollospora maritima, Kallichroma tethys, Lulworthia grandispora, Torpedospora radiata, Verruculina enalia* and *Halenospora varia* in Minicoy Island. Other islands surveyed for mangrove fungi include the Seychelles, Maldives and Andaman-Nicobar Islands (Hyde and Jones 1989; Chinnaraj 1993a, 1993b).

Vertical zonation

Vertical zonation of mangrove fungi is common in different mangroves (e.g. Sierra Leone – Aleem 1980; Belize – Kohlmeyer et al. 1995; Seychelles – Hyde and Jones 1988; Andaman and Nicobar Islands – Chinnaraj 1993b; Malaysia – Jones and Tan 1987; Thailand – Hyde et al. 1990, 1993; Brunei – Hyde 1988; and Philippines – Jones et al. 1988). Fungal zonation was seen in different mangrove plant species, e.g. *Rhizophora apiculata* (Hyde 1988; Hyde et al. 1990, 1993), *R. mucronata* (Hyde 1988) and *Sonneratia griffithii* (Hyde et al. 1990). Some fungi have uniform distribution throughout the tidal range (e.g. *Hydea pygmea, Leptosphaeria australiensis* and *Lulworthia grandispora*), while others are confined to the upper littoral zone (e.g. *Cytospora rhizophorae, Halorosellinia oceanica* and *Savoryella lignicola*), mid-littoral zone (e.g. *Caryosporella rhizophorae*) and lower littoral zone (e.g. *Antennospora quadricornuta* and *Thalassogena sphaerica*) (Aleem 1980; Hyde 1988; Hyde and Jones 1988; Hyde 1989c). On immersed wood in the mangroves of Belize, many fungi sporulated below the water line (2.0–2.4 m from the surface or 0.6–1.2 m above the sediment; Kohlmeyer et al. 1995). However, some fungi fruited on wood close to the sediment (e.g. *Antennospora quadricornuta, Lulworthia* sp. and *Phoma* sp.). Some preferred above the high water line (*Halocyphina villosa, Lulworthia* sp. and *Trichocladium alopallonellum*). *Nia vibrissa* preferred the submerged part of the wood.

There seems to be structural adaptation of fungi to the different zones (or niches) of mangroves. Based on the fungal vertical zonation in mangroves, Hyde and Lee (1995) commented on their diversity that:

(i) mean tide level possesses a high diversity of fungi;
(ii) fungi possessing carbonaceous ascomata, those with active spore release, and those with colored or ornamented ascospores with superficial ascomata are confined to above mean tide levels;
(iii) fungi possess passive spore dispersal, those with membranous walls and with immersed ascomata occur throughout the tidal range.

The zone between the high intertidal region and terrestrial region supports many unusual fungi, which are adapted to decay the mangrove wood without being severely impacted by salinity. It is interesting to study fungal zonation in narrow, as well as deep mangroves: Belize (0.15 m), Florida (0.8 m), South China Sea (2.6 m), Brunei (3.0 m) and the northern territory of Australia (7.0 m) (Hyde and Lee 1995; Kohlmeyer et al. 1995).

Exposure of test blocks/substrata in mangroves

Fungal colonization and succession on different substrata after exposure in selected mangroves has been studied by many individuals in different geographical locations: Belize (Kohlmeyer et al. 1995), Brunei (Hyde 1991), India (Maria and Sridhar 2004; Sridhar and Maria 2006), Malaysia (Alias and Jones 2000b), and Singapore (Tan et al. 1989; Leong et al. 1991). These studies noted different sequences in the sporulation of fungi on the selected substrata, as only fruit body succession was observed. They did not isolate fungi from the substrata or determine the presence of unculturable fungi (Hyde and Jones 2002). However, the investigation established patterns in occurrence with time, e.g. early, intermediate or late colonizers (or other terms), with the time scale varying with each study and the substratum examined (Jones and Hyde 2002). Other terms used to characterize the fungal community include common, frequent, infrequent or rare, with the percentage frequency of occurrence provided.

The basic pattern of fungal colonization was early, intermediate, and late (persistent; Tan et al. 1989; Leong et al. 1991; Alias and Jones 2000a; Maria and Sridhar 2004). In Table 14.2, we compare the colonization sequence of fungi on exposed substrata in selected mangroves in India, Malaysia and Singapore. Two studies employed twigs ~1 cm in diameter and 10–15 cm in length, while the Malaysian study used decorticated *Avicennia* spp. test blocks that were 5 × 1 × 1 cm (Table 14.2). The exposure periods varied from 60 to 90 weeks and the number of fungi per site were 23 (Singapore), 33 (India) and 46 (Malaysia). The Indian site was subject to a monsoon season, when the salinity of the water declined resulting in the appearance of a number of terrestrial fungi. Given the variations in protocol employed at the different sites, a number of conclusions can be drawn:

(i) initial colonization was by relatively few fungi;
(ii) the greatest diversity was during the intermediate stage;
(iii) species diversity declined with time of exposure;
(iv) many colonizers were common to the three studies: *Lignincola laevis, Lulworthia* sp., *Verruculina enalia* and *Halosarpheia marina*; and,
(v) few species were persistent throughout the entire exposure period, while some appear late in the succession e.g. *Aigialus mangrovei, Savoryella paucispora* (Alias and Jones 2000a; Maria and Sridhar 2003b).

Tan et al. (1989) reported a similar pattern of colonization of *Avicennia alba* and *A lanata* twigs at Mandai mangrove, Singapore; as did Alias and Jones (2000a) on *A. marina* and *Bruguiera parviflora* test blocks in Kuala Selangor mangrove, Malaysia. However, differences in the pattern of colonization of *A. officinalis* and *Rhizophora mucronata* twigs immersed in Udyavara mangrove, India were observed. This highlights the fact that

Table 14.2 Comparison of the five most frequent fungi at each stage of fungal colonization.

Avicennia alba		Avicennia officinalis		Avicennia marina	
Number of fungal species: 23		Number of fungal species: 33		Number of fungal species: 46	
Exposure time: 60 weeks		Exposure time: 72 weeks		Exposure time: 90 weeks	
Leong et al. (1991)		Maria and Sridhar (2003b)		Alias and Jones (2000a)	
Stage	Species	Stage	Species	Stage	Species
Early:	*Halosarpheia marina*	4–12 weeks	*Hydea pygmea*	6–18 weeks	*H. marina*
6–18 weeks	*Lignincola laevis*		*L. laevis*		*L. laevis*
	Luworthia sp.		*Lulworthia* sp.		*Neptunella longirostris*
	Payosphaeria minuta		*Matsusporium tropicale*		*Saagaromyces ratnagiriensis*
	Verruculina enalia		*Trichocladium achrasporum*		*V. enalia*
Intermediate:	*H. marina*	16–32 weeks	*H. pygmea*	26–54 weeks	*H. marina*
22–32 weeks	*Halocyphina villosa*		*L. laevis*		*L. laevis*
	Lulworthia sp.		*Lulworthia* sp.		*N. longirostris*
	P. minuta		*M. tropicale*		*Natantispora retorquens*
	V. enalia		*Tirispora* sp.		*S. lignicola*
Late:	*Aigialus parvus*	36–72 weeks	*H. pygmea*	73–90 weeks	*Dictyosporium pelagicum*
37–60 weeks	*H. villosa*		*Lulworthia* sp.		*H. marina*
	Lulworthia sp.		*Savoryella lignicola*		*N. longirostris*
	P. minuta		*Tirispora* sp.		*S. lignicola*
	V. enalia		*V. enalia*		*V. enalia*

our knowledge of mangrove fungal ecology is still fragmentary and molecular techniques are required to determine the precise colonization of fungi and the effect they exert on each other over time. Tan et al. (1995) have shown that the sporulation of a fungus can be affected by the presence of another species. This could result in some species not being detected using conventional methods for enumerating fungal communities. Fungal diversity on wood was low at all stages of decomposition in Belize and Brunei (Hyde 1991; Kohlmeyer et al. 1995) compared to the mangroves of Singapore and Malaysia (Tan et al. 1989; Leong et al. 1991; Alias and Jones 2000b).

Biogeography and biodiversity

A comparison of the mangroves in 12 geographical areas revealed wide differences in the mean number of fungi colonizing each wood sample (Table 14.3). In mangroves in Bahamas, Brunei and Egypt, the lowest number of fungi per wood sample was seen (0.9), while it was highest (up to 2.2) in the mangroves in Singapore. The upper limit was the lowest in Egypt (1.3) and the highest was in Hong Kong (4.2). The highest number of fungi per sample was seen on wood blocks of *Avicennia marina* and *Bruguiera parviflora* immersed in a Malaysian mangrove. This was higher during the intermediate (6.0–8.0) rather than the early (1.8–4.2) and late (4.0–7.0) stages of immersion. A similar result was found in study of a Singapore mangrove (Tan et al. 1989). Comparison of 13 woody mangrove substrates revealed the highest number of fungal species per wood sample on *Acanthus ilicifolius* (4.2), followed by decorticated wood blocks of *Avicennia marina* and woody litter of *R. mucronata* (3.4). The number of species was the lowest on the roots and trunks of *Bruguiera parviflora* with intact bark (1.0), see Table 14.4.

Mangrove leaves

Many ecological groups of fungi are associated with the leaves of mangrove plants: phylloplane (Lee and Hyde 2002), endophytic (see Chapter 16) and marine and terrestrial saprophytic fungi. Fungi associated with mangrove leaves in marine environments can either be endophytes of the plants or secondary colonizers in the seawater. Rodrigues

Table 14.3 Mean number of fungal species recovered per wood sample at different mangroves.

Mangroves	Mean number of fungal species/wood sample	References
Bahamas	0.9–1.9	Jones and Abdel-Wahab (2005)
Brunei	0.9–1.6	Hyde (1989b), Hyde (1990a)
Egypt	0.9–1.3	Abdel-Wahab (2005)
Hong Kong	1.2–4.2	Vrijmoed et al. (1994), Sadaba et al. (1995)
India	1–3.3	Borse (1988), Borse et al. (2000), Patil and Borse (2001), Maria and Sridhar (2002, 2003b, 2004), Ananda and Sridhar (2003, 2004)
Macau	1.7	Vrijmoed et al. (1994)
Malaysia	1.3–3.3	Jones and Tan (1987), Jones and Agerer (1992), Tan and Leong (1992), Alias et al. (1995), Alias and Jones (2000a, 2000b)
Mauritius	1.1–3.4	Poonyth et al. (1999, 2001)
Seychelles	1.1–1.5	Hyde and Jones (1988)
Singapore	2.2–2.6	Tan et al. (1989)
Sumatra	1.5	Hyde (1989c)
Thailand	1.8	Hyde (1989d)

Table 14.4 Mean number of fungal species recovered per wood sample of different mangrove plant species.

Plant species	Mean number of fungal species/wood sample	References
Acanthus ilicifolius L.	2.1–4.2	Sadaba et al. (1995), Maria and Sridhar (2003b)
Aegiceras corniculatum (L.) Blanco	2	Hyde (1989d)
Avicennia alba Blume	2.2–2.7	Tan et al. (1989), Hyde (1990b)
Avicennia lanata Ridley	2.6	Tan et al. (1989)
Avicennia marina (Forsk.) Vierh.	3.3	Alias and Jones (2000a)
Avicennia officinalis L.	2.6–3.3	Maria and Sridhar (2003b)
Bruguiera gymnorrhiza (L.) Lamk.	2.9–3	Maria and Sridhar (2003b)
Bruguiera parviflora (L.) Lamk.	1.0–2.6	Hyde (1989d), Alias and Jones (2000a)
Rhizophora apiculata Bl.	1.3–1.6	Hyde (1990b), Jones and Agerer (1992), Alias and Jones (2000b)
Rhizophora mucronata Lamk.	1.1–3.4	Hyde (1989d), Hyde (1990b), Poonyth et al. (1999), Poonyth et al., 2001), Maria and Sridhar (2003b, 2004)
Sonneratia alba Smith	1.8	Hyde (1990b)
Sonneratia caseolaris (L.) Engl.	2.4–2.8	Maria and Sridhar (2003b)
Xylocarpus granatum Koen.	1.1–1.7	Hyde (1989d, 1990b)

and Petrini (1997) suggested that endophytic fungi can become saprophytes with the death/senescence of the plant (or part of the plant), giving them an advantage in resource capture over other non-endophytic saprophytes. In a study on the endophytic and saprophytic fungi associated with *Magnolia liliifera*, similar fungi (in terms of phylogenetic affinity) were isolated from both ecological groups of fungi. The endophytic fungi also produced the same degrading enzymes as the saprophytic fungi, suggesting that endophytes can switch lifestyle to become saprobes (Promputtha et al. 2010). In mangrove habitats, however, the endophytes have to overcome the salinity barrier to be a degrader when the leaves fall onto the floor. Preliminary results suggest that some endophytes of *Avicennia marina* and *Kandelia obovata* are capable of producing pectinases in a seawater medium (YF Huang and KL Pang, unpublished results).

Lanceispora amphibia, an ascomycete, also occurs on both senescent yellow leaves on the trees and fallen leaves of *Bruguiera gymnorrhiza*, suggesting that this fungus has a terrestrial origin (Nakagiri et al. 1997). Other fungi occurring on fallen mangrove leaves are mostly the asexual fungi of known terrestrial taxa, including *Aspergillus* spp., *Cladosporium* spp. and *Phoma* spp. (Nakagiri et al. 1997; Ananda et al. 2008). Whether these taxa are truly marine and involved in the degradation of leaf litter requires further studies. Typical marine mangrove fungi were also identified from mangrove leaves, again predominantly asexual taxa, e.g. *Hydea pygmea, Periconia prolifica* and *Trichocladium alopallonellum*, and ascomycetes, e.g. *Lulworthia* spp. (Nakagiri et al. 1997; Ananda et al. 2008; Sridhar 2009). Other groups of true fungi including Basidiomycota, Zygomycota and Chytridiomycota (see Chapter 7) have rarely been isolated from fallen mangrove leaves. *Clathrus* c.f. *crispus* (Basidiomycota: Clathraceae) was reported to occur on the upper mangrove soil and decomposing leaves of *Rhizophora mangle* in Puerto Rico (Maldonado-Ramírez et al. 2005).

Marine fungal-like organisms, Oomycota and Labyrinthulomycota, are common inhabitants of fallen mangrove leaves. In particular, *Halophytophthora* species have been reported from the early to late stages of mangrove leaf decomposition (Leaño et al. 2000). Raghukumar et al. (1994) studied the degradation of *Rhizophora apiculata* detritus and found that *Aurantiochytrium mangrovei* and *Halophytophthora vesicula* were capable of producing degradative enzymes, such as cellulases, xylanases and pectic enzymes, suggesting their involvement in the degradation of leaf detritus.

Endophytes

The majority of foliar and root endophytic fungi reported from mangrove plant species belong to terrestrial-like fungi (Suryanarayanan et al. 1998; Suryanarayanan and Kumaresan 2000; Kumaresan and Suryanarayanan 2001; Ananda and Sridhar 2002; Maria and Sridhar 2003a; Pang et al. 2008; Chaeprasert et al. 2010). Common fungi include *Sporormiella minima, Acremonium* spp., *Phomopsis* spp. and *Phyllosticta* spp., while the last two genera are cosmopolitan in distribution (Suryanarayanan and Kumaresan 2000). Endophytic *Acremonium* and *Colletotrichum* were dominant in senescent standing stems of *Acanthus ilicifolius* and they also serve as saprophytes on senescence of mangrove leaves (Kumaresan and Suryanarayanan 2002). Several mangrove endophytic fungi have multiple ecological roles as saprotrophs or opportunistic pathogens (e.g. *Chaetomium globosum* and *Paecilomyces variotii*; Ananda and Sridhar 2002; Arnold et al. 2007; Naik et al. 2007; Hyde and Soytong 2008; Vega et al. 2008). Plant pathogenic fungi (e.g. *Alternaria alternata, Curvularia clavata* and *Drechslera halodes*) and toxigenic fungi (e.g. *Aspergillus flavus, A. ochraceus* and *Trichoderma harzianum*) are endophytic in mangrove plant species (Anita and Sridhar 2009; Anita et al. 2009). Some of the endophytic fungi in mangrove plant species are also entomopathogenic (e.g. *Paecilomyces* spp.; Ananda and Sridhar 2002; Maria and Sridhar 2003a). Endophytic fungi of seagrasses are documented by Sakayaroj et al. (2010) and in Chapter 16.

Further ecological observations

A literature search reveals that approximately 80 mangrove fungi are dominant and are regarded as a core group on woody litter. Nearly 40 of them are confined to woody litter of *Rhizophora* spp. (Sarma and Hyde 2001; Maria 2003). Several fungi are common and confined to bark of *Rhizophora apiculata* (e.g. *Hypophloeda rhizospora, Phomopsis mangrovei* and *Rhizophila marina*), while some are only reported from woody tissues (e.g. *Caryosporella rhizophorae*) (Hyde et al. 1993). A higher number of fungi were identified on the proproots of *R. apiculata* compared to other substrates (seedlings and roots) in India (Ravikumar and Vittal 1996; Sarma and Vittal 2000). Overall, woody substrates of *Rhizophora* spp. possess a higher species richness and diversity of mangrove fungi compared to other host species (Sarma and Vittal 2000). It is also interesting to note that a number of typical mangrove fungi (e.g. *Aigialus mangrovei, Hydea pygmea, Matsusporium tropicale, Oceanitis cincinnatula, Lulworthia grandispora, Passeriniella mangrovei, Verruculina enalia* and *Zalerion maritima*) on immersed woody litter (*Avicennia officinalis* and *Rhizophora mucronata*) belonged to the core group, even during monsoon season. This shows their ability to colonize and decompose the woody litter under conditions of low salinity (Sridhar and Maria 2006).

Concluding remarks

The importance and vital role of mangroves in tropical coastal ecosystems is beyond measure. Examples of their value are many, including:

(i) A reduction in coastal erosion, to dampen storm surges such as tsunamis, a reduction in flooding of coastal planes and lining the banks of rivers to prevent erosion (Saenger et al. 1983).
(ii) Providing spawning and nursery areas for many marine species with microorganisms, enhancing the polyunsaturated fatty acid content of detritus for their growth.
(iii) Export of particulate and dissolved nutrients to the open ocean. In this, fungi play a vital role in the decomposition of recalcitrant material in mangroves.
(iv) Mangroves, along with salt marshes and sea grasses, act as the Earth's blue carbon sink, capturing some 55% of biological carbon.

Mangroves are reported to be the most productive of any forest ecosystem. Therefore the maintenance and conservation of mangroves is essential because of their multifunctional role in marine ecosystems.

Acknowledgements

We thank Professor Gareth Jones for valuable comments. K.L. Pang would like to thank National Science Council of Taiwan for a research grant (NSC98–2321-B-019–004) and providing financial support.

References

Abdel-Wahab MA. Diversity of marine fungi from Egyptian Red Sea mangroves. Bot Mar. 2005;48:348–355.

Aleem AA. Distribution and ecology of marine fungi in Sierra Leone (Tropical West Africa). Bot Mar. 1980;23:679–688.

Alias SA, Jones EBG. Colonization of mangrove wood by marine fungi at Kuala Selangor mangrove stand, Malaysia. Fungal Divers. 2000a;5:9–21.

Alias SA, Jones EBG. Vertical distribution of marine fungi on *Rhizophora apiculata* at Morib mangrove, Selangor, Malaysia. Mycoscience 2000b;41:431–436.

Alias SA, Jones EBG. Marine fungi from mangroves of Malaysia. IOES Monograph Series 8, Kuala Lumpur: University of Malaya, 2009.

Alias SA, Kuthubutheen AJ, Jones EBG. Frequency of occurrence of fungi on wood in Malaysian mangroves. Hydrobiol. 1995;295:97–106.

Alias SA, Zainuddin N, Jones EBG. Biodiversity of marine fungi in Malaysian mangroves. Bot Mar. 2010;53:545–554.

Alongi DM. Present state and future of the world's mangrove forests. Environ Conserv. 2002; 29:331–349.

Alongi DM. Paradigm shifts in mangrove ecology. In: Perillo GME, Wolanski E, Cahoon DDR, Brinson MM eds. Coastal wetlands: an integrated ecosystem approach. Amsterdam: Elsevier, 2009:615–640.

Ananda K, Sridhar KR. Diversity of endophytic fungi in the roots of mangrove species on west coast of India. Can J Microbiol. 2002;48:871–878.

Ananda K, Sridhar KR. Filamentous fungal assemblage of two island mangroves of India. In: Madhyastha MN, Sridhar KR, Lakshmi A eds. Prospects and problems of environment across the millennium. New Delhi: Daya Publishing House, 2003:35–44.

Ananda K, Sridhar KR. Diversity of filamentous fungi on decomposing leaf and woody litter of mangrove forests of southwest coast of India. Curr Sci. 2004;87:1431–1437.

Ananda K, Sridhar KR, Raviraja NS, Bärlocher F. Breakdown of fresh and dried *Rhizophora mucronata* leaves in a mangrove of Southwest India. Wetlands Ecol Manage. 2008;16:1–9.

Anita DD, Sridhar KR. Assemblage and diversity of fungi associated with mangrove wild legume *Canavalia cathartica*. Trop Subtrop Agroecosys. 2009;10:225–235.

Anita DD, Sridhar KR, Bhat R. Diversity of fungi associated with mangrove legume *Sesbania bispinosa* (Jacq.) W. Wight (Fabaceae). Livestock research for rural development 2009;21: Article # 67. (Accessed April 11, 2012 at http://www.lrrd.org/lrrd21/5/cont2105.htm.)

Arnold AE, Henk DA, Eells RL, Lutzoni F, Vilgalys R. Diversity and phylogenetic affinities of foliar fungal endophytes in loblolly pine inferred by culturing and environmental PCR. Mycologia 2007;99:185–206.

Bandaranayake WM. Survey of mangrove plants from Northern Australia for phytochemical constituents and UV-absorbing compounds. Current Topics Phytochem. 1995;14:69–78.

Bandaranayake WM. Traditional and medicinal uses of mangroves. Mangroves Salt Marshes. 1998;2:133–148.

Bandaranayake WM. Bioactivities, bioactive compounds and chemical constituents of mangrove plants. Wetl Ecol Manag. 2002;10:421–452.

Bernini E, Rszende CE. Litterfall in a mangrove in Southeast Brazil. Pan-Am J Aquat Sci. 2010;5:508–519.

Besitulo A, Moslem MA, Hyde KD. Occurrence and distribution of fungi in a mangrove forest on Siargao Island, Philippines. Bot Mar. 2010;53:535–543.

Borse BD. Frequency of occurrence of marine fungi from Maharashtra coast, India. Indian J Mar Sci. 1988;17:165–167.

Borse BD, Kelkar DJ, Patil AC. Frequency of occurrence of marine fungi from Pirotan Island (Gujarat), India. Geobios. 2000;27:145–148.

Bunt JS. Continental scale patterns in mangrove litter fall. Hydrobiol. 1995;259:135–140.

Chaeprasert S, Piapukiew J, Whalley AJS, Sihanonth P. Endophytic fungi from mangrove plant species of Thailand: their antimicrobial and anticancer potentials. Bot Mar. 2010;53:555–564.

Chapman VJ. Mangrove vegetation. Liechtenstein, Germany: Cramer Vadyz, 1976.

Chen L, Zan Q, Li M, Shen J, Liao W. Litter dynamics and forest structure of the introduce *Sonneratia caseolaris* mangrove forest in Shenzhen, China. East Coast Shelf Sci. 2009;85:241–246.

Chinnaraj S. Manglicolous fungi from atolls of Maldives, Indian Ocean. Indian J Mar Sci. 1993a;22:141–142.

Chinnaraj S. Higher marine fungi from mangroves of Andaman and Nicobar Islands. Sydowia 1993b;45:109–115.
Clough BF. Primary productivity and growth of mangrove forests. In: Robertson AI, Alongi DM, eds. Tropical mangrove ecosystems. Washington DC: American Geophysical Union, 1992: 225–249.
Day JW Jr, Coronado-Molina C, Vera-Herrera FR, Twilley R, Rivera-Monroy VH, Alvarez-Guillen H, Day R, Conner W. A 7 year record of above-ground net primary production in a southeastern Mexican mangrove forest. Aqut Bot. 1996;55:39–60.
DeLonge WP, DeLonge PJ. An appraisal of factors controlling the latitudinal distribution of mangrove (*Avicennia marina* var. *resinifera*). NZ J Coastal Res. 1994;10:539–548.
Duke NC. Phenological trends with latitude in the mangrove tree *Avicennia marina*. J Ecol. 1990;78:113–133.
Duke NC. Mangrove floristics and biogeography. In: Robertson AI, Alongi DM, eds. Tropical mangrove ecosystems. Washington DC: American Geophysical Union, 1992:53–100.
Ellison AM. Managing mangroves with benthic biodiversity in mind: Moving beyond roving banditry. J Sea Res. 2008;59:2–15.
Field C. Journey amongst mangroves. Okinawa, Japan: International Society for Mangrove Ecosystems, 1995.
Ghosh PB, Singh BN, Chakrabarty C, Saha A, Das RL, Choudhury A. Mangrove litter production in a tidal creek of Lothian Island of Sunderbans, India. Indian J Mar Sci. 1990;19:292–293.
Hossain M, Hoque AKF. Litter production and decomposition in mangroves – a review. Indian J Forestry 2008;31:227–238.
Hyde KD. Studies on the tropical marine fungi of Brunei. II. Notes on five interesting species. Trans Mycol Soc Jpn. 1988;29:161–171.
Hyde KD. Intertidal fungi from mangrove fern, *Acrostichum speciosum*, including *Massarina acrostichi* sp. nov. Mycol Res. 1989a;93:435–438.
Hyde KD. Ecology of tropical marine fungi. Hydrobiol. 1989b;178:199–208.
Hyde KD. Ecology of tropical marine fungi from North Sumatra. Can J Bot. 1989c;67:3078–3082.
Hyde KD. *Caryospora mangrovei* sp. nov. and notes on marine fungi from Thailand. Trans Mycol Soc Jpn. 1989d;30:333–341.
Hyde KD. A study of vertical zonation of intertidal fungi on *Rhizophora apiculata* at Kampong Kapok mangrove, Brunei. Aquat Bot. 1990a;36:255–262.
Hyde KD. A comparison of the intertidal mycota of five mangrove tree species. Asian Mar Biol. 1990b;7:93–107.
Hyde KD. Fungal colonization of *Rhizophora apiculata* and *Xylocarpus granatum* poles in Kampong Kapok mangrove, Brunei. Sydowia 1991;43:31–38.
Hyde KD, Chalermongse A, Boonthavikoon T. The distribution of intertidal fungi on *Rhizophora apiculata*. In: Morton JB ed. The marine biology of South China Sea. Hong Kong: Hong Kong University Press, 1993:643–652.
Hyde KD, Chalermpongse A, Boonthavikoon T. Ecology of intertidal fungi at Ranong mangrove, Thailand. Trans Mycol Soc Jpn. 1990;31:17–28.
Hyde KD, Farrant CA, Jones EBG. Marine fungi from Seychelles. III. *Aniptodera mangrovei* sp. nov. from mangrove wood. Can J Bot. 1986;64:2989–2992.
Hyde KD, Jones EBG. Marine mangrove fungi. PSZNI Mar Ecol. 1988;9:15–33.
Hyde KD, Jones EBG. Ecological observations on marine fungi from the Seychelles. Bot J Linn Soc. 1989;100:237–254.
Hyde KD, Jones EBG. Introduction to fungal succession. Fungal Divers. 2002;10:1–4.
Hyde KD, Lee SY. Ecology of mangrove fungi and their role in the nutrient cycling: what gaps occur in our knowledge? Hydrobiol. 1995;295:107–118.

Hyde KD, Sarma VV. Biodiversity and ecological observations on filamentous fungi of mangrove palm *Nypa fruticans* Wurumb. (Loiliopsida-Arecales) along the Tutong River, Brunei. Ind J Mar Sci. 2006;35:297–307.

Hyde KD, Soytong K. The fungal endophyte dilemma. Fungal Divers. 2008;33:163–173.

Imbert D, Ménard S. Structure de la végétation et production primarie dans la mangrove de la Baie de Forte-de-France, Martinique (FWI). Biotropica 1997;29:413–425.

Imgraben S, Dittmann S. Leaf litter dynamics and litter consumption in two temperate South Australian mangrove forests. J Sea Res. 2008;59:83–93.

Jones EBG. Fungal adhesion. Mycol Res. 1994;98:961–981.

Jones EBG. Ultrastructure and taxonomy of the aquatic ascomycetous order Halosphaeriales. Mycol Res. 1995;73(Suppl 1):S790–S801.

Jones EBG. Marine fungi: some factors influencing biodiversity. Fungal Divers. 2000;4:53–73.

Jones EBG, Agerer R. *Calathella mangrovei* sp nov and observations on the mangrove fungus *Halocyphina villosa*. Bot Mar. 1992;35:259–265.

Jones EBG, Tan TK. Observations on manglicolous fungi from Malaysia. Trans Br Mycol Soc. 1987;89:390–392.

Jones EBG, Abdel-Wahab MA. Marine fungi from Bahamas Islands. Bot Mar. 2005;48: 356–364.

Jones EBG, Hyde KD. Succession: where do we go from here? Fungal Divers. 2002;10:241–253.

Jones EBG, Uyenco FR, Follosco M. Fungi on driftwood collected in the intertidal zone from the Philippines. Asian Mar Biol. 1988;5:103–106.

Jones EBG, Vrijmoed LLP. Observations on subtropical fungi on driftwood from mangroves and sandy beaches in the Peral River Estuary. In: Janardhanan KK, Rajedran C, Natarajan K, Hawksworth DL, eds. Tropical mycology. USA: Science Publishers Inc, 1997:51–59.

Kathiresan K, Bingham BL. Biology of mangrove ecosystem. Adv Mar Biol. 2001;40:81–251.

Khafaji AK, Mandura AS, Saifullah SM, Sambas AZ. Litter production in two mangrove stands of the Southern Red Sea coast of Saudi Arabia (Jizan). JKAU Mar Sci. 1991;2:93–100.

Kohlmeyer J, Bebout B, Volkmann-Kohlmeyer B. Decomposition of mangrove wood by marine fungi and Teredinids in Belize. PSZNI Mar Ecol. 1995;16:27–39.

Kohlmeyer J, Kohlmeyer E. Marine mycology – the higher fungi. New York: Academic Press, 1979.

Kumar IJN, Sajish PR, Kumar RN, Basil G, Shailendra V. Nutrient dynamics in an *Avicennia marina* (Forsk.) Vierh. Mangrove forest in Vamleshwar, Gujrat, India. Not Sci Biol. 2011;3: 51–56.

Kumaresan V, Suryanarayanan TS. Occurrence and distribution of endophytic fungi in a mangrove community. Mycol Res. 2001;105:1388–1391.

Kumaresan V, Suryanarayanan TS. Endophyte assemblage in young, mature and senescent leaves of *Rhizophora apiculata*: evidence for the role of endophytes in mangrove litter degradation. Fungal Divers. 2002;9:81–91.

Leaño EM, Jones EBG, Vrijmoed LLP. Why are *Halophytophthora* species well adapted to mangrove habitats? Fungal Divers. 2000;5:131–135.

Lee OHK, Hyde KD. Phylloplane fungi in Hong Kong mangroves: evaluation of study methods. Mycologia 2002;94:596–606.

Leong WF, Tan TK, Jones EBG. Fungal colonization of submerged *Bruguiera cylindrica* and *Rhizophora apiculata* wood. Bot Mar. 1991;34:69–76.

Mackey AP, Smail G. The decomposition of mangrove litter in a subtropical mangrove forest. Hydrobiol. 1996;332:93–98.

Maldonado-Ramírez SL, Torres-Pratts H. First report of *Clathrus* cf. *crispus* (Basidiomycota: Clathraceae) occurring on decomposing leaves of *Rhizophora mangle* in Puerto Rico. Caribbean J Sci. 2005;41:357–359.

Maria GL. Studies on the mangrove mycoflora of west coast of India, PhD Dissertation, India: Mangalore University, 2003.

Maria GL, Sridhar KR. Richness and diversity of filamentous fungi on woody litter of mangroves along the west coast of India. Curr Sci. 2002;83:1573–1580.

Maria GL, Sridhar KR. Endophytic fungal assemblage of two halophytes from west coast mangrove habitats, India. Czech Mycol. 2003a;55:241–251.

Maria GL, Sridhar KR. Diversity of filamentous fungi on woody litter of five mangrove plant species from the southwest coast of India. Fungal Divers. 2003b;14:109–126.

Maria GL, Sridhar KR. Fungal colonization of immersed wood in mangroves of the southwest coast of India. Can J Bot. 2004;82:1409–1418.

Mepham RH, Mepham JS. The flora of tidal forest – a rationalization of the use of the term "mangrove". S Afr J Bot. 1985;51:77–99.

Mfilinge PL, Meziane T, Bachok Z, Tsuchiya M. Litter dynamics and particulate organic matter outwelling from a subtropical mangrove in Okinawa Island, South Japan. Est Coast Shelf Sci. 2005;63:301–313.

Miller JD, Whitney NJ. Fungi from the Bay of Fundy I: Lignicolous marine fungi. Can J Bot. 1981;59:1128–1133.

Naik BS, Shashikala J, Krishnamurthy YL. Study on the diversity of endophytic communities from rice (*Oryza sativa*L.) and their antagonistic activities *in vitro*. Microbiol Res. 2007;164:90–296.

Nakagiri A. Intertidal fungi from Iriomote Island. Inst Ferment Res Commun. 1993;16:24–62.

Nakagiri A, Okane I, Ito T, Katumoto K. *Lanceispora amphibia* gen *et* sp nov, a new amphisphaeriaceous ascomycete inhabiting senescent and fallen leaves of mangrove. Mycoscience 1997;38:207–213.

Newell SY. Establishment and potential impacts of eukaryotic mycelial decomposers in marine/terrestrial ecotones. J Exp Mar Biol Ecol. 1996;200:187–206.

Newell SY, Fell JW. Ergosterol content of living and submerged decaying leaves and twigs of red mangrove. Can J Bot. 1992;38:979–982.

Newell SY, Fell JW. Competition among mangrove oomycetes and between oomycetes and other microbes. Aquat Microb Ecol. 1997;12:21–28.

Othman S. The rate of litter production in mangrove forest at Siar Beach, Lundu, Sarawak. Pertanika 1989;12:47–51.

Pang KL, Jheng JS, Jones EBG. Marine mangrove fungi of Taiwan. Keelung: National Taiwan Ocean University Press, 2011.

Pang KL, Vrijmoed LLP, Goh TK, Plaingam N, Jones EBG. Fungal endophytes associated with *Kandelia candel* (Rhizophoraceae) in Mai Po Nature Reserve, Hong Kong. Bot Mar. 2008;51:171–178.

Patil KB, Borse BD. Studies on higher marine fungi from Gujarat coast (India). Geobios. 2001;28:41–44.

Poon MOK, Hyde KD. Biodiversity of intertidal estuarine fungi on *Phragmites* at Mai Po marshes, Hong Kong. Bot Mar. 1998a;41:141–155.

Poon MOK, Hyde KD. Evidence for the vertical distribution of saprophytic fungi on senescent *Phragmites australis* culms at Mai Po marshes. Bot Mar. 1998b;41:285–292.

Poonyth AD, Hyde KD, Peerally A. Intertidal fungi in Mauritian mangroves. Bot Mar. 1999;42: 243–252.

Poonyth AD, Hyde KD, Peerally A. Colonization of *Bruguiera gymnorrhiza* and *Rhizophora mucronata* wood by marine fungi. Bot Mar. 2001;44:75–80.

Promputtha I, Hyde KD, McKenzie EHC, Peberdy JF, Lumyong S. Can leaf degrading enzymes provide evidence that endophytic fungi becoming saprobes? Fungal Divers. 2010;41:89–99.

Qasim SZ, Wafar MVM. Marine resources in the tropics. Res Manag Optim. 1990;7:141–169.

Raghukumar S, Sharma S, Raghukumar C, Sathe-Patak V, Chandramohan D. Thraustochytrids and fungal components of marine detritus: IV. Laboratory studies on decomposition of leaves of the mangrove *Rhizophora apiculata* Blume. J Exp Mar Biol Ecol. 1994;183:113–131.

Raveendran K, Manimohan R. Marine fungi of Kerala – a preliminary floristic and ecological study. India: Malabar Natural History Society, 2007.

Ravikumar DR, Vittal BPR. Fungal diversity on decomposing biomass of mangrove plant *Rhizophora* in Pichavaram estuary, east coast of India. Indian J Mar Sci. 1996;25:142–144.

Read SJ, Jones EBG, Moss ST, Hyde KD. Ultrastructure of asci and ascospores of two mangrove fungi: *Swampomyces armeniacus* and *Marinosphaera mangrovei*. Mycol Res. 1995;99: 1465–1471.

Robertson AI, Alongi DM, Boto KG. Concluding remarks: research and mangrove conservation. In: Robertson AI, Alongi DM eds. Tropical mangrove ecosystem. Washington DC: American Geophysical Union, 1992:293–326.

Rodrigues KF, Petrini O. Biodiversity of endophytic fungi in tropical regions. In: Hyde KD ed. Biodiversity of tropical microfungi. Hong Kong: Hong Kong University Press, 1997:57–69.

Sadaba RB, Hodgkiss LJ, Vrijmoed LLP, Jones EBG. Observations on vertical distribution of fungi associated with standing senescent *Acanthus ilicifolius* stems at Mai Po mangrove, Hong Kong. Hydrobiol. 1995;295:119–126.

Saenger P, Hegerl EJ, Davie JDS. Global status of mangrove ecosystems. IUCN Commission on Ecology Papers No. 3. Switzerland: Gland, 1983.

Saenger P, Snedaker SC. Pantropical trends in mangrove above-ground biomass and annual litterfall. Oecologia 1993;96:293–299.

Sakayaroj J, Jones EBG, Chatmala I, Phongpaichit S. Marine fungi. In: Jones EBG, Tantichareon M, Hyde KD, eds. Thai fungal diversity. Thailand: BIOTEC, 2004:107–117.

Sakayaroj J, Preedanon S, Supaphon O, Jones EBG, Phongpaichit S. Phylogenetic diversity of endophyte assemblages associated with the tropical seagrass *Enhalus acoroides* in Thailand. Fungal Divers. 2010;42:27–45.

Sarma VV, Hyde KD. A review on frequently occurring fungi in mangroves. Fungal Divers. 2001;8:1–34.

Sarma VV, Vittal BPR. Biodiversity of mangrove fungi on different substrata of *Rhizophora apiculata* and *Avicennia* spp. from Godavari and Krishna deltas, east coast of India. Fungal Divers. 2000;5:23–41.

Schmit JP, Shearer CA. A checklist of mangrove-associated fungi, their geographical distribution and known host plants. Mycotaxon 2003;53:423–477.

Shearer CA, Descals E, Kohlmeyer B, Kohlmeyer J, Marvanova L, Padgett DE, Porter D, Raja HA, Schmit JP, Thorton HA, Voglymayr H. Fungal biodiversity in aquatic habitats. Biodivers Conserv. 2007;16:49–67.

Slim FJ, Gwada PM, Kodjo M, Hemminga MA. Biomass and litterfall of *Ceriops tagal* and *Rhizophora mucronata* in the mangrove forest of Gazi Bay, Kenya. Mar Freshw Res. 1996;47: 999–1007.

Spalding M, Blasco F, Field C. World mangrove atlas. Cambridge, UK: Cambridge Samara Pub Co, 1997.

Sridhar KR. Mangrove fungi of the Indian Peninsula. In: Sridhar KR ed. Frontiers in fungal ecology, diversity and metabolites. New Delhi: IK International Publishing House Pvt Ltd, 2009:28–50.

Sridhar KR, Karamchand KS, Sumathi P. Fungal colonization and breakdown of sedge (*Cyperus malaccensis* Lam.) in a southwest mangrove, India. Bot Mar. 2010;53:525–533.

Sridhar KR, Maria GL. Fungal diversity on woody litter of *Rhizophora mucronata* in a southwest Indian mangrove. Indian J Mar Sci. 2006;35:318–325.

Suryanarayanan TS, Kumaresan V. Endophytic fungi of some halophytes from an estuarine mangrove forest. Mycol Res. 2000;104:1465–1467.

Suryanarayanan TS, Kumaresan V, Johnson JA. Foliar fungal endophytes from two species of the mangrove *Rhizophora*. Can J Microbiol. 1998;44:1003–1006.

Tan TK, Leong WF, Jones EBG. Succession of fungi on wood of *Avicennia alba* and *A. lanata* in Singapore. Can J Bot. 1989;67:2687–2691.

Tan TK, Teng CL, Jones EBG. Substrate and microbial interactions as factors affecting ascocarp formation by mangrove fungi. Hydrobiol. 1995;295:127–134.

Tan TK, Leong WF. Lignicolous fungi of tropical mangrove wood. Mycol Res. 1992;96:413–414.

Tomlinson PB. The botany of mangroves. Cambridge, UK: Cambridge University Press, 1986.

Twilley RR, Chen R, Hargis T. Carbon sinks in mangroves and their implications to carbon budget of tropical ecosystems. Water Air Soil Pollut. 1992;64:265–288.

Vega FE, Posada F, Aime MC, Pava-Ripoll M, Infante F, Rehner SA. Entomopathogenic fungal endophytes. Biol Control. 2008;46:72–82.

Volkmann-Kohlmeyer B, Kohlmeyer J. Biogeographic observations on Pacific marine fungi. Mycologia 1993;85:337–346.

Vrijmoed LLP. Isolation and culture of higher filamentous fungi. In: Hyde KD, Pointing SB, eds. Marine mycology – a practical approach. Hong Kong: Fungal Diversity Press, 2000:1–20.

Vrijmoed LLP, Hyde KD, Jones EBG. Observations on mangrove fungi from Macau and Hong Kong, with the description of two new ascomycetes: *Diaporthe salsuginosa* and *Aniptodera haispora*. Mycol Res. 1994;98:699–704.

Wafar S, Untawale AG, Wafar M. Litterfall and energy flux in a mangrove ecosystem. Est Coast Shelf Sci. 1997;44:111–124.

Wilkie ML, Fortuna S. (Status and trends in mangrove area extent worldwide. Forest Resources Assessment Working Paper # 63. Rome: Forest Resources Division, FAO, 2003.

Wong MKM, Poon MOK, Hyde KD. *Phragmitensis marina* gen. et sp. nov., an intertidal saprotroph from *Phragmites australis* in Hong Kong. Bot Mar. 1998;41:379–382.

15 Biodiversity of fungi on the palm *Nypa fruticans*

Apilux Loilong, Jariya Sakayaroj, Nattawut Rungjindamai, Rattaket Choeyklin and E.B. Gareth Jones

Mangrove forests play a major role in the ecology of coastal tropical/subtropical waters as they serve as hatchery and nursery habitats for marine organisms, and cover some 198,818 km^2 of the Earth's surface (Fisher and Spalding 1993). *Nypa fruticans* is a brackish water palm that grows in mangroves (at salinities ranging from 0–35‰), in soft mud and slow-moving tidal and river waters that bring in nutrients. The palm can be found as far inland as the tide can deposit the floating nuts. It is common on coasts and rivers flowing into the Indian and Pacific Oceans, ranging from Bangladesh to the Pacific Islands (Tomlinson 1986).

Many new taxa are reported when new substrata are surveyed for fungi, for example 123 species have been reported from temperate and tropical salt marshes, with *Spartina* spp. being the most common host (Gessner and Kohlmeyer 1976; Barata 2002; Calado and Barata, Chapter 18). The grass *Phragmites australis* and the sedge *Schoenoplectus littoralis* were shown to support 61 (17 ascomycetes and 44 asexual taxa) and 31 (six ascomycetes and 25 asexual taxa) species, respectively, in Mai Po Marshes, Hong Kong (Poon and Hyde 1998a, 1998b; Wong et al. 1998), with many being new species. *Juncus roemerianus* yielded 107 species (44 obligate, 25 facultative and 38 halotolerant/terrestrial), of which 48 are new species belonging to seven orders, 20 families and 44 genera (Kohlmeyer and Volkmann-Kohlmeyer 2001; Jones 2011). Similarly, many new taxa have been described from *Nypa fruticans*, and these have been well documented by Hyde (1992a, 1992b), Hyde and Alias (1999), Hyde et al. (1999), Alias and Jones (2009) and Besitulo et al. (2010). Currently, 139 fungal taxa have been recorded on *N. fruticans* from South East Asian countries including Brunei (Hyde 1988a, 1988b, 1988c, 1992a; Hyde and Sarma 2006), Malaysia (Hyde et al. 1999; Hyde and Alias 1999, 2000), the Philippines (Besitulo et al. 2002), Papua New Guinea (Hyde 1992b) and Thailand (Hyde and Nakagiri 1989; Hyde 1992b; Pilantanapak et al. 2005).

Our objectives in this chapter are to:

(i) review the diversity of fungi reported from *N. fruticans*; and
(ii) present data on a recent biodiversity survey undertaken in Thailand.

In Thailand, *N. fruticans* forms extensive forests in many regions, i.e. the eastern, central and southern parts. Each region differs in water salinity, ranging from freshwater to brackish and marine habitats.

Materials and methods

Sample collection

Decaying intertidal petioles and fronds of *N. fruticans* were collected at six locations in Thailand: 1) Trang, Satun and Surat Thani provinces in the south; and 2) Samut Sakhon, Samut Prakarn and Chachoengsao provinces in the central region. Water salinity was measured at each site and *Nypa* samples placed in polyethylene bags and returned to the mycology laboratory at BIOTEC, Thailand. Samples were incubated at room temperature on tissue paper soaked with tap water in plastic boxes for up to three months.

Fungal diversity analysis

The frequency of the fungal occurrence of each taxon collected from was analyzed by the formula:

Frequency of occurrence (FoO) = occurrence of each taxon × 100/number of samples examined.

Results

Frequency of occurrence of saprobic fungi on *Nypa fruticans* in Thailand

Table 15.1 lists the fungi identified on 331 samples collected from the six localities yielding 71 fungi (48 ascomycetes, 21 asexual fungi and two basidiomycetes). *Linocarpon appendiculatum* (12.0–47.4% FoO), *Astrosphaeriella striatispora* (10.3–40.0%) and *Trichocladium nypae* (8.8–76.3% FoO) were the most frequently collected taxa on the *Nypa* palm. *Linocarpon* was the most speciose genus with five species, followed by *Aniptodera* and *Savoryella* (four and three species, respectively). Most taxa were marine, with 44 ascomycetes such as *Aniptodera, Astrosphaeriella, Linocarpon, Savoryella* and *Tirisporella* and asexual fungi *Cirrenalia macrocephala, Halenospora varia* and *Trichocladium* spp. Many of the asexual fungi were typical freshwater species, such as *Helicoma hyalonema, Helicosporium hongkongense*, and *Thozetella nivea* (Table 15.1).

The fungal community on *N. fruticans* at different salinities was compared. At two low salinity (2–5‰) sites (in Trang and Surat Thani provinces), most of the fungal taxa were freshwater species, e.g. *Acrogenospora, Helicoma* and *Helicosporium* spp. In contrast, at the higher (15–26‰) salinity locations (in Samutsakorn, Satun, Samutprakarn and Chachoengsao provinces), the fungal taxa were typically marine fungi (*Halosarpheia, Helicascus, Lulworthia, Saagaromyces,* and *Savoryella* species). The basidiomycete *Grammothele fuligo* was first recorded on *Nypa* fronds by Dr Rattaket Choeyklin (Jones et al. 2009) and has also been collected in the present study. This fungus appears to be widespread on *Nypa* bases in low-salinity estuarine waters. *Cosmospora* sp. and *Arecophila* sp. were recorded for the first time on *Nypa* (Table 15.1, Figures 15.1–15.20), and may be new taxa that will be described at a later stage.

In a separate study a further nine basidiomycetes were recorded on *Nypa* collected in the freshwater intertidal zone of the estuary, with *Grammothele fuligo* being the most common species (Table 15.2, Figures 15.21–15.29). Further studies are warranted to determine their role in the breakdown of this palm in estuarine habitats.

Table 15.1 Frequency of occurrence of saprobic fungi on *Nypa fruticans* from six locations in Thailand.

Fungi	Frequency of occurrence (%)					
	South Thailand			Central Thailand		
	Trang	Satun	Surat Thani	Samut Prakarn	Chachoengsao	Samut Sakorn
Salinity (psu ‰)	5	26	2	15	17	15
Ascomycetes						
Aniptodera chesapeakensis	3.4	1.1	2.5	3.9	2.6	1.1
Aniptodera longispora	2.3					
Aniptodera nypae	9.2	13.9	2.5	13.7	15.8	8.0
Aniptodera intermedia	2.3	2.3		3.9		
Anthostomella nypae	3.4	1.1		1.96	5.3	
Arecophila sp.	1.1			5.9		
Astrosphaeriella nypae	3.4	1.1		5.9	10.5	
Astrosphaeriella striatispora	10.3	9.2	40.0	7.8	10.5	
Carinispora nypae	4.6	1.1				
Carinispora velatispora						3.4
Cosmospora sp.	2.3		15.0			
Dactylospora haliotrepha		1.1				
Fasciatispora nypae		6.9		1.96	7.9	
Panorbis viscosus		2.3		7.8		
Halomassarina thalassiae	1.1					
Halosarpheia marina		1.1				
Helicascus nypae	8.0		2.5			
Kallichroma tethys		1.1				
Leptosphaeria australiensis				1.96		
Leptosphaeria nypicola	2.3	19.5	2.5	1.96	2.6	3.4
Leptosphaeria sp.	5.7	4.6	7.5	1.96	2.6	
Lignincola laevis	1.1	2.3	5.0	15.7	2.6	
Lignincola nypae	3.4	12.9	2.5			1.1
Linocarpon appendiculatum	19.5	24.7	32.5	21.6	47.4	12
Linocarpon bipolaris	2.3		2.5			
Linocarpon nypae			15.0	3.9	23.7	3.4
Linocarpon angustatum	1.1	2.3	5.0			2.3
Linocarpon longisporum		2.3				
Lophiostoma mangrovei				15.7		
Lophiostoma sp.	1.1					

Table continued on next page.

Table 15.1 (Continued)

Fungi	Frequency of occurrence (%)					
	South Thailand			Central Thailand		
	Trang	Satun	Surat Thani	Samut Prakarn	Chachoengsao	Samut Sakorn
Lulworthia grandispora	6.9	1.1		9.8		1.1
Marinosphaera mangrovei		2.3				
Massarina sp.	1.1					
Halomassarina thalassie	1.1					
Natantispora retorquens	2.3		2.5			
Neptunella longirostris	1.1	4.6		1.96		1.1
Oceanitis cincinnatula	2.3	4.6				
Oxydothis nypae	2.3	2.3		9.8	2.6	
Phomatospora nypicola		4.6	5.0	1.96		4.6
Saagaromyces ratnagiriensis			2.5			
Saccadoella cf. *mangrovei*			5.0			
Savoryella cf. *aquatica*	1.1	1.1				
Savoryella paucispora		4.6		1.96	2.6	1.1
Savoryella lignicola		5.7	22.5	13.7	2.6	
Tirisporella beccariana	5.7	5.7	7.5	1.96	2.6	4.6
Trematosphaeria mangrovei	1.1					
Vibrissea nypicola	4.6	25.6		7.8	13.2	
Verruculina enalia		1.1				
Tirispora unicaudata		1.1				
Asexual fungi						
Acrogenospora sphaerocephala	1.1					
Cirrenalia macrocephala		2.3				
Cumulospora sp.		1.1		5.9		
Dactylaria sp.	1.1					1.1
Dictyosporium sp.			5.0	1.96	7.9	
Helicoma sp.		1.1				
Helicoma hyalonema			2.5			
Helicorhoidion nypicola	3.4	26.9	5.0	7.8	26.3	5.7
Helicosporium pannosum			2.5			
Helicosporium hongkongense			2.5			
Monodictys sp.		1.1	2.5	2.0		
Halenospora varia		1.1		2.0	2.6	
Monacrosporium sp.						

Table continued on next page.

Table 15.1 (Continued)

Fungi	Frequency of occurrence (%)					
	South Thailand			Central Thailand		
	Trang	Satun	Surat Thani	Samut Prakarn	Chachoengsao	Samut Sakorn
Sporidesmium sp.		6.9	2.5			
Thozetella nivea			2.5			
Trichocladium achrasporum		3.4				
Trichocladium nypae	8.8	40.9	62.5	64.7	76.3	13.0
Vanakripa sp.	2.3	3.4	5.0	19.6		
Diplodia sp.		1.1	2.5	1.96		
Phoma sp.		2.3				
Phomopsis sp.		2.3				
Basidiomycetes						
Grammothele fuligo	1.1			1.96		
Halocyphina villosa			2.5			
Average number of fungi per sample	2.41	1.97	1.3	1.53	1.9	1.38
Total taxa	36	47	31	33	20	16
Total 72 taxa						

Review of fungi found on *Nypa fruticans*

Table 15.3 lists all of the fungi recorded from *Nypa* fronds in Asia with 135 taxa (90 Ascomycota, three Basidiomycota and 42 asexual taxa) of which 97 are described species. Studies of the fungal diversity on the intertidal brackish water palm *N. fruticans* have been undertaken by Pilantanapak et al. (2005) and Hyde and Sarma (2006), while previous studies focused on describing new taxa (Hyde 1988d, 1992a, 1992b; Jones et al. 1996; Hyde et al. 1999; Hyde and Alias 2000). Pilantanapak et al. (2005) collected 81 species on *Nypa* in Thailand, while Hyde and Sarma (2006) documented 46 species from the Tutong River in Brunei. However, there are few quantitative data on the fungi occurring on this palm. Of the five surveys of fungi on *Nypa*, the most common species were *Linocarpon appendiculatum* (in all studies, 20–53% FoO), *L. nypae* (in four studies, 17.5–32.5% FoO), *Oxydothis nypae* (in all five studies, 12–26% FoO) and *Astrosphaeriella striatispora* (in four studies, 18.0–49.5% FoO) (Hyde 1992a, 1992b; Hyde and Alias 2000; Besitulo et al. 2002; Pilantanapak et al. 2005; Hyde and Sarma 2006; Besitulo et al. 2010). Table 15.4 lists the ten most frequent fungi found on *Nypa* palm in three studies (Pilantanapak et al. 2005; Hyde and Sarma 2006; Besitulo et al. 2010). *Linocarpon appendiculatum*, *Astrosphaeriella striatispora* and *Helicorhoidion nypicola* were frequent in all three studies, while *Aniptodera nypae*, *Lignincola nypae* and *Oxydothis nypae* featured in two studies. The FoO varied greatly between the three studies, ranging from 2.1–13.0% (Hyde and Sarma 2006) to 3.0–53.0% in collections in the Philippines (Besitulo et al. 2010).

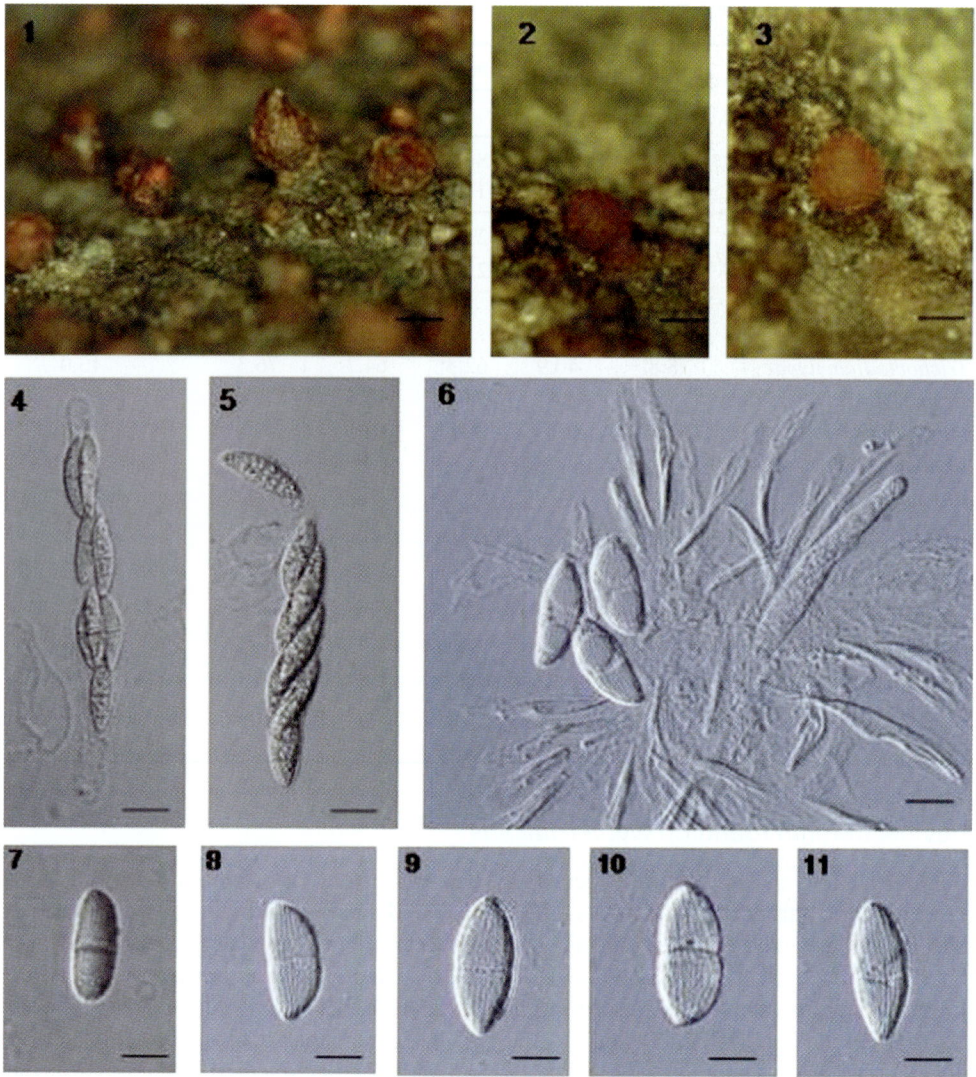

Figures 15.1–15.11 *Cosmospora* sp. Figures 15.1–15.3 Ascomata on base and petiole of *Nypa fruticans*. Figure 15.4 Ascus with immature ascospores. Figure 15.5 Mature ascus. Figure 15.6 Young asci. Figures 15.7–15.11 Striated ascospores. Scale bars 1–3 = 100 µm; 4–6 = 20 µm; 7–11 = 10 µm.

The fungi occurring on *Nypa* can be categorized into four groups based on their known occurrence:

(i) Typically marine/mangrove, e.g. *Aniptodera chesapeakensis, A. longispora, Kallichroma tethys, Marinosphaera mangrovei, Lignincola laevis, Lulworthia grandispora, Savoryella paucispora, Saagaromyces ratnagiriensis* and *Verruculina enalia*.

(ii) Host-specific, e.g. *Aniptodera nypae, A. intermedia, Anthostomella nypae, Fasciatispora nypae, Helicascus nypae, Lignincola nypae, Linocarpon appendiculatum,*

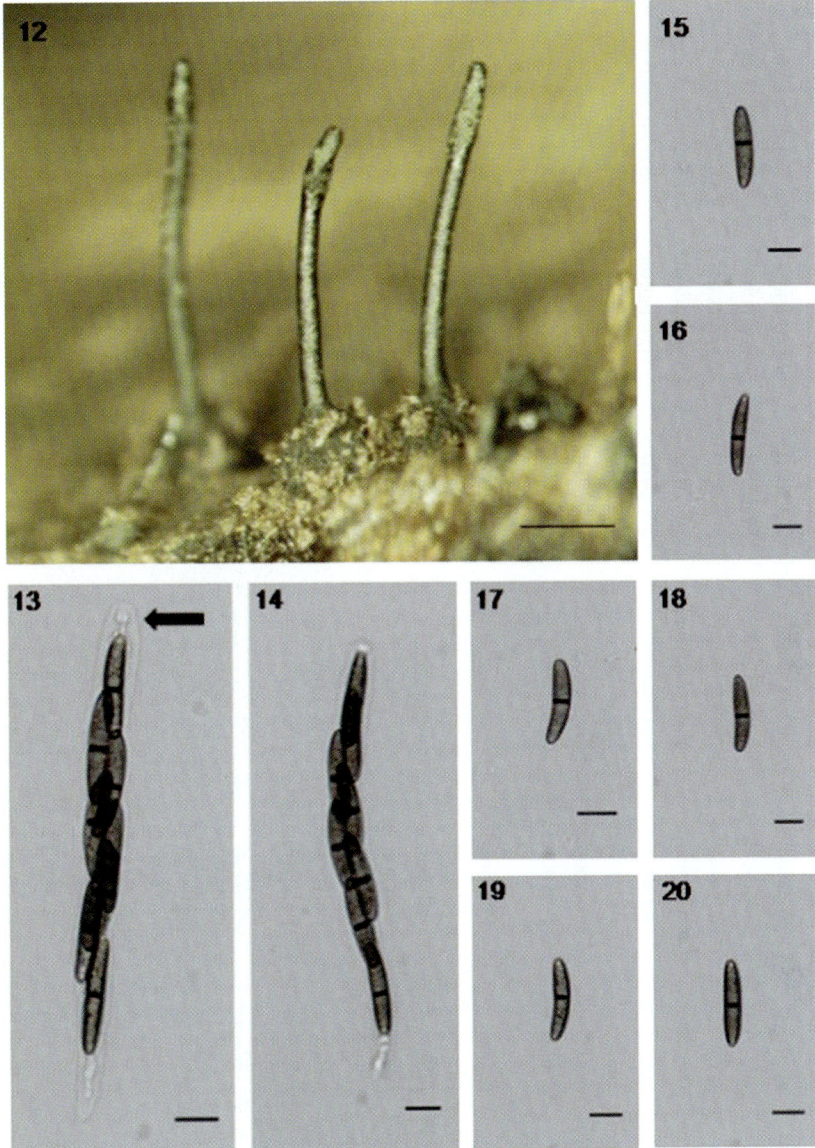

Figures 15.12–15.20 *Arecophila* sp. Figure 15.12 Ascomata on petiole base of *Nypa fruticans*. Figures 15.13–15.14 Ascus with J-subapical ring (arrow). Figures 15.15–15.20 Ascospores. Scale bars 12 = 200 μm; 13–14 = 20 μm; 15–20 = 10 μm.

L. bipolaris, L. angustatum, Oxydothis nypae, Tirisporella beccariana, Trichocladium nypae, and *Helicorhoidion nypicola.*

(iii) New species e.g. *Arecophila* sp., *Cosmospora* sp. and *Dactylaria* sp.
(iv) Freshwater species, e.g. *Helicoma hyalonema, Helicosporium pannosum, H. hongkongense* and *Thozetella nivea.*

Table 15.2 Lists of basidiomycetes collected from *Nypa fruticans* from several provinces in Thailand.

Original Code	Genus	Epithet	Order	Family	Collection Date	Substrate	Province
RCK00214	*Marasmiellus*	sp. 1	Agaricales	Marasmiaceae	15/9/2010	*Nypa* leaf	Samut Sakorn
RCK00215	*Psathyrella*	sp. 1	Agaricales	Psathyrellaceae	15/9/2010	*Nypa* petiole	Samut Sakorn
1(25/1/2010)	*Psathyrella*	sp. 2	Agaricales	Psathyrellaceae	25/1/2010	*Nypa* petiole	Surat Thani
RCK00226	*Schizophyllum*	*commune*	Agaricales	Schizophyllaceae	15/9/2010	*Nypa* petiole	Samut Sakorn
RCK00228	*Coprinus*	sp.	Agaricales	Coprinaceae	15/9/2010	*Nypa* petiole	Samut Sakorn
RCK00230	*Marasmiellus*	sp. 2	Agaricales	Marasmiaceae	15/9/2010	*Nypa* petiole	Samut Prakarn
1(24/1/2010)	*Grammothele*	*fuligo*	Polyporales	Polyporaceae	24/1/2010	*Nypa* petiole	Ranong
1(28/4/2010)	*Hyphoderma*	*sambuci*	Polyporales	Meruliaceae	28/4/2009	*Nypa* petiole	Nakhon Sri Thammarat
3(28/4/2010)	*Grammothele*	*fuligo*	Polyporales	Polyporaceae	28/4/2009	*Nypa* petiole	Nakhon Sri Thammarat
RCK00227	*Grammothele*	*fuligo*	Polyporales	Polyporaceae	15/9/2010	*Nypa* petiole	Samut Sakorn
RCK00232	*Grammothele*	*fuligo*	Polyporales	Polyporaceae	15/9/2010	*Nypa* petiole	Samut Prakarn
RCK00236	*Grammothele*	*fuligo*	Polyporales	Polyporaceae	15/9/2010	*Nypa* petiole	Samut Prakarn
RCK00238	*Grammothele*	*fuligo*	Polyporales	Polyporaceae	15/9/2010	*Nypa* petiole	Samut Prakarn
RCK00240	*Grammothele*	*fuligo*	Polyporales	Polyporaceae	15/9/2010	*Nypa* petiole	Samut Prakarn
RCK00241	*Grammothele*	*fuligo*	Polyporales	Polyporaceae	15/9/2010	*Nypa* sheath	Samut Prakarn
RCK00242	*Grammothele*	*fuligo*	Polyporales	Polyporaceae	15/9/2010	*Nypa* sheath	Samut Prakarn
RCK00243	*Grammothele*	*fuligo*	Polyporales	Polyporaceae	15/9/2010	*Nypa* sheath	Samut Prakarn
RCK00233	*Phanerochaete*	sp.	Polyporales	Phanerochaetaceae	15/9/2010	*Nypa* petiole	Samut Prakarn
2(28/4/2010)	Unidentified	Unidentified	Unidentified	Unidentified	24/4/2010	*Nypa* petiole	Nakhon Sri Thammarat
RCK00218	Unidentified	Unidentified	Unidentified	Unidentified	15/9/2010	*Nypa* petiole	Samut Sakorn
RCK00231	Unidentified	Unidentified	Unidentified	Unidentified	15/9/2010	*Nypa* petiole	Samut Prakarn

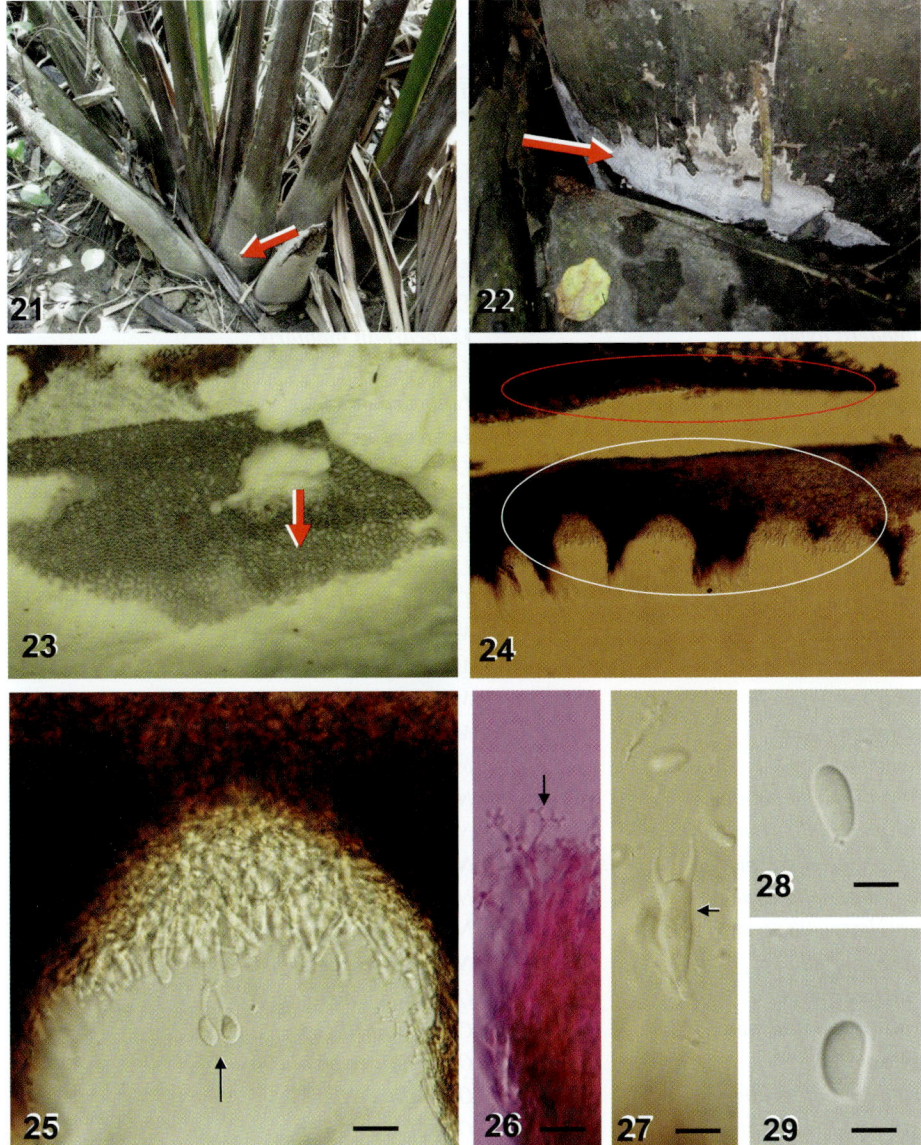

Figures 15.21–15.29 Figures 15.21–15.22 Basidiocarp in its natural habitat. Figure 15.23 Pores surface. Figure 15.24 Cross-section of a hymenium (white circle). Figure 15.25 Basidiospores 21–22: on the basidium. Figure 15.26 Dendrohyphidia stained with phloxine. Figure 15.27 Basidium. Figures 15.28–15.29 Basidiospore. Scale bars: 25–29 = 10 µm.

Vertical distribution of marine fungi

Vertical distribution of mangrove fungi on woody tissue has been demonstrated in a number studies, while many species are distributed throughout the tidal range (Hyde 1988a, 1988b; Sadaba et al. 1995; Poon and Hyde 1998b; Alias and Jones 2000). However, Besitulo et al. (2010) found no evidence of vertical distribution of fungi on the palm *Nypa fruticans*

Table 15.3 List of the fungi documented as growing on *Nypa fruticans* from South East Asia (species in bold are known only from *Nypa*).

Taxa	Taxa
Ascomycetes	
Aniptodera chesapeakensis	*Lignincola laevis*
Aniptodera intermedia	***Lignincola nypae***
Aniptodera **cf.** *limnetica*	*Lignincola tropica*
Aniptodera longispora	*Linocarpon angustatum*
Aniptodera mangrovei	***Linocarpon appendiculatum***
Aniptodera nypae	***Linocarpon bipolaris***
Anthostomella eructans	*Linocarpon livistonae*
Anthostomella nypicola	***Linocarpon nypae***
Anthostomella nypae	*Linocarpon angustatum*
Anthostomella nypensis	***Linocarpon longisporum***
Anthostomella sp.	*Lophiostoma mangrovei*
Anthostomella cf. *remii*	*Lophiostoma* sp.
Annulatascus velatispora	*Lulworthia grandispora* (500–1000 × 3.7 µm)
Annulatascus palmietensis	*Lulworthia grandispora* (462–500 × 3.7 µm)
Annulatascus sp. 1	*Manglicola guatemalensis*
Annulatascus sp. 2	*Marinosphaera mangrovei*
Apioclypea nypicola	*Massarina* sp. 1
Arecophila nypae	*Massarina* sp. 2
Arecophila sp.	*Massarina* sp. 3
Astrosphaeriella aquaticus	*Natantispora retorquens*
Astrosphaeriella cf. *mangrovis*	*Neolinocarpon globosicarpum*
Astrosphaeriella nipicola	***Neolinocarpon* cf. *nypicola***
Astrosphaeriella nypae	*Neptunella longirostris*
Astrosphaeriella striatispora	***Nipicola carbonispora***
Astrosphaeriella sp.	***Nipicola selangorensis***
Carinispora nypae	*Oceanitis cincinnatula*
Carinispora velatispora	*Ophiostomella* sp.
Cosmospora sp.	***Oxydothis nypicola***
Cucullosporella mangrovei	***Oxydothis nypae***
Dactylospora haliotrepha	*Oxydothis* sp. 1
Didymella sp.	*Oxydothis* sp. 2
Fasciatispora lignicola	*Panorbis viscosa*
Fasciatispora nypae	*Phomatospora nypae*
Frondicola tunitricuspis	*Phomatospora nypicola*
Fasciatispora petrakii	*Phomatospora* sp.
Halomassarina thalassiae	*Rhipidocarpon javanicum*
Halorosellinia oceanica	*Saagaromyces abonnis*
Halosarpheia marina	*Saagaromyces ratnagiriensis*
Halosarpheia sp.	*Saccardoella* cf. *mangrovei*
Helicascus nypae	*Savoryella* cf. *aquatica*

Table continued on next page.

Table 15.3 (Continued)

Taxa	Taxa
Savoryella lignicola	*Helicosporium hongkongense*
Savoryella paucispora	*Hydea pygmea*
Swampomyces sp.	*Monodictys* sp. 1
Tirisporella beccariana	*Monacrosporium* sp.
Trematosphaeria mangrovei	***Nypaella frondicola***
Tirispora unicaudata	*Papulospora* sp.
Trematosphaeria lineolatispora	*Phialogeniculata* sp.
Tubeufia sp.	*Phialophorma litoralis*
Verruculina enalia	*Phoma* sp.
Vibrissea nypicola	*Phomopsis* sp.
	Phomatospora nypae
Asexual fungi	*Pestalotiopsis* sp.
Acrogenospora sphaerocephala	***Plectophomella nypae***
Agerita sp.	*Pleospora* sp.
Cancellidium applanatum	***Pleurophomopsis nypae***
Chaetospermum sp.	*Sporidesmium crassisporum*
Cirrenalia macrocephala	*Sporidesmium* sp.
Matsusporium tropicale	*Tetraploa* cf. *aristata*
Cumulospora sp.	*Thozettella nivea*
Dactylaria sp. 1	*Trichocladium alopallonellum*
Dictyosporium elegans	*Trichocladium achrasporum*
Dictyochaeta sp.	***Trichocladium nypae?***
Dictyosporium sp.	*Vanakripa* sp.
Diplodia sp.	*Xylomyces* cf. *rhizophorae*
Endophragmia sp.	
Halenospora varia	**Basidiomycetes**
Helicoma sp.	*Grammothele fuligo*
Helicoma hyalonema	*Halocyphina villosa*
Helicorhoidion nypicola	*Halocyphina* sp.
Helicosporium pannosum	

Table 15.4 Frequency of occurrence of the top ten marine fungi on *Nypa* fronds from various studies (bold case indicates data from the three studies).

Fungus	Pilantanapak et al. 2005	Hyde and Sarma 2006	Besitulo et al. 2010
Linocarpon appendiculatum	**34.0**	**12.5**	**53.0**
Astrosphaeriella striatispora	**26.4**	**6.2**	**18.0**
Helicorhoidion nypicola	**20.4**	**2.6**	**4.0**
Oxydothis nypae	26.8	10.4	
Trichocladium nypae	34.8	6.2	
Aniptodera nypae	14.0	2.1	
Neolinocarpon globosicarpum		3.6	
Sporidesmium crassisporum		3.1	
Linocarpon bipolaris		13.0	
Lignincola nypae		6.2	6.0
Microthyrium sp.			25.0
Oxydothis nypicola			12.0
Halocyphina villosa			8.0
Didymella sp.			7.0
Aniptodera intermedia			4.0
Massarina sp.			3.0
Lulworthia grandispora	30.8		
Lignincola laevis	8.8		
Dictyosporium elegans	6.4		
Anthostomella cf. *rehmii*	6.0		

in Siargao Island, Philippines, although at the same locality zonation occurred on the mangrove trees *Rhizophora apiculata* and *Xylocarpus granatum*. *Cucullosporella mangrovei*, *Morosphaeria ramunculicola*, and *Marinosphaera mangrovei* were found in the upper level with *Acrocordiopsis patilii* at the lower level only. Hyde and Sarma (2006) noted vertical zonation and found that more fungi occurred on the submerged parts of the palm compared to those in the intertidal or terrestrial parts. A few fungi have been reported from the aerial parts of *Nypa*, with only *Linocarpon nypae* being common to terrestrial and intertidal parts (Hyde and Alias 2000).

Comparison of common fungi on *Nypa* and selected freshwater palms

Fungi occurring on palms have been intensively studied by Hyde and his coworkers (see Hyde et al. 1997). Taylor and Hyde (2003) estimate the number of terrestrial and aquatic fungi from palms to be 1,522 (592 ascomycetes, 270 basidiomycetes, 400 hyphomycetes and 260 coelomycetes). Most palm fungi are ascomycetes and asexual fungi, with fewer basidiomycetes reported. Endophytes have also been isolated from a number of palms, including the oil palm *Elaeis guineensis* (Rungjindamai et al. 2008).

Pinnoi et al. (2006) and Pinruan et al. (2007) conducted a detailed study on the occurrence of saprobic fungi on four palms (*Eleiodoxa conferta*, *Licuala longecalycata*, *Metroxylon sagus* and *Nenga pumila*) at the Sirindhorn peat swamp forest in Narathiwat, Thailand and the ten most common species on each palm are listed in Tables 15.5 and 15.6. None of these species have been reported on *N. fruticans*, despite many studies in Brunei, Malaysia, the Philippines, Thailand, Papua New Guinea, Indonesia and India (Hyde et al. 1999; Pilantanapak et al. 2005; Hyde and Sarma 2006; Besitulo et al. 2010). Equally, few of the fungi reported from *N. fruticans* (Table 15.3) have been found on the peat swamp palms.

Table 15.5 The ten most common fungi reported from four freshwater palms.

Eleiodoxa conferta	*Licuala longecalycata*	*Metroxylon sagus*	*Nenga pumila*
Cancellidium applanatum	*Microthyrium* sp.	*Anthostomella bispapillata*	*Diplococcium stoveri*
Xylomyces aquaticus	*Phaeoisaria clematidis*	*Nawawia filiformis*	*Dinemasporium* sp.
Astrosphaeriella sp.	*Annulatascus velatisporus*	*Oxydothis*-like	*Arecomyces epigeni*
Stilbohypoxylon elaeicola	*Massarina bipolaris*	*Apioclypea* sp.	*Linocarpon* sp. 4
Lophiostoma frondisubmersa	**Phruensis brunneispora**	*Apiospora* sp.	*Lophodermium* sp.
Microthyrium sp.	*Solheimia costaspora*	*Dinemasporium lanatum*	*Dactylaria palmae*
Morenoina palmicola	**Thailandiomyces bisetulosus**	*Tetraploa aristata*	*Lophiostoma* sp.
Phaeoisaria clematidis	*Nectria* sp. 1	*Ornatispora* sp.	*Oxydothis* sp. 8
Jahnula appendiculata	*Helicoma* sp. 1	*Massarina bipolaris*	*Spadicoides* sp. 4
	Astrosphaeriella malayensis	*Acrogenospora sphaerocephala*	*Jahnula appendiculata*

Table 15.6 The ten most common fungi reported from four peat swamp palms.

Fungi	Records	Percentage presence
Microthyrium sp.	40	16.5
Cancellidium applanatum	35	14.4
Diplococcium stoveri	33	13.6
Xylomyces aquaticus	32	13.2
Phaeoisaria clematidis	29	12.0
Astrosphaeriella aquatica-like	26	10.7
Stilbohypoxylon elaeicola	26	10.7
Jahnula appendiculata	25	10.3
Lophiostoma frondisubmersa	24	9.9
Morenoina palmicola	21	8.6

A comparison of the fungi colonizing the peat swamp palm *Eleiodoxa conferta* and *N. fruticans* (Pinnoi et al. 2006) shows that there are few species/genera in common: *Astrosphaeriella*, *Linocarpon* and *Oxydothis*. However, the genera *Carinispora*, *Fasciatispora*, *Halocyphina*, *Helicascus*, *Lignincola* and *Lulworthia*, which are common on *Nypa*, have not been recorded on *E. conferta*. These genera are more commonly found on substrata in marine habitats (Poonyth et al. 1999) and may require salt for growth, while those on *E. conferta* may not be salt tolerant. The latter may be more tolerant to acidic waters, while marine fungi tend to occur in more alkaline waters.

When a comparison is made with other peat swamp palms, such as *Licuala longicalycata* from a freshwater site (Pinruan et al. 2007), again the fungi found differed from those on *Nypa*, although some freshwater species occured, including *Helicoma*, *Helicosporium* and *Thozetella*, on *Nypa*. Thus, most of the fungi found on the *Nypa* palm are intertidal and do not appear to occur on other palms. There are only three other species of mangrove palms that may support these intertidal palm fungi, *Calamus erinaceus*, *Oncosperma tigillarium* and *Phoenix* sp. (Tomlinson 1986), and these have yet to be examined in greater detail for fungi.

Comparison of common fungi on mangrove wood and *Nypa* fronds

Many of the fungi colonizing *Nypa* fronds are unique to this host and may be host-specific, differing markedly from those common on other mangrove substrata. Table 15.3 lists 135 fungi documented as growing on *Nypa*, of which 39 are only known from *Nypa* (Hyde and Alias 2000; Pilantanapak et al. 2005; Hyde and Sarma 2006). In contrast, the common mangrove fungi on woody tissue are *Dactylospora haliotrepha*, *Halocyphina villosa*, *Halosarpheia marina*, *Halorosellinia oceanica*, *Hydea pygmea*, *Kallichroma tethys*, *Leptosphaeria australiensis* and *Verruculina enalia* (Alias and Jones 2009). There is therefore a clear difference in the fungi colonizing *Nypa* fronds when compared with woody mangrove substrata (Alias and Jones 2009). This is also well illustrated in the study by Besitulo et al. (2010), who compared the fungi colonizing four mangrove substrata (*N. fruticans*, *Rhizophora apiculata*, *Xylocarpus granatum* and driftwood) in Siargao mangroves in the Philippines. Sixteen species on *Nypa* (out of a total of 23) did not occur on the other substrata, but this could be attributed to the lower-salinity habitat of the palm. Hyde and Sarma (2006) studied the horizontal distribution of fungi on *Nypa* along the estuary of the Tutong River in Brunei at four sites ranging from marine to brackish to freshwater. They reported no significant difference in the very frequent and frequent fungi observed, even at the freshwater site. However, the freshwater fungi *Dictyochaeta* sp., *Anthostomella eructans* and *Annulatascus velatispora* were found at the freshwater site.

Molecular studies

Many fungal genera and species have been well-documented from *N. fruticans* (Pilantanapak et al. 2005; Hyde and Sarma 2006). However, there are few molecular studies of fungi associated with *N. fruticans*. Suetrong et al. (2011) have undertaken a phylogenetic study of *Tirisporella beccariana*, once classified in the Dothideomycetes. Molecular data place the species in the Sordariomycetes, Sordariomycetidae, order Diaporthales, with strains grouping in a distinct clade with *Thailandiomyces bisetulosus*, a freshwater ascomycete isolated from a senescent palm submerged in a peat swamp (Pinruan et al. 2008). To understand population dynamics of the fungal community on *N. fruticans*, different fungal groups, including pathogens, saprobes, endophytes, epiphytes and unculturable fungi, should be studied and compared. Some of these fungal groups cannot be identified based on morphology alone due to the failure to produce reproductive structures in culture. Molecular analysis is able to provide the missing link for this community (Guo et al. 2003).

New taxa on *Nypa* fronds

Cosmospora sp. (Figures 15.1–15.11)

Description: *Ascomata* 350–500 μm high and 250–350 μm in diameter, superficial, aggregate, gregarious, coriaceous, globose to obpyriform, orange to red color, collapse when dry changing color in 3% KOH to deeply red, ostioles obtuse. *Asci* with paraphyses: 75–100 × 30–37.5 μm, eight-spored, clavate, thin-walled, unitunicate, deliquesce early, lack an apical ring. *Ascospores* 32.5–37.5 × 10–15 μm, biseriate to multiseriate, hyaline, two-celled, ellipsoid, striate, not constricted or slightly constricted at the septum, with round or conical ends, no mucilaginous sheath or appendages.

No asexual state observed

Habitat: Saprobic on petiole and base of *Nypa fruticans*.

Specimen examined: NF00232 NF30218 *Cosmospora* sp.

Status: Teleomorph

Collected by: A. Loilong; E. B. G. Jones; N. Rungjindamai

Isolated by: A. Loilong 25/1/2010, on SCMA

Site: Thailand, Surat Thani, Mueang

Note: Some 65 *Cosmospora* spp. have been attributed to the genus, with *C. coccinea* as the type species, and *Acremonium*-like anamorphs often occurring on other fungi. Gräfenhan et al. (2011) have undertaken an extensive phylogenetic assessment of the genus and have confirmed that it is polyphyletic, accepting eight species. Species with other anamorphs have been assigned to other genera. Our species differs in that no anamorph has been observed in nature or in culture, and it is from a marine habitat. The *Nypa* species has ellipsoidal ascospores that are one-septate with striations on the spore wall. Five *Cosmospora* species have ellipsoidal ascospores and *C. diminuta* appears to be the most similar to the *Nypa* taxon. Other species have smaller ascospores (*C. matuoi, C. marellina, C. japonica* and *C. zealandica*). However none of these are included in *Cosmospora* as defined by Gräfenhan et al. (2011).

Arecophila sp. (Figures 15.12–15.20)

Description: *Ascomata* 150–200 μm in diameter, immersed, becoming superficial, globose, black carbonaceous, solitary or gregarious. Ostiole central, papillate. *Necks* up to 350 μm long. ***Paraphyses*** absent. ***Asci*** 100–125 × 12.5–15 μm, eight-spored, cylindrical, unitunicate, short pedicellate, apical rounded, with J-subapical ring, ***Ascospores*** 25–30 × 5 μm, biseriate, ellipsoidal, two-celled, pale brown to brown, not constricted at the septum, smooth wall and not surrounded by any mucilaginous sheath or apical appendage.

Habitat: Saprobic on petiole and base of *Nypa fruticans*.

Specimen examined: Trang and Samut Prakarn provinces, Thailand. On petioles and bases of *Nypa fruticans*.

Anamorph: Unknown

Note: Currently 15 *Arecophila* species have been described, many from palms or bamboo (Hyde and Alias 2000). The genus is assigned to the Cainiaceae, Xylariales (Taylor and Hyde 2003). Most species have immersed ascomata and a short papillate neck, often produced under a clypeus. The above *Arecophila* sp. differs in having ascomata that are immersed to erumpent with long necks up to 350 μm. *Arecophila nypae* was found growing on pinnae of *N. fruticans* collected at Selangor mangrove, Malaysia, but differs from our species with its dark brown ascospores, a prominent sheath, minute striations of the spore wall and smaller spores: 19–26 × 7–8 μm (Hyde 1996). *Arecophila calamicola* differs from *A. nypae* in that it is terrestrial with larger ascospores (24–33 × 5.0–5.9 μm), also with a verrucose spore wall and a gelatinous sheath.

Acknowledgements

This research was supported by a research grant from the Biodiversity and Training Program (BRT R_352112) of Thailand. We would like to thank Professor Morakot Tanticharoen, Dr Kanyawim Kirtikara and Dr Lily Eurwilaichitr at BIOTEC for their continued interest and support. Thanks go to Ms Sujinda Sommai for field assistance.

References

Alias SA, Jones EBG. Vertical distribution of marine fungi on *Rhizophora apiculata* at Morib mangrove, Selangor, Malaysia. Mycoscience 2000;41:431–436.
Alias SA, Jones EBG. Marine fungi from mangroves of Malaysia. Inst Ocean Earth Studies. 2009;8:109.
Barata M. Fungi on the halophyte *Spartina maritima* in salt marshes. In: Hyde KD, ed. Fungi in marine environments. Hong Kong: Fungal Diversity Press, 2002:179–193.
Besitulo A, Sarma VV, Hyde KD. Mangrove fungi from Siargao Islands, Philippines. In: Hyde KD, ed. Fungi in marine environments. Hong Kong: Fungal Diversity Press, 2002:267–283.
Besitulo A, Moslem MA, Hyde KD. Occurrence and distribution of fungi in a mangrove forest on Siargao Island, Philippines. Bot Mar. 2010;54:535–544.
Fisher P, Spalding MD. Protected areas with mangrove habitat. Cambridge, UK: Draft Report World Conservation Monitoring Centre, 1993.
Gessner RV, Kohlmeyer J. Geographical distribution and taxonomy of fungi from salt marsh *Spartina*. Can J Bot. 1976;54:2023–2037.
Gräfenhan T, Schroers HJ, Nirenberg HI, Seifert KA. An overview of the taxonomy, phylogeny, and typification of nectriaceous fungi in *Cosmospora, Acremonium, Fusarium, Stilbella* and *Volutella*. Stud Mycol. 2011;68:79–113.

Guo LD, Huang GR, Wang Y, He WH, Zheng WH, Hyde KD. Molecular identification of white morphotype strains of endophytic fungi from *Pinus tabulaeformis*. Mycol Res. 2003;107:680–688.

Hyde KD. A study of the vertical zonation of intertidal fungi on *Rhizophora apiculata* at Kampong Kapok mangrove, Brunei. Aquat Bot. 1988a;36:255–262.

Hyde KD. Observation on the vertical distribution of marine fungi on *Rhizophora* spp. at Kg. Danau mangrove, Brunei. Asian Mar Biol. 1988b;5:77–81.

Hyde KD. Studies on the tropical marine fungi of Brunei. II. Notes on five interesting species. Trans Mycol Soc Jpn. 1988c;29:161–171.

Hyde KD. The genus *Linocarpon* from the mangrove palm *Nypa fruticans*. Trans Mycol Soc Jpn. 1988d;29:338–350.

Hyde HD. Fungi from decaying intertidal fronds of *Nypa fruticans*, including three new genera and four new species. Bot J Linn Soc. 1992a;110:95–110.

Hyde KD. Fungi from *Nypa fruticans*: *Nipicola carbospora* gen. *et* sp. nov. (Ascomycotina). Crypt Bot. 1992b;2:330–332.

Hyde KD. Fungi from palms. XXIX. *Arecophila* gen. nov. (Amphisphaeriaceae, Ascomycota), with five new species and two new combinations. Nova Hedw. 1996;63:81–100.

Hyde KD, Alias SA. *Linocarpon angustatum* sp. nov., and *Neolinocarpon nypicola* sp. nov. from petioles of *Nypa fruticans*, and a list of fungi from aerial parts of *Nypa* palm. Mycoscience 1999;40:145–149.

Hyde KD, Alias SA. Biodiversity and distribution of fungi associated with decomposing *Nypa* palm. Biodivers Conserv. 2000;9:393–402.

Hyde KD, Nakagiri A. A new species of *Oxydothis* from the mangrove palm *Nypa fruticans*. Trans Mycol Soc Jpn. 1989;30:69–75.

Hyde KD, Sarma VV. Biodiversity and ecological observations on filamentous fungi of mangrove palm *Nypa fruticans* Wurumb (Liliosida-Arecales) along the Tutong River, Brunei. India J Mar Sci. 2006;35:297–307.

Hyde KD, Fröhlich J, Taylor JE. Diversity of ascomcyeons on palms in the tropics. In: Hyde KD, ed Biodiversity of tropical microfungi. Hong Kong: Hong Kong University Press, 1997:141–156.

Hyde KD, Goh TK, Lu BS, Alias SA. Eleven new intertidal fungi from *Nypa fruticans*. Mycol Res. 1999;103:1409–1422.

Jones EBG. Are there more marine fungi to be described? Bot Mar. 2011;54:343–354.

Jones EBG, Hyde KD, Read SJ, Moss ST, Alias SA. *Tirisporella* gen. nov., an ascomycete from the mangrove palm *Nypa fruticans*. Can J Bot. 1996;74:1487–1495.

Jones EBG, Sakayaroj J, Suetrong S, Somrithipol S, Pang KL. Classification of marine Ascomycota, anamorphic taxa and Basidiomycota. Fungal Divers. 2009;35:1–187.

Kohlmeyer J, Volkmann-Kohlmeyer B. The biodiversity of fungi on *Juncus roemerianus*. Mycol Res. 2001;105:1411–1412.

Pilantanapak A, Jones EBG, Eaton RA. Marine fungi on *Nypa fruticans* in Thailand. Bot Mar. 2005;48:365–373.

Pinnoi A, Lumyong S, Hyde KD, Jones EBG. Biodiversity of fungi on the palm *Eleiodoxa conferta* in Sirindhorn peat swamp forest, Narathiwat, Thailand. Fungal Divers. 2006;22:205–208.

Pinruan U, Hyde KD, Lumyong S, McKenzie EHC, Jones EBG. Occurrence of fungi on tissues of the peat swamp palm *Licuala longicalycata*. Fungal Divers. 2007;25:157–173.

Pinruan U, Sakayaroj J, Hyde KD, Jones EBG. *Thailandiomyces bisetulosus* gen. *et* sp. nov. (Diaporthales Sordariomycetidae, Sordariomycetes) and its anamorphs *Craspedodidymum*, is described based on nuclear SSU and LSU rDNA sequences. Fungal Divers. 2008;29:89–98.

Poon MOK, Hyde KD. Biodiversity of intertidal estuarine fungi on *Phragmites* at Mai Po marshes, Hong Kong. Bot Mar. 1998a;41:141–155.

Poon MK, Hyde KD. Evidence for the vertical distribution of saprophytic fungi on senescent *Phragmites australis* culms at Mai Po Marshes, Hong Kong. Bot Mar. 1998b;41:285–292.

Poonyth AD, Hyde KD, Peerally A. Intertidal fungi in Mauritius mangroves. Bot Mar. 1999;42: 285–292.

Rungjindamai N, Pinruan U, Choeyklin R, Hattori T, Jones EBG. Molecular characterization of basidiomycetous endophytes isolated from leaves, rachis and petioles of the oil palm, *Elaeis guineensis,* in Thailand. Fungal Divers. 2008;33:139–161.

Sadaba RB, Vrijmoed LLP, Jones EBG, Hodgkiss IJ. Observations on vertical distribution of fungi associated with standing senescent *Acanthus ilicifolius* stems at Mai Po mangrove, Hong Kong. Hydrobiol. 1995;295:119–126.

Suetrong S, Klaysuban A, Loilong A, Sakayaroj J, Phongpaichit S, Jones EBG. A re-appraisal of the systematic status of marine ascomycetes: *Tirisporella beccariana* from *Nypa fruticans* based on molecular characters. In: Abstract 12th International Marine and Freshwater Mycology Symposium, Incheon, Korea: Asian Mycological Congress, 2011:436.

Taylor JE, Hyde KD. Microfungi of tropical and temperate palms. Hong Kong: Fungal Diversity Press, 2003.

Tomlinson PB. The botany of mangroves. London: Cambridge University Press, 1986.

Wong MKM, Poon MOK, Hyde KD. *Phragmitensis marina* gen. *et* sp. nov., an intertidal saprotroph from *Phragmites australis* in Hong Kong. Bot Mar. 1998;41:379–382.

16 Diversity of endophytic and marine-derived fungi associated with marine plants and animals

Jariya Sakayaroj, Sita Preedanon, Souwalak Phongpaichit, Jirayu Buatong, Prapaipit Chaowalit and Vatcharin Rukachaisirikul

There are many definitions of endophytes, endophytic fungi or fungal endophytes, as recently reviewed by Hyde and Soytong (2008). Among the definitions proposed, the most commonly used is that of by Petrini (1991): "All organisms inhabiting plant organs that at some time in their life, can colonize internal plants tissues without causing apparent harm to the host". Schulz and Boyle (2005) opine that fungal endophytes are fungi that colonize a plant without causing visible disease symptoms at any specific moment. However, with the advent of molecular data the definition of fungal endophytes has been defined based on molecular phylogenetic affinity. The most recent is by Arnold (2007): "a polyphyletic group of highly diverse, primarily ascomycetous fungi defined functionally by their occurrence within asymptomatic tissues of plants".

Endophytic fungi have been isolated from a wide selection of plant hosts, ranging from temperate conifers (Ganley and Newcombe 2006; Arnold et al. 2007; Higgins et al. 2007) to tropical trees and plants (Oses et al. 2008; Tao et al. 2008) to lichens (Li et al. 2007) to terrestrial grasses (Sánchez Márquez et al. 2008). Fungal endophytes are also found in the above-ground tissues of liverworts, hornworts, mosses, lycophytes, equisetopsids, ferns and seed plants from the Arctic to the tropics and from agricultural fields to the most biologically-diverse tropical forests (Arnold 2007).

Studies on the biodiversity of marine and mangrove fungal endophytes

Mangroves are a large group of plant families that are adapted to the tropical intertidal environment. They are defined as trees, shrubs, palms or ground ferns, generally exceeding 50 cm in height, and which normally grow above the mean sea level in the intertidal zone of marine coastal environments or estuarine margins (Tomlinson 1986; Kathiresan and Bingham 2001; Duke 2006). Endophytic fungi are also common in mangrove and estuarine plants (Raghukumar 2008; Table 16.1).

Kobayashi and Ishibashi (1993) are of the opinion that marine endophytes may have the potential to produce novel structural compounds that are different from those of their terrestrial counterparts. Therefore nowadays, researchers are attracted to the study of fungal endophytes due to their importance in ecology, which includes an array of benefits to their hosts, such as tolerance to heavy metals, increased drought resistance, reduced herbivory, defense against pathogens, enhanced growth and competitive ability (Saikkonen et al. 1998). They are also interested in them as potential sources of secondary

Table 16.1 Dominant endophytic fungi isolated from mangrove, saltmarsh, seagrass and marine-associated plant hosts.

Plant hosts		Source	No. of isolates obtained	Density of colonization* (%)	Dominant endophytes (%)	Reference	Research focus
Family	**Genus, species**						
Acanthaceae	*Acanthus ilicifolius*	India	36	NA	*Phomopsis* spp. (3.4) *Ampullifera* sp. (2.3)	Suryanarayanan and Kumaresan (2000)	Species diversity
		India	NA	NA	*Cylindrocarpon* sp. (23.3) *Phoma* sp. (10.0)	Ananda and Sridhar (2002)	Species diversity
		India	40	NA	Sterile fungus # 4 (NA)	Kumaresan and Suryanarayanan (2001)	Species diversity
		India	160	74.5	*Colletotrichum* sp. (26.0) *Cytospora* sp. (6.5)	Maria and Sridhar (2003)	Species diversity
		Thailand	250	83.3	*Phyllosticta* sp. 1 (40.0)	Chaeprasert et al. (2010)	Antimicrobial screening
Aizoaceae	*Sesuvium portulacastrum*	India	36	NA	*Acremonium* sp. (6.3) *Colletotrichum* sp. 2 (4.0)	Suryanarayanan and Kumaresan (2000)	Species diversity
Amaranthaceae	*Arthrocnemum indicum*	India	36	NA	*Camarosporium propinquum* (29.6) *Phyllosticta* sp. (23.3)	Suryanarayanan and Kumaresan (2000)	Species diversity
	Arthrocnemum macrostachum	Egypt	10	NA	*Conoplea* spp. (NA)	El-Morsy (2000)	Species diversity, Enzyme screening

Family	Host	Country	#	%	Dominant fungi	Reference	Study type
	Halocnemum strobilaceum***	Egypt	16	NA	Penicillium purpurogenum (NA) Nigrospora sphaerica (NA)	El-Morsy (2000)	Species diversity, Enzyme screening
	Suaeda maritima	India	36	NA	Camarosporium palliatum (11.7) Sporormiella minima (6.0)	Suryanarayanan and Kumaresan (2000)	Species diversity
Arecaceae	Nypa fruticans	Malaysia	17	NA	NA	Ariffin et al. (2011)	Cytotoxic and antimicrobial screening
Avicenniaceae	Avicennia alba	Thailand	44	14.7	Phyllosticta sp. 1 (34.1)	Chaeprasert et al. (2010)	Antimicrobial screening
		Thailand	34	NA (IR** = 8.5)	Diaporthaceae sp. (NA) Phomopsis sp. (NA)	Chaowalit (2009), Buatong (2010), Buatong et al. (2011)	Antimicrobial screening, phylogenetic diversity
	Avicennia marina	India	NA	32.3	Phoma sp. (15.3) Colletotrichum sp. (2.7)	Kumaresan and Suryanarayanan (2001)	Species diversity
		Egypt	16	NA	Cladosporium cladosporioides (NA) Stachybotrys chartarum (NA)	El-Morsy (2000)	Species diversity, enzyme screening
	Avicennia officinalis	India	47	18.7	Paecilomyces sp. (6.0) Sporormiella minima (2.0)	Kumaresan and Suryanarayanan (2001)	Species diversity
		India	NA	NA	Allescheriella crocea (6.7) Fusarium oxysporum (6.7)	Ananda and Sridhar (2002)	Species diversity

Table continued on next page.

Table 16.1 (Continued)

Plant hosts		Source	No. of isolates obtained	Density of colonization* (%)	Dominant endophytes (%)	Reference	Research focus
		Thailand	17	NA (IR = 3.5)	*Diaporthe* sp. (NA)	Chaowalit (2009), Buatong (2010), Buatong et al. (2011)	Antimicrobial screening, phylogenetic diversity
Brassicaceae	*Zilla coccineum****	Egypt	6	NA	*Alternaria alternata* (NA)	El-Morsy (2000)	Species diversity, enzyme screening
	*Zilla spinosa****	Egypt	12	NA	*Acremonium* spp. (NA)	El-Morsy (2000)	Species diversity, enzyme screening
Ceratopteridaceae	*Acrostichum aureum*	India	158	77.5	*Penicillium* sp. (14.5) *Acremonium* sp. (14.5)	Maria and Sridhar (2003)	Species diversity
Combretaceae	*Lumnitzera littorea*	Thailand	34	NA (IR = 17)	*Pleosporales* sp. (NA) *Hypoxylon* sp. (NA)	Chaowalit (2009), Buatong (2010), Buatong et al. (2011)	Antimicrobial screening, phylogenetic diversity
		Thailand	240	80	*Phyllosticta* sp. 1 (72.9)	Chaeprasert et al. (2010)	Antimicrobial screening
	Lumnitzera racemosa	India	NA	52.1	*Phyllosticta* sp. (11.7) *Phomopsis* sp. (10.3)	Kumaresan and Suryanarayanan (2001)	Species diversity

Family	Species	Country	No.	% Colonization	Dominant species (%)	Reference	Study focus
Euphorbiaceae	*Excoecaria agallocha*	India	NA	31.3	*Glomerella* sp. (7.7) *Sporormiella minima* (5.3)	Kumaresan and Suryanarayanan (2001)	Species diversity
Loranthaceae	*Dendrophthoe falcate*	India	NA	77.7	*Phyllosticta* sp. MG123 (40.3) *Alternaria* sp. MG94 (10.3)	Kumaresan et al. (2002)	Species diversity
Malvaceae	*Heritiera littoralis*	Thailand	11	NA (IR = 11.0)	NA	Chaowalit (2009), Buatong (2010), Buatong et al. (2011)	Antimicrobial screening, phylogenetic diversity
	Thespesia populneoides	Thailand	182	60.66.0	*Phoma* sp. (44.5)	Chaeprasert et al. (2010)	Antimicrobial screening
Meliaceae	*Xylocarpus granatum*	Thailand	29	NA (IR = 7.3)	*Lophiostoma* sp. *Guignardia* sp.	Chaowalit (2009), Buatong (2010), Buatong et al. (2011)	Antimicrobial screening, phylogenetic diversity
		Thailand	166	55.32.0	*Colletotrichum* sp. 3 (20.8)	Chaeprasert et al. (2010)	Antimicrobial screening
	Xylocarpus moluccensis	Thailand	43	NA (IR = 10.8)	NA	Chaowalit (2009), Buatong (2010), Buatong et al. (2011)	Antimicrobial screening, phylogenetic diversity
		Thailand	180	60.0	*Phyllosticta* sp. 2 (26.7)	Chaeprasert et al. (2010)	Antimicrobial screening

Table continued on next page.

Table 16.1 (Continued)

Plant hosts		Source	No. of isolates obtained	Density of colonization* (%)	Dominant endophytes (%)	Reference	Research focus
Pandanaceae	*Pandanus amaryllifolius*	Malaysia	32	NA	NA	Ariffin et al. (2011)	Cytotoxic and antimicrobial screening
Myrsinaceae	*Aegiceras corniculatum*	India	NA	6.2	Sterile fungus MG242 (2.3)	Kumaresan and Suryanarayanan (2001)	Species diversity
		Thailand	4	NA (IR = 4.0)	*Pestalotiopsis* sp. (NA)	Chaowalit (2009), Buatong (2010), Buatong et al. (2011)	Antimicrobial screening, phylogenetic diversity
Plumbaginaceae	*Limoniastrum monopetalum****	Egypt	7	NA	*Papulaspora immersa* (NA)	El-Morsy (2000)	Species diversity, enzyme screening
Rhizophoraceae	*Bruguiera cylindrical*	India	NA	38.3	*Colletotrichum gloeosporioides* (34.0) *Sporormiella minima* (3.7)	Kumaresan and Suryanarayanan (2001)	Species diversity
		Thailand	21	NA (IR = 21.0)	NA	Chaowalit (2009), Buatong (2010), Buatong et al. (2011)	Antimicrobial screening, phylogenetic diversity

Diversity of endophytic and marine-derived fungi

Host	Country	Isolates	CR (%)	Dominant species	Reference	Study focus
Bruguiera gymnorrhiza	Japan	296	NA	*Colletotrichum* sp. (35.0–60.0) *Phyllosticta* sp. (15.0–35.0)	Okane et al. (2001a)	Species diversity
	Thailand	2	NA (IR = 2.0)	NA	Chaowalit (2009), Buatong (2010), Buatong et al. (2011)	Antimicrobial screening, phylogenetic diversity
Bruguiera parviflora	Thailand	48	NA (IR = 9.6)	Xylariaceae sp. (NA)	Chaowalit (2009), Buatong (2010), Buatong et al. (2011)	Antimicrobial screening, phylogenetic diversity
Bruguiera sexangula var. *rhynchopetala*	China	91	NA	*Aspergillus* sp. 1 (14.2) *Trichoderma* sp. (8.5) *Pestalotiopsis oxyanthi* (7.6)	Xing and Guo (2011)	Species diversity
Rhizophoraceae						
Ceriops decandra	India	NA	47.5	*Sporormiella minima* (2.0) *Acremonium* sp. (1.0)	Kumaresan and Suryanarayanan (2001)	Species diversity
	Thailand	25	8.33	*Phyllosticta* sp. 1 (64.0)	Chaeprasert et al. (2010)	Antimicrobial screening
	Thailand	8	NA (IR = 4.0)	NA	Buatong (2010), Buatong et al. (2011)	Antimicrobial screening, phylogenetic diversity
Ceriops tagal	Thailand	5	NA (IR = 2.5)	NA	Buatong (2010), Buatong et al. (2011)	Antimicrobial screening, phylogenetic diversity

Table continued on next page.

Table 16.1 (Continued)

Plant hosts	Source	No. of isolates obtained	Density of colonization* (%)	Dominant endophytes (%)	Reference	Research focus
	China	67	NA	*Pestalotiopsis microspora* (14.2) *Aureobasidium* sp. (9.4) *Phomopsis* sp. 1 (6.4)	Xing and Guo (2011)	Species diversity
Kandelia candel	Hong Kong	880	NA	*Phomopsis* sp. (NA) *Pestalotiopsis* sp. (NA) *Guignardia* sp. (NA)	Pang et al. (2008)	Species diversity
Rhizophora apiculata	India	NA	NA	*Sporormiella minima* (2.0–16.7) *Acremonium* sp. (1.7–4.0) *Cladosporium* sp. (0.3–7.3)	Suryanarayanan et al. (1998)	Species diversity
	India	NA	20.0–98.0	*Phyllosticta* sp. (5.5–38.0) *Sporormiella minima* (2.0–10.0)	Kumaresan and Suryanarayanan (2002)	Species diversity
	Thailand	200	NA (IR = 16.6)	*Acremonium* sp. (NA) *Cladosporium* sp. (NA) *Phomopsis* sp. (NA)	Chaowalit (2009), Buatong (2010), Buatong et al. (2011)	Antimicrobial screening, phylogenetic diversity
	Thailand	439	23.0–69.01.0	*Phyllosticta* sp. 2 (16.0–19.0) *Cladosporium* sp. (2.1)	Chaeprasert et al. (2010)	
	China	80	NA	*Pestalotiopsis oxyanthi* (11.0) *Fusarium solani* (6.2) *Pestalotiopsis clavispora* (5.6)	Xing and Guo (2011)	Species diversity

Diversity of endophytic and marine-derived fungi

Family	Host plant	Country	No.	CR	Dominant species	Reference	Purpose
	Rhizophora mucronata	India	NA	NA	*Sporormiella minima* (15.7–19.3) *Acremonium* sp. (2.3–2.7) *Cladosporium* sp. (4.7)	Suryanarayanan et al. (1998)	Species diversity
		India	49	NA	Sterile fungus #3 (NA)	Kumaresan and Suryanarayanan (2001)	Species diversity
		India	NA	NA	*Sporormiella minima* (NA) *Phyllosticta* sp. MG90 (NA)	Kumaresan and Suryanarayanan (2002)	Species diversity
		India	NA	NA	*Aspergillus* sp. (13.3) *Fusariella obstipa* (6.7)	Ananda and Sridhar (2002)	Species diversity
		Thailand	84	NA (IR = 9.3)	NA	Chaowalit (2009), Buatong (2010), Buatong et al. (2011)	Antimicrobial screening, phylogenetic diversity
		Thailand	264	34.66.0–53.33.0	*Phyllosticta* sp. 2 (12.0) *Pestalotiopsis* sp. 1 (17.0)	Chaeprasert et al. (2010)	Antimicrobial screening
	Rhizophora stylosa	China	57	NA	*Cladosporium cladosporioides* (15.7) *Fusarium solani* (11.2) *Pestalotiopsis virgatula* (10.7)	Xing and Guo (2011)	Species diversity
Rubiaceae	*Scyphiphora hydrophyllacea*	Thailand	22	NA (IR = 7.3)	*Leptosphaerulina* sp. (NA) *Didymella* sp. (NA) *Pestalotiopsis* sp. (NA)	Chaowalit (2009), Buatong (2010), Buatong et al. (2011)	Antimicrobial screening, phylogenetic diversity

Table continued on next page.

Table 16.1 (Continued)

Plant hosts		Source	No. of isolates obtained	Density of colonization* (%)	Dominant endophytes (%)	Reference	Research focus
Sonneratiaceae	Sonneratia alba	Thailand	158	52.61.0	Phyllosticta sp. 1 (29.0)	Chaeprasert et al. (2010)	Antimicrobial screening
	Sonneratia apetala	China	104	NA	Mycosphaerella sp. 3 (43.0) Phomopsis liquidambari (33.7)	Xing et al. (2010)	Species diversity
	Sonneratia caseolaris	India	31	NA	Phomopsis sp. 3 (40.7) Sterile fungus # 1 (NA)	Kumaresan and Suryanarayanan (2001)	Species diversity
		India	NA	NA	Cytospora abietis (6.7)	Ananda and Sridhar (2002)	Species diversity
		Thailand	34	NA (IR = 8.5)	Pestalotiopsis sp. (NA)	Chaowalit (2009), Buatong (2010), Buatong et al. (2011)	Antimicrobial screening, phylogenetic diversity
		China	58	NA	Phomopsis sp. 1 (43.6) Fusarium sp. 1 (40.0) Stemphylium solani (32.0)	Xing et al. (2010)	Species diversity
	Sonneratia griffithii	Thailand	4	NA (IR = 4.0)	NA	Buatong (2010)	Antimicrobial screening
	Sonneratia hainanensis	China	65	NA	Mycosphaerella sp. 2 (42.4) Diaporthe sp. (28.5) Glomerella cingulata (42.5)	Xing et al. (2010)	Species diversity
	Sonneratia ovata	Thailand	19	NA (IR = 9.5)	NA	Chaowalit (2009), Buatong (2010), Buatong et al. (2011)	

Family	Species	Location	#	%	Dominant fungi	Reference	Focus
	Sonneratia paracaseolaris	China	82	NA	Glomerella cingulata (38.0) Cytospora rhizophorae (23.9) Glomerella sp. (44.0)	Xing et al. (2010)	Species diversity
		China	82	NA	Fusarium sp. 2 (40.0) Mycosphaerella sp. 4 (33.0) Phoma sp. (34.0)	Xing et al. (2010)	Species diversity
Tamaricaceae	Tamarix nilotica***	Egypt	5	NA	Trimmatostroma sp. (NA)	El-Morsy (2000)	Species diversity, enzyme screening
Zygophyllaceae	Zygophyllum album***	Egypt	8	NA	Alternaria alternata (NA) Penicillium chrysogenum (NA)	El-Morsy (2000)	Species diversity, enzyme screening
	Zygophyllum simplex***	Egypt	5	NA	Alternaria alternata (NA)	El-Morsy (2000)	Species diversity, enzyme screening
Seagrasses							
Cymodoceaceae	Halodule bermudensis	Bermuda	9	NA	Penicillium sclerotiorum (17.2–26.4) Aspergillus granulosus (5.2)	Wilson (1998)	Species diversity
	Syringodium filiforme	Bermuda	9	NA	Penicillium sclerotiorum (1.0) Candida utilis (4.0)	Wilson (1998)	Species diversity
Hydrocharitaceae	Enhalus acoroides	Thailand	42	11.0	Penicillium spp. (2.2) Hypocreales sp. (2.8) Cladosporium spp. (1.1)	Sakayaroj et al. (2010)	Phylogenetic diversity

Table continued on next page.

Table 16.1 (Continued)

Plant hosts		Source	No. of isolates obtained	Density of colonization* (%)	Dominant endophytes (%)	Reference	Research focus
	Halophila ovalis	India	24	NA	*Cladosporium cladosporioides* (0.6) *Phialophora* sp. (0.4) *Talaromyces* sp. (0.6)	Devarajan et al. (2002)	Species diversity
	Thalassia testudinum	Bermuda	9	NA	*Aspergillus granulosus* (15.4) *Penicillium sclerotiorum* (69.2) *Sporobolomyces holsaticus* (34.6)	Wilson (1998)	Species diversity
		Hong Kong	17 sp.	NA	*Gamsia* sp. (5.6) *Arthrinium arundinis* (3.8) *Fusarium* sp. (2.5)	Alva et al. (2002)	Species diversity
		Puerto Rico	13 sp.	NA	*Acremonium* spp. (NA) *Cladosporium* spp. (NA) *Penicillium* spp. (NA)	Rodríguez (2008)	Antimicrobial screening
Zosteraceae	*Zostera japonica*	Hong Kong	16 sp.	NA	*Alternaria* sp. (2.5) *Cladosporium* sp. (2.5) *Penicillium sclerotiorum* (1.3)	Alva et al. (2002)	Species diversity
	Zostera marina	Hong Kong	8 sp.	NA	*Cladosporium* sp. (4.0) *Paecilomyces* sp. (3.1)	Alva et al. (2002)	Species diversity

*Density of colonization (%) = Total number of segments colonized by an endophyte divided by total number of segments screened × 100 (Hata and Futai 1995; Suryanarayanan et al. 2010).
**Isolation rate (IR %)=Total number of isolates yielded by a given sample divided by total number of leaf segments in that sample × 100 (Petrini et al. 1982).
***Salt affected land plant.
NA = Not applicable.

metabolites of pharmaceutical importance (Chen et al. 2007; Liu et al. 2007a; Huang et al. 2008; Jones et al. 2008; Aly et al. 2010; Chaeprasert et al. 2010; Buatong et al. 2011; Debbab et al. 2011).

Marine endophytic fungi have also been proven to be important sources of enzymes. El-Morsy (2000) reported the ability of endophytes isolated from the endorhizosphere of haloplants to produce pectinase, xylanase and phenolase. Maria et al. (2005) tested seven mangrove endophytic fungi isolated from *Acanthus ilicifolius* and *Acrostichum aureum* for the production of extracellular enzymes. Cellulase and lipase activities were detected in all fungi, with amylase and protease present in a few. No fungus exhibited chitinase, laccase or tyrosinase activities.

Over the past 13 years, the common and dominant marine, mangrove and estuarine plant species have been shown to support fungal endophytes. More than 52 plant host species from 23 plant families, as well as the salt-affected land plants, have been investigated for the presence of endophytic fungi. Most of the studies have been undertaken in India, with some reports from Bermuda, China, Egypt, Hong Kong (China), Japan, Malaysia, Philippines, Puerto Rico and Thailand. The occurrence, distribution, density of colonization, species diversity and number of fungi per segment were frequently recorded (Table 16.1).

Most of the early studies focused on the abundance and presence of fungi based on the morphological characterization of isolated fungal endophytes. Nevertheless, many of the fungi failed to sporulate or produce reproductive structures, therefore the traditional identification approach proved to be difficult. These non-sporulating cultures were generally referred to as Mycelia sterilia (Guo et al. 2000). The use of rDNA sequence data has been reported to be helpful in comparing sequence divergence and taxonomic identities within phylogenetically-referenced databases of recognized species (Arnold 2007). Recently there have been several studies undertaken using rDNA sequences, especially the ITS rDNA region, to identify the phylogenetic diversity of endophytic fungi from marine and mangrove plant hosts (Sakayaroj et al. 2010; Xing et al. 2010; Buatong et al. 2011; Xing and Guo 2011).

Sakayaroj et al. (2010) investigated the presence of endophyte assemblages from the tropical seagrass *Enhalus acoroides* in Thailand. Little information is available for fungi associated with healthy seagrasses, especially fungal endophytes (Raghukumar 2008). There are only a few reports on seagrass endophytes, including those by Wilson (1998), Alva et al. (2002), Devarajan et al. (2002) and Rodríguez (2008), see Table 16.1. The study by Sakayaroj and colleagues aimed to investigate the phylogenetic diversity of fungal endophytes in *E. acoroides* based on LSU, ITS1, ITS2 and 5.8S rDNA sequence analyses. The majority of endophytes belonged to the Ascomycota (98%), with the Basidiomycota accounting for only 2%. The predominant ascomycetous orders were Hypocreales, followed by Eurotiales and Capnodiales, respectively. The most representative endophytic fungi from *E. acoroides* were unidentified species in Hypocreales, with *Aspergillus*, *Penicillium* and *Cladosporium* being the most frequent genera found. They are also the most commonly reported in every seagrass species so far studied, including *Thalassia testudinum, Halophila ovalis, Halodule bermudensis, Syringodium filiforme, Zostera marina* and *Z. japonica* (Sakayaroj et al. 2010).

Xing et al. (2010) reported that there were fungal endophytes associated with mangrove plants in China. Five *Sonneratia* species – *S. caseolaris, S. hainanensis,*

S. ovata, S. paracaseolaris and *S. apetala* – were examined for the presence of fungal endophytes in roots, stems and leaves. Thirty-nine out of 391 morphotypes were selected for a molecular identification using ITS rDNA sequence analysis. The dominant fungal taxa comprised the genera *Cytospora, Diaporthe, Fusarium, Glomerella, Mycosphaerella, Phoma, Phomopsis* and *Stemphylium*. Most recovered species were found on more than one host species, but the dominant species differed according to the host and tissue type. Some fungi found in this study have also been reported to be endophytes of seagrass species (*Z. marina, T. testudinum* and *E. acoroides*), such as species of *Pestalotiopsis, Nigrospora, Aspergillus, Cladosporium, Phoma, Alternaria* and *Colletotrichum* (Alva et al. 2002; Sakayaroj et al. 2010).

Recently, Xing and Gao (2011) investigated the taxonomic identities and diversity of fungal endophytes isolated from four Rhizophoraceae mangrove plant species (*Ceriops tagal, Rhizophora apiculata, R. stylosa* and *Bruguiera sexangula* var. *rhynchopetala*) using a combination of morphological and molecular approaches. Thirty-eight out of 295 fungal isolates were identified to various taxonomic levels based on ITS rDNA sequences. The genera *Aspergillus, Pestalotiopsis* and *Phomopsis* were the most frequent endophytes within and among the four host species. Some of the endophytes exhibited host and tissue specificity. The colonization frequencies of endophytic fungi in the stems of the four host plants were evidently higher than in the roots. The similarity in the endophytic fungal community between the four Rhizophoraceae mangrove plants was low. The result from this study was in concordance with an earlier study by Xing et al. (2010) stating that mangrove trees of close phylogenetic affinity do not necessarily harbor similar fungal communities.

Research on the biodiversity of mangrove endophytes in Thailand

Thailand is a biodiversity-rich country supporting vast coastal areas with coral reefs, significant mangrove stands and healthy seagrass beds (Jones and Hyde 2004). It is endowed with coastal and marine resources with 2,600 km of coastline on the Gulf of Thailand and along the Andaman Sea, covering 23 coastal provinces. The estimated mangrove area in Thailand was 240,000 ha in 2002 (RFD 2004). There are 34 species of mangrove plants in Thailand (FAO 2007), with Rhizophoraceae, Sonneratiaceae and Avicenniaceae being the dominant families (Aksornkoae 1993). There are only a few reports on species composition and the biodiversity of endophytic fungi from Thai mangrove plants (Chaowalit 2009; Buatong 2010; Chaeprasert et al. 2010; Sakayaroj et al. 2010; Buatong et al. 2011). Most studies are restricted to investigations of lignicolous fungi in marine and mangrove habitats (Jones et al. 2006; Sakayaroj et al. 2011).

Chaeprasert et al. (2010) investigated the occurrence of endophytes from various mangrove plants (*Acanthus ilicifolius, Avicennia alba, Ceriops decandra, Rhizophora apiculata, R. mucronata, Lumnitzera littorea, Sonneratia alba, Thespesia populneoides, Xylocarpus granatum* and *X. moluccensis*) from eastern, central and southern locations in Thailand. One-thousand-nine-hundred-twenty-two fungal isolates were obtained, with the genera *Cladosporium, Colletotrichum, Phomopsis* and *Xylaria* being the most common. The results showed that there were differences in the endophytic fungal flora among the three locations, and that the species diversity and species evenness of endophytes varied between mangrove hosts.

Molecular identification of representative fungal endophytes isolated from mangrove plants in southern Thailand

The results presented here are the continuation from studies by Chaowalit (2009), Buatong (2010) and Buatong et al. (2011). We isolated fungal endophytes from 18 mangrove plant species from southern Thailand. This included leaves and branches of *Aegiceras corniculatum, Avicennia alba, A. officinalis, Bruguiera cylindrica, B. gymnorrhiza, B. parviflora, Ceriops decandra, C. tagal, Heritiera littoralis, Lumnitzera littorea, Rhizophora apiculata, R. mucronata, Sonneratia caseolaris, S. griffithii, S. ovata, Scyphiphora hydrophyllacea, Xylocarpus granatum* and *X. moluccensis*. The objective of this study was to characterize the presence of endophytic fungi in selected host plants, screen the crude extracts for antimicrobial activity and identify the strains that produce antimicrobial substances.

A total of 619 endophytic fungi were obtained, with the highest isolation rate from *B. cylindrica*. Crude extracts of 150 selected fungal endophytes were evaluated against potential human pathogens (*Staphylococcus aureus* ATCC25923, a clinical isolate of methicillin-resistant *S. aureus* [MRSA] SK1, *Escherichia coli* ATCC25922, *Pseudomonas aeruginosa* ATCC27853, *Candida albicans* ATCC90028, *Cryptococcus neoformans* ATCC90112 and *Microsporum gypseum* MU-SH4). Twenty-six representative isolates exhibited interesting antimicrobial activity and/or displayed high enzyme activity and were selected for a molecular identification (Table 16.2 and Figure 16.1). Combining both the morphological characters and rDNA sequence data, we evaluated the taxonomic identities of these endophytes to generic and species levels or at least to the family level. ITS1, ITS2, and 5.8S rDNA sequences were incorporated into one data matrix comprising representatives from the Ascomycota and Basidiomycota. Results from the phylogenetic analyses revealed that all the active producing endophytic fungi belonged to the Sordariomycetes (Diaporthales, Hypocreales and Xylariales), the Dothideomycetes (Botryosphaeriales, Capnodiales and Pleosporales) and the Basidiomycota (Agaricales).

Six isolates (MA81, MA96, MA60, MA194, MA1 and MA2) grouped well in a subclade of the Diaporthales, comprising the genus *Diaporthe* and its asexual state *Phomopsis*. Isolate MA81 had close affinity with the strains EF488434 (*Diaporthe* sp.) and EU395772 (*Phomopsis* sp.), isolated from pharmaceutical plants as endophytes (Lin et al. 2007; Yuan et al. 2008). Due to the lack of fruiting structures in the culture of MA81, it can only be identified as a member of the Diaporthaceae. Endophyte MA96 clustered well in the *Phomopsis* subclade with sequences of *P. phyllanthicola* (AY620819 and EF488375) and *Phomopsis* sp. (EU330616). Isolate MA60 gave the highest nucleotide similarity (99%) with *P. eucommii* (AY601921), an isolate from woody plants; thus MA60 can be identified as *P. eucommii*. MA194 formed a distinct monophyletic clade with *Phomopsis* spp. AB247183 and FJ037768, with relatively high similarity (98%), which had been isolated from terrestrial and mangrove plants, respectively. MA2 and MA1 formed a monophyletic basal subclade in the family Diaporthaceae without any closely related taxa, while LSU sequence analysis also confirmed its placement in *Diaporthe* (tree not shown).

The endophytic isolate MA70 grouped with EU164804 (an unidentified sequence from a terrestrial plant species) with good support and formed a sister group to *Acremonium crotocinigenum* (DQ882846 and AJ621773) with 91–100% ITS sequence similarity.

Table 16.2 Summary of the molecular identification of endophytic fungi isolated from mangrove plants whose crude extracts produced antimicrobial and/or enzyme activities.

Fungal endophyte species	Source	Bioactivity of crude extracts, secondary metabolite production and/or enzyme production
Diaporthales		
Diaporthaceae sp. MA81	*Avicennia alba*	Weak activity against CA and MG (MIC 128–200 µg/ml)
Phomopsis sp. MA96	*Rhizophora apiculata*	Moderate activity against SA, MRSA and MG (MIC 32–200 µg/ml)
P. eucommii MA60	*Avicennia alba*	Strong activity against CA, CN and MG (MIC 16–32 µg/ml)
Phomopsis sp. MA194	*Rhizophora apiculata*	Strong activity against SA, MRSA, CA, CN and MG (MIC 8–32 µg/ml; Buatong et al. 2011)
Diaporthe sp. MA2	*Avicennia officinalis*	Strong activity against SA, MRSA and MG (MIC 32–64 µg/ml); lipase production (EPR = 4.75)
Diaporthe sp. MA1	*Avicennia officinalis*	Moderate activity against SA, MRSA, CN and MG (MIC 32–200 µg/ml); lipase production (EPR = 4.95)
Hypocreales		
Acremonium sp. MA70	*Rhizophora apiculata*	Moderate activity against SA, MRSA, CA, CN and MG (MIC 32–200 µg/ml)
Xylariales		
Pestalotiopsis sp. MA177	*Scyphiphora hydrophyllacea*	Moderate activity against SA, CA, CN and MG (MIC 32–200 µg/ml)
Pestalotiopsis sp. MA92	*Rhizophora apiculata*	Three new α-pyrones, pestalotiopyrones A–C, and one known compound (Rukachaisirikul et al. 2011)
Pestalotiopsis sp. MA119*	*Rhizophora apiculata*	Two new seiricuprolides, pestalotioprolides A and B, together with one known compound (Rukachaisirikul et al. 2011)
Pestalotiopsis sp. MA129	*Rhizophora apiculata*	Weak activity against SA and MRSA (MIC 200 µg/ml)
Pestalotiopsis sp. MA165	*Aegiceras corniculatum*	Weak activity against SA, CN and MG (MIC 64–200 µg/ml)
Pestalotiopsis sp. MA21	*Sonneratia caseolaris*	Moderate activity against SA and MG (MIC 32–200 µg/ml)
Hypoxylon sp. MA156	*Lumnitzera littorea*	Moderate activity against SA, CA, CN and MG (MIC 8–128 µg/ml; Buatong et al. 2011)
Xylariaceae sp. MA105	*Bruguiera parviflora*	Weak activity against SA and MRSA (MIC 200 µg/ml)

Table continued on next page.

Table 16.2 (Continued)

Fungal endophyte species	Source	Bioactivity of crude extracts, secondary metabolite production and/or enzyme production
Capnodiales		
Cladosporium sp. MA111	*Rhizophora apiculata*	Lipase production (EPR = 3.04)
Mycosphaerellaceae sp. MA12	*Sonneratia caseolaris*	Strong activity against CN (MIC 8 µg/ml)
Botryosphaeriales		
Guignardia sp. MA140	*Xylocarpus granatum*	Lipase production (EPR = 3.55)
Guignardia sp. MA127	*Xylocarpus granatum*	Lipase production (EPR = 3.16)
Pleosporales		
Pleosporales sp. MA150	*Lumnitzera littorea*	Cellulase production (EPR = 1.92)
Pleosporales sp. MA158	*Lumnitzera littorea*	Cellulase production (EPR = 1.66)
Pleosporales sp. MA122	*Rhizophora mucronata*	Cellulase production (EPR = 1.54)
Lophiostoma sp. MA130	*Xylocarpus granatum*	Cellulase production (EPR = 1.96)
Leptosphaerulina chartarum MA164	*Scyphiphora hydrophyllacea*	Moderate activity against CN and MG (MIC 64–128 µg/ml)
Didymella bryoniae MA71	*Scyphiphora hydrophyllacea*	Moderate activity against SA and MRSA (MIC 64–200 µg/ml)
Agaricales, Basidiomycota		
Schizophyllum commune MA56	*Sonneratia ovata*	Lipase production (EPR = 3.25)

SA: *Staphylococcus aureus* ATCC25923, MRSA: methicillin-resistant *S. aureus* SK1, PA: *Pseudomonas aeruginosa* ATCC27853, EC: *Escherichia coli* ATCC25922, CA: *Candida albicans* ATCC90028, CN: *Cryptococcus neoformans* ATCC90112, MG: *Microsporum gypseum* MU-SH4, MIC: minimum inhibitory concentration, EPR: extracellular production ratios (Choi et al. 2005).
*Identified by morphology only.

Six *Pestalotiopsis* isolates (MA21, MA92, MA119, MA129, MA165 and MA177) based on morphological and molecular identifications grouped in the Xylariales (tree not shown). The data indicate that *Pestalotiopsis* strains sequenced from this study constituted distinct clades with sequences of other pathogenic and endophytic *Pestalotiopsis* species. MA177 was closely related to *P. longisetula* AF409971 (100% nucleotide similarity), while isolate MA92 had close relationships with sequences of *Pestalotiopsis* spp.

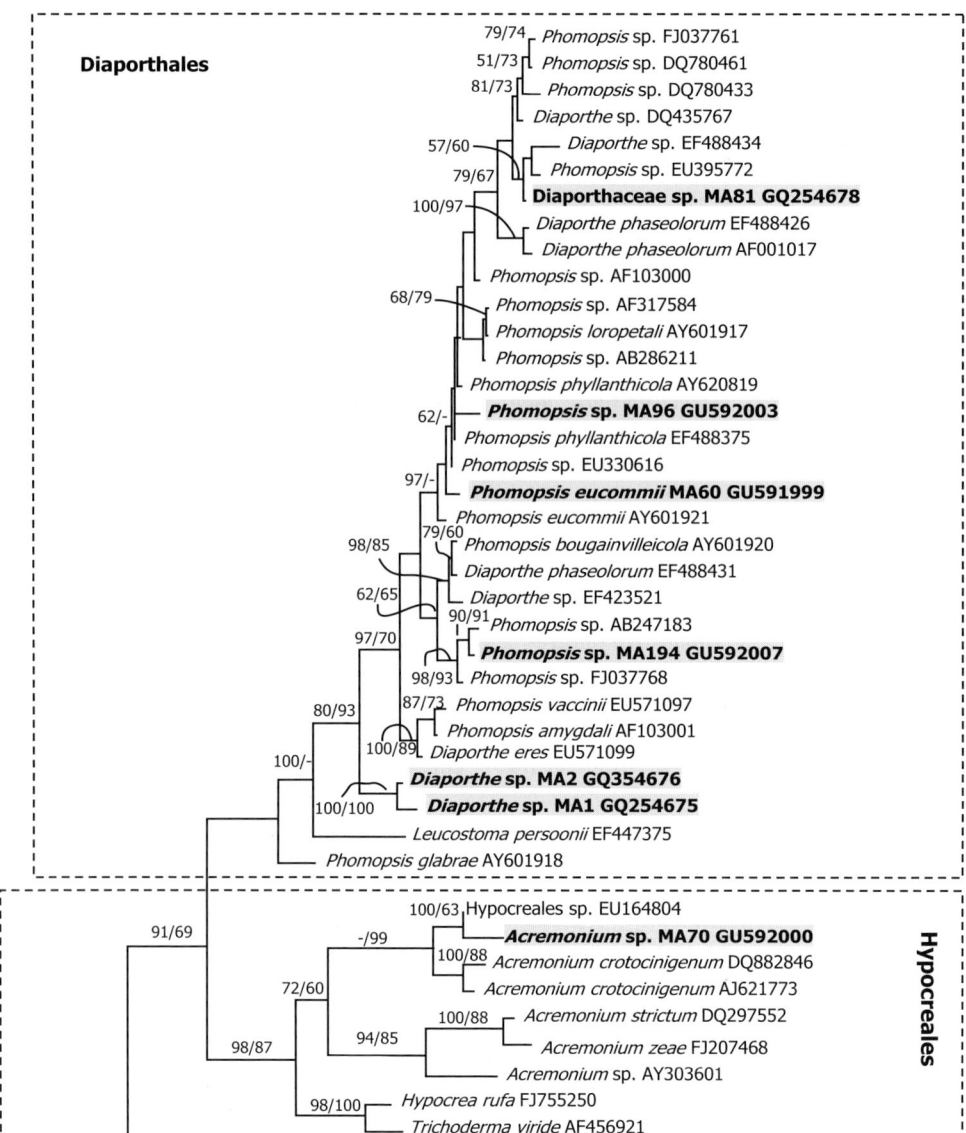

Figure 16.1 (Continued)

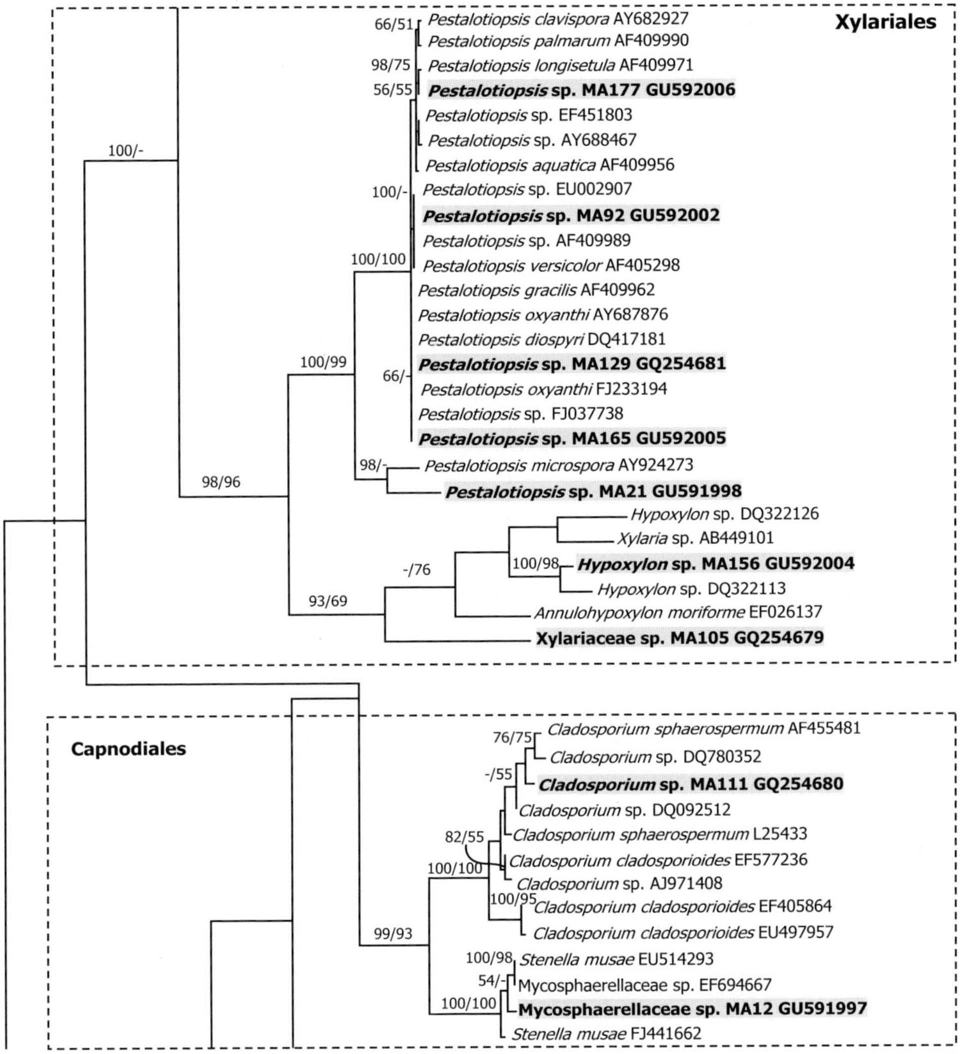

Figure 16.1 (Continued)

Figure 16.1 (Continued)

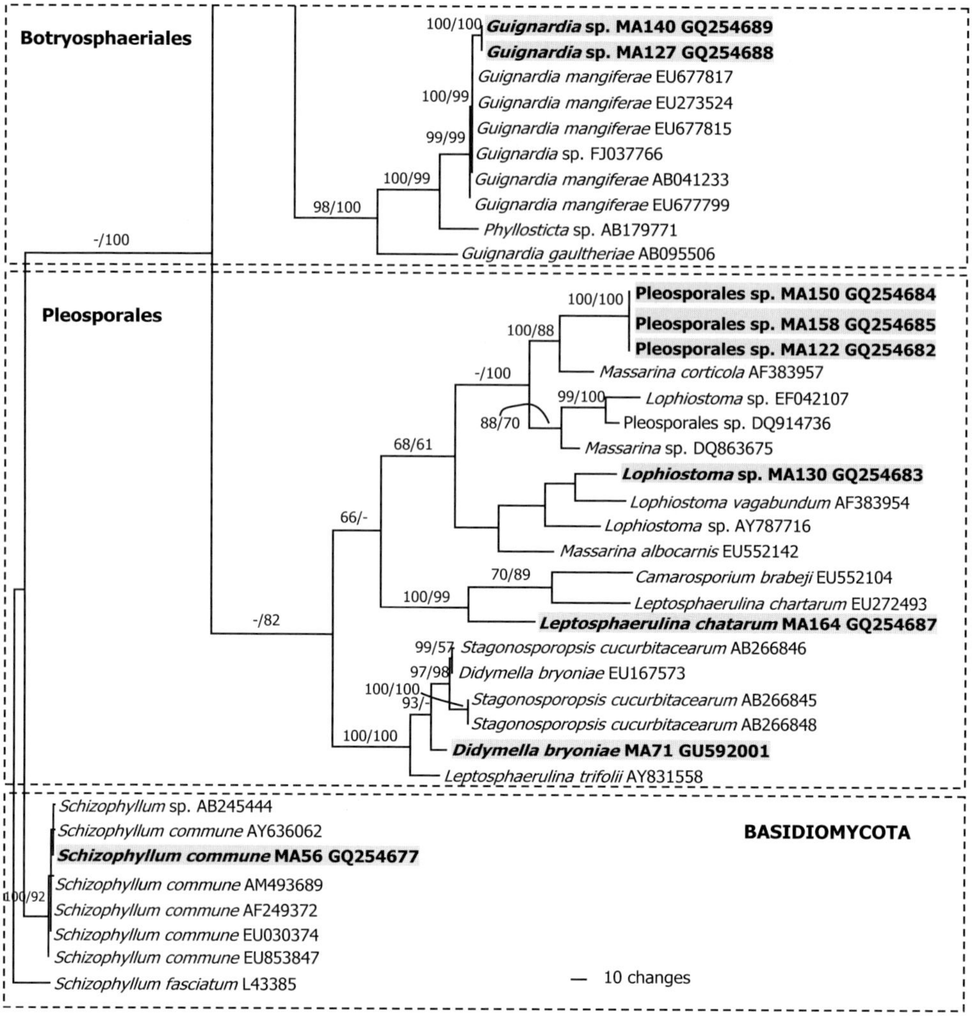

Figure 16.1 One of the most parsimonious trees generated from maximum parsimony analysis of ITS1, ITS2 and 5.8S rDNA sequences from mangrove endophytes isolated from southern Thailand. Bootstrap support values from neighbor joining and maximum parsimony analyses are shown above each branch, respectively. Tree length = 3,302 changes, consistency index = 0.4028, and retention index = 0.8159. Highlighted fonts represent sequences generated from this study. Branch lengths are proportional to the number of character state changes.

(EU002907 and AF409989) and *P. versicolor* (AF405298) with 100% sequence similarity. MA129 and MA165 showed affinity with *P. diospyri* (DQ417181), *P. oxyanthi* (FJ233194), and *Pestalotiopsis* sp. (FJ037738); while MA21 formed a monophyletic clade with *P. microspora* (AY92427), although it had a long branch length.

Endophyte MA156 formed a monophyletic clade with a *Hypoxylon* sp. (DQ322113) in the Xylariales and was the first report of a *Hypoxylon* species found as an endophyte in a mangrove plant (Buatong et al. 2011). Additionally, MA105 could tentatively be identified as a member in the Xylariaceae, as it formed a long branch length distinct at a basal clade within the family without any closely-related species.

Two of the endophyte isolates grouped in the Capnodiales, with strain MA111 identified as a *Cladosporium* sp. based on its morphological and ITS sequence similarity. Isolate MA12 was closely related with a sequence in the Mycosphaerellaceae (EF694667), with a weak bootstrap support.

The Botryosphaeriales were represented by the two monophyletic endophytes (MA140 and MA127), and formed a well-supported clade with sequences of *Guignardia mangiferae*. Therefore, these two endophytes could be identified as *G. mangiferae*.

Six endophyte assemblages (MA150, MA158, MA122, MA130, MA164 and MA71) clustered within the Pleosporales. Endophytes MA150, MA158 and MA122 were monophyletic and possibly the same species; but they had no closely related taxon, and can only be tentatively classified as members in the Pleosporales. Endophyte MA130 had affinities with *Lophiostoma*, with *L. vagabundum* (AF383954) and *Lophiostoma* sp. (AY787716) being the closest taxa. Nevertheless, their branch length indicated nucleotide differences; therefore MA130 could not be identified at the generic level. Endophyte MA164 can be identified as *Leptosphaerulina chartarum* (EU272493), as it shared identical LSU and ITS sequences. Additionally, a sporulating fungus MA71 was closely related to the genera *Didymella* and *Stagonosporopsis*. Based on morphological features and ITS sequence analyses, this fungus could be named as *Didymella bryoniae*. One isolate (MA56) formed a well-supported clade within the *Schizophyllum commune* group (Basidiomycota), with ITS sequence similarity among this group of 99–100%.

The dominant endophytic fungal genera found from marine and mangrove plants

The identification of fungal endophytes from various plant species, based on traditional and molecular approaches, revealed a highly diverse taxonomic community (Stone et al. 2004). Most belong to the Ascomycota, and are restricted to the Dothideomycetes and the Sordariomycetes, as the major foliar endophyte species. These classes account for more than 75% of endophytes in sites ranging from the Arctic to the tropics (Arnold 2007). The most common genera found from marine and mangrove plant species belong to the ascomycetous orders Botryosphaeriales, Capnodiales, Chaetothyriales, Diaporthales, Dothideales, Eurotiales, Glomerellales, Hypocreales, Pleosporales, Sordariales, Trichosphaeriales and Xylariales (Table 16.3). This might indicate that they have independently evolved into an endophytic mode several times (Stone et al. 2004). Moreover, these frequently-found mangrove endophytes were mainly "terrestrial" in origin, commonly found from the foliage of woody perennial plants (Arnold 2007), as well as from other marine sources, such as algae, sponges, sea fans, corals, solar salterns and hypersaline

Table 16.3 Genera of endophytic fungi commonly isolated from marine and mangrove plants (see references in Table 16.1).

Order	Genera
Ascomycota	
Apiosporaceae, Order *incertae sedis*	*Arthrinium*
Botryosphaeriales	*Camarosporium, Guignardia, Phyllosticta*
Capnodiales	*Cladosporium, Mycosphaerella*
Chaetothyriales	*Phialophora*
Diaporthales	*Cytospora, Diaporthe, Phomopsis*
Dothideales	*Aureobasidium*
Eurotiales	*Aspergillus, Penicillium, Talaromyces*
Glomerellales	*Colletotrichum, Glomerella*
Hypocreales	*Acremonium, Cylindrocarpon, Fusarium, Paecilomyces, Stachybotrys, Trichoderma*
Pleosporales	*Alternaria, Didymella, Leptosphaerulina, Lophiostoma, Phoma, Sporormiella, Stemphylium*
Saccharomycetales	*Candida*
Sordariales	*Papulaspora*
Trichosphaeriales	*Nigrospora*
Xylariales	*Hypoxylon, Pestalotiopsis*
Asexual Pezizomycotina	*Ampullifera, Gamsia, Fusariella*
Basidiomycota	
Cantharellales	*Allescheriella*
Sporidiobolales	*Sporobolomyces*

environments (Koh et al. 2002; Morrison-Gardiner 2002; Cantrell et al. 2006; Phongpaichit et al. 2006; Preedanon 2007; Zalar et al. 2007; Wang et al. 2008; Li and Wang 2009; Suryanarayanan et al. 2010).

Results from the present study indicate that endophytic fungi from mangrove plants were diverse, and distributed in broad taxonomic orders of the Ascomycota, although not all endophytes obtained were sequenced for their taxonomic identity. The predominant genera obtained from this present study were *Diaporthe, Phomopsis, Guignardia, Pestalotiopsis* and members in the Pleosporales. This is in congruence with other studies by Huang et al. (2008), Pang et al. (2008) and Chaeprasert et al. (2010).

The genus *Diaporthe* and its asexual state (*Phomopsis*) are generally well known as plant pathogens and frequent colonizers of various hosts, from woody dicotyledonous to herbaceous monocots (Lodge et al. 1996; Eriksson and Yue 1998). They are the most prevalent endophytic fungi isolated from tropical and temperate woody plants (Rossman et al. 2007; Santos and Phillips 2009). They occur as common endophytic genera in mangrove plant species such as *A. ilicifolius* (Suryanarayanan and Kumaresan 2000), *L. racemosa* (Kumaresan and Suryanarayanan 2001), *C. tagal* (Xing and Guo 2011), *K. obovata* (as *K. candel* in Pang et al. 2008), *S. apetala, S. caseolaris, S. hainanensis* (Xing et al. 2010), *A. alba, A. officinalis* and *R. apiculata* (present study). The molecular identification of these genera at species level has proved to be difficult; therefore multi-gene phylogenetic data, e.g. ITS rDNA, in combination with the translation elongation factor-1 alpha and actin genes were used (Castlebury and Mengistu 2006).

The genus *Guignardia* and its asexual state (*Phyllosticta*) are cosmopolitan plant parasites (van der Aa 1973; Baayen et al. 2002; Okane et al. 2003; Pandey et al. 2003). Many species are commonly isolated as endophytes from several plant hosts and may remain latent in their hosts for a prolonged period of time (Rodrigues et al. 2004). *Guignardia* species are commonly found as endophytes from *K. obovata* (Pang et al. 2008) and *X. granatum* (present study). *Phyllosticta* species were found on wider range of hosts, including *Acanthus ilicifolius, Avicennia alba, L. littorea, C. decandra, S. alba* (Chaeprasert et al. 2010), *L. racemosa, Arthrocnemum indicum* (Suryanarayanan and Kumaresan 2000), *D. falcate* (Kumaresan et al. 2002), *B. gymnorrhiza* (Okane et al. 2001a), *R. apiculata* and *R. mucronata* (Kumaresan and Suryanarayanan 2002).

Pestalotiopsis species are ubiquitous, occurring on a wide range of substrata. Most species are plant pathogens (Zhu et al. 1991; Zhang et al. 2003) while some are saprobes in soil (Agarwal and Chauhan 1988). They have frequently been isolated as endophytes from various plants (Hu et al. 2007; Liu et al. 2007b; Wei et al. 2007). In mangrove plants, *Pestalotiopsis* species have frequently been found on *B. sexangula* var. *rhynchopetala, C. tagal, R. apiculata, R. stylosa* (Xing and Guo 2011), *K. obovata* (Pang et al. 2008), *R. mucronata* (Chaeprasert et al. 2010), *A. corniculatum, Scypiphora hydrophyllacea* and *Sonneratia caseolaris* (present study). *Pestalotiopsis* species are of particular interest as they are a good source of secondary metabolites (Strobel 2002). Nevertheless, the taxonomy at species level has proved to be difficult; therefore interrelationships were recently inferred using the combination of nuclear ITS rDNA and *beta-tubulin 2* gene (Liu et al. 2010).

Members within the Dothideomycetes are found predominantly as foliar, bark and shoot endophytes from diverse angiosperm and gymnosperm plant species. They are dominated by genera such as *Alternaria, Curvularia, Pleospora, Sporormia, Sporormiella, Stemphylium* (Pleosporales; Stone et al. 2004; Arnold 2007; Sieber 2007). This is in agreement with the list of common pleosporaceous genera found from mangrove plants (Table 16.3). *Phoma* spp. have frequently been found from several mangrove plant species such as *A. ilicifolius* (Ananda and Sridhar 2002), *A. marina* (Kumaresan and Suryanarayanan 2001), *T. populneoides* (Chaeprasert et al. 2010) and *S. paracaseolaris* (Xing et al. 2010). The genus *Sporormiella* has been commonly found from *S. maritima* (Suryanarayanan and Kumaresan 2000), *A. officinalis, E. agallocha, B. cylindrical, C. decandra* (Kumaresan and Suryanarayanan 2001), *R. apiculata* and *R. mucronata* (Suryanarayanan et al. 1998).

In general, Basidiomycota are reported much less frequently than ascomycetous endophytes. Basidiomycetous endophytes occur more often in woody tissues than in foliage in a variety of terrestrial plant taxa including orchids, liverworts, monocotyledonous and dicotyledonous plants (Arnold 2007; Rungjindamai et al. 2008). They include members of the Agaricomycotina and Pucciniomycotina (Rodriguez et al. 2009). Basidiomycetous endophytes have rarely been isolated from any marine or mangrove plants; however, Sakayaroj et al. (2010) found one taxon, *Peniophora* sp., as a seagrass endophyte. Moreover *Schizophyllum commune* was reported for the first time from *Sonneratia ovata* in the current study.

Our results, and those obtained in earlier studies (Table 16.1), indicate that the numbers of dominant endophytic species detected from marine host plants are usually low (Jones 2011) compared to the endophytic fungi of other plant species (Stone et al. 2004; Sieber 2007). The number of endophytic species may depend on many biotic, abiotic and experimental factors, e.g. the host species, type and phase disposition of the plant organ, climatic conditions, the isolation procedure and the number and size of samples (Sieber 2007).

Investigations of endophytic fungi from terrestrial plants have frequently yielded novel taxa. For example, Strobel et al. (2001) isolated a volatile compound-producing fungus (*Muscodor albus*), which was a new endophytic species of *Cinnamomum zeylanicum*; while another new species (*Muscodor cinnamomi*) was described as an endophyte within the leaf tissues of *Cinnamomum bejolghota* by Suwannarach et al. (2010). Peterson et al. (2005) described a novel endophyte, *Penicillium coffeae*, from a *Coffea arabica* plant in Hawaii. The new endophyte *Preussia mediterranea* was described from a Mediterranean plant species based on morphological and molecular characters (Arenal et al. 2007). Additionally, Zhao et al. (2009) discovered a new taxol-producing fungus, *Aspergillus niger* var. *taxi*, from *Taxus cuspidata* in China. Recently, Chaverri et al. (2011) described a new species *Trichoderma amazonicum*, an endophytic fungus isolated from living sapwood and leaves of the *Hevea* tree. In contrast, few new endophytic taxa have been recovered from marine plants. Okane et al. (2001b) described a new fungal endophyte, *Surculiseries rugispora* (Xylariales), from the leaves of *B. gymnorrhiza*.

It is well known that mangrove plants are physiologically tolerant of high sea salt levels. They regulate salt concentration in the plant tissue through a combination of salt exclusion, salt excretion and salt accumulation. Moreover, mangrove trees are rich in polyphenols and tannins, although the levels of these substances vary seasonally (Kathiresan and Qasim 2005). Therefore, the restricted colonization of fungal endophytes in mangrove plant hosts, compared to terrestrial plants, could be due to these factors.

Molecular characterization of representative sea fan-derived fungi isolated from the Andaman Sea, Thailand

Studies by Raghukumar and Raghukumar (1991), Koh et al. (2000), Petes et al. (2003) and Toledo-Hernández et al. (2007) have recovered fungi from gorgonian sea fans. In preliminary studies, fungi have been isolated from sea fans at Mu Ko Similan National Park, Andaman Sea, Thailand, by Phongpaichit et al. (2006) and Preedanon (2007). They reported fungal infections in sea fan tissues after the 2004 tsunami and identified nine out of 51 fungal strains based on their morphological characteristics. Since then, the numbers of fungi were periodically examined from diseased and healthy sea fans using morphological and molecular data and screened for novel secondary metabolites. Thirty-eight fungal strains that produced antimicrobial activity and/or had interesting nuclear magnetic resonance profiles (Table 16.4) could be referred to three Ascomycota classes (Dothideomycetes, Eurotiomycetes and Sordariomycetes) and six orders (Eurotiales, Pleosporales, Capnodiales, Xylariales, Trichosphaeriales and Hypocreales) based on morphological and ITS rDNA sequence data (Figure 16.2).

Four active fungal taxa (F36, F33, F78 and F154), which produced *Aspergillus*-type conidia, were confirmed by morphological and molecular data. Isolates F33 and F36 had affinity with a sequence of an uncultured fungus (GQ999178) and several strains of *A. versicolor*, with 100% ITS sequence similarity. Strain F154 was identified as *A. sydowii* (tree not shown) and is the first report of this fungus from a gorgonian sea fan from

Table 16.4 Summary of the molecular identification of marine-derived fungi isolated from sea fans (*Annella* sp.) whose crude extracts produced antimicrobial activities.

Sea fan-derived fungi	Bioactivity of crude extracts and/or secondary metabolite production
Eurotiales	
Aspergillus versicolor F36	Moderate activity against SA, MRSA and MG (MIC 32–128 µg/ml)
A. versicolor F33	Moderate activity against MG (MIC 32 µg/ml)
Aspergillus aculeatus F78	Moderate activity against MG (MIC 20 µg/ml)
Aspergillus sydowii F154*	Three new sesquiterpenes, named aspergillusenes A and B and (+)-(7S)-7-O-methylsydonic acid; and 2 new hydrogenated xanthone derivatives, named aspergillusones A and B; with 10 known compounds (Trisuwan et al. 2011a)
Penicillium sp. F40*	Nine new fungal metabolites: penicisochromans A–E, penicipyrone, penicipyranone, peniciphenol, and penicisoquinoline; with 5 known compounds (Trisuwan et al. 2010b)
Pezizomycotina incertae sedis	
Scolecobasidium sp. F118	Moderate activity against PA (MIC 20 µg/ml)
Scolecobasidium sp. F69	Weak activity against SA and MRSA (MIC 160–320 µg/ml)
Scolecobasidium sp. F77	Moderate activity against EC (MIC 40 µg/ml)
Capnodiales	
Cladosporium sp. F3	Weak activity against all test microorganisms (MIC >256 µg/ml)
Cladosporium sp. F32	Moderate activity against MG (MIC 32 µg/ml)
Cladosporium cladosporioides F98	Moderate activity against PA (MIC 80 µg/ml)
Cladosporium sp. F128	Weak activity against EC, PA (MIC 80 ≥ 1,280 µg/ml)
Pleosporales	
Alternaria sp. F54	Moderate activity against MG (MIC 32 µg/ml)
Curvularia affinis F22	Three new metabolites: curvulapyrone, curvulalide and curvulalic acid; with 6 known compounds (Trisuwan et al. 2011b)
Ascomycota sp. F46	Strong activity against MG (MIC 16 µg/ml)
Pleosporales sp. F4	Weak activity against all test microorganisms (MIC > 256 µg/ml)
Letendraea helminthicola F158	Moderate activity against PA (MIC 40–80 µg/ml)
L. helminthicola F10	Weak activity against all test microorganisms (MIC > 256 µg/ml)

Table continued on next page.

Table 16.4 (Continued)

Sea fan-derived fungi	Bioactivity of crude extracts and/or secondary metabolite production
Hypocreales	
Fusarium sp. F14	One new metabolite: fusaranthraquinone; with 9 known compounds (Trisuwan et al. 2010a)
Fusarium sp. F135	Four new metabolites: fusarnaphthoquinones A–C and fusarone; with 9 known compounds (Trisuwan et al. 2010a)
Fusarium sp. F37	Strong activity against SA, MRSA, CN and MG (MIC 4–16 μg/ml)
Gibberella sp. F7	Moderate activity against MG (MIC 32 μg/ml)
Fusarium sp. F156	Strong activity against MRSA (MIC 20 μg/ml)
Fusarium sp. F63	Strong activity against SA and MRSA (MIC 10–20 μg/ml)
Trichoderma aureoviride F6	Moderate activity against MG (MIC 32 μg/ml)
T. aureoviride F95	Moderate activity against SA, MRSA and MG (MIC 40–80 μg/ml)
Trichosphaeriales	
Nigrospora sp. F57	Interesting nuclear magnetic resonance profile
Nigrospora sp. F5	Four new metabolites: nigrospoxydons A–C and nigrosporapyrone; together with 9 known compounds (Trisuwan et al. 2008)
Ascomycete sp. F13	Strong activity against MG (MIC 1 μg/ml)
Nigrospora sp. F2	Weak activity against MG (MIC 128 μg/ml)
Nigrospora sp. F105	Weak activity against PA (MIC 160 μg/ml)
Nigrospora sp. F18	Four new pyrones, named nigrosporapyrones A–D (Trisuwan et al. 2009)
Nigrospora sp. F11*	Two new compounds: nigrosporanenes A and B; one known compound (Rukachaisirikul et al. 2010)
Nigrospora sp. F12*	Five known compounds (Rukachaisirikul et al. 2010)
Xylariales	
Xylariales sp. F80	Weak activity against SA and MRSA (MIC 160 μg/ml)
Hypoxylon sp. F159	Weak activity against PA (MIC 80 μg/ml)
Annulohypoxylon nitens F151	Strong activity against PA (MIC 10 μg/ml)
Xylaria sp. F100	New cytochalasin derivative and xylarisin (Rukachaisirikul et al. 2009)

SA: *Staphylococcus aureus* ATCC25923, MRSA: methicillin-resistant *S. aureus* SK1, PA: *Pseudomonas aeruginosa* ATCC27853, EC: *Escherichia coli* ATCC25922, CA: *Candida albicans* ATCC90028, CN: *Cryptococcus neoformans* ATCC90112, MG: *Microsporum gypseum* MU-SH4, MIC: minimum inhibitory concentration.
*Identified by morphology only.

Thailand. Another fungal isolate, F78, nestled within a subclade comprising *A. aculeatus* strains with high bootstrap support and 99% ITS sequence similarity. Additionally, isolate *Penicillium* sp. F40 (tree not shown) was characterized based on its morphological features and shown to produce new fungal secondary metabolites (Trisuwan et al. 2010b), see Table 16.4.

Diversity of endophytic and marine-derived fungi | 317

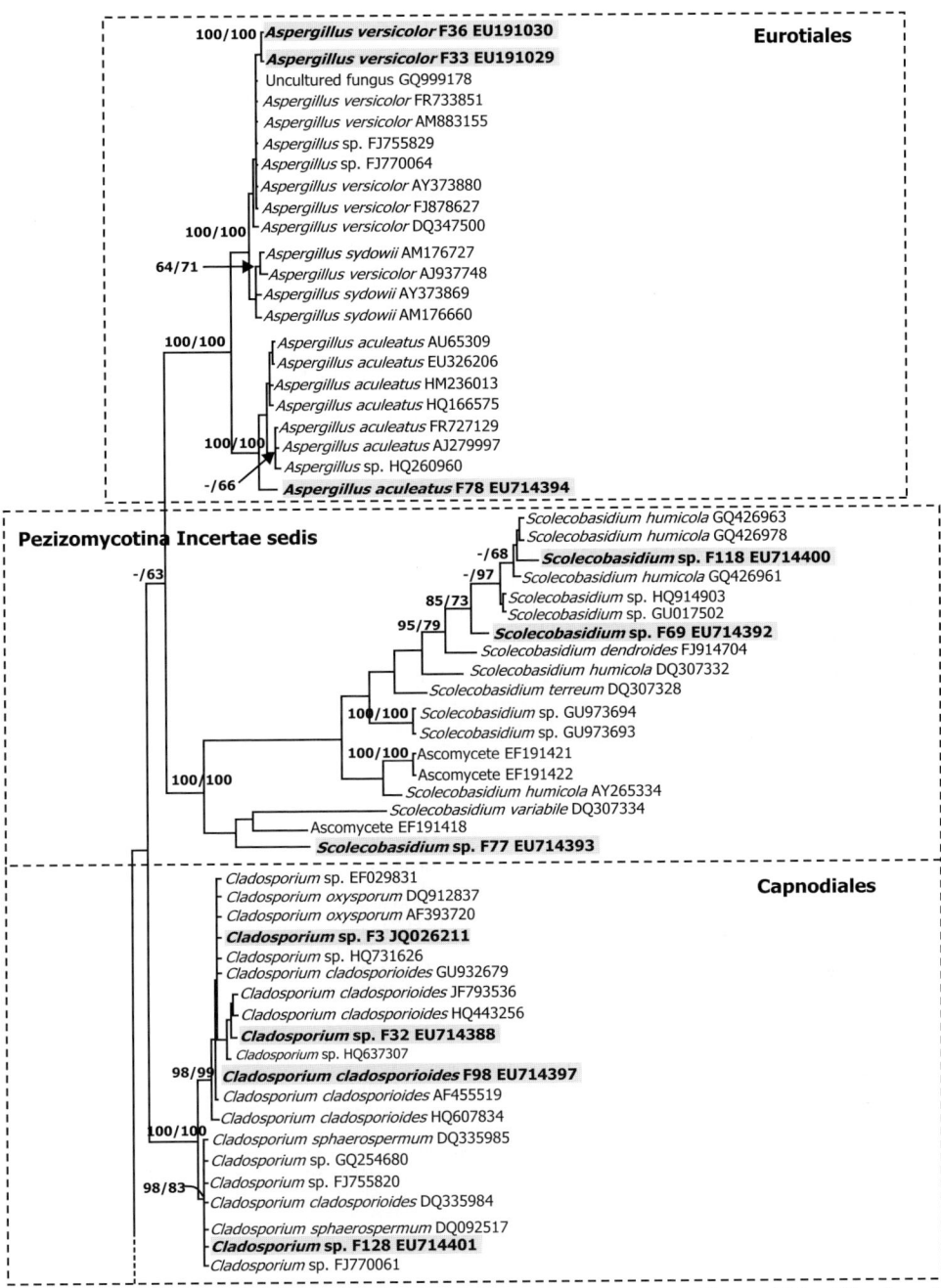

Figure 16.2 (Continued)

Figure 16.2 (Continued)

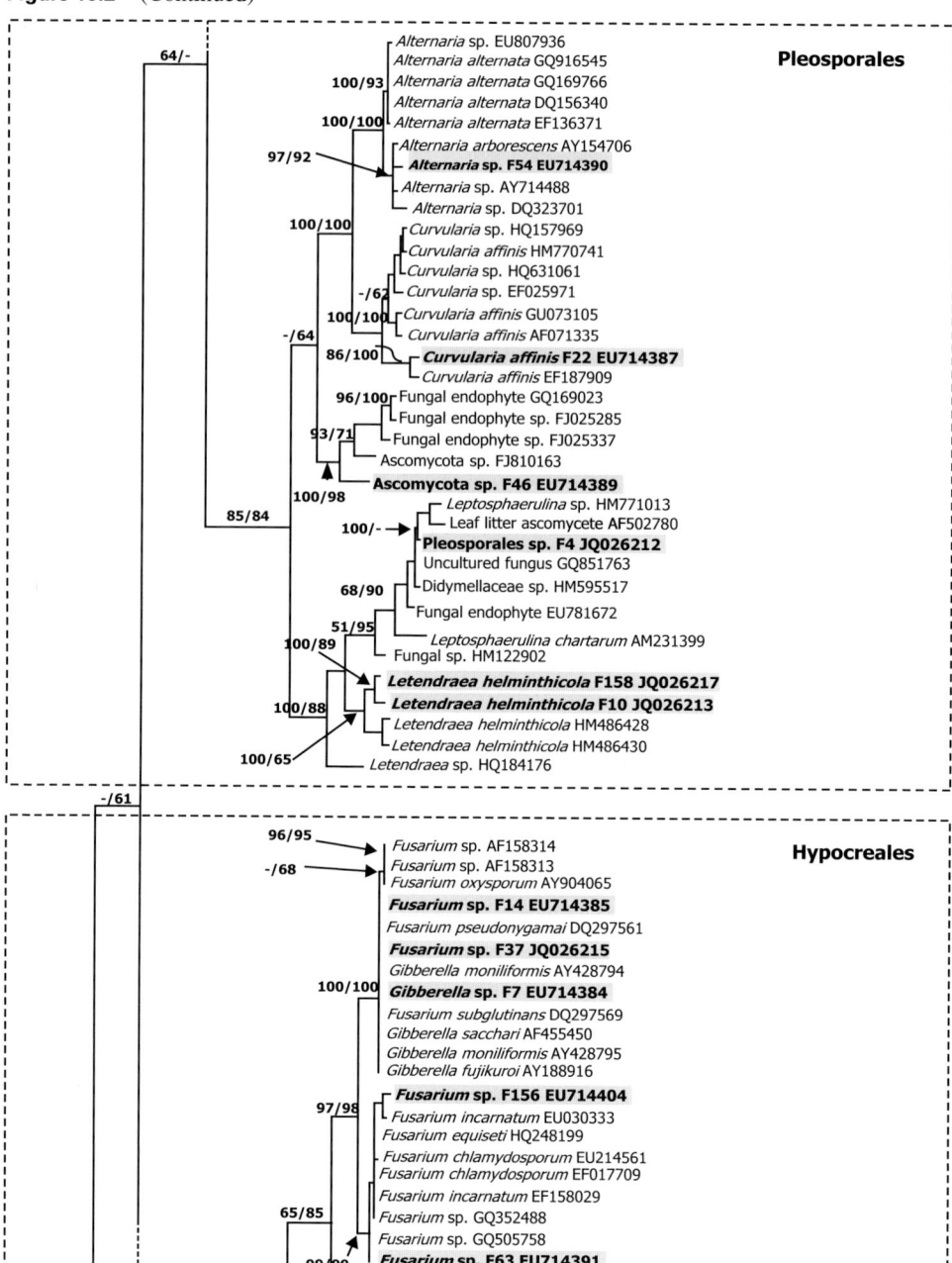

Figure 16.2 (Continued)

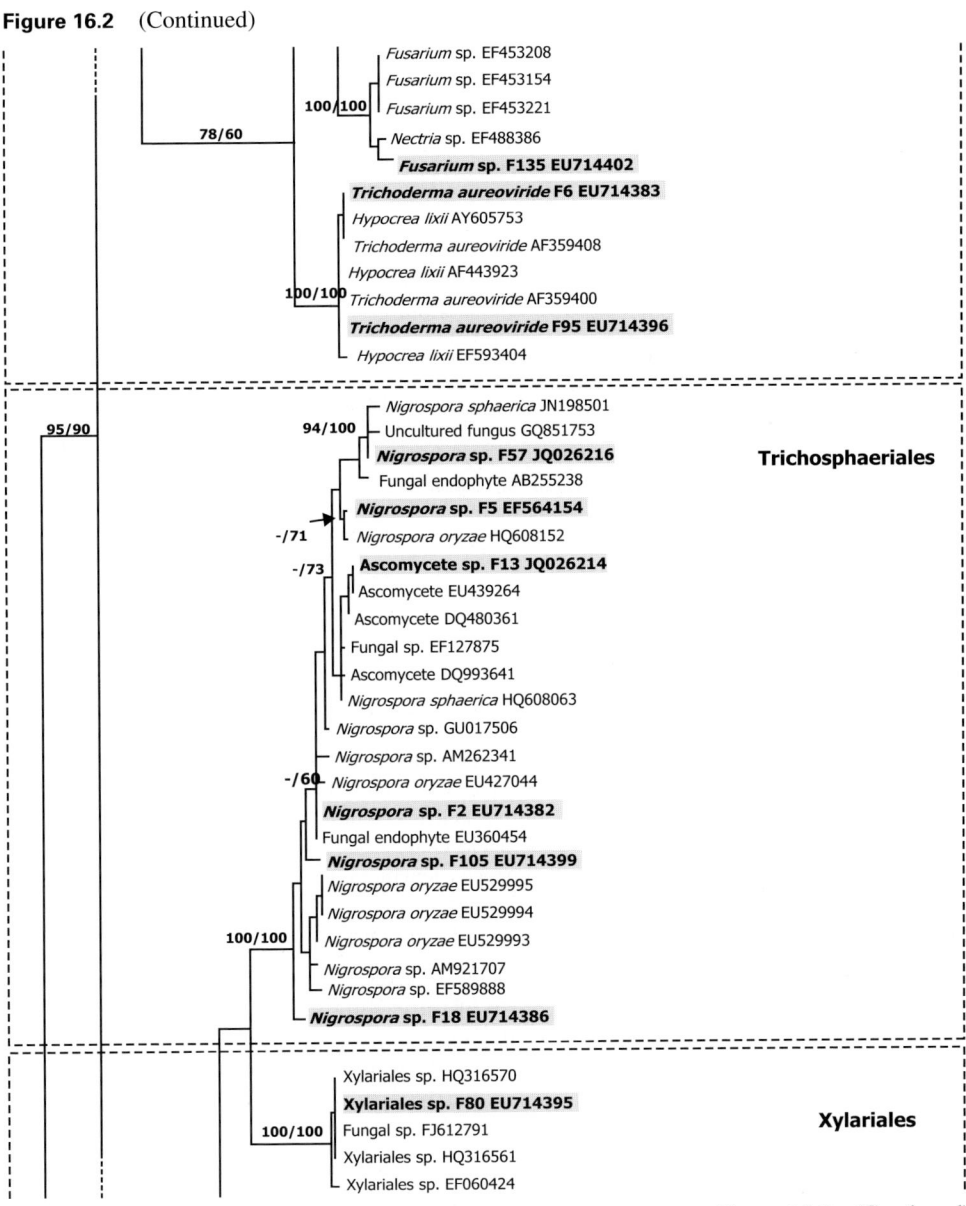

Figure 16.2 (Continued)

Figure 16.2 (Continued)

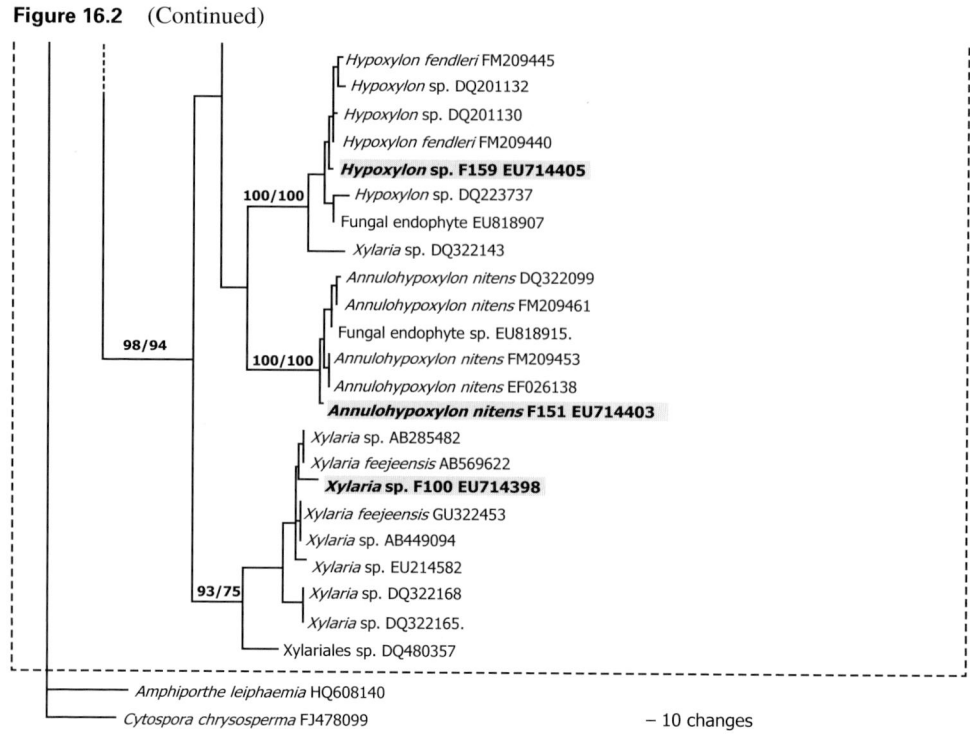

Figure 16.2 One of the most parsimonious trees generated from maximum parsimony analysis of ITS1, ITS2 and 5.8S rDNA sequences from sea fan-derived fungi isolated from southern Thailand. Bootstrap support values from neighbor joining and maximum parsimony analyses are shown above each branch, respectively. Tree length = 5,178 changes, consistency index = 0.389, retention index = 0.833. Highlighted fonts represent sequences generated from this study. Branch lengths are proportional to the number of character state changes.

Three fungal strains (F118, F69 and F77) were closely related to the genus *Scolecobasidium*. Isolate F118 had a close affinity with *S. humicola* (GQ426963 and GQ426978); while F69 was related to *Scolecobasidium* sp. (GU017532), which has also been isolated from a seagrass (Sakayaroj et al. 2010). Strain F77 was distantly placed in a basal clade of the group, with ascomycete EF191418 and *S. variabile* (DQ307334) as sister taxa.

Four strains (F3, F32, F98 and F128) isolated in our study had an affinity with the genus *Cladosporium* (Capnodiales). Isolate F3 had an ITS sequence identical to *C. oxysporum* (DQ912837 and AF393720) and *Cladosporium* sp. (HQ731626) with 98–100% similarity; while isolates F32 and F98 were closely related to several *C. cladosporioides* strains, with high sequence similarity. In addition, isolate F128 clustered within a subclade comprising various *Cladosporium* species at the base of the Capnodiales. Therefore, the fungal isolates obtained from sea fan were identified as *Cladosporium* sp. F3, *Cladosporium* sp. F32, *C. cladosporioides* F98 and *Cladosporium* sp. F128.

Six active isolates (F54, F22, F46, F4, F158 and F10) could be referred to various genera in the Pleosporales. Fungus F54 was identified as an *Alternaria* sp. as it grouped with identical nucleotide sequences of *A. arborescens* (AY154706), and *Alternaria* sp. (AY714488 and DQ323701). *Curvularia affinis* (F22) was monophyletic with another strain (EF187909); however, identification of isolates F46 and F4 could not be resolved. Isolates F158 and F10 were monophyletic and formed a clade with a sequence of *Letendraea helminthicola* with strong support.

The Hypocreales were represented by eight strains (F14, F37, F7, F156, F63, F135, F6 and F95) of sea fan-derived fungi. Five isolates had affinity with the genera *Fusarium* and *Gibberella*. Isolates F14 and F37 had sequences identical to various *Fusarium* species within the upper clade of a tree (97–99% nucleotide similarity). Isolate F7 was placed in a subclade comprising the genera *Gibberella* and *Fusarium* species, with *G. miniliformis* (AY428794) as the closest taxon. Isolates F156 and F63 clustered with several *Fusarium* species, while F135 was referred to the genus based on its morphology. Isolates F6 and F95 formed a basal clade of the Hypocreales with 100% sequence similarity to the *T. aureoviride-Hypocrea lixii* grouping.

There were nine active isolates distributed within the Trichosphaeriales (F11, F12, F57, F2, F105, F18, F13, F57 and F5) and two (F11 and F12) identified as *Nigrospora* spp. based on morphological features (Table 16.4). Strain F57 clustered closely with an uncultured fungus (GQ851753) and an unidentified fungal endophyte (AB255238), and showed an affinity within the *Nigrospora* group. Isolate F5 was closely related to *N. oryzae* (HQ608152), with 95% sequence similarity. Additionally, strains F2, F105 and F18 had affinities with various *N. oryzae* and *Nigrospora* species, while strain F13 clustered within a clade comprising several unidentified ascomycetes. Finally, within the Xylariales we found four isolates (F80, F159, F151 and F100) that had affinity with the genera *Hypoxylon, Annulohypoxylon* and *Xylaria*. Isolate F80 grouped well with a sequence of an unidentified xylariaceous fungus (HQ316570) and fungal sp. FJ612791 isolated from marine environments. F159 grouped within a *Hypoxylon* grouping; while F151 was identified as *Annulohypoxylon nitens*, as it nestled with several *A. nitens* strains in a well-supported clade. Isolate F100 grouped with *X. feejeensis* (AB569622) and the *Xylaria* species subclade supported by moderate bootstrap values.

Diversity and dominant sea fan-derived fungi from the gorgonian sea fan *Annella* sp.

A wide variety of fungal genera have been isolated from *Annella* sp. sea fans. These fungi belong to diverse genera and orders within the Ascomycota, and are mainly dematiaceous anamorphs. Dominant genera isolated from this study were *Aspergillus, Cladosporium, Fusarium, Nigrospora, Scolecobasidium* and also pleosporaceous taxa (Figure 16.4). These fungi, especially the genera *Aspergillus, Cladosporium, Fusarium* and *Nigrospora*, have been commonly associated with diseased corals from the Caribbean, Australia, Singapore and the Andaman Islands (Kendricks et al. 1982; Raghukumar and Raghukumar 1991; Geiser et al. 1998; Koh et al. 2000; Yarden et al. 2007). A recent study by Toledo-Hernàndez et al. (2007) also documented *Aspergillus, Penicillium, Cladosporium, Gloeotinia, Rhodotorula, Stachybotrys* and *Xylaria* from the gorgonian sea fan *Gorgonia ventalina*.

Here we report, for the first time, the isolation of several ubiquitous fungal species from a gorgonian sea fan from Thailand. The detection of various fungi in sea fans is of concern because several are known to produce toxic metabolites (Geiser et al. 1998;

Toledo-Hernàndez et al. 2007). Pure cultures of the pathogenic sea fan *A. sydowii* produced unique compounds that had not been found in nonpathogenic strains (Geiser et al. 1998). Similarly, *Stachybotrys* species isolated from diseased sea fans also produce mycotoxins such as trichothecenes and stachybotrin (Pietra 1997; Toledo-Hernàndez et al. 2007). Interestingly, the fungi isolated in this study have shown many interesting activities against other microbial pathogens. It is possible therefore that these fungi may produce some unique compounds that could play a role during the disease process. It has yet to be determined whether these fungi produce such compounds in their natural marine environments.

Conclusion

The mangrove endophytic and marine-derived fungi highlighted here are shown to be a diverse group. They are distributed in various genera and orders, and most have originally been described from terrestrial habitats. Jones (2011) is of the opinion that the greater number of endophytes of marine plants, seaweeds, and marine animals could reach ~6,000 species; therefore, studies of endophytic fungi and marine-derived fungi are needed to further provide fundamental information for evaluating global fungal diversity and distribution. However it is important to document their diversity by molecular techniques.

Fungi isolated from this present study are able to produce unique secondary metabolites with interesting structural features and promising antimicrobial and biological activities. Our results are in concordance with previous reports by Jones et al. (2008), Schulz et al. (2008) and Debbab et al. (2011). It is believed that the natural products pattern of endophytic fungi might play an important role in endophyte–host communication and for adaptation of the hosts to environmental challenges (Gunatilaka 2006) while it is possible that sea fan-derived fungi may produce some unique compounds that could play a role during the disease process (Geiser et al. 1998; Toledo-Hernàndez et al. 2007). The knowledge obtained from these studies is a basis for further understanding how marine ecosystems function.

Acknowledgements

This work was supported by the TOTAL Corporate Foundation, TOTAL E&P Thailand, and the TRF/BIOTEC Special Program for Biodiversity Research and Training Grant BRT R_351004, the Commission on Higher Education, the Thailand Research Fund (TRF Senior Research Scholar Grant No. RTA5180007), and the Prince of Songkla University. We acknowledge Professor Morakot Tanticharoen, Dr Kanyawim Kirtikara and Dr Lily Eurwilaichitr for continued support. We thank Professor E. B. Gareth Jones and Dr Ka-Lai Pang for advice and kindly editing this chapter.

References

Agarwal AK, Chauhan S. A new species of the genus *Pestalotiopsis* from Indian Soil. Indian Phytopath. 1988;41:625–627.

Aksornkoae S. Ecology and management of mangrove, Bangkok: IUCN, 1993.

Alva P, Mckenzie EHC, Pointing SP, Pena-Murala R, Hyde KD. Do seagrasses harbour endophytes? In: Hyde KD, ed. Fungi in marine environments. Hong Kong: Fungal Diversity Press, 2002:167–178.

Aly AH, Debbab A, Kjer J, Proksch P. Fungal endophytes from higher plants: a prolific source of phytochemicals and other bioactive natural products. Fungal Divers. 2010;41:1–16.

Ananda K, Sridhar KR. Diversity of endophytic fungi in the roots of mangrove species on the west coast of India. Can J Microbiol. 2002;48:871–878.

Arenal F, Platas G, Peláez F. A new endophytic species of *Preussia* (*Sporormiaceae*) inferred from morphological observations and molecular phylogenetic analysis. Fungal Divers. 2007;25:1–17.

Ariffin SA, Davis P, Ramasamy K. Cytotoxic and antimicrobial activities of Malaysian marine endophytic fungi. Bot Mar. 2011;54:95–100.

Arnold AE. Understanding the diversity of foliar endophytic fungi: progress, challenges, and frontiers. Fung Biol Rev. 2007;21:51–66.

Arnold AE, Henk DA, Eells RL, Lutzoni F, Vilgalys R. Diversity and phylogenetic affinities of foliar fungal endophytes in loblolly pine inferred by culturing and environmental PCR. Mycologia 2007;99:185–206.

Baayen RP, Bonants PJM, Verkley G, Carroll GC, van der Aa HA, de Weerdt M, van Brouwershaven IR, Schutte GC, Maccheroni W, de Blanco C, Azevedo JL. Nonpathogenic isolates of the *Citrus* black spot fungus, *Guignardia citricarpa*, identified as a cosmopolitan endophyte of woody plants, *G. mangiferae* (*Phyllosticta capitalensis*). Phytopathol. 2002;92:464–477.

Buatong J. Endophytic fungi producing antimicrobial substances from mangrove plants in the south of Thailand. Prince of Songkla University, Thailand, MSc thesis, 2010.

Buatong J, Phongpaichit S, Rukachaisirikul V, Sakayaroj J. Antimicrobial activity of crude extracts from mangrove fungal endophytes. World J Microbiol Biotechnol. 2011;27:3005–3008.

Cantrell SA, Casillas-Martínez L, Molina M. Characterization of fungi from hypersaline environments of solar salterns using morphological and molecular techniques. Mycol Res. 2006;110:962–970.

Castlebury LA, Mengistu A. Phylogenetic distinction of *Diaporthe/Phomopsis* isolates from soybeans. Inoculum 2006;57:13.

Chaeprasert S, Piapukiew J, Whalley AJS, Sihanonth P. Endophytic fungi from mangrove plant species of Thailand: their antimicrobial and anticancer potentials. Bot Mar. 2010;53:555–564.

Chaowalit P. Screening of endophytic fungi from mangrove plants which produce lipase, cellulase, amylase or protease enzymes. Prince of Songkla University, Thailand, MSc thesis, 2009.

Chaverri P, Gazis RO, Samuels GJ. *Trichoderma amazonicum*, a new endophytic species on *Hevea brasiliensis* and *H. guianensis* from the Amazon basin. Mycologia 2011;103:139–151.

Chen G, Zhu Y, Wang H-Z, Wang S-J, Zhang R-Q. The metabolites of a mangrove endophytic fungus, *Penicillium thomi*. J Asian Nat Prod Res. 2007;9:159–164.

Choi YW, Hodgkiss IJ, Hyde KD. Enzyme production by endophytes of *Brucea javanica*. J Agric Technol. 2005;1:55–66.

Debbab A, Aly AH, Proksch P. Bioactive secondary metabolites from endophytes and associated marine derived fungi. Fungal Divers. 2011;49:1–12.

Devarajan PT, Suryanarayanan TS, Geetha V. Endophytic fungi associated with the tropical seagrass *Halophila ovalis* (Hydrocharitaceae). Indian J Mar Sci. 2002;31:73–74.

Duke NC. Australia's mangroves. Brisbane, Australia: University of Queensland, 2006.

El-Morsy EM. Fungi isolated from the endorhizosphere of halophytic plants from the Red Sea Coast of Egypt. Fungal Divers. 2000;5:43–54.

Eriksson OE, Yue JZ. Bambusicolous pyrenomycetes, an annotated check list. Myconet 1998;1:25–78.

FAO. Proceedings of the workshop on coastal area planning and management in Asian tsunami-affected countries, 27–29 September 2006, Bangkok, Thailand. RAP Publication 2007/06: Bangkok, 2007. (Accessed April 17, 2012 at www.fao.org/forestry/site/35734/en.)

Ganley RJ, Newcombe G. Fungal endophytes in seeds and needles of *Pinus monticola*. Mycol Res. 2006;110:318–327.

Geiser DM, Taylor JW, Ritchie KB, Smith GW. Cause of sea fan death in the West Indies. Nature 1998;394:137–138.

Gunatilaka AAL. Natural products from plant-associated microorganisms: Distribution, structural diversity, bioactivity, and implication of their occurrence. J Nat Prod. 2006;69:509–526.

Guo LD, Hyde KD, Liew ECY. Identification of endophytic fungi from *Livistona chinensis* based on morphology and rDNA sequences. New Phytol. 2000;147:617–630.

Hata K, Futai K. Endophytic fungi associated with healthy pine needles and needles infested by the pine needle gall midge, *Thecodiplosis japonensis*. Can J Bot. 1995;73:384–390.

Higgins KL, Arnold AE, Miadlikowska J, Sarvate SD, Lutzoni F. Phylogenetic relationships, host affinity, and geographic structure of boreal and arctic endophytes from three major plant lineages. Mol Phylogenet Evol. 2007;42:543–555.

Hu HL, Jeewon R, Zhou DQ, Zhou TX, Hyde KD. Phylogenetic diversity of endophytic *Pestalotiopsis* species in *Pinus armandii* and *Ribes* spp.: evidence from rDNA and β-tubulin gene phylogenies. Fungal Divers. 2007;24:1–22.

Huang WY, Cai YZ, Hyde KD, Corke H, Sun M. Biodiversity of endophytic fungi associated with 29 traditional Chinese medicinal plants. Fungal Divers. 2008;33:61–75.

Hyde KD, Soytong K. The fungal endophyte dilemma. Fungal Divers. 2008;33:163–173.

Jones EBG. Are there more marine fungi to be described? Bot Mar. 2011;54:343–354.

Jones EBG, Hyde KD. Introduction to Thai fungal diversity. In: Jones EBG, Tanticharoen M, Hyde KD, eds. Thai fungal diversity. Thailand: BIOTEC, 2004:7–35.

Jones EBG, Pilantanapak A, Chatmala I, Sakayaroj J, Phongpaichit S, Choeyklin R. Thai marine fungal diversity. Songklanakarin J Sci Technol. 2006;28:687–715.

Jones EBG, Stanley SJ, Pinruan U. Marine endophyte sources of new chemical natural products: a review. Bot Mar. 2008;51:163–170.

Kathiresan K, Bingham BL. Biology of mangrove and mangrove ecosystem. Adv Mar Biol. 2001;40:81–251.

Kathiresan K, Qasim SZ. Biodiversity of mangrove ecosystems. New Delhi: Hindustan Publishing Corporation, 2005.

Kendricks B, Risk MJ, Michaelides J, Bergman K. Amphibious microborers: bioeroding fungi isolated from live corals. Bull Mar Sci. 1982;32:862–867.

Kobayashi J, Ishibashi M. Bioactive metabolites of symbiotic marine microorganisms. Chem Rev. 1993;93:1753–1769.

Koh LL, Tan TK, Chou LM, Goh NKC. Fungi associated with gorgonians in Singapore. Proc 9th Int Coral Reef Symp. 2000;1:521–526.

Koh LL, Tan TK, Chou LM, Goh NKC. Antifungal properties of Singapore gorgonians: a preliminary study. J Exp Mar Biol Ecol. 2002;273:121–130.

Kumaresan V, Suryanarayanan TS. Occurrence and distribution of endophytic fungi in a mangrove community. Mycol Res. 2001;105:1388–1391.

Kumaresan V, Suryanarayanan TS. Endophyte assemblages in young, mature and senescent leaves of *Rhizophora apiculata*: evidence for the role of endophytes in mangrove litter degradation. Fungal Divers. 2002;9:81–91.

Kumaresan V, Suryanarayanan TS, Johnson JA. Ecology of mangrove endophytes. In: Hyde KD, ed. Fungi in marine environments. Hong Kong: Fungal Diversity Press, 2002:145–166.

Li Q, Wang G. Diversity of fungal isolates from three Hawaiian marine sponges. Microbiol Res. 2009;163:233–241.

Li WC, Zhou J, Guo SY, Guo LD. Endophytic fungi associated with lichens in Baihua mountain of Beijing, China. Fungal Divers. 2007;25:69–80.

Lin X, Huang Y, Zheng Z. Endophytic fungi from pharmaceutical plant, *Annona squamosa* L.: Isolation, bioactivity, identification and diversity of polyketide synthase gene. Fungal Divers. 2007;41:41–51.

Liu AR, Xu T, Guo LD. Molecular and morphological description of *Pestalotiopsis hainanensis* sp. nov., a new endophyte from a tropical region of China. Fungal Divers. 2007a;24:23–36.

Liu X, Dong M, Chen X, Jiang M, Yan G. Antioxidant activity and phenolics of endophytic *Xylaria* sp. from *Ginkgo biloba*. Food Chem. 2007b;105:548–554.

Liu AR, Chen SC, Wu SY, Xu T, Guo LD, Jeewon R, Wei JG. Cultural studies coupled with DNA based sequence analyses and its implication on pigmentation as a phylogenetic marker in *Pestalotiopsis* taxonomy. Mol Phylogenet Evol. 2010;57:528–535.

Lodge DJ, Fisher PJ, Sutton BC. Endophytic fungi of *Manilkara bidentata* leaves in Puerto Rico. Mycologia 1996;88:733–792.

Maria GL, Sridhar KR. Endophytic fungal assemblage of two halophytes from west coast mangrove habitats, India. Czech Mycol. 2003;55:3–4.

Maria GL, Sridhar KR, Raviraja NS. Antimicrobial and enzyme activity of mangrove endophytic fungi of southwest coast of India. J Agric Technol. 2005;1:67–77.

Morrison-Gardiner S. Dominant fungi from Australian reefs. Fungal Divers. 2002;9:105–121.

Okane I, Nakagiri A, Ito I. Assemblages of endophytic fungi on *Bruguiera gymnorrhiza* in the Shira River Basin, Iriomote Is. IFO Res Comm. 2001a;20:41–49.

Okane I, Nakagiri A, Ito T. *Surculiseries rugispora* gen. *et* sp. nov., a new endophytic mitosporic fungus leaves of *Bruguiera gymnorrhiza*. Mycoscience 2001b;42:115–122.

Okane I, Lumyong S, Nakagiri A, Ito T. Extensive host range of an endophytic fungus, *Guignardia endophyllicola* (anamorph: *Phyllosticta capitalensis*). Mycoscience 2003;44:353–363.

Oses R, Valenzuela S, Freer J, Sanfuentes E, Rodríguez J. Fungal endophytes in xylem of healthy Chilean trees and their possible role in early wood decay. Fungal Divers. 2008;33:77–86.

Pandey AK, Reddy MS, Suryanarayanan TS. ITS-RFLP and ITS sequence analysis of a foliar endophytic *Phyllosticta* from different tropical trees. Mycol Res. 2003;107:439–444.

Pang KL, Vrijmoed LLP, Goh TK, Plaingam N, Jones EBG. Fungal endophytes associated with *Kandelia candel* (Rhizophoraceae) in Mai Po Nature Reserve, Hong Kong. Bot Mar. 2008;51:171–178.

Peterson SW, Vega FE, Posada F, Nagai C. *Penicillium coffeae*, a new endophytic species isolated from a coffee plant and its phylogenetic relationship to *P. fellutanum*, *P. thiersii* and *P. brocae* based on parsimony analysis of multilocus DNA sequences. Mycologia 2005;97:659–666.

Petes LE, Harvell CD, Peters EC, Webb MAH, Mullen KM. Pathogens compromise reproduction and induce melanization in Caribbean sea fans. Mar Ecol Prog Ser. 2003;264:167–171.

Petrini O, Stone J, Carroll FE. Endophytic fungi in evergreen shrubs in western Oregon: A preliminary study. Can J Bot. 1982;60:789–796.

Petrini O. Fungal endophytes of tree leaves. In: Andrews JH, Hirano SS, eds. Microbial ecology of leaves. Germany: Springer, 1991:179–197.

Phongpaichit S, Preedanan S, Rungjindamai N, Sakayaroj J, Benzies C, Chuaypat J, Plathong S. Aspergillosis of the gorgonian sea fan *Annella* sp., after the 2004 tsunami at Mu Ko Similan National Park, Andaman Sea, Thailand. Coral Reefs 2006;25:296.

Pietra F. Secondary metabolites from marine microorganisms: bacteria, protozoa, algae and fungi. Achievements and prospects. Nat Prod Rep. 1997;14:453–464.

Preedanon S. Screening and identification of sea fan-derived fungi that produce antimicrobial substances. Prince of Songkla University, Thailand, MSc thesis, 2007.

Raghukumar C. Marine fungal biotechnology: an ecological perspective. Fungal Divers. 2008;31:19–35.

Raghukumar C, Raghukumar S. Fungal invasion of massive corals. PSZNI Mar Ecol. 1991;12:251–260.

RFD (Royal Forest Department). Forest statistics. Royal Forest Department 61: Phaholyathin, Ladyao, Bangkok, Thailand, 2004. (http://www.phuketcampground.com/mangrove.htm.)

Rodríguez GM. Potential of fungal endophytes from *Thalassia testudinum* Bank ex K.D. Koenig as producers of bioactive compounds. University of Puerto Rico, Puerto Rico, MSc thesis, 2008.

Rodriques KF, Seiber TN, Grünig CR, Holdenrieder O. Characterization of *Guignardia mangiferae* isolated from tropical plants based on morphology, ISSR-PCR amplifications and ITS1–5.8S-ITS2 sequences. Mycol Res. 2004;108:45–52.

Rodriguez RJ, White JF Jr, Arnold AE, Redman RS. Fungal endophytes: diversity and functional roles. New Phytol. 2009;182:314–330.

Rossman AY, Farr DF, Castlebury LA. A review of the phylogeny and biology of the Diaporthales. Mycoscience 2007;48:135–144.

Rukachaisirikul V, Khamthong N, Sukpondma Y, Pakawatchai C, Phongpaichit S, Sakayaroj J, Kirtikara K. An [11] Cytochalasin derivative from the marine-derived fungus *Xylaria* sp. PSU-F100. Chem Pharm Bull. 2009;57:1409–1411.

Rukachaisirikul V, Khamthong N, Sukpondma Y, Phongpaichit S, Hutadilok-Towatana N, Graidist P, Sakayaroj J, Kirtikara K. Cyclohexene, diketopiperazine, lactone and phenol derivatives from the sea fan-derived fungi *Nigrospora* sp. PSU-F11 and PSU-F12. Arch Pharm Res. 2010;33:375–380.

Rukachaisirikul V, Rodglin A, Phongpaichit S, Buatong J, Sakayaroj J. α-Pyrone and seiricuprolide derivatives from the mangrove-derived fungi *Pestalotiopsis* spp. PSU-MA92 and PSU-MA119. Phytochem Lett. 2011;5:13–17.

Rungjindamai N, Pinruan U, Choeyklin R, Hattori T, Jones EBG. Molecular characterization of basidiomycetous endophytes isolated from leaves, rachis and petioles of the oil palm, *Elaeis guineensis*, in Thailand. Fungal Divers. 2008;33:139–161.

Sakayaroj J, Preedanon S, Supaphon O, Jones EBG, Phongpaichit S. Phylogenetic diversity of endophyte assemblages associated with the tropical seagrass *Enhalus acoroides* in Thailand. Fungal Divers. 2010;42:27–45.

Sakayaroj J, Supaphon O, Jones EBG, Phongpaichit S. Diversity of higher marine fungi at Hat Khanom-Mu Ko Thale Tai National Park, Southern Thailand. Songklanakarin J Sci Technol. 2011;33:15–22.

Sánchez Márquez S, Bills GF, Zabalgogeazcoa I. Diversity and structure of the fungal endophytic assemblages from two sympatric coastal grasses. Fungal Divers. 2008;33:87–100.

Schulz B, Boyle C. The endophyte continuum. Mycol Res. 2005;109:661–686.

Schulz B, Draeger S, Edison dela Cruz T, Rheinheimer J, Siems K, Loesgen S, Bitzer J, Schloerke O, Zeeck A, Kock I, Hussain H, Dai J, Krohn K. Screening strategies for obtaining novel, biologically active, fungal secondary metabolites from marine habitats. Bot Mar. 2008;51:219–234.

Saikkonen K, Faeth SH, Helander M, Sullivan TJ. Fungal endophytes: a continuum of interactions with host plants. Ann Rev Ecol Syst. 1998;29:319–343.

Santos JM, Phillips AJL. Resolving the complex of *Diaporthe* (*Phomopsis*) species occurring on *Foeniculum vulgare* in Portugal. Fungal Divers. 2009;34:111–125.

Sieber TN. Endophytic fungi in forest trees: are they mutualists? Fung Biol Rev. 2007;21:75–89.

Stone JK, Polishook JD, White Jr JF. Endophytic fungi. In: Mueller GM, Bills GF, Foster MS, eds. Biodiversity of fungi inventory and monitoring method. Amsterdam: Elsevier Academic Press, 2004:241–270.

Strobel GA. Microbial gifts from rain forests. Can J Plant Pathol. 2002;24:14–20.

Strobel GA, Dirkse E, Sears J, Markworth C. Volatile antimicrobials from *Muscodor albus*, a novel endophytic fungus. Microbiol. 2001;147:2943–2950.

Suryanarayanan TS, Kumaresan V. Endophytic fungi of some halophytes from an estuarine mangrove forest. Mycol Res. 2000;104:1465–1467.

Suryanarayanan TS, Kumaresan V, Johnson JA. Foliar fungal endophytes from two species of the mangrove *Rhizophora*. Can J Microbiol. 1998;44:1003–1006.

Suryanarayanan TS, Venkatachalam A, Thirunavukkarasu N, Ravishankar JP, Doble M, Geetha V. Internal mycobiota of marine macroalgae from the Tamilnadu coast: distribution, diversity and biotechnological potential. Bot Mar. 2010;53:457–468.

Suwannarach N, Bussaban B, Hyde KD, Lumyong S. *Muscodor cinnamomi*, a new endophytic species from *Cinnamomum bejolghota*. Mycotaxon 2010;114:15–23.

Tao G, Liu ZY, Hyde KD, Lui XZ, Yu ZN. Whole rDNA analysis reveals novel and endophytic fungi in *Bletilla ochracea* (Orchidaceae). Fungal Divers. 2008;33:101–122.

Toledo-Hernández C, Bones-González A, Ortiz-Vázquez OE, Sabat AM, Bayman P. Fungi in the sea fan *Gorgonia ventalina*: diversity and sampling strategies. Coral Reefs 2007;26:725–730.

Tomlinson PB. The botany of mangroves. Cambridge, UK: Cambridge Univ Press, 1986.

Trisuwan K, Rukachaisirikul V, Sukpondma Y, Preedanon S, Phongpaichit S, Rungjindamai N, Sakayaroj J. Epoxydons and a pyrone from the marine-derived fungus *Nigrospora* sp. PSU-F5. J Nat Prod. 2008;71:1323–1326.

Trisuwan K, Rukachaisirikul V, Sukpondma Y, Preedanon S, Phongpaichit S, Sakayaroj J. Pyrone derivatives from the marine-derived fungus *Nigrospora* sp. PSU-F18. Phytochem. 2009;70: 554–557.

Trisuwan K, Khamthong N, Rukachaisirikul V, Phongpaichit S, Preedanon S, Sakayaroj J. Anthraquinone, Cyclopentanone, and Naphthoquinone derivatives from the sea fan-derived fungi *Fusarium* spp. PSU-F14 and PSU-F135. J Nat Prod. 2010a;73:1507–1511.

Trisuwan K, Rukachaisirikul V, Sukpondma Y, Preedanon S, Phongpaichit S, Sakayaroj J. Furo [3,2-h] isochroman, furo [3,2-h] isoquinoline, isochroman, phenol, pyranone, and pyrone derivatives from the sea fan-derived fungus *Penicillium* sp. PSU-F40. Tetrahedron 2010b;66: 4484–4489.

Trisuwan K, Rukachaisirikul V, Kaewpet M, Phongpaichit S, Hutadilok-Towatana N, Preedanon S, Sakayaroj J. Sesquiterpene and Xanthone derivatives from the sea fan-derived fungus *Aspergillus sydowii* PSU-F154. J Nat Prod. 2011a;74:1663–1667.

Trisuwan K, Rukachaisirikul V, Preedanon S, Phongpaichit S, Sakayaroj J. Modiolide and Pyrone derivatives from the sea fan-derived fungus *Curvularia* sp. PSU-F22. Arch Pharm Res. 2011b;34:709–714.

van der Aa HA. Studies in *Phyllosticta*. Stud Mycol. 1973;5:1–110.

Wang G, Li Q, Zhu P. Phylogenetic diversity of culturable fungi associated with the Hawaiian sponges *Suberites zeteki* and *Gelliodes fibrosa*. Anton Leeuw Int J. 2008;93:163–174.

Wei JG, Xu T, Guo LD, Liu AR, Zhang Y, Pan XH. Endophytic *Pestalotiopsis* species associated with plants of Podocarpaceae, Theaceae and Taxaceae in southern China. Fungal Divers. 2007;24:55–74.

Wilson WL. Isolation of endophytes from seagrasses from Bermuda. The University of New Brunswick, Canada, MSc thesis, 1998.

Xing X, Guo S. Fungal endophyte communities in four Rhizophoraceae mangrove species on the south coast of China. Ecol Res. 2011;26:403–409.

Xing XK, Chen J, Xu MJ, Lin WH, Guo SX. Fungal endophytes associated with *Sonneratia* (Sonneratiaceae) mangrove plants on the south coast of China. Forest Pathol. 2010;41: 334–340.

Yarden O, Ainsworth TD, Roff G, Leggat W, Fine M, Hoegh-Guldberg O. Increased prevalence of ubiquitous ascomycetes in an acropoid coral (*Acropora formosa*) exhibiting symptoms of brown band syndrome and skeletal eroding band. Appl Environ Microbiol. 2007;73:2755–2757.

Yuan L, Zhao PJ, Wang T. New chemical constituents from endophyte *Phomopsis* species Lz42 cultivated on *Maytenus hookeri*. Chem J Chinese Univers. 2008;30:78–81.

Zalar P, de Hoog GS, Schroers H-J, Crous PW, Groenewalf JZ, Gunde-Cimernam N. Phylogeny and ecology of the ubiquitous saprobe *Cladosporium sphaerospermum*, with descriptions of seven new species from hypersaline environments. Stud Mycol. 2007;58:157–183.

Zhao K, Ping W, Li Q, Hao S, Zhao L, Gao T, Zhou D. *Aspergillus niger* var. *taxi*, a new species variant of taxol-producing fungus isolated from *Taxus cuspidate* in China. J Appl Microbiol. 2009;107:1202–1207.

Zhang JX, Xu T, Ge QX. Notes on *Pestalotiopsis* from Southern China. Mycotaxon 2003;85: 91–92.

Zhu P, Ge Q, Xu T. The perfect stage of *Pestalosphaeria* from China. Mycotaxon 1991;50: 129–140.

17 Fungi from marine algae

E.B. Gareth Jones, Ka-Lai Pang and Susan J. Stanley

Some of the earliest records of marine fungi were of species growing on marine algae, e.g. *Leptosphaeria chondri* (Rostrup 1889), *Zignoella calospora* (Patouillrad 1897), and *Mycosphaerella ascophylli* (Cotton 1909), with many described from herbarium material (Cribb and Cribb 1960). Sutherland (1915, 1916a, 1916b) described numerous algicolous fungi from collections made in Scotland and the south coast of England at Lulworth Cove. Bugni and Ireland (2004) commented that fungi isolated or growing on algae were the second largest source of marine fungi. They include parasites, saprobes and endophytes of seaweeds and planktonic taxa, and most are ascomycetes (Zuccaro and Mitchell 2005). Algal genera that have been shown to support fungi include *Ascophyllum* (Sutherland 1915, Webber 1959, 1967), *Ballia* (Kohlmeyer 1967), *Chondrus* (Schatz 1984), *Dilsea* (Stanley 1992), *Fucus* (Zuccaro et al. 2003, 2004, 2008), *Laminaria* (Kohlmeyer 1968), and *Sargassum* (Kohlmeyer 1971), to name but a few.

Although much has been written on algicolous fungi (Kohlmeyer 1973; Kohlmeyer and Demoulin 1981; Schatz 1983, 1984; Kohlmeyer and Volkmann-Kohlmeyer 2003), it is largely taxonomic-orientated or concerned with the marine origin of marine fungi (Demoulin 1974; Kohlmeyer and Kohlmeyer 1979). Zuccaro and Mitchell (2005) list some 79 marine fungi growing on seaweeds as parasites or saprobes, with only a few new algicolous species described over the past two decades (Zuccaro et al. 2004; Janson et al. 2005; Mantle et al. 2006; Jones et al. 2009a, 2009b).

Recently there has been a resurgence of interest in fungi growing on marine algae. Loque et al. (2009) studied the filamentous fungi and yeasts associated with the marine algae *Adenocystis utricularis*, *Desmarestia anceps* and *Palmaria decipiens* from Antarctica. Seventy-five species were isolated (27 filamentous fungi and 48 yeasts) belonging to the genera *Geomyces*, *Antarctomyces*, *Oidiodendron*, *Penicillium*, *Phaeosphaeria*, *Aureobasidium*, *Cryptococcus*, *Leucosporidium*, *Metschnikowia* and *Rhodotorula*. Chytrids and the Chromistan (Straminipiles) oomycetes parasitic on algae have also attracted interest (Gachon et al. 2006; Sekimoto et al. 2008a, 2008b; Strittmatter et al. 2009; Gachon et al. 2010), see Chapters 7 and 9–13). Many of these studies have been concerned with commercially-important algae, e.g. *Porphyra* in the production of nori, and the resultant economic losses (Gachon et al. 2010). Filamentous marine algae, e.g. *Pylaiella*, have also been examined for marine oomycetes (Sekimoto et al. 2008a).

A few studies have followed the formation of ascomata in marine algae, with Schatz (1983, 1984) reporting on *Phycomelaina laminariae* and *Lautitia danica*. There is no modern study, however, on the detailed structure of algicolous ascomycetes.

Jones (2011) has drawn attention to the large number of algae that have yet to be explored for the occurrence of fungi. Not only are algae very numerous in marine habitats (there are 9,200–12,500 described seaweeds) but they also cover vast areas of the sea bottom, e.g. ~30% of the bottom surface in the Maritime Antarctica are algal beds, yielding an estimated 74,000 tons of wet biomass (Nedzarek and Rakusa-Suszczewski 2004). In Japan and Korea, 655,000 and 777,090 tons of wet weight of seaweed are harvested, respectively. This again indicates the potential source for marine fungi (Ohno and Largo 1998; Sohn 1998). An estimate of the standing crop of kelp bed biomass for British Columbia was 651,697 wet tons, of which 130,340 wet tons was harvested for various products (Lindstrom 1998). These extensive standing seaweed crops are in urgent need of more intensive surveys for marine fungi.

Fungal taxonomic groups on seaweeds

Algicolous fungi are an ecological group and belong to a wide range of taxonomical groups (Table 17.1). Currently some 79 fungi have been listed as growing on algae, but no one taxonomic unit is more prevalent. The genera with the most marine species are the lichens *Verrucaria* (24) and *Collemopsidium* (six) and genera growing parasitically on various seaweeds: *Pontogeneia* (eight), *Chadefaudia* (six), *Haloguignardia* (five) and *Spathulospora* (five). Common hosts are the larger Phaeophyta (e.g. *Fucus, Laminaria*, and *Sargassum* species). The newly-described lichen genera *Hydropunctaria* (with two species) and *Wahlenbergiella* (two species) were based on species previously referred to *Verrucaria*.

Marine-derived fungi from seaweeds

Fungi on algae constitute the second largest group of marine fungi, with most being isolated from their hosts (Bugni and Ireland 2004). These have been isolated in the continued search for new antimicrobial compounds (Jensen and Fenical 2002), and most are asexual fungi (Table 17.2). The greater number of species thus isolated remain unidentified or are merely referred to a genus. Whether all of the fungi isolated are truly marine species remains to be resolved (Jones 2011).

Fungal associations

Fungi form various associations with algae: endophytes, parasites, primitive marine lichens, and mycophycobioses (Kohlmeyer and Kohlmeyer 1979). However, demarcation can be somewhat arbitrary: generally *Haloguignardia* species are considered to be parasites (Kohlmeyer and Kohlmeyer 1979); while Harvey and Goff (2010) referred to the association as an endophytic one. Few studies have examined the relationship between a marine fungus and its host. Stanley (1991) studied the relationship between three marine fungi: the basidiomycetes *Mycaureola dilsea* and *Dilsea carnosa*; the ascomycetes *Mycophycias ascophylli* and *Ascophyllum nodosum* and *Pelvetia canalicaulata*; and *Lautitia danica* and *Chondrus crispus*. *Mycophycias ascophylli* hyphae did not penetrate the cells of *A. nodosum* (Jones et al. 2008b), but *M. dilsea* and *L. danica* did (Figures 17.1–17.10). The hyphae of many species can be recovered from all parts of the algal host (*M. ascophylli*, Webber 1967; *Mycophycias ascophylli*, Jones et al. 2008b), while *Spathulospora* sp. may invade only a single cell (Figures 17.11–17.14). Sanders et al. (2004, 2005) have also examined the relationship between photobiont and mycobiont in the lichen *Verrucaria tavaresiae* at the light microscope and ultrastructure levels.

Table 17.1 Parasitic, saprophytic and lichen fungi on algal hosts.

Phylum	Order	Family	Genus (No. of species)	Reference
Basidiomycota	*Physalacria* clade	Physalacriaceae	*Mycaureola* (1)	Porter and Farnham (1986)
Ascomycota	Capnodiales	Mycosphaerellaceae	*Pharcidia* (3)	Kohlmeyer (1973)
	Dothideales *incertae sedis*	Botryosphaeriaceae	*Thalassoascus* (3)	Kohlmeyer (1981)
	Pleosporales	Lophiostomataceae	*Massarina* (1)	Kohlmeyer and Kohlmeyer (1979)
		Phaeosphaeriaceae	*Lautitia* (1)	Schatz (1984)
		Pleosporaceae	*Pleospora* (2)	Sutherland (1915)
	Pleosporales *incertae sedis*	*incertae sedis*	*Didymella* (3)	Feldmann (1958)
	Pyrenulales	Pyrenulaceae	*Xenus* (1)	Kohlmeyer and Volkmann-Kohlmeyer (1992)
		Xanthopyreniaceae	*Collemopsidium* (6)	Grube and Ryan (2002)
	Capnodiales		*Mycophycias* (2)	Toxopeus et al. (2011)
	Verrucariales	Verrucariaceae	*Verrucaria* (20)	McCarthy (2008)
			Hydropunctaria (2)	McCarthy (2008)
			Mastodia (1)	McCarthy (2008)
			Wahlenbergiella (2)	McCarthy (2008)
	Helotiales	Dermateaceae	*Laetinaevia* (1)	Kirk and Spooner (1984)
	Lichinales	Lichinaceae	*Lichina* (2)	Jones et al. (2009a)
	Arthoniales	Roccellaceae	*Halographis* (1)	Kohlmeyer and Volkmann-Kohlmeyer (1988)
	Hypocreales	Bionectriaceae	*Pronectria* (1)	Lowen (1990)
	Microascales	Halosphaeriaceae	*Chadefaudia* (6)	Kohlmeyer (1973)
			Corollospora (4)	Jones et al. (2009a)
			Trailia (1)	Sutherland (1915)
	Lulworthiales	Lulworthiaceae	*Lulworthia* (3)	Kohlmeyer and Volkmann-Kohlmeyer (1991)
			Lindra (1)	Orpurt et al. (1964)
			Kohlmeyeriella (1)	Jones et al. (1983)

Table continued on next page.

Table 17.1 (Continued)

Phylum	Order	Family	Genus (No. of species)	Reference
	Koralionastetales	Koralionastetaceae	*Haloguignardia* (5)	Inderbitzin et al. (2004)
			Spathulospora (5)	Kohlmeyer (1973)
	Phyllachorales	Phyllachoraceae	*Pontogenia* (8)	Kohlmeyer (1975)
	Phyllachorales *incertae sedis*	*incertae sedis*	*Polystigma* (1)	Kohlmeyer and Demoulin (1981)
			Phycomelaina (1)	Kohlmeyer (1968)
	Untunicate *incertae sedis*	Hispidicarpomycetaceae	*Hispidicarpomyces* (1)	Nakagiri (1993)
		Spathulosporaceae	*Retrostium* (1)	Nakagiri and Ito (1997)
			Turgidosculum (1)	Kohlmeyer and Kohlmeyer (1972b)
		Mastodiaceae	*Mastodia* (1)	Kohlmeyer and Kohlmeyer (1979)
			Orcadia (1)	Sutherland (1915)
		incertae sedis	*Crinigera* (1)	Schmidt (1969)
Anamorphic fungi	NA	NA	*Acremonium* (2)	Zuccaro et al. (2004)
			Amorosia (1)	Mantle et al. (2006)
			Blodgettia (1)	Kohlmeyer and Kohlmeyer (1979)
			Cladosporium (1)	Kohlmeyer and Kohlmeyer (1979)
			Dendryphiella (2)	Jones et al. (2008a)
			Halosigmoidea (2)	Jones et al. (2009b)
			Penicillium (1)	Janson et al. (2005)
			Scopulariopsis (1)	Tubaki (1973)
			Septoria (1)	Malnik and Petrov (1966)
			Stagonospora (1)	Kohlmeyer (1973)
			Stemphylium (1)	Kohlmeyer and Kohlmeyer (1979)
			Gloeosporidina (1)	Jones et al. (2009a)
			Phoma (1)	Kohlmeyer and Kohlmeyer (1979)
			Varicosporina (1)	Meyers and Kohlmeyer (1965)

Table 17.2 Selected list of marine-derived fungi from algae.

Species	Algal host/substrate	Reference
Amorosia littoralis	NA	Mantle et al. (2006)
Antarctomyces psychrotrophilus	*Adenocystis utricularis*	Loque et al. (2009)
Aspergillus niger	*Ulva lactuca*	Suryanarayanan et al. (2010)
Aspergillus vesicular	*Penicillus capitatus*	Belofsky et al. (1998), Suryanarayanan et al. (2010)
Aureobasidium pullulans	*Caulerpa racemosa*	Suryanarayanan et al. (2010)
Cryptococcus carnescens	*Palmaria decipiens*	Loque et al. (2009)
Curvularia lunata	Ulvaceae	Suryanarayanan et al. (2010)
Curvularia tuberculata	Ulvaceae	Suryanarayanan et al. (2010)
Drechslera papendorfii	Halymeniaceae	Suryanarayanan et al. (2010)
Emericella nidulans	*Dictyota dichotoma*	Suryanarayanan et al. (2010)
Geomyces pannorum	*Adenocystis utricularis*	Loque et al. (2009)
	Desmarestia anceps	Loque et al. (2009)
Gliocladium roseum	NA	Omura et al. (1999)
Phaeosphaeria herpotrichoides	*Adenocystis utricularis*	Loque et al. (2009)
Penicillium dravuni	NA	Janson et al. (2005)
Penicillium waksmanii	NA	Amagata et al. (1998)
Metschnikowia australis	*Adenocystis utricularis*	Loque et al. (2009)
	Desmarestia anceps	Loque et al. (2009)
	Palmaria decipiens	Loque et al. (2009)
Rhodotorula mucilaginosa	*Adenocystis utricularis*	Loque et al. (2009)

NA: algal source not identified.

Parasites

Pathogenic fungi probably constitute one-third of those documented from algae, see Table 17.1 (Kohlmeyer and Kohlmeyer 1979); see also Chapters 9–13 for fungal-like organisms. Symptoms of fungal attack extend from no significant change in the host morphology (*Chadefaudia balliae* in *Ballia callitricha*), to those causing marked discoloration of the thalli (*L. danica* on *C. crispus*, Figures 17.7–17.8), to those causing malformation galls of the host (*Haloguignardia* spp. on the brown algae *Cystoseira, Halidrys* and *Sargassum*; Kohlmeyer and Volkmann-Kohlmeyer 2003).

Lichens

Mycobionts of the lichens include *Collemopsidium, Lichina* and *Verrucaria* species, with most being found in intertidal temperate waters. In *Verrucaria*, the photobionts include the green algae *Coccobotrys, Desmococcus, Dilabifilum* and *Myremecia* and the xanthophyte *Heterococcus*. *Verrucaria tavaresiae* is unique in having a brown alga, *Petroderma maculiforme*, as its photobiont (Moe 1997).

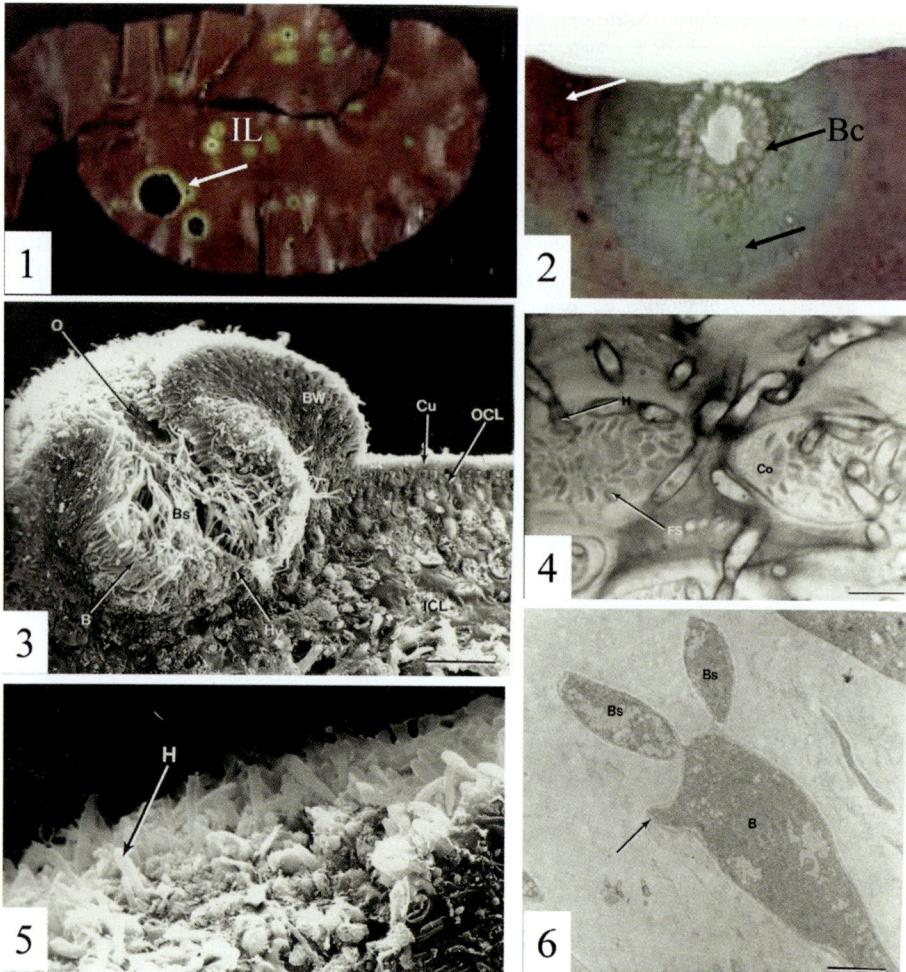

Figures 17.1–17.6 *Mycaureola dilsea*. Figure 17.1. Thallus of *Dilsea carnosa* infected by *M. dilsea*, immature lesions (IL) and a mature lesion (arrowed). Figure 17.2. Higher magnification of hyaline basidiomes (Bc), clearly-defined boundary between healthy (red) and infected (green) surrounded by green areas infected by the fungus. Figure 17.3. Scanning electron micrograph of basidiome with ostiole (arrowed O), basidia (B), basidiospores (Bs) and basal hymenium (Hy) partly immersed in algal tissue. Cu = cuticle, OCL = outer cortical layer, and ICL = inner cortical layer. Figure 17.4. Longitudinal section of infected algal tissue, penetrating hypha into cortical cell (H), starch grains (FS), cortical cell (Co). Figure 17.5. SEM surface of basidiome with peridial hairs (H). Figure 17.6. Transmission electron micrograph of a basidium (B), basidiospores (Bs) and a developing basidiospore (arrow). Scale bars: 3 = 50 µm; 4, 5 = 10 µm; 6 = 2 µm.

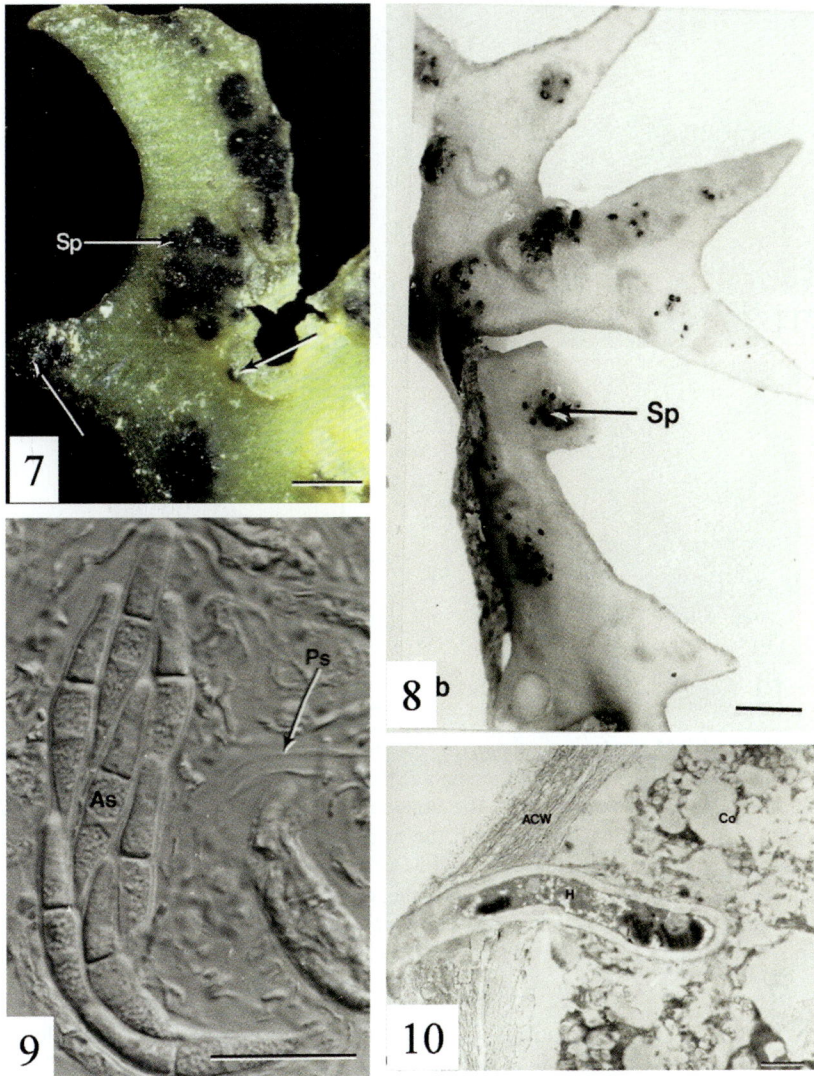

Figures 17.7–17.10 *Lautitia danica*. Figures 17.7–17.8. *Chondrus crispus* thallus infected with *L. danica*, tetrasporangial pustules with black ascomata (Sp). Arrowed areas grazed. Figure 17.9. Ascus with bicelled hyaline ascospores (As) and pseudoparaphyses (Ps). Figure 17.10. Penetration of algal cortical cell (Co) by hypha (H), no intact Floridean starch grains remain after infection. ACW = algal cell wall. Scale bars: 7–8 = 2 mm; 9 = 20 µm; 10 = 1 µm.

Mycophycobioses

Mycophycobioses are obligately symbiotic associations between a systemic marine fungus and a marine macroalga in which the habit of the alga dominates (Kohlmeyer and Kohlmeyer 1979). A classic example is *Mycophycias ascophylli* (= *Mycosphaerella ascophylli*) on the marine alga *Ascophyllum nodosum* and *Pelvetia canaliculata* (Figures 17.11–17.13). (Sutherland 1915; Kohlmeyer and Volkmann-Kohlmeyer 1998;

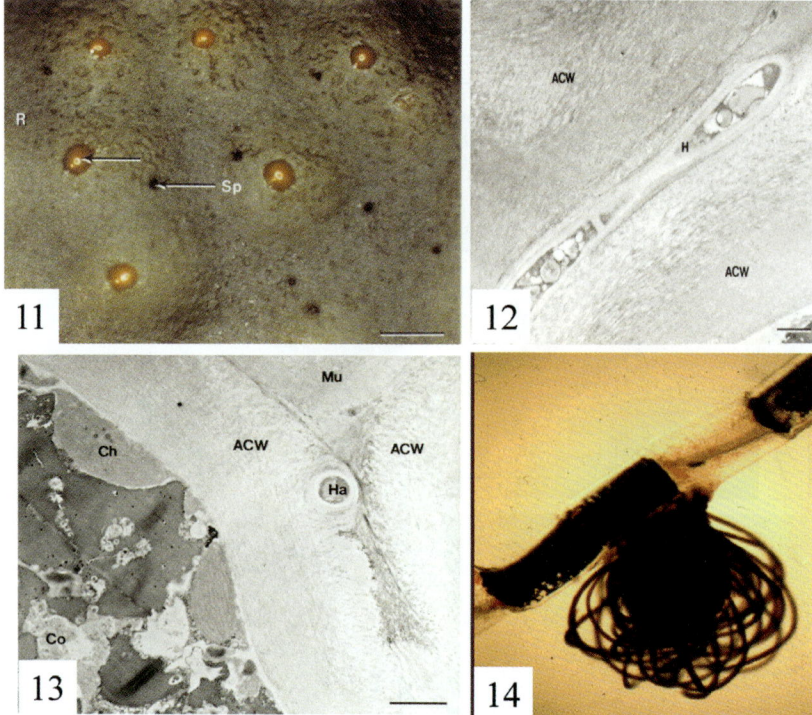

Figures 17.11–17.14 *Mycophycias ascophylli*. Figure 17.11. Surface view of algal receptacles of *Ascophyllum nodosum* (R), with black ascomata (Sp), close to conceptacle openings. Mucilaginous mass containing algal gametes around the conceptacle opening (arrow). Figure 17.12. TEM of intercellular hypha (H) associated with the adjacent algal cell walls (ACW). Hypha does not penetrate the algal cells. Figure 17.13. TEM of transverse section of cortical host cell (Co) with intra-membranous haustorium (Ha) in the outer cell wall (ACW). Ch = chloroplast. Figure 17.14. *Spathulospora* sp. Ascomata with mycelium infecting a single cell of the algal host (*Ballia* sp.). Scale bars: 11 = 2 mm; 12 = 1 µm; 13 = 2 µm.

Toxopeus et al. 2011). The relationship between mycobiont and photobiont leads to the question of whether *Ascophyllum* is a lichen (Kohlmeyer and Kohlmeyer 1972a; Garbary 2009). This relationship has been the subject of much research (Garbary and Gautam 1989; Garbary and MacDonald 1995), with Garbary and London (1995) suggesting that it confers desiccation resistance on *A. nodosum*.

Kingham and Evans (1986) have also investigated the relationship between *M. ascophylli* and *P. canaliculata*, including morphological studies of mycelium distribution in the host thallus and physiological studies. They reported a network of mycelium from just a few cells from the apex of the alga to the attaching holdfasts. No haustoria were reported, but intra-membranous structures (? haustoria) occurred within the algal cell walls. Physiological observations indicated that there was fungal dependence on the alga, but there was no evidence of any benefits to the alga from the presence of the fungus.

Endophytes

Marine algae are known to harbor endophytes (Jones et al. 2008b), but only two studies have surveyed algae for endophytic/endosymbiotic fungi in any detail: Zuccaro et al. (2003) and Suryanarayanan et al. (2010).

Zuccaro et al. (2003) examined the fungal–algal interactions of the brown alga *Fucus serratus* by isolating fungi from live (washed thalli) and dead plants, sterilized live plants and direct DNA extraction from the alga. Thirty-six and 67 isolates were made from dead thalli and live washed thalli, respectively, and most species were asexual species of the genera *Cladosporium, Fusarium* and *Penicillium*, with 20 sterile strains. These isolates represent only 13 endophytes, with one ascomycete (*Chaetomium* sp.), eight anamorphic fungi and four sterile cultures. While many species were isolated onto media, few were detected at the molecular level. The fungal sequences recovered belonged to the Halosphaeriaceae, Hypocreales and Lulworthiales. Environmental sequences aligned with *Corollospora angusta* (two sequences), *Sigmoidea* (now *Halosigmoidea*) *marina* (one), *Hypocrea lutea* (one), *Engyodontium aranearum* (one), *Emericellopsis/Acremonium* clade (one) and Lulworthiales clade (four sequences).

In a more extensive study on the Tamilnadu coast of India, Suryanarayanan et al. (2010) examined eleven, six and eight brown, green and red algae, respectively, for fungal endophytes. Each algal species yielded endophytes and a total of 72 fungi were isolated, while the density varied with the algal host. Brown algae supported the highest diversity, with 25 species from a *Turbinaria* sp. The majority of the fungi isolated were asexual species, with five sterile and nine unidentified, of which six were yeasts. Most species were typical terrestrial fungi, with the genus *Aspergillus* dominant, and 37 species were present at a low frequency and confined to a few algae.

Suryanarayanan et al. (2010) also screened 54 endophyte isolates for bioactive compounds, with autobiograms showing that 29 fungi produced metabolites that showed antialgal or antifungal activity or both. Some were also shown to produce antioxidants and insecticidal metabolites.

Further studies are warranted to screen seaweeds for endophytes, especially the larger seaweeds: tropical *Sargassum* spp. and the extensive kelp beds of temperate waters. Molecular techniques must also be used to identify the isolated endophytes to determine whether they are truly marine species. Most marine fungal endophytes are asexual and few novel taxa have been documented to date (Janson et al. 2005).

Molecular studies of algicolous fungi

Studies of the phylogenetic relationships of algicolous fungi have been few until recently, when those forming lichen associations were investigated (Lumbsch and Lindemuth 2001; Lumbsch et al. 2005; Gueidan et al. 2007, 2009; Harvey and Goff 2010; Pérez-Ortega et al. 2010). These studies fall into two categories:

(i) use of sequence data to resolve phylogenetic relationships; and
(ii) biodiversity studies of unculturable fungi and those addressing algal–fungal symbiosis.

Taxonomic studies

Mycaureola dilsea was originally classified as an ascomycete (Maire and Chemin 1922), but following a scanning and transmission electron microscopy (SEM and TEM) study, Porter and Farnham (1986) showed that hyphae had dolipore septa with perforate parenthosomes, which placed it in the Basidiomycota. This is an unusual organism as it is the only marine basidiomycete parasitic on an alga and occurs exclusively on the red alga *Dilsea carnosa*. Sporulation of the basidiomycete is restricted to the colder months of the year in temperate climates, when the temperature falls below 15°C (Stanley 1991). Binder et al. (2006) undertook a molecular study to determine its taxonomic affinities, and showed that it formed a second independent lineage of marine fungi in the eugarics clade. Furthermore, it is not related to other cyphelloid marine basidiomycetes (e.g. *Halocyphina villosa*, see Chapter 4) (Binder et al. 2001; Hibbett and Binder 2001). Maximum likelihood analysis of ITS data places *Gloiocephala phormiorum* as a sister group of *M. dilsea*, a genus that contains a freshwater species (*G. aquatica*) growing on submerged culms in a lake (Desjardin et al. 1995). Binder et al. (2006) conclude that the "unique ecology of mangroves may have promoted multiple transitions from terrestrial to marine habitats".

Taxonomic studies of marine lichens are detailed in Chapter 2 on the phylogeny of marine Dothideomycetes. Studies have focused on the multigene analysis of the phylogeny of species in the Verrucariales (Gueidan et al. 2007, 2009; Pérez-Ortega et al. 2010) and the reclassification of *Mycophycias ascophylli* in the Capnodiales (Toxopeus et al. 2011).

Biodiversity studies

Most studies of algicolous fungi have been morphological, involving descriptions of new taxa or their relationships with their phycobiont (Kohlmeyer 1973; Kohlmeyer and Demoulin 1981; Kohlmeyer and Volkmann-Kohlmeyer 2003). Few molecular studies have been undertaken to detect fungi associated with algae (Zuccaro et al. 2003; Loque et al. 2009; Harvey and Goff 2010). Zuccaro et al. (2003) used molecular techniques to survey ascomycetes associated with the *Fucus serratus* to "assess fungal–algal interactions directly, with and without fungal cultivation, so that key, or novel fungi might be identified". Environmental sequences included four main ascomycete groups: Halosphaeriaceae, Hypocreales, Lulworthiales and Pleosporales. However, few species were detected molecularly, indicating low colonization of fronds, with no novel taxa.

In a subsequent study, Zuccaro et al. (2008) examined the filamentous fungi associated with healthy and decaying *F. serratus* thalli using molecular techniques, such as, 28S rRNA gene polymerase chain reaction-denaturing gradient gel electrophoresis (DGGE) and real-time polymerase chain reaction analysis. The genera *Lindra*, *Lulworthia*, *Engyodontium*, *Sigmoidea/Corollospora* complex, and *Emericellopsis/Acremonium*-like ribotypes were recovered from healthy *F. serratus*. *Sigmoidea marina* (= *Halosigmoidea marina*) was the most common species reported, accounting for 60% of the DGGE bands detected. They failed to recover *S. marina* from decaying material and concluded that it cannot survive in the environment without the protection of the alga. This conflicts with the results of the study by Haythorn et al. (1980) that isolated the fungus from decaying brown algae in the UK. This study also revealed that many sequences were of members of the Pleosporales (*Phaeosphaeria* and *Didymella*), which did not match those of known taxa, suggesting the existence of new fungal lineages. Zuccaro et al. (2008) also concluded that there was a change in the mycoflora between healthy and decaying algal material. The study also emphasized how little is known about marine fungi inhabiting marine algae, which corresponds with the view expressed by Jones (2011) on the diversity of marine fungi.

Loque et al. (2009) used sequence data to identify fungi associated with the marine algae *Adenocystis utricularis, Desmarestia anceps* and *Palmaria decipiens*. Twenty-seven filamentous fungi and 48 yeasts were isolated and identified, some based on sequence data. The most frequently isolated were *Geomyces pannorum* and the yeast *Metschnikowia australis*, neither of which are obligate marine fungi, although both have been isolated from marine habitats (Fell and Hunter 1968; Zuccaro et al. 2008). Most of the fungi isolated were anamorphic taxa and yeasts, and differed significantly from those recovered by Zuccaro et al. (2008) from *F. serratus*. As molecular techniques improve, especially the development of fungal-specific primers, a better understanding of the endophytes and unculturable fungi in marine algae should result in a better comprehension of their occurrence.

The genus *Haloguignardia* comprises five species, all occurring on the large brown algae *Cystoseira, Halidrys* and *Sargassum*, often forming galls of the thalli. The genus is widely distributed usually in cooler oceanic waters (Jones et al. 2009a). Initially classified in the Phyllachoraceae, Phyllachorales, it has been shown to cluster within the *Lulworthia* complex in the family Lulworthiaceae (Inderbitzin et al. 2004). Harvey and Goff (2010) investigated the genetic covariation of *Haloguignardia irritans* on the west coast of North America, with its two brown algal hosts *Cystoseira* (*C. osmundaceae, C. neglecta*) and *Halidrys dioica*. Based on neighbor-joining analysis of the fungal ITS rDNA, four groups were identified:

(i) eight strains on *C. osmundaceae* from Santa Rosa Island;
(ii) five strains from La Jolla and Still Water Cove, on *C. osmundaceae*;
(iii) five strains from Cabrillo National Monument, on *H. dioica*; and
(iv) 17 strains from central northern California and two strains form La Jolla, on *C. osmundaceae*.

It was concluded that the degree of ITS sequence variation within *H. irritans* was up to 6%, suggesting that more than one cryptic, phylogenetic species exists in the population sampled (Harvey and Goff 2010).

Ecological and physiological studies of algicolous fungi

Detailed studies on the seasonal occurrence of marine pathogenic fungi on seaweeds are few (Webber 1967; Stanley 1992) and are a topic worthy of greater research. Stanley (1992) reported that *Mycaureola dilseae* sporulated on *Dilsea carnosa* during the months of September and October, while ascomata of *Lautitia danica* could be found on *Chondrus crispus* from September through to January. *Mycophycias ascophylli* occurred on *Ascophyllum nodosum* from December to May, which corresponds to the observations made by Webber (1959, 1967) and Kingham (1976). Cotton (1908) and Sutherland (1915) suggested that the sporulation of *M. ascophylli* coincided with the development and release of *A. nodosum* oospores, although the data recorded by Webber (1959) did not support this hypothesis.

Closing remarks

Although about one-third of all described marine fungi are algicolous, they have received little attention in recent years – with the exception of the fungal-like organisms where there has been a resurgence of interest in their study (see Chapters 9–13). Few algae have been studied for their endophytes, especially the existence of unculturable species and novel taxa. Equally, much needs to be done to resolve the phylogenetic relationship of key species and the discovery of new lineages of marine fungi, e.g. *Lautita danica, Phycomelaina laminariae, Hispidicarpomyces*, and *Retrostium*. Seaweeds in storage, prior to the extraction of phycocolloids, are often subject to deterioration by anamorphic fungi, such as *Penicillium* and *Trichoderma* (A.T. Critchley and E.B.G. Jones, unpublished data). This requires further research. Finally, more information is required on pathogenic marine fungi, their seasonal occurrence and distribution, and how submerged algal hosts are infected by basidiospores and ascospores.

Acknowledgements

EBGJ thanks Dr Aisyah Alias for logistical support and the Institute of Ocean and Earth Sciences, University of Malaya for financial support.

References

Amagata T, Usami Y, Minoura K, Ito T, Numata A. Cytotoxic substances produced by a fungal strain from a sponge: physico-chemical properties and structures. J Antib. 1998;51:33–40.

Belofsky GN, Jensen PR, Renner MK, Fenical W. New cytotoxic sesquiterpenoid nitrobenzoyl esters from a marine isolate of the fungus *Aspergillus versicolor*. Tetrahedron 1998;54:1715–1724.

Binder M, Hibbett DS, Molitoris HP. Phylogenetic relationships of the marine gasteromycete *Nia vibrissa*. Mycologia 2001;93:679–688.

Binder M, Hibbett DS, Wang Z, Farnham WF. Evolutionary relationships of *Mycaureola dilseae* (Agaricales), a basidiomycetes pathogen of a subtidal Rhodophyte. Am J Bot. 2006; 93:547–556.

Bugni TS, Ireland CM. Marine-derived fungi: a chemically and biologically diverse group of microorganisms. Nat Prod Rep. 2004;21:143–163.

Cotton AD. Notes on marine Pyrenomycetes. Trans Br Mycol Soc. 1909;3:96.

Cotton AD. Notes on marine Pyrenomycetes. Ann Bot. 1908;2:399–400.

Cribb AB, Cribb JW. Some marine fungi on algae in European herbaria. Univ Queensl Pap Dep Bot. 1960;4:45–48.

Demoulin V. The origin of Ascomycetes and Basidiomycetes. The case for a red algal ancestry. Bot Rev. 1974;40:315–345.

Desjardin DS, Martinez-Peck L, Rachenberg M. An unusual psychrophilic aquatic agaric from Argentina. Mycologia 1995;87:547–550.

Feldmann G. Un nouvel Ascomycéte parasite d'une algue marine: *Chadefaudia marina*. Rev Gen Bot. 1958;66:140–152.

Fell JW, Hunter IL. Isolation of heterothallic yeast strains of *Metschnikowia kamienski* and their mating reaction with *Chlamydozyma wickerham*. Antonie Leeuwenhoek 1968;34:365–376.

Gachon MM, Küpper H, Küpper FC, Setlik I. Single-cell chlorophyll fluorescence kinetic microscopy of *Pylaiella littorlais* (Phaeophyceae) infected by *Chytridium polysiphoniae* (Chytridiomycota). Eur J Phycol. 2006;41:395–403.

Gachon MM, Sime-Ngando T, Strittmatter M, Chambouvet A, Kim GH. Algal diseases: spotlight on a black box. Trends Plant Sci. 2010;15:633–640.

Garbary D. Why is *Ascophyllum* not a lichen? Int Lichenol News 2009;41:34–35.

Garbary D, Gautam A. The *Ascophyllum/Polysiphonia/Mycosphaerella* symbiosis 1. Population ecology of *Mycosphaerella* from Nova Scotia. Bot Mar. 1989;32:181–186.

Garbary D, London J. The *Ascophyllum/Polysiphonia/Mycosphaerella* symbiosis V. Fungal infection protects *A. nodosum* from desiccation. Bot Mar. 1995;38:529–533.

Garbary D, MacDonald K. The *Ascophyllum/Polysiphoni/Mycosphaerella* symbiosis IV. Mutualism in the *Ascophyllum/Mycosphaerella* interaction. Bot Mar. 1995;38:221–225.

Grube M, Ryan BD. *Colemopsidium*. In: Nash TH, Ryan BD, Gries C, Bungarts F, eds. Lichen flora of the Greater Sonoran Desert Region. Temp: Lichyenc Unlimited, 2002:1162–1164.

Gueidan C, Roux C, Lutzoni F. Using a multigene phylogenetic analysis to assess generic delineation and character evolution in Verrucariaceae (Verrucariales, Ascomycota). Mycol Res. 2007;111:1145–1168.

Gueidan C, Savié S, Thüs H, Roux C, Keller C, Tibell L, Prieto M, Heiomarsson S, Breuss O, Orange A, Fröbeerg L, Wynns AA, Navarro-Rosinés P, Krzewicka B, Pykälä J, Grube M, Lutzoni F. Generic classification of the Verrucariaceae (Ascomycota) based on molecular and morphological evidence: recent progress and remaining challenges. Taxon 2009;58:184–208.

Harvey JBJ, Goff LJ. Genetic covariation of the marine fungal symbiont *Haloguignardia irritans* (Ascomycota, Pezizomycotina) with its algal hosts *Cystoseira* and *Halidrys* (Phaeophyceae, Fucales) along the west coast of North America. Fungal Biol. 2010;114:82–95.

Haythorn JM, Jones EBG, Harrison JL. Observations on marine algicolous fungi, including the hyphomycete *Sigmoidea marina* sp. nov. Trans Br Mycol Soc. 1980;74:615–6234.

Hibbett DS, Binder M. Evolution of marine mushrooms. Biol Bul. 2001;201:319–322.

Inderbitzin P, Lim SR, Volkmann-Kohlmeyer B, Kohlmeyer J. The phylogenetic position of *Spathulospora* based on DNA sequences from dried herbarium material. Mycol Res. 2004;108:737–748.

Janson JE, Bernan VS, Greenstein M, Bugni TS, Ireland CM. *Penicillium dravuni*, a new marine-derived species from an alga in Fiji. Mycologia 2005;97:444–453.

Jensen PR, Fenical W. Secondary metabolites from marine fungi. Fungal Divers Res Ser. 2002;7:293–315.

Jones EBG. Are there more marine fungi to be described? Bot Mar. 2011;54:391–402.

Jones EBG, Johnson RG, Moss ST. Taxonomic studies of the Halosphaeriaceae: *Corollospora* Werdmann. Bot J Linn Soc. 1983;87:193–212.

Jones EBG, Chatmala I, Klasuban A, Pang KL. Ribosomal DNA phylogeny of marine anamorphic fungi: *Cumulospora varia*, *Dendryphiella* species and *Orbimyces spectabilis*. Raffles Bull Zool Suppl. 2008a;19:11–18.

Jones EBG, Stanely SJ, Pinruan U. Marine endophytes: sources of new chemical natural products: a review. Bot Mar. 2008b;51:163–170.

Jones EBG, Sakayaroj J, Suetrong S, Somrithipol S, Pang KL. Classification of marine Ascomycota, anamorphic taxa and Basidiomycota. Fungal Divers. 2009a;35:1–187.

Jones EBG, Zuccaro A, Nakagiri A, Mitchell JL, Pang KL. Phylogenetic relationships of the genus *Sigmoidea* and a new genus *Halosigmoidea* gen. nov. Bot Mar. 2009b;52:349–359.

Kingham DL. Studies relating to the Fucacean endomycobiont *Mycosphaerella ascophylli*. University of Leeds, UK, PhD thesis, 1976.

Kingham DL, Evans LV. The *Pelvetia–Mycosphaerella* interrelationship. In: Moss ST, ed. The biology of marine fungi. Cambridge: Cambridge University Press, 1986:177–188.

Kirk PM, Spooner BM. An account of the fungi of Arran, Gigha and Kintyre. Kew Bull. 1984;38:503–597.

Kohlmeyer J. Intertidal and phycophilous fungi from Tenerife (Canary Islands). Trans Br Mycol Soc. 1967;50:137–147.

Kohlmeyer J. Revisions and descriptions of algicolous marine fungi. Phytopathol Z. 1968;63: 341–363.

Kohlmeyer J. Fungi from the Sargasso Sea. Mar Biol. 1971;8:344–350.

Kohlmeyer J. Fungi from marine algae. Bot Mar. 1973;16:201–215.

Kohlmeyer J. Revision of algicolous *Zignoella* spp. and description of *Pontogeneia* gen. nov. (Ascomycetes) Bot Jahrb. 1975;96:200–211.

Kohlmeyer J. Marine fungi from Easter Island and notes on *Thalassoascus*. Mycologia 1981;73: 833–843.

Kohlmeyer J, Demoulin V. Parasitic and symbiotic fungi on marine algae. Bot Mar. 1981;24:9–18.

Kohlmeyer J, Kohlmeyer E. Is *Ascophyllim nodsum* lichenised? Bot Mar. 1972a;15:109–113.

Kohlmeyer J, Kohlmeyer E. A new genus of marine Ascomycetes on *Ulva vexata* Setch. *et* Gard. Bot Jahrb. 1972b;92:429–432.

Kohlmeyer J, Kohlmeyer E. Marine mycology – the higher fungi. New York: Academic Press, 1979.

Kohlmyer J, Volkmann-Kohlmeyer B. *Halographis* (Opegraphales). A new endolithic lichenoid from corals and snails. Can J Bot. 1988;66:1138–1141.

Kohlmeyer J, Volkmann-Kohlmeyer B. Illustrated key to the filamentous marine fungi. Bot Mar. 1991;34:1–61.

Kohlmeyer J, Volkmann-Kohlmeyer B. Two Ascomycotina from coral reefs in the Caribbean and Australia. Crypto Bot. 1992;2:367–374.

Kohlmeyer J, Volkmann-Kohlmeyer B. *Mycophycias*, a new genus for the mycobionts of *Apophlaea, Ascophyllum* and *Pelvetia*. Syst Ascomycet. 1998;16:1–7.

Kohlmeyer J, Volkmann-Kohlmeyer B. Marine ascomycetes from algae and animal hosts. Bot Mar. 2003;46:285–306.

Lindstrom S. The seaweed resources of British Columbia, Canada. In: Critchley AT, Ohno M, eds. Seaweed resources of the world. Japan: Japan Inter Coop Agency, 1998:266–272.

Loque CP, Medeiros AO, Pellizzari FM, Oliveira EC, Rosa CA, Rosa LH. Fungal community associated with marine macroalgae from Antarctica. Polar Biol. 2009;33:641–648.

Lowen R. New combinations in *Pronectria*. Mycotaxon 1990;39:461–463.

Lumbsch HT, Lindemuth R. Major lineages of Dothideomycetes (Ascomycota) inferred from SSU and LSU rDNA sequences. Mycol Res. 2001;105:901–908.

Lumbsch HT, Schmitt I, Lindemuth R, Miller A, Mangold A, Fernandez F, Huhndorf S. Performance of four ribosomal DNA regions to infer higher-level phylogenetic relationships of inoperculate euascomycetes (Leotiomyceta). Mol Phylogenet Evol. 2005;34:512–524.

Maire R, Chemin E. Un nouveau Pyrenomycete marin. Compt rend hebd Séan l'Acad Sci (Paris). 1922;175:319–321.

Malnik VA, Petrov YUE. Novyi vid griba s morskoi buroi Vodorosli *Ascophyllum nodosum* (L.) Le Jolis. Nov Sist Niz Rast. 1966:211–212.

Mantle PG, Hawksworth DL, Pazoutova S, Collinson LM, Rassing BR. *Amorosia littoralis* gen. sp. nov., a new genus and species name for the scorpinone and caffeine-producing hyphomycetes from the littoral zone in The Bahamas. Mycol Res. 2006;110:371–1378.

McCarthy PM. A new species and new combination of Australian Verrucariaceae. Aust Lichenol. 2008;63:17–18.

Meyers SP, Kohlemeyer J. *Varicosporina ramulosa* gen. nov. sp. nov., an aquatic Hyphomycete from marine areas. Can J Bot. 1965;43:915–921.

Moe R. *Verrucaria tavaersiae* sp. nov., a marine lichen with a brown algal photobiont. Bull Cal Lichen Soc. 1997;4:7–11.

Nakagiri A. A new ascomycetes in Sapthulosporales *Hispidicarpomyces galauaricola* gen. *et* sp. nov. (Hispidicarpomyetceae fam. nov.), inhabiting a red alga *Galaxaura falcata*. Mycologia 1993;85:638–652.

Nakagiri A, Ito T. *Retrostium amphiroae* gen. *et* sp. nov. inhabiting a marine red alga, *Amphiroa zonata*. Mycologia 1997;89:484–493.

Nedzarek A, Rakusa-Suszczewski S. Decomposition of macro-algae and the release of nutrient in Admiralty Bay, King George Island, Antarctica. Polar Biosc. 2004;17:16–35.

Ohno M, Largo DB. The seaweed resources of Japan. In: Critchley CT, Ohno M, eds. Seaweed resources of the world. Japan: Japan Inter Coop Agency, 1998:1–14.

Omura S, Tomoda H, Tabata N, Ohyama Y, Abe T, Namikoshi M. Roselipins, novel fungal metabolites having a highly methylated fatty acid modified with a mannose and an arabitol. J Anti. 1999;52:586–859.

Orpurt PA, Meyeers SP, Boral LL, Simms J, Thalassiomycetes V. A new species of *Lindra* on turtle grass, *Thalassia testudinum* Köning. Bull Mar Sci Gulf Caribb. 1964;14:405–417.

Patouillard N. *Zignoella calospora* Patouillard, n. sp. J Bot Paris. 1897;11:242.

Pérez-Ortega S, De los Rios A, Crespo A, Sancho LG. Symbiotic lifestyle and phylogenetic relationships of the bionts of *Mastodia tessellata* (Ascomycota, *Incertae sedis*). Am J Bot. 2010;97:738–752.

Porter D, Farnham WF. *Mycaureola edulis*, a marine basidiomycete parasite of the red alga, *Dilsea carnosa*. Trans Br Mycol Soc. 1986;87:575–582.

Rostrup E. Mykologiske Meddeleser. Bot Tidsskr. 1890;17:234.

Sanders WB, Moe RL, Ascaso C. The intertidal marine lichen formed by the pyrenomycete fungus *Verrucaria tavaersiae* (Ascomycotina) and the brown alga *Petroderma maculiforme* (Phaeophyceae): thallus organisation and symbiont interaction. Am J Bot. 2004;90:511–522.

Sanders WB, Moe RL, Ascaso C. Ultrastructure study of the brown alga *Petroderma maculiforme* (Phaeophyceae) in the free-living state and in lichen symbiosis with the intertidal marine fungus *Verrucaria tavaersiae* (Ascomycotina). Eur J Phycol. 2005;40:353–361.

Schatz S. The developmental morphology and life history of *Phycomelaina laminariae*. Mycologia 1983;77:762–77.

Schatz S. The life history, developmental morphology, and taxonomy of *Lautitia danica* gen. nov., comb. nov. Can J Bot. 1984;62:28–32.

Schmidt I. *Corollospora intermedia*, nov. spec., *Carbosphaerella leptosphaerioides*, nov. spec., und *Crinigera maritima*, nov. spec., 3 neue marine Pilzarten von der ostseeküste. Nat Naturschutz Mecklenburg. 1969;7:5–14.

Sekimoto S, Beakes W, Gachon CMM, Müller DG, Küpper FC, Honda D. The developmental, ultrastructural cytology, and molecular phylogeny of the basal Oomycete *Eurychasma dicksonii*, infecting the filamentous Phaeophyta aglae *Ectocarpus siliculosus* and *Pylaiella littoralis*. Protist 2008a;159:299–318.

Sekimoto S, Yoko K, Kawamura Y, Honda D. Taxonomy, molecular phylogeny, and ultrastructure of *Olpidiopsis porphyrae* sp. nov. (Oomycetes, straminipiles), a unicellular obligate endoparasite of *Bangia* and *Porphyra* spp. Bangiales, Rhodophyta). Mycol Res. 2008b;12:361–374.

Sohn CH. The seaweed resources of Korea. In: Critchley AT, Ohno M, eds. Seaweed resources of the world. Japan: Japan Inter Coop Agency, 1998:15–33.

Stanley SJ. The autecology and ultrastructure interaction between *Mycosphaerella ascophylli* Cotton, *Lautitia danica* (Berlese) Schatz, *Mycaureola dilsea* Maire *et* Chemin and their respective marine algal hosts. PhD thesis, University of Portsmouth, 1991.

Stanley SJ. Observations on the seasonal occurrence of marine endophytic and parasitic fungi. Can J Bot. 1992;70:2089–2096.

Strittmatter M, Gachon CMM, Küpper FC. Ecology of lower Oomycetes. In: Lamour K, Kamoun S, eds. Oomycete genetics and genomics: Diversity, interactions and research tools. New Jersey: John Wiley and Sons Inc., 2009:25–46.

Suryanarayanan TS, Venkatachalam A, Thirunavukkarasu N, Ravishankar JP, Doble M, Geetha V. Internal mycobiota of marine macroalgae from the Tamilnadu coast: distribution, diversity and biotechnological potential. Bot Mar. 2010;53:456–468.

Sutherland GK. New marine fungi on *Pelvetia*. New Phytol. 1915;14:33–42.

Sutherland GK. Marine Fungi Imperfecti. New Phytol. 1916a;15:35–48.

Sutherland GK. Additional notes on marine pyrenomycetes. Trans Br Mycol Soc. 1916b;5:257–263.

Toxopeus J, Kozeraa CJ, O'Leary JB, Garbary DJ. A reclassification of *Mycophycias ascophylli* (Ascomycota) based in nuclear large ribosomal subunit DNA sequences. Bot Mar. 2011;54:325–334.

Tubaki K. An undescribed halophilic species of *Scopulariopsis*. Trans Mycol Soc Jpn. 1973;14:367–369.

Webber FC. Marine fungi. PhD thesis, University of Wales, 1959.

Webber FC. Observations on the structure, life history and biology of *Mycosphaerella ascophylli*. Trans Br Mycol Soc. 1967;50:583–601.

Zuccaro A, Mitchell JI. Fungal communities of seaweeds. In: Dighton J, White JF, Oudeman P, eds. The fungal community: its organization and role in the ecosystem. 3rd ed. New York: CRC Press, Boca Raton CRC, Taylor and Francis, 2005:533–579.

Zuccaro A, Schulz B, Mitchell JI. Molecular detection of ascomycetes associated with *Fucus serratus*. Mycol Res. 2003;107:1451–1466.

Zuccaro A, Summerbell RC, Gams W, Schroers H-F, Mitchell JI. A new *Acremonium* species associated with *Fucus* spp, and its affinity with a phylogenetically distinct marine *Emericellopsis* clade. Stud Mycol. 2004;50:283–297.

Zuccaro A, Schoch CL, Spatafora JW, Kohlmeyer J, Dreager S, Mitchelll JI. Detection and identification of fungi intimately associated with the brown seaweed *Fucus serratus*. Appl Environ Microbiol. 2008;74:931–941.

18 Salt marsh fungi

Maria da Luz Calado and Margarida Barata

Introduction: Salt marsh ecosystem functioning and the importance of the microbial community

Salt marsh ecosystem

Salt marshes represent coastal marine ecosystems that occur mainly in temperate and high-latitude estuaries (Allen and Pye 1992; Simas et al. 2001) and are exposed to low hydrodynamic conditions and periodic tidal flooding (Simas et al. 2001). They are plastic, dynamic systems created by the combined action of water, sediments and vegetation, and constitute a typical example of open ecosystems (Chapman 1977; Boorman 1999). Salt marshes have long been recognized as being one of the most productive ecosystems in the world (Kohlmeyer and Kohlmeyer 1979; McLusky and Elliott 2004) due to their high primary production rates (Bouchard and Lefeuvre 2000; McLusky and Elliott 2004). A number of emergent macrophytes, in particular *Spartina* spp., *Juncus roemerianus* and *Phragmites australis*, are grass-like plants that thrive in such an environment and represent one of the main sources of nutrients and organic matter (Teal 1962; Christian et al. 1990; Newell et al. 1996b; Van Ryckegem et al. 2006). The primary production of these macrophytes is essentially composed of highly refractory lignocellulosic compounds, such as lignin, hemicellulose and cellulose (Maccubbin and Hodson 1980; Benner et al. 1984a, 1984b; Torzilli and Andykovitch 1986; Newell et al. 1996b; Lyons et al. 2010), and hence only a small fraction is consumed as living tissue (Teal 1962; Maccubbin and Hodson 1980); most of the production is actually converted into detritus, which either remains in the salt marsh or is transported to coastal waters (Teal 1962; Asaeda et al. 2002). For these emergent macrophytes, the decay process is initiated in the standing crops, and continues after abscission and deposition of dead plant material onto the marsh surface (Fell and Hunter 1979; Newell and Fallon 1989; Newell et al. 1989; Christian et al. 1990; Samiaji and Barlocher 1996; Newell et al. 1998; Gessner 2001; Van Ryckegem et al. 2006; Menéndez and Sanmartí 2007). Much of the decay of marsh grass tissue takes place above the sediment (Newell and Porter 2002).

Decomposer microbial community

The decomposer microbial community, including fungi and bacteria, assumes a fundamental ecological role in the degradation of plant material, which is enriched with structural polymers, and in the consequent release of nutrients that are essential to the metabolism of a wide marine community (Benner et al. 1984b; Boorman 1999; Newell and Porter 2002; Lyons et al. 2005). Though the role of fungi in this process has been long neglected, several studies have highlighted the importance of the metabolic activities of these saprobic microorganisms on the biogeochemical carbon and nutrients cycles, and in the energy fluxes within these ecotonal marine ecosystems (Gessner and

Goos 1973; Torzilli and Andrykovitch 1986; Newell 1996; Newell et al. 1996b; Hyde et al. 1998; Gessner et al. 2007).

Saprobic fungi that colonize standing-dead tissues of salt marsh grasses initiate the decay process (Torzilli and Andrykovitch 1986; Barlocher 1996; Lyons et al. 2005), and represent the main secondary producers of the microbial community (Newell and Fallon 1989; Newell et al. 1989; Newell 1996; Newell et al. 1996a, 1996b; Castro and Freitas 2000; Newell et al. 2000b; Gessner 2001; Newell 2001a; Findlay et al. 2002; Van Ryckegem et al. 2006, 2007). Bacteria may become more active in the latter phase of decomposition (i.e. when the plant material collapses onto the marsh sediment surface; Benner et al. 1984b; Newell et al. 1989; Newell and Porter 2002). However, Buchan et al. (2003) and Lyons et al. (2005) demonstrated that metabolically-active bacteria and fungi co-occur on *Spartina* detritus, which contradicts this idea of temporally segregated interventions during the decay process, but apparently without establishing species-specific ecological associations.

In the microbial community associated with standing-decaying tissues of emergent macrophytes, there is a clear dominance of fungi over bacteria. This is expressed in biomass and productivity. This dominance occurs because the morphological and physiological characteristics of the saprobic fungi confer an adaptive advantage on the use and degradation of this substrate. In fact, in addition to their ability to tolerate a wide range of environmental conditions, fungi can degrade the most resistant substrates in a more efficient manner than bacteria. Filamentous fungi are well suited to penetrate substrates due to their rigid cell walls, apical growth and ability to produce lignocellulose-degrading enzymes (Torzilli 1982; Newell 1996; Newell et al. 1996b; Raghukumar 2004). Saprobic fungi act on the surface or within the tissues of macrophytes by the production of lignocellulolytic enzymes and physical penetration of the host cell walls, bringing about the decomposition of senescent tissues (Newell and Porter 2002). Additionally, saprobic fungi have the ability to retain and convert inorganic nitrogen into fungal biomass during the initial phases of plant tissue decomposition (Findlay et al. 2002; Van Ryckegem et al. 2006, 2007) and immobilize this nutrient from the surrounding environment (Newell 1996; Van Ryckegem et al. 2006). The incorporation of nitrogen into fungal biomass, together with the extracellular enzymes produced during the process, results in a nutritive enrichment of substrates, which in turn becomes more palatable to several animal consumers (Raghukumar 2004). The fungal community associated with the decomposition of macrophytes is composed mainly of ascomycetes (Gessner and Kohlmeyer 1976; Newell et al. 1996a, 2000b; Newell 2001a, 2001b; Barata 2002; Buchan et al. 2003; Van Ryckegem and Verbeken, 2005a, 2005b, 2005c).

Mycota of salt marshes: biotic and abiotic factors affecting community structure

Despite similar general biophysical characteristics, salt marsh ecosystems present some environmental and ecological variations that will reflect on the composition and dynamics of the fungal community. The marine fungal community in salt marshes, as in other ecosystems, is composed of ubiquitous species, which occur on a broad range of substrates and environmental conditions, and other species that appear to be strictly associated with particular ecological niches (Gessner and Kohlmeyer 1976). The presence of a given fungus in the ecosystem depends on an appropriate combination of various biotic and abiotic factors, which vary according to species. These diverse factors include:

(i) degree of host/substrate specificity (Apinis and Chesters 1964; Newell and Porter 2002; Blum et al. 2004; Torzilli et al. 2006; Lyons et al. 2010);
(ii) ability to interact and compete with other microorganisms (Torzilli and Andrykovitch 1986; Buchan et al. 2003; Lyons et al. 2005);
(iii) vulnerability/resistance to predation (Newell and Wasowsky 1995; Newell 2001a, 2001b); and
(iv) ecological requirements, such as water (Newell et al. 1996a; Poon and Hyde 1998a), oxygen availability (Wong and Hyde 2002; Menéndez and Sanmartí 2007), dissolved organic nutrients (Newell et al. 1996a, 2000b, Newell 2001b; Newell and Porter 2002), salinity (Van Ryckegem and Verbeken 2005a), and temperature (Castro and Freitas 2000; Van Ryckegem et al. 2007).

Host/substrate specificity

Among the intrinsic biological and environmental factors mentioned, the host/substrate specificity – which is related to the chemical and structural composition of plant tissues – appears to be primarily responsible for determining fungal community composition and productivity (Fell and Hunter 1979; Newell et al. 2000b; Newell and Porter 2002; Blum et al. 2004; Torzilli et al. 2006; Van Ryckegem et al. 2006; 2007; Lyons et al. 2010). This specificity occurs during the selection process of the host plant species to be colonized, but also in the choice of the plant tissue.

Host plant and associated fungal diversity

Studies of salt marsh fungi associated with diverse host plants reveal no overlap between the fungal-decay communities, which emphasizes the general high-level specificity with the chemical and structural characteristics of each plant (Newell and Porter 2002; Blum et al. 2004; Torzilli et al. 2006). Torzilli et al. (2006) compared the mycota associated with four salt marsh plants – *Spartina alterniflora*, *Juncus roemerianus*, *Distichlis spicata* and *Sarcocornia perennis* – and concluded that the greater the similarity between the type of plant tissues, the greater is the similarity between the associated fungal communities. The same conclusion was reached by Lyons et al. (2010), who found the same major ascomycetes on various species of *Spartina* (*S. alterniflora*, *S. foliosa*, *S. alterniflora* × *S. foliosa* and *S. densiflora*).

Walker and Campbell (2010) inventoried the fungal community associated with *S. alterniflora* and *Juncus roemerianus* using morphological and molecular approaches, and obtained different results. The morphological analyses revealed different species on host plants, but terminal-restriction fragment length polymorphism community profiles showed that more than 50% of the fungal terminal-restriction fragments were found on both plants. The authors suggested that the absence of fruiting structures of the same species on *S. alterniflora* and *J. roemerianus* might indicate that some fungi are able to colonize but not sporulate on both hosts, and thus might be host-specific to complete their lifecycle.

A comparison of the species composition of fungal communities associated with the main primary producers in marsh ecosystems (Table 18.1) confirms the above observation. The overlap of fungal community between all possible host combinations

Table 18.1 Filamentous fungi associated with *Juncus roemerianus*, *Spartina* spp. and *Phragmites australis* in marsh ecosystems.

Fungi	Host Plant		
	Juncus roemerianus	*Spartina* spp.	*Phragmites australis*
Ascomycota			
Amauroascus albicans (Apinis) Arx		Barata (2002)	
Amphisphaeria culmicola Sacc.		Barata (2002)	
Aniptodera chesapeakensis Shearer et M.A. Mill.*		Kohlmeyer and Volkmann-Kohlmeyer (2002)	Poon and Hyde (1998a)
Aniptodera juncicola Volkm.-Kohlm. et Kohlm.*	Volkmann-Kohlmeyer and Kohlmeyer (1994); Jones (2011)		
Anthostomella atroalba Kohlm., Volkm.-Kohlm. et O.E. Erikss.	Kohlmeyer et al. (1998b); Jones (2011)		
Anthostomella poecila Kohlm., Volkm.-Kohlm. et O.E. Erikss.*	Kohlmeyer et al. (1995b); Walker and Campbell (2010)		
Anthostomella punctulata (Roberge ex Desm.) Sacc.			Van Ryckegem et al. (2007)
Anthostomella semitecta Kohlm., Volkm.-Kohlm. et O.E. Erikss.	Kohlmeyer et al. (1995b); Jones (2011)		
Anthostomella spissitecta Kohlm. et Volkm.-Kohlm.*		Kohlmeyer and Volkmann-Kohlmeyer (2002)	
Anthostomella torosa Kohlm. et Volkm.-Kohlm.*	Kohlmeyer and Volkmann-Kohlmeyer (2002); Jones (2011)		
Anthostomella sp.		Barata (2002)	
Apiospora montagnei Sacc.			Van Ryckegem and Verbeken (2005b)
Aposphaeria sp.			Van Ryckegem and Verbeken (2005b, 2005c)

Table continued on next page.

Table 18.1 (Continued)

Fungi	Host Plant		
	Juncus roemerianus	*Spartina* spp.	*Phragmites australis*
Aquamarina speciosa Kohlm., Volkm.-Kohlm. et O.E. Erikss.*	Kohlmeyer et al. (1995d); Jones (2011)		
Aropsiclus junci (Kohlm. et Volkm.-Kohlm.) Kohlm. et Volkm.-Kohlm.*	Kohlmeyer and Volkmann-Kohlmeyer (1994); Jones (2011)		
Atkinsonella hypoxylon (Peck) Diehl		Barata (2002)	
Atrotorquata lineata Kohlm. et Volkm.-Kohlm.*	Kohlmeyer and Volkmann-Kohlmeyer (1993b); Jones (2011)		
Belonium heteromorphum (Ellis et Everh.) Seaver		Barata (2002)	
Botryosphaeria festucae (Lib.) Arx et E. Müll.			Van Ryckegem and Verbeken (2005b, 2005c)
Brunnipila palearum (Desm.) Baral		Barata (2002)	
Buergenerula spartinae Kohlm. et R.V. Gessner*		Barata (2002); Buchan et al. (2002); Kohlmeyer and Volkmann-Kohlmeyer (2002); Buchan et al. (2003); Walker and Campbell (2010)	
Byssothecium obiones (P. Crouan et H. Crouan) M.E. Barr*		Barata (2002); Kohlmeyer and Volkmann-Kohlmeyer (2002)	
Ceratosphaeria sp.	Fell and Hunter (1979)		
Ceriosporopsis halima Linder*		Barata (2002); Kohlmeyer and Volkmann-Kohlmeyer (2002)	

Table continued on next page.

Table 18.1 (Continued)

Fungi	Host Plant		
	Juncus roemerianus	*Spartina* spp.	*Phragmites australis*
Chaetomium crispatum (Fuckel) Fuckel		Barata (2002)	
Chaetomium funicola Cooke		Barata (2002)	
Chaetomium globosum Kunze		Barata (2002)	Poon and Hyde (1998a)
Chaetomium thermophilum La Touche		Barata (2002)	
Chaetomium sp.	Fell and Hunter (1979)		
Cistella fugiens (W. Phillips) Matheis			Van Ryckegem and Verbeken (2005c)
Claviceps purpurea (Fr.) Tul.		Barata (2002)	
Claviceps sp.		Barata (2002)	
Corollospora maritima Werderm.*		Barata (2002); Kohlmeyer and Volkmann-Kohlmeyer (2002)	
Corynascus sepedonium (C.W. Emmons) Arx		Barata (2002)	
Decorospora gaudefroyi (Pat.) Inderb., Kohlm. *et* Volkm.-Kohlm.*		Barata (2002)	
Didymella glacialis Rehm			Van Ryckegem and Verbeken (2005b, 2005c); Van Ryckegem et al. (2007)
Didymella sp.		Barata (2002)	Van Ryckegem and Verbeken (2005b); Van Ryckegem et al. (2007)
Didymosphaeria lignomaris Strongman *et* J.D. Mill.*		Barata (2002); Kohlmeyer and Volkmann-Kohlmeyer (2002)	

Table continued on next page.

Table 18.1 (Continued)

Fungi	Host Plant		
	Juncus roemerianus	*Spartina* spp.	*Phragmites australis*
Discostroma sp.			Van Ryckegem and Verbeken (2005b)
Ellisiodothis inquinans (Ellis *et* Everh.) Theiss.		Barata (2002)	
Gaeumannomyces graminis var. *graminis* (Sacc.) Arx *et* D.L. Olivier		Buchan et al. (2003)	
Gaeumannomyces medullaris Kohlm., Volkm.-Kohlm. *et* O.E. Erikss. (anamorph *Trichocladium medullare* Kohlm. *et* Volkm.-Kohlm.)*	Kohlmeyer et al. (1995c); Jones (2011)		
Gaeumannomyces sp.			Poon and Hyde (1998a)
Gibberella gordonii C. Booth		Barata (2002)	
Gibberella zeae (Schwein.) Petch			Van Ryckegem and Verbeken (2005b, 2005c)
Gibberella sp.		Barata (2002)	
Gloeotinia granigena (Quél.) T. Schumach.	Walker and Campbell (2010)		
Glomerobolus gelineus Kohlm. *et* Volkm.-Kohlm.	Kohlmeyer and Volkmann-Kohlmeyer (1996a); Jones (2011)		
Gnomonia salina E.B.G. Jones*		Barata (2002); Kohlmeyer and Volkmann-Kohlmeyer (2002)	
Guignardia sp.	Fell and Hunter (1979)		
Haematonectria haematococca (Berk. *et* Broome) Samuels *et* Rossman			Poon and Hyde (1998a)

Table continued on next page.

Table 18.1 (Continued)

Fungi	Host Plant		
	Juncus roemerianus	*Spartina* spp.	*Phragmites australis*
Haligena elaterophora Kohlm.*		Barata (2002); Kohlmeyer and Volkmann-Kohlmeyer (2002)	
Halosarpheia culmiperda Kohlm., Volkm.-Kohlm. *et* O.E. Erikss.*	Kohlmeyer et al. (1995c); Jones (2011)		
Halosarpheia phragmiticola Poon *et* K.D. Hyde*			Poon and Hyde (1998a)
Heleiosa barbatula Kohlm., Volkm.-Kohlm. *et* O.E. Erikss.*	Kohlmeyer et al. (1996); Jones (2011)		
Helicascus kanaloanus Kohlm.*		Kohlmeyer and Volkmann-Kohlmeyer (2002)	
Hydropisphaera arenula (Berk. *et* Broome) Rossman *et* Samuels			Van Ryckegem and Verbeken (2005b)
Hydropisphaera erubescens (Roberge ex Desm.) Rossman *et* Samuels		Buchan et al. (2002, 2003); Kohlmeyer and Volkmann-Kohlmeyer (2002)	
Julella herbatilis Kohlm., Volkm.-Kohlm. *et* O.E. Erikss.*	Kohlmeyer et al. (1997); Jones (2011)		
Juncigena adarca Kohlm., Volkm.-Kohlm. *et* O.E. Erikss. (anamorph *Cirrenalia adarca* Kohlm., Volkm.-Kohlm. *et* O.E. Erikss.)*	Kohlmeyer et al. (1997); Jones (2011)		
Kananascus sp.		Buchan et al. (2003)	
Keissleriella rara Kohlm., Volkm.-Kohlm. *et* O.E. Erikss.	Kohlmeyer et al. (1995d); Jones (2011)		
Keissleriella sp.	Fell and Hunter (1979)		

Table continued on next page.

Table 18.1 (Continued)

Fungi	Host Plant		
	Juncus roemerianus	*Spartina* spp.	*Phragmites australis*
Lachnum spartinae S.A. Cantrell		Kohlmeyer and Volkmann-Kohlmeyer (2002); Buchan et al. (2003)	
Lautospora simillima Kohlm., Volkm.-Kohlm. *et* O.E. Erikss.*	Kohlmeyer et al. (1995a); Jones (2011)		
Lentithecium arundinaceum (Sowerby) K.D. Hyde, J. Fourn. *et* Yin. Zhang		Barata (2002)	Van Ryckegem and Verbeken (2005b, 2005c); Van Ryckegem et al. (2007)
Lentithecium fluviatile (Aptroot *et* Van Ryck.) K.D. Hyde, J. Fourn. *et* Yin. Zhang			Van Ryckegem and Verbeken (2005b, 2005c); Van Ryckegem et al. (2007)
Lentithecium lineare (E. Müll. ex Dennis) K.D. Hyde, J. Fourn. *et* Yin. Zhang			Van Ryckegem and Verbeken (2005c)
Leptosphaeria albopunctata (Westend.) Sacc.		Barata (2002)	
Leptosphaeria australiensis (Cribb *et* J.W. Cribb) G.C. Hughes*	Fell and Hunter (1979)	Barata (2002); Kohlmeyer and Volkmann-Kohlmeyer (2002)	
Leptosphaeria lacustris (Fuckel) Wint.		Barata (2002)	
Leptosphaeria marina Ellis *et* Everh.		Barata (2002); Kohlmeyer and Volkmann-Kohlmeyer (2002)	
Leptosphaeria orae-maris Linder		Barata (2002); Kohlmeyer and Volkmann-Kohlmeyer (2002)	

Table continued on next page.

Table 18.1 (Continued)

Fungi	Host Plant		
	Juncus roemerianus	*Spartina* spp.	*Phragmites australis*
Leptosphaeria pelagica E.B.G. Jones*		Barata (2002); Kohlmeyer and Volkmann-Kohlmeyer (2002); Walker and Campbell (2010)	
Leptosphaeria sp.	Walker and Campbell (2010)		Poon and Hyde (1998a)
Lewia infectoria (Fuckel) M.E. Barr *et* E.G. Simmons			Van Ryckegem and Verbeken (2005b)
Lignincola laevis Höhnk*		Barata (2002); Kohlmeyer and Volkmann-Kohlmeyer (2002)	Poon and Hyde (1998a)
Lophiostoma arundinis (Pers.) Ces. *et* De Not.			Van Ryckegem and Verbeken (2005c)
Lophiostoma semiliberum (Desm.) Ces. *et* De Not.			Van Ryckegem and Verbeken (2005c)
Lophodermium arundinaceum (Schrad.) Chevall.			Van Ryckegem and Verbeken (2005b, 2005c)
Loratospora aestuarii Kohlm. *et* Volkm.-Kohlm.*	Kohlmeyer and Volkmann-Kohlmeyer (1993b); Jones (2011)		
Lulworthia floridana Meyers*		Barata (2002)	
Lulworthia medusa (Ellis *et* Everh.) Cribb *et* J.W. Cribb*		Jones (1963); Barata (2002); Kohlmeyer and Volkmann-Kohlmeyer (2002)	
Lulworthia spp.		Barata (2002)	
Magnisphaera spartinae (E.B.G. Jones) J. Campb., J.L. Anderson *et* Shearer*		Barata (2002); Kohlmeyer and Volkmann-Kohlmeyer (2002)	Van Ryckegem and Verbeken (2005c)

Table continued on next page.

Table 18.1 (Continued)

Fungi	Host Plant		
	Juncus roemerianus	*Spartina* spp.	*Phragmites australis*
Massariella sp.		Barata (2002)	
Massarina carolinensis Kohlm., Volkm.-Kohlm. *et* O.E. Erikss.	Kohlmeyer et al. (1995d); Jones (2011)		
Massarina phragmiticola Poon *et* K.D. Hyde*			Poon and Hyde (1998a)
Massarina ricifera Kohlm., Volkm.-Kohlm. *et* O.E. Erikss.*	Kohlmeyer et al. (1995c); Walker and Campbell (2010); Jones (2011)		
Massarina spp.	Fell and Hunter (1979)		Van Ryckegem and Verbeken (2005b)
Massariosphaeria erucacea Kohlm., Volkm.-Kohlm. *et* O.E. Erikss.*	Kohlmeyer et al. (1996); Jones (2011)		
Massariosphaeria scirpina (G. Winter) Leuchtm.		Barata (2002)	
Massariosphaeria typhicola (P. Karst.) Leuchtm.*		Barata (2002); Kohlmeyer and Volkmann-Kohlmeyer (2002)	
Massariosphaeria sp.			Van Ryckegem and Verbeken (2005c)
Meliola spartinae (Ellis *et* Everh.) Berl. *et* Voglino		Barata (2002)	
Micronectriella agropyri Apinis *et* Chesters		Barata (2002)	
Microthecium levitum Udagawa *et* Cain		Barata (2002)	
Microthyrium microscopicum Desm.		Barata (2002)	
Mollisia atriella Cooke		Barata (2002)	
Mollisia cf. *palustris* (Roberge ex Desm.) P. Karst.			Van Ryckegem and Verbeken (2005b)

Table continued on next page.

Table 18.1 (Continued)

Fungi	Host Plant		
	Juncus roemerianus	*Spartina* spp.	*Phragmites australis*
Mollisia hydrophila (P. Karst.) Sacc.			Van Ryckegem and Verbeken (2005b)
Mollisia retincola (Rabenh.) P. Karst.			Van Ryckegem and Verbeken (2005b, 2005c)
Morenoina phragmitis J.P. Ellis			Van Ryckegem and Verbeken (2005b, 2005c)
Mycosphaerella eurypotami Kohlm., Volkm.-Kohlm. *et* O.E. Erikss.	Kohlmeyer et al. (1999); Jones (2011)		
Mycosphaerella lineolata (Roberge ex Desm.) J. Schröt.			Van Ryckegem and Verbeken (2005b, 2005c)
Mycosphaerella salicorniae (Rabenh.) Lindau*		Barata (2002)	
Mycosphaerella spp.	Fell and Hunter (1979); Walker and Campbell (2010)	Barata (2002); Buchan et al. (2002, 2003); Lyons et al. (2010); Walker and Campbell (2010)	
Nais inornata Kohlm.*		Barata (2002); Kohlmeyer and Volkmann-Kohlmeyer (2002)	
Natantispora retorquens (Shearer *et* J.L. Crane) J. Campb., J.L. Anderson *et* Shearer*		Barata (2002)	
Nectria sp.	Fell and Hunter (1979)		
Ommatomyces coronatus Kohlm., Volkm.-Kohlm. *et* O.E. Erikss.*	Kohlmeyer et al. (1995c); Jones (2011)		

Table continued on next page.

Table 18.1 (Continued)

Fungi	Host Plant		
	Juncus roemerianus	*Spartina* spp.	*Phragmites australis*
Orbilia junci Kohlm., Baral *et* Volkm.-Kohlm. (anamorph *Dwayaangam junci* Kohlm., Baral *et* Volkm.-Kohlm.)	Kohlmeyer et al. (1998a); Jones (2011)		
Otthia sp.	Fell and Hunter (1979)		
Panorbis viscosus (I. Schmidt) J. Campb., J.L. Anderson *et* Shearer*		Barata (2002); Buchan et al. (2003)	
Papulosa amerospora Kohlm. *et* Volkm.-Kohlm.*	Kohlmeyer and Volkmann-Kohlmeyer (1993a); Jones (2011)		
Paraphaeosphaeria apicicola Kohlm., Volkm.-Kohlm. *et* O.E. Erikss. (anamorph *Coniothyrium* sp.)	Kohlmeyer et al. (1999); Jones (2011)		
Paraphaeosphaeria michotii (Westend.) O.E. Erikss.			Van Ryckegem and Verbeken (2005b)
Paraphaeosphaeria pilleata Kohlm., Volkm.-Kohlm. *et* O.E. Erikss. (anamorph *Coniothyrium* sp.)	Kohlmeyer et al. (1995d); Jones (2011)		
Phaeosphaeria anchiala Kohlm., Volkm.-Kohlm. *et* K.M. Tsui	Kohlmeyer et al (2005); Jones (2011)		
Phaeosphaeria caricinella (P. Karst.) O.E. Erikss.		Barata (2002)	
Phaeosphaeria culmorum (Auersw. ex Rehm) Leuchtm.			Van Ryckegem and Verbeken (2005b); Van Ryckegem et al. (2007)

Table continued on next page.

Table 18.1 (Continued)

Fungi	Host Plant		
	Juncus roemerianus	*Spartina* spp.	*Phragmites australis*
Phaeosphaeria eustoma (Fuckel) L. Holm			Van Ryckegem and Verbeken (2005b, 2005c); Van Ryckegem et al. (2007)
Phaeosphaeria gessneri Shoemaker *et* C.E. Babc.*		Kohlmeyer and Volkmann-Kohlmeyer (2002)	
Phaeosphaeria halima (T.W. Johnson) Shoemaker *et* C.E. Babc.*		Barata (2002); Buchan et al. (2002, 2003); Kohlmeyer and Volkmann-Kohlmeyer (2002); Lyons et al. (2010); Walker and Campbell (2010)	
Phaeosphaeria herpotrichoides (De Not.) L. Holm		Barata (2002)	
Phaeosphaeria juncina (Auersw.) L. Holm	Fell and Hunter (1979)		
Phaeosphaeria luctuosa (Niessl ex Sacc.) Otani *et* Mikawa			Van Ryckegem and Verbeken (2005b, 2005c)
Phaeosphaeria macrosporidium (E.B.G. Jones) Shoemaker *et* C.E. Babc.*		Barata (2002)	
Phaeosphaeria neomaritima (R.V. Gessner *et* Kohlm.) Shoemaker *et* C.E. Babc.*		Barata (2002); Kohlmeyer and Volkmann-Kohlmeyer (2002)	
Phaeosphaeria nodorum (E. Müll.) Hedjar.		Buchan et al. (2003)	

Table continued on next page.

Table 18.1 (Continued)

Fungi	Host Plant		
	Juncus roemerianus	*Spartina* spp.	*Phragmites australis*
Phaeosphaeria olivacea Kohlm., Volkm.-Kohlm. *et* O.E. Erikss.*	Kohlmeyer et al. (1997); Jones (2011)		
Phaeosphaeria pontiformis (Fuckel) Leuchtm.			Van Ryckegem and Verbeken (2005b, 2005c); Van Ryckegem et al. (2007)
Phaeosphaeria roemeriani Kohlm., Volkm.-Kohlm. *et* O.E. Erikss.*	Kohlmeyer et al. (1998b); Walker and Campbell (2010); Jones (2011)		
Phaeosphaeria spartinae (Ellis *et* Everh.) Shoemaker *et* C.E. Babc.*		Barata (2002); Kohlmeyer and Volkmann-Kohlmeyer (2002)	
Phaeosphaeria spartinicola Leuchtm.*		Barata (2002); Buchan et al. (2002); Kohlmeyer and Volkmann-Kohlmeyer (2002); Buchan et al. (2003); Lyons et al. (2010); Walker and Campbell (2010)	
Phaeosphaeria typharum (Desm.) L. Holm*		Barata (2002)	
Phaeosphaeria vagans (Niessl) O.E. Erikss.			Van Ryckegem and Verbeken (2005b)
Phaeosphaeria spp.			Van Ryckegem and Verbeken (2005b, 2005c); Van Ryckegem et al. (2007)

Table continued on next page.

Table 18.1 (Continued)

Fungi	Host Plant		
	Juncus roemerianus	*Spartina* spp.	*Phragmites australis*
Phomatospora bellaminuta Kohlm., Volkm.-Kohlm. *et* O.E. Erikss.*	Kohlmeyer et al. (1995b); Jones (2011)		
Phomatospora berkeleyi Sacc.			Van Ryckegem and Verbeken (2005b, 2005c); Van Ryckegem et al. (2007)
Phomatospora dinemasporium J. Webster		Barata (2002)	Van Ryckegem and Verbeken (2005c)
Phomatospora phragmiticola Poon *et* K.D. Hyde*			Poon and Hyde (1998a)
Phomatospora spp.	Fell and Hunter (1979)		Van Ryckegem and Verbeken (2005b, 2005c)
Phragmitensis marina M.K.M. Wong, Poon *et* K.D. Hyde*			Poon and Hyde (1998a)
Phyllachora cynodontis Niessl		Barata (2002)	
Phyllachora graminis var. *graminis* (Pers.) Fuckel		Barata (2002)	
Phyllachora sylvatica Sacc. *et* Speg.		Barata (2002)	
Physalospora citogerminans Kohlm., Volkm.-Kohlm. *et* O.E. Erikss.	Kohlmeyer et al. (1995b); Jones (2011)		
Pleospora abscondita Sacc. *et* Roum.			Van Ryckegem and Verbeken (2005b)
Pleospora herbarum (Pers.) Rabenh.		Barata (2002)	

Table continued on next page.

Table 18.1 (Continued)

Fungi	Host Plant		
	Juncus roemerianus	*Spartina* spp.	*Phragmites australis*
Pleospora pelagica T.W. Johnson*		Barata (2002); Kohlmeyer and Volkmann-Kohlmeyer (2002); Buchan et al. (2003); Lyons et al. (2010)	
Pleospora spartinae (J. Webster *et* M.T. Lucas) Apinis *et* Chesters*		Barata (2002); Kohlmeyer and Volkmann-Kohlmeyer (2002); Buchan et al. (2003)	Poon and Hyde (1998a)
Pleospora vagans var. *vagans* Niessl		Barata (2002)	
Preussia funiculata (Preuss) Fuckel		Barata (2002)	
Pseudohalonectria falcata Shearer			Poon and Hyde (1998a)
Pseudohalonectria halophila Kohlm. *et* Volkm.-Kohlm.*	Kohlmeyer et al. (2005); Jones (2011)		
Remispora hamata (Höhnk) Kohlm.	Fell and Hunter (1979)	Barata (2002)	Van Ryckegem and Verbeken (2005b, 2005c); Van Ryckegem et al. (2007)
Rivilata ius Kohlm., Volkm.-Kohlm. *et* O.E. Erikss.	Kohlmeyer et al. (1998b); Jones (2011)		
Schizothecium hispidulum (Speg.) N. Lundq.			Van Ryckegem and Verbeken (2005b)
Scirrhia annulata Kohlm., Volkm.-Kohlm. *et* O.E. Erikss.*	Kohlmeyer et al. (1996); Jones (2011)		
Sordaria fimicola (Roberge ex Desm.) Ces. *et* De Not.		Barata (2002)	

Table continued on next page.

Table 18.1 (Continued)

Fungi	Host Plant		
	Juncus roemerianus	*Spartina* spp.	*Phragmites australis*
Sphaerulina albispiculata Tubaki*		Barata (2002)	
Sphaerulina orae-maris Linder*		Jones (1963); Barata (2002)	
Sphaerulina sp.	Fell and Hunter (1979)		
Splanchnonema sp.	Fell and Hunter (1979)		
Sporormia sp.	Fell and Hunter (1979)		
Sporormiella intermedia (Auersw.) S.I. Ahmed *et* Cain ex Kobayasi		Barata (2002)	
Stictis sp.			Van Ryckegem and Verbeken (2005b)
Thelebolus crustaceus (Fuckel) Kimbr.		Barata (2002)	
Tremateia halophila Kohlm., Volkm.-Kohlm. *et* O.E. Erikss.*	Kohlmeyer et al. (1995a); Jones (2011)	Barata (2002)	
Trichodelitschia bisporula (P. Crouan *et* H. Crouan) Munk		Barata (2002)	
Zopfiella latipes (N. Lundq.) Malloch *et* Cain*			Poon and Hyde (1998a)
BASIDIOMYCOTA			
Halocyphina villosa Kohlm. *et* E. Kohlm.*		Barata (2002)	
Merismodes bresadolae (Grélet) Singer			Van Ryckegem and Verbeken (2005c)
Nia vibrissa R.T. Moore *et* Meyers*		Barata (2002); Kohlmeyer and Volkmann-Kohlmeyer (2002)	
Puccinia magnusiana Körn.			Van Ryckegem and Verbeken (2005b); Van Ryckegem et al. (2007)

Table continued on next page.

Table 18.1 (Continued)

Fungi	Host Plant		
	Juncus roemerianus	*Spartina* spp.	*Phragmites australis*
Puccinia phragmitis (Schumach.) Körn.			Van Ryckegem et al. (2007)
Puccinia seymouriana Arthur		Barata (2002)	
Puccinia sparganioides Ellis *et* Tracy		Barata (2002)	
Sporobolomyces sp.			Van Ryckegem and Verbeken (2005b); Van Ryckegem et al. (2007)
Tremella spicifera Van Ryck., Van de Put *et* P. Roberts			Van Ryckegem and Verbeken (2005b, 2005c)
Uromyces acuminatus Arthur		Barata (2002)	
Uromyces argutus F. Kern		Barata (2002)	
ASEXUAL FUNGI			
Acremonium sp.	Fell and Hunter (1979)		
Alternaria alternata (Fr.) Keissl.	Fell and Hunter (1979)	Barata (2002)	Van Ryckegem and Verbeken (2005b); Van Ryckegem et al. (2007)
Alternaria maritima G.K. Sutherl.		Barata (2002)	
Arthrinium phaeospermum (Corda) M.B. Ellis		Barata (2002)	Van Ryckegem and Verbeken (2005b, 2005c); Van Ryckegem et al. (2007)
Arthrinium sp. (state of *Apiospora montagnei*)			Poon and Hyde (1998a)
Arthrinium sp. (state of *Apiospora* sp.)			Poon and Hyde (1998a)

Table continued on next page.

Table 18.1 (Continued)

Fungi	Host Plant		
	Juncus roemerianus	*Spartina* spp.	*Phragmites australis*
Arthrinium sp.	Fell and Hunter (1979)		
Arthrobotrys sp.			Poon and Hyde (1998a)
Ascochyta cf. *arundinariae* Tassi			Van Ryckegem and Verbeken (2005b); Van Ryckegem et al. (2007)
Ascochyta cf. *leptospora* (Trail) Hara			Van Ryckegem and Verbeken (2005b)
Ascochyta spartinae Trel.		Barata (2002)	
Ascochyta sp.		Barata (2002)	Van Ryckegem and Verbeken (2005b)
Aspergillus nidulans (Eidam) G. Winter		Barata (2002)	
Aspergillus niger Tiegh.	Fell and Hunter (1979)		
Aspergillus ustus (Bainier) Thom *et* Church		Walker and Campbell (2010)	
Aspergillus spp.	Fell and Hunter (1979)	Barata (2002)	
Asteromyces cruciatus Moreau *et* M. Moreau ex Hennebert*		Kohlmeyer and Volkmann-Kohlmeyer (2002)	
Aureobasidium sp.	Fell and Hunter (1979)		
Bactrodesmium atrum M.B. Ellis			Van Ryckegem and Verbeken (2005c)
Botryodiplodia sp.	Fell and Hunter (1979)		
Botrytis cinerea Pers.		Barata (2002)	
Camarosporium feurichii Henn.			Van Ryckegem and Verbeken (2005b)
Camarosporium sp.			Van Ryckegem and Verbeken (2005b, 2005c)

Table continued on next page.

Table 18.1 (Continued)

Fungi	Host Plant		
	Juncus roemerianus	*Spartina* spp.	*Phragmites australis*
Chaetasbolisia sp.			Poon and Hyde (1998a)
Chaetospermum camelliae Agnihothr.			Poon and Hyde (1998a)
Cirrenalia macrocephala (Kohlm.) Meyers *et* R.T. Moore*		Barata (2002); Kohlmeyer and Volkmann-Kohlmeyer (2002)	
Cirrenalia pseudomacrocephala Kohlm.*	Fell and Hunter (1979)		
Cladosporium algarum Cooke *et* Massee*		Barata (2002)	
Cladosporium cladosporioides (Fresen.) G.A. de Vries	Fell and Hunter (1979)		
Cladosporium sphaerospermum Penz.	Fell and Hunter (1979)		
Cladosporium spp.	Fell and Hunter (1979)	Barata (2002)	Van Ryckegem and Verbeken (2005b); Van Ryckegem et al. (2007)
Cochliobolus hawaiiensis Alcorn	Fell and Hunter (1979)		Poon and Hyde (1998a)
Cochliobolus tuberculatus Sivan.	Fell and Hunter (1979)		
Colletotrichum sp.			Poon and Hyde (1998a)
Coniothyrium spp.	Fell and Hunter (1979)		
Cremasteria cymatilis Meyers *et* R.T. Moore	Fell and Hunter (1979)		
Cumulospora marina (nomen dubium) I. Schmidt*		Kohlmeyer and Volkmann-Kohlmeyer (2002)	
Curvularia protuberata R.R. Nelson *et* Hodges	Fell and Hunter (1979)		
Curvularia sp.	Fell and Hunter (1979)		
Cytoplacosphaeria phragmiticola Poon *et* K.D. Hyde*			Poon and Hyde (1998a)

Table continued on next page.

Table 18.1 (Continued)

Fungi	Host Plant		
	Juncus roemerianus	*Spartina* spp.	*Phragmites australis*
Cytoplacosphaeria rimosa Petr.			Van Ryckegem and Verbeken (2005b, 2005c)
Cytoplea sp.			Poon and Hyde (1998a)
Deightoniella roumeguerei (Cavara) Constant.			Van Ryckegem et al. (2007)
Dendrostilbella sp.			Poon and Hyde (1998a)
Dictyosporium oblongum (Fuckel) S. Hughes			Van Ryckegem and Verbeken (2005c); Van Ryckegem et al. (2007)
Dictyosporium pelagicum (Linder) G.C. Hughes ex E.B.G. Jones*		Barata (2002); Kohlmeyer and Volkmann-Kohlmeyer (2002)	
Dictyosporium toruloides (Corda) Guég.		Jones (1963)	
Drechslera sp.	Fell and Hunter (1979)		
Dumortieria sp.	Fell and Hunter (1979)		
Epicoccum nigrum Link	Fell and Hunter (1979)	Barata (2002)	
Flagellospora sp.	Fell and Hunter (1979)		
Floricola striata Kohlm. *et* Volkm.-Kohlm.	Kohlmeyer and Volkmann-Kohlmeyer (2000); Jones (2011)		
Fusarium incarnatum (Desm.) Sacc.		Walker and Campbell (2010)	
Fusarium spp.	Fell and Hunter (1979)		Van Ryckegem and Verbeken (2005b, 2005c)
Geniculosporium sp.	Fell and Hunter (1979)		
Gliocladium sp.	Fell and Hunter (1979)		

Table continued on next page.

Table 18.1 (Continued)

Fungi	Host Plant		
	Juncus roemerianus	*Spartina* spp.	*Phragmites australis*
Gliomastix spp.	Fell and Hunter (1979)		Poon and Hyde (1998a)
Halenospora varia (Anastasiou) E.B.G. Jones*		Barata (2002); Kohlmeyer and Volkmann-Kohlmeyer (2002)	
Haplobasidion lelebae Sawada ex M.B. Ellis	Fell and Hunter (1979)		
Hendersonia culmicola Sacc.		Barata (2002)	
Hendersonia culmiseda Sacc.			Van Ryckegem and Verbeken (2005b); Van Ryckegem et al. (2007)
Humicola sp.	Fell and Hunter (1979)		
Hymenopsis chlorothrix Kohlm. *et* Volkm.-Kohlm.*	Kohlmeyer and Volkmann-Kohlmeyer (2001a); Jones (2011)		
Hyphopolynema juncatile Kohlm. *et* Volkm.-Kohlm.	Kohlmeyer and Volkmann-Kohlmeyer (1999); Jones (2011)		
Khuskia oryzae H.J. Huds.	Fell and Hunter (1979)		
Kolletes undulatus Kohlm. *et* Volkm.-Kohlm.	Kohlmeyer et al. (2005); Jones (2011)		
Koorchaloma galateae Kohlm. *et* Volkm.-Kohlm.*	Kohlmeyer and Volkmann-Kohlmeyer (2001b); Jones (2011)		
Koorchaloma spartinicola V.V. Sarma, S.Y. Newell *et* K.D. Hyde*		Buchan et al. (2003)	
Leptosphaerulina chartarum Cec. Roux	Fell and Hunter (1979)		
Memnoniella echinata (Rivolta) Galloway	Fell and Hunter (1979)		

Table continued on next page.

Table 18.1 (Continued)

Fungi	Host Plant		
	Juncus roemerianus	*Spartina* spp.	*Phragmites australis*
Microsphaeropsis spp.			Poon and Hyde (1998a); Van Ryckegem and Verbeken (2005b); Van Ryckegem et al. (2007)
Monodictys austrina Tubaki	Fell and Hunter (1979)		
Myrothecium roridum Tode	Fell and Hunter (1979)		
Neottiospora sp.	Fell and Hunter (1979)		
Neottiosporina australiensis B. Sutton *et* Alcorn			Van Ryckegem and Verbeken (2005b, 2005c); Van Ryckegem et al. (2007)
Paecilomyces sp.	Fell and Hunter (1979)		
Papulaspora halima Anastasiou	Fell and Hunter (1979)		
Penicillium spp.	Fell and Hunter (1979)	Barata (2002)	Poon and Hyde (1998a)
Periconia cookei E.W. Mason *et* M.B. Ellis	Fell and Hunter (1979)		Van Ryckegem et al. (2007)
Periconia digitata (Cooke) Sacc.	Fell and Hunter (1979)		Van Ryckegem and Verbeken (2005b)
Periconia echinochloae (Bat.) M.B. Ellis	Fell and Hunter (1979)		
Periconia minutissima Corda	Fell and Hunter (1979)		Van Ryckegem and Verbeken (2005b)
Periconia sp.	Fell and Hunter (1979)		
Pestalotia planimi Vize		Barata (2002)	
Pestalotia sp.	Fell and Hunter (1979)		
Pestalotiopsis juncestris Kohlm. *et* Volkm.-Kohlm.	Kohlmeyer and Volkmann-Kohlmeyer (2001b); Jones (2011)		
Phaeoseptoria sp.			Van Ryckegem and Verbeken (2005b)

Table continued on next page.

Table 18.1 (Continued)

Fungi	Host Plant		
	Juncus roemerianus	*Spartina* spp.	*Phragmites australis*
Phaeosiaria sp.			Poon and Hyde (1998a)
Phialophorophoma litoralis Linder*		Barata (2002)	
Phialophorophoma sp.			Van Ryckegem and Verbeken (2005b, 2005c)
Phoma glomerata (Corda) Wollenw. *et* Hochapfel*		Barata (2002)	
Phoma suaedae Jaap*		Barata (2002)	
Phoma spp.	Fell and Hunter (1979)	Barata (2002)	Poon and Hyde (1998a); Van Ryckegem and Verbeken (2005b, 2005c); Van Ryckegem et al. (2007)
Phomopsis spp.	Fell and Hunter (1979)	Barata (2002)	Poon and Hyde (1998a)
Phyllosticta spartinae Brunaud		Barata (2002)	
Phyllosticta sp.		Barata (2002)	
Piricauda pelagica T. Johnson* (Nomen dubium)		Barata (2002); Kohlmeyer and Volkmann-Kohlmeyer (2002)	
Pithomyces atro-olivaceus (Cooke *et* Harkn.) M.B. Ellis	Fell and Hunter (1979)		
Pithomyces maydicus (Sacc.) M.B. Ellis	Fell and Hunter (1979)		Poon and Hyde (1998a)
Prathoda longissima (Deighton *et* MacGarvie) E.G. Simmons	Fell and Hunter (1979)		
Psammina sp.	Fell and Hunter (1979)		

Table continued on next page.

Table 18.1 (Continued)

Fungi	Host Plant		
	Juncus roemerianus	*Spartina* spp.	*Phragmites australis*
Pseudorobillarda phragmitis (Cunnell) M. Morelet*			Poon and Hyde (1998a)
Pseudorobillarda sp.		Barata (2002)	
Pseudoseptoria donacis (Pass.) B. Sutton			Van Ryckegem and Verbeken (2005b); Van Ryckegem et al. (2007)
Pycnodallia dupla Kohlm. *et* Volkm.-Kohlm.	Kohlmeyer and Volkmann-Kohlmeyer (2001a); Jones (2011)		
Pyrenochaeta sp.	Fell and Hunter (1979)		
Rhinocladiella spp.	Fell and Hunter (1979)		Poon and Hyde (1998a)
Scolecobasidium arenarium (Nicot) M.B. Ellis* (= *Dendryphiella arenaria*)		Barata (2002); Kohlmeyer and Volkmann-Kohlmeyer (2002)	
Scolecobasidium humicola G.L. Barron *et* L.V. Busch	Fell and Hunter (1979)		
Scolecobasidium salinum (G.K. Sutherl.) M.B. Ellis (= *Dendryphiella salina*)		Barata (2002); Kohlmeyer and Volkmann-Kohlmeyer (2002)	
Scopulariopsis spp.	Fell and Hunter (1979)		
Selenophoma sp.	Fell and Hunter (1979)		
Septogloeum spartinae (Ellis *et* Everh.) Wollenw. *et* Reinking		Barata (2002)	
Septonema secedens Corda	Fell and Hunter (1979)		
Septoria sp.	Fell and Hunter (1979)		Van Ryckegem and Verbeken (2005b)
Septoriella phragmitis Oudem.			Van Ryckegem and Verbeken (2005b, 2005c)

Table continued on next page.

Table 18.1 (Continued)

Fungi	Host Plant		
	Juncus roemerianus	*Spartina* spp.	*Phragmites australis*
Septoriella unigalerita Kohlm. *et* Volkm.-Kohlm.	Kohlmeyer and Volkmann-Kohlmeyer (2000); Jones (2011)		
Septoriella sp.			Poon and Hyde (1998a); Van Ryckegem and Verbeken (2005b, 2005c); Van Ryckegem et al. (2007)
Setosphaeria rostrata K.J. Leonard	Fell and Hunter (1979)	Barata (2002); Kohlmeyer and Volkmann-Kohlmeyer (2002)	
Spegazzinia tessarthra (Berk. *et* M.A. Curtis) Sacc.	Fell and Hunter (1979)		Poon and Hyde (1998a)
Sphaeronaema sp.	Fell and Hunter (1979)		
Sporothrix sp.	Fell and Hunter (1979)		
Stachybotrys chartarum (Ehrenb.) S. Hughes	Fell and Hunter (1979)		
Stachybotrys kampalensis Hansf.	Fell and Hunter (1979)		
Stachybotrys nephrospora Hansf.	Fell and Hunter (1979)		
Stachybotrys sp.	Fell and Hunter (1979)	Buchan et al. (2003)	Poon and Hyde (1998a)
Stachylidium bicolor Link	Fell and Hunter (1979)		
Stagonospora abundata Kohlm. *et* Volkm.-Kohlm.	Kohlmeyer and Volkmann-Kohlmeyer (2000); Jones (2011)		
Stagonospora cylindrica Gunnell			Van Ryckegem and Verbeken (2005c)
Stagonospora elegans (Berk.) Sacc. *et* Traverso			Van Ryckegem and Verbeken (2005b, 2005c)

Table continued on next page.

Table 18.1 (Continued)

Fungi	Host Plant		
	Juncus roemerianus	*Spartina* spp.	*Phragmites australis*
Stagonospora spp.	Fell and Hunter (1979)	Barata (2002); Buchan et al. (2003)	Poon and Hyde (1998a); Van Ryckegem and Verbeken (2005b, 2005c)
Stauronema sp.			Poon and Hyde (1998a)
Stemphylium lycopersici (Enjoji) W. Yamam.	Fell and Hunter (1979)		
Stemphylium maritimum T.W. Johnson*		Barata (2002)	
Stemphylium vesicarium (Wallr.) E.G. Simmons	Fell and Hunter (1979)		
Tetranacriella papillata Kohlm. *et* Volkm.-Kohlm.	Kohlmeyer and Volkmann-Kohlmeyer (2001b); Jones (2011)		
Tetranacrium sp.			Poon and Hyde (1998a)
Tetraplosphaeria tetraploa (Scheuer) Kaz. Tanaka *et* K. Hiray.	Fell and Hunter (1979)		Poon and Hyde (1998a)
Tiarosporella halmyra Kohlm. *et* Volkm.-Kohlm.*	Kohlmeyer and Volkmann-Kohlmeyer (1996b); Jones (2011)		
Tracyella spartinae (PK.) Tassi		Barata (2002)	
Trichoderma viride Pers.	Fell and Hunter (1979)		
Trichoderma sp.			Poon and Hyde (1998a)
Tubercularia sp.		Buchan et al. (2003)	
Veronaea sp.	Fell and Hunter (1979)		
Virgaria nigra (Link) Nees	Fell and Hunter (1979)		

Table continued on next page.

Table 18.1 (Continued)

Fungi	Host Plant		
	Juncus roemerianus	*Spartina* spp.	*Phragmites australis*
Xepicula leucotricha (Peck) Nag Raj	Fell and Hunter (1979)		
Zalerion maritima (Linder) Anastasiou*		Barata (2002); Kohlmeyer and Volkmann-Kohlmeyer (2002)	
Zythia spp.	Fell and Hunter (1979)		
Total taxa	**136**	**132**	**109**

*Marine fungi (based on Jones et al. 2009, 2011).
The list includes all the taxa associated with these plant hosts, mentioned in the following studies: Jones, 1963 (three taxa); Fell and Hunter 1979 (88 taxa); Kohlmeyer and Volkmann-Kohlmeyer 1993a (one taxa); Kohlmeyer and Volkmann-Kohlmeyer 1993b (two taxa); Kohlmeyer and Volkmann-Kohlmeyer 1994 (one taxa); Volkmann-Kohlmeyer and Kohlmeyer 1994 (one taxa); Kohlmeyer et al. 1995a (two taxa); Kohlmeyer et al. 1995b (four taxa); Kohlmeyer et al. 1995c (four taxa); Kohlmeyer et al. 1995d (four taxa); Kohlmeyer and Volkmann-Kohlmeyer 1996a (one taxa); Kohlmeyer and Volkmann-Kohlmeyer 1996b (one taxa); Kohlmeyer et al. 1996 (three taxa); Kohlmeyer et al. 1997 (three taxa); Kohlmeyer et al. 1998a (one taxa); Kohlmeyer et al. 1998b (three taxa); Poon and Hyde 1998a (40 taxa); Kohlmeyer and Volkmann-Kohlmeyer 1999 (one taxa); Kohlmeyer et al. 1999 (two taxa); Kohlmeyer and Volkmann-Kohlmeyer 2000 (three taxa); Kohlmeyer and Volkmann-Kohlmeyer 2001a (two taxa); Kohlmeyer and Volkmann-Kohlmeyer 2001b (three taxa); Barata 2002 (115 taxa); Buchan et al. 2002 (five taxa); Kohlmeyer and Volkmann-Kohlmeyer 2002 (40 taxa); Buchan et al. 2003 (16 taxa); Kohlmeyer et al. 2005 (three taxa); Van Ryckegem and Verbeken 2005b (58 taxa); Van Ryckegem and Verbeken 2005c (40 taxa); Van Ryckegem et al. 2007 (27 taxa); Lyons et al. 2010 (four taxa); Walker and Campbell 2010 (12 taxa); Jones 2011 (45 taxa). The names of the taxa follow the Index Fungorum (http://www.indexfungorum.org), except *Byssothecium obiones*.

(*Spartina* spp./*J. roemerianus*, *J. romerianus*/*Ph. australis*, *Spartina* spp./*Ph. australis* and *Spartina* spp./*J. roemerianus*/*Ph. australis*), is very low (<5%), which means that each host plant supports a distinct mycota. In fact, from the 332 taxa found on *Spartina* spp., *J. roemerianus* and/or *Ph. australis*, 89% are exclusively associated with one, 9% associated with two and 2% associated with the three host species, like *Remispora hamata* (doubtful species) and *Alternaria alternata* (Table 18.1).

Spartina species

Spartina appears to be conducive for the growth of saprobic fungi, given its high lignocellulose content and non-lignin cinnamyl phenols (Newell et al. 1996b). A total of 132 taxa of higher filamentous fungi have been documented from *Spartina* spp. (Table 18.1). Among the various species associated with this standing marsh grass, *Phaeosphaeria spartinicola* and *Mycosphaerella* sp. 2 are ubiquitous and dominant in the community, and exhibit high spore expulsion rates (Newell and Wasowski 1995; Newell and Zakel 2000; Newell 2001a; Buchan et al. 2002, 2003; Lyons et al. 2010). Additional common species in this community are *Phaeosphaeria halima* and *Buergenerula spartinae* (Newell and Zakel 2000; Newell 2001a; Buchan et al. 2002, 2003; Lyons et al. 2010; Walker and Campbell 2010).

Juncus roemerianus

Another salt marsh plant, *Juncus roemerianus*, supports a surprisingly high number of fungal taxa (Table 18.1). Most of the fungi associated with this host have been collected and identified by the Kohlmeyers and their colleagues in marshes located on the east coast of the United States. Jones (2011) has recently updated the list of fungi documented by Kohlmeyer and Volkmann-Kohlmeyer (2001c). One-hundred-thirty-six fungi have been identified on *J. roemerianus* (Table 18.1). Newell and Porter (2002) have pointed out that *Loratospora aestuarii*, *Papulosa amerospora*, *Aropsiclus junci*, *Anthostomella poecila*, *Physalospora citogerminans*, *Scirrhia annulata*, *Massarina ricifera* and *Tremateia halophila* are the most common fungi occurring on this substrate.

Phragmites australis

Phragmites australis, another host that has been extensively surveyed for fungi, is a cosmopolitan grass that has a widespread worldwide distribution, colonizing not only intertidal marshes but also freshwater and terrestrial habitats in temperate and subtropical regions. Over 300 fungi have been reported for this plant (Wong and Hyde 2001, 2002), of which 109 species were detected in intertidal marshes in Hong Kong (Table 18.1).

Poon and Hyde (1998a) identified 41 species associated with *Ph. australis* in an intertidal subtropical marsh in Hong Kong, from which *Massarina phragmiticola*, *Phomatospora phragmiticola* and *Cytoplacosphaeria phragmiticola* were described as new species (Table 18.1). Wong et al. (1998) described *Phragmitensis marina* from the same locality, while Poon and Hyde (1998a) noted that *Lignincola laevis*, *Trichoderma* sp., *Halosarpheia phragmiticola* and *Colletotrichum* sp. were the most common species on this host.

Host tissue preference

In addition to host-specificity, the majority of the saprobic fungi also exhibit other ecological requirements that determine its occurrence on different parts of the host plant (Newell and Wasowsky 1995; Newell et al. 1996b; Kohlmeyer and Volkmann-Kohlmeyer 2001c; Newell 2001a). This pattern may be a result of a nutritive preference for a particular substrate and/or interactions of mutualism, parasitism or competition that are established among the different species of fungi.

Phaeosphaeria spartinicola and *Mycosphaerella* sp. 2, for example, occur preferably on the leaf blades of *Spartina* spp. and are involved in lignocellulose degradation (Bergbauer and Newell 1992; Newell et al. 1996b; Newell and Porter 2002); both species seem to establish a highly-efficient mutualistic relationship that suppresses potential competitors (Newell 2001a; Newell and Porter 2002; Buchan et al. 2003). *Buergenerula spartinae* is dominant on leaf sheaths of *Spartina* spp., occurring in non-melanized patches (Newell 2001a), but it can also colonize the leaf blades, developing characteristic melanotic patches (Buchan et al. 2002). These findings support the hypothesis proposed by Newell and Porter (2002) that the melanization of leaf blades by *B. spartinae* is the result of competition between this species and the complex of *Ph. spartinicola* and *Mycosphaerella* sp. 2.

In a given host plant, the colonization of the diverse plant tissues by saprobic fungi may or may not proceed simultaneously. Van Ryckegem and Verbeken (2005c) have observed that fungal sexual-reproductive structures in *Ph. australis* stems only appeared after three months of senescence, when nearly 50% of the leaf sheath tissue was decomposed. This time lag in the colonization process was attributed by the fact that the stems were more recalcitrant and consequently less susceptible to fungal breakdown than the sheaths (i.e. there were fewer stomata, more sclerenchymatous tissue and thicker cuticle).

Species composition of the fungal community associated with a particular plant tissue does not generally remain unchanged, since the plant material undergoes physical and chemical changes during decomposition and different fungi require specific nutrients for their metabolism. Fell and Hunter (1979) directly compared the fungal communities colonizing *J. roemerianus* leaves of different physiological states and distinguished fungi that occurred mainly or exclusively in living, senescent or dead-standing leaves. Buchan et al. (2002, 2003) reached the same conclusions, having found fungi present in early- or late-decay blades of *S. alterniflora* and other species in both decay stages. Van Ryckegem and Verbeken (2005c) argued that the reduction of carbon resources may lead to an overlap of ecological niches, thereby contributing to an increase in competition between species; in this context, the species with higher antibiotic activity and with a broader spectrum of enzymes would certainly be favored.

Thus, in a particular decaying plant tissue, we can frequently detect patterns in the succession of fungi, with total or partial replacement of the colonizing species (Kohlmeyer and Volkmann-Kohlmeyer 2001c; Buchan et al. 2002; 2003; Van Ryckegem and Verbeken 2005b, 2005c; Van Ryckegem et al. 2007). Van Ryckegem and Verbeken (2005b) followed the decay and simultaneous colonization process of the leaf sheaths of *Ph. australis* by saprobic fungi, and characterized three successive phases:

(i) The process begins with a pioneer community composed of few species, like *Sporobolomyces* sp., *Alternaria alternata*, *Cladosporium* sp., *Septoriella* sp., *Phoma* sp. and *Phaeosphaeria* sp. (weak pathogens, biotrophic species and opportunistic saprotrophs). These exhibit a low tolerance to stress.
(ii) The second phase includes a more closed and mature community, with a high diversity of more competitive species, such as *Stictis* sp. and *Lophodermium arundinaceum*.
(iii) The third phase presents an impoverished community dominated by a few stress-tolerant species and/or highly competitive taxa, like *Lentithecium arundinaceum* and *Mycosphaerella lineolata*.

The presence of dominant fungi in plant material, however, seems to be independent of the degree of decay of plant tissue (Buchan et al. 2002), which apparently only has effects on the spore-expulsion rate (Newell 2001a; Buchan et al. 2003).

In the latter stages of decomposition, the dead plant tissues finally detach and reach the marsh sediment surface, which offers different microenvironmental conditions. Under these new abiotic conditions, a shift in the species composition of the fungal community occurs with an alteration of the productivity of the community (Newell et al. 1989; Kohlmeyer and Volkmann-Kohlmeyer 2001c; Van Ryckegem and Verbeken 2005b, 2005c; Van Ryckegem et al. 2006, 2007).

Ecological requirements

Tidal regime

In marsh ecosystems fungi usually display patterns of vertical distribution on the colonized emergent macrophytes, which go beyond their nutritive requirements and ability to compete with other fungi but also reflect their tolerance limits to environmental conditions, particularly tidal flooding (Gessner 1977; Barata 1997; Poon and Hyde 1998b; Kohlmeyer and Volkmann-Kohlmeyer 2001c; Barata 2002; Van Ryckegem and Verbeken 2005a, 2005b, 2005c; Van Ryckegem et al. 2006, 2007). Vertical distribution of fungi may distinguish between obligate and facultative marine species. Gessner (1977), Kohlmeyer and Volkmann-Kohlmeyer (2001c) and Barata (2002) characterized the mycota associated with *S. alterniflora, S. maritima* and *J. roemerianus*. They found that in general terrestrial halotolerant fungi occurred at the tips of the leaves, which rarely or never become submerged. Obligate marine fungi occupy the lower periodically-inundated portions of the leaves, and facultative marine fungi overlap both plant portions.

Seasonality

Environmental fluctuations inherent to seasonality do not seem to interfere with the species composition of fungal communities associated with a particular marsh grass (Torzilli et al. 2006). Nevertheless, seasonality has significant impact on the differentiation of reproductive structures (Newell 2001a) and the abundance of several species (Buchan et al. 2003), as well as on the biomass and productivity of the entire fungal community (Samiaji and Barlocher 1996; Castro and Freitas 2000; Newell 2001b; Newell and Porter 2002).

Even though Castro and Freitas (2000) and Newell (2001b) have shown that fungal biomass and productivity increased during winter and spring and diminished during the summer and fall periods, the potential reasons for these phenomena were different. Castro and Freitas (2000) regarded the decrease of fungal biomass during summer with higher air temperatures and salinity resulted in less favorable moisture conditions for fungal survival. Contrarily, Newell (2001b) suggested that the higher tides and higher rainfall that occurred in the late summer and fall during that particular study, could have contributed to a reduction of fungal biomass and productivity because the leaves had become more accessible to mycophagous invertebrates and/or bacterial competitors, and had lost more nutrients through leaching.

Nitrogen

The availability of nitrogen in plant tissues and the surrounding environment is likely to be one of the key factors that regulate decomposer activity (Torzilli and Andykovitch 1986) and fungal productivity (Newell et al. 1996a, 2000b; Newell 2001b). In fact, the studies previously mentioned have demonstrated that when nitrogen increases, fungal productivity also increases; this observation seems to suggest that nitrogen might be the limiting factor in these ecosystems.

Water supply

Newell et al. (1996a) have demonstrated that the fungal community is more productive in a normal regime of repeated drying/wetting episodes, being incapable of achieving higher production when water supply is constant, which proves that this community is well adapted to these environments. On the other hand, Poon and Hyde (1998b) and Wong and Hyde (2002) compared the fungal community associated with *Ph. australis* in two intertidal areas subjected to distinct water-availability regimes, and observed that fungal diversity was higher under periodic submersion than under a permanent one. Barata (2006) also confirmed this finding in a study performed with *S. maritima* baits in two zones of a salt marsh that were exposed to different periods of submergence; the intertidal zone was the most favorable place for colonization by a high diversity of marine fungi. Newell (1995) added that when the plant tissues become dry during a dry season, the colonizing fungi can interrupt their metabolic activity and resume when humid conditions are restored, without any loss of biomass.

Pollution

The impact of anthropogenic pollution on fungal community dynamics is still poorly understood (Pointing and Hyde 2000). However, some studies carried out in contaminated salt marshes found that the fungal communities remain unchanged, which suggests a higher resistance and resilience of fungal and plant communities to these pollutants (Newell and Wall 1998; Newell et al. 2000a; Wall et al. 2001).

Conclusion

Despite diverse studies of salt marsh fungi, it is quite possible that the total fungal diversity associated with these ecosystems is not yet known. Neither are the abiotic and biotic factors that determine the presence of a given fungal species. The continuing investment in this field of investigation is therefore fundamental, and should encompass different geographic regions and/or other host plants in order to fill this knowledge gap.

References

Allen JRL, Pye K. Coastal saltmarshes: their nature and importance. In: Allen JRL, Pye K, eds. Saltmarshes: morphodynamics, conservation, and engineering significance. Cambridge: Cambridge University Press, 1992:1–18.

Apinis AE, Chesters CGC. Ascomycetes of some salt marshes and sand dunes. Trans Brit Mycol Soc. 1964;47:419–435.

Asaeda T, Nama LH, Hietz P, Tanaka N, Karunaratne S. Seasonal fluctuations in live and dead biomass of *Phragmites australis* as described by a growth and decomposition model: implications of duration of aerobic conditions for litter mineralization and sedimentation. Aquat Bot. 2002;73:223–239.

Barata M. Fungos marinhos superiores associados a *Spartina maritima* em estuários da costa Portuguesa. PhD thesis, Faculdade de Ciências da Universidade de Lisboa, Lisbon, 1997.

Barata M. Fungi on the halophyte *Spartina maritima* in salt marshes. In: Hyde KD, ed. Fungi in marine environments. Hong Kong: Fungal Diversity Press, 2002:179–193.

Barata M. Marine fungi from Mira river salt marsh in Portugal. Rev Iberoam Micol. 2006;23: 179–184.

Benner R, MacCubbin E, Hodson RE. Preparation, characterization and microbial degradation of specifically radiolabeled [14C] lignocellulose from marine and freshwater macrophytes. Appl Environ Microbiol. 1984a;47:381–389.

Benner R, Newell SY, Maccubbin AE, Hodson RE. Relative contributions of bacteria and fungi to rates of degradation of lignocellulosic detritus in salt-marsh sediments. Appl Environ Microbiol. 1984b;48:36–40.

Bergbauer M, Newell SY. Contribution to lignocellulose degradation and DOC formation from a salt marsh macrophyte by the ascomycete *Phaeosphaeria spartinicola*. FEMS Microbiol Ecol. 1992;86:341–348.

Blum LK, Roberts MS, Garland JL, Mills AL. Distribution of microbial communities associated with the dominant high marsh plants and sediments of the United States East Coast. Microb Ecol. 2004;48:375–388.

Boorman LA. Salt marshes – present functioning and future change. Mangroves and Salt Marshes 1999;3:227–241.

Bouchard V, Lefeuvre JC. Primary production and macro-detritus dynamics in a European salt marsh: carbon and nitrogen budgets. Aquat Bot. 2000;67:23–42.

Buchan A, Newel SY, Moreta JIL, Moran MA. Analysis of internal transcribed spacer (ITS) regions of rRNA genes in fungal communities in southeastern U.S. Salt Marsh. Microb Ecol. 2002;43:329–340.

Buchan A, Newell SY, Butler M, Biers EJ, Hollibaugh JT, Moran MA. Dynamics of bacterial and fungal communities on decaying salt marsh grass. Appl Environ Microbiol. 2003;69: 6676–6687.

Castro P, Freitas H. Fungal biomass and decomposition in *Spartina maritima* leaves in the Mondego salt marsh (Portugal). Hydrobiol. 2000;428:171–177.

Chapman VJ. Wet coastal ecosystems – Ecosystems of the world. New York: Elsevier, 1977.

Christian RR, Bryant WL, Brinson MM. *Juncus roemerianus* production and decomposition along gradients of salinity and hydroperiod. Mar Ecol Prog Ser. 1990;68:137–145.

Fell JW, Hunter IL. Fungi associated with the decomposition of the black rush, *Juncus roemerianus*, in south Florida. Mycologia 1979;71:322–342.

Findlay SEG, Dye S, Kuehn KA. Microbial growth and nitrogen retention in litter of *Phragmites australis* compared to *Typha angustifolia*. Wetlands 2002;22:616–625.

Gessner MO. Mass loss, fungal colonization and nutrient dynamics of *Phragmites australis* leaves during senescence and early decay. Aquat Bot. 2001;69:325–339.

Gessner RV. Seasonal occurrence and distribution of fungi associated with *Spartina alterniflora* from Rhode Island estuary. Mycologia 1977;69:477–491.

Gessner RV, Goos RD. Fungi from decomposing *Spartina alterniflora*. Can J Bot. 1973;51:51–55.

Gessner RV, Kohlmeyer J. Geographical distribution and taxonomy of fungi from salt marsh *Spartina*. Can J Bot. 1976;54:2023–2037.

Gessner MO, Gulis V, Kuehn K, Chauvet E, Suberkropp K. Fungal decomposers of plant litter in aquatic ecosystems. In: Kubicek C, Druzhinia I, eds. Environmental and microbial relationships: The Mycota, Volume 4. Berlin: Springer, 2007:301–324.

Hyde KD, Jones EBG, Leano E, Pointing SB, Poonyth AD, Vrijmoed LLP. Role of fungi in marine ecosystems. Biodivers Conserv. 1998;7:1147–1161.

Jones EBG. Marine fungi: II. Ascomycetes and deuteromycetes from submerged wood and drift Spartina. Trans Br Mycol Soc. 1963;46:135–144.

Jones EBG. Fifty years of marine mycology. Fungal Divers. 2011;50:73–112.

Kohlmeyer J, Kohlmeyer E. Marine mycology – The higher fungi. New York: Academic Press, 1979.

Kohlmeyer J, Volkmann-Kohlmeyer B. Two new genera of Ascomycotina from saltmarsh *Juncus*. Syst Ascomycet. 1993a;11:95–106.

Kohlmeyer J, Volkmann-Kohlmeyer B. *Atrotorquata* and *Loratospora*: new ascomycete genera on *Juncus roemerianus*. Syst Ascomycet. 1993b;12:7–22.

Kohlmeyer J, Volkmann-Kohlmeyer B. *Aropsiclus* nom. nov. (Ascomycotina) to replace *Sulcospora* Kohlm. & Volkm.-Kohlm. Syst Ascomycet. 1994;13:24.

Kohlmeyer J, Volkmann-Kohlmeyer B. Fungi on *Juncus roemerianus*. 6. *Glomerobolus* gen. nov., the first ballistic member of Agonomycetales. Mycologia 1996a;88:328–337.

Kohlmeyer J, Volkmann-Kohlmeyer B. Fungi on *Juncus roemerianus*. 7. *Tiarosporella halmyra* sp. nov. Mycotaxon 1996b;59:79–83.

Kohlmeyer J, Volkmann-Kohlmeyer B. Fungi on *Juncus roemerianus*. 13. *Hyphopolynema juncatile* sp. nov. Mycotaxon 1999;70:489–495.

Kohlmeyer J, Volkmann-Kohlmeyer B. Fungi on *Juncus roemerianus*. 14. Three new coelomycetes, including *Floricola*, anam.-gen. nov. Bot Mar. 2000;43:385–392.

Kohlmeyer J, Volkmann-Kohlmeyer B. Fungi on *Juncus roemerianus*: new coelomycetes with notes on *Dwayaangam junci*. Mycol Res. 2001a;105:500–505.

Kohlmeyer J, Volkmann-Kohlmeyer B. Fungi on *Juncus roemerianus*. 16. More new coelomycetes, including *Tetranacriella*, gen. nov. Bot Mar. 2001b;44:147–156.

Kohlmeyer J, Volkmann-Kohlmeyer B. The biodiversity of fungi on *Juncus roemerianus*. Mycol Res News 2001c;105:1411–1412.

Kohlmeyer J, Volkmann-Kohlmeyer B. Fungi on *Juncus* and *Spartina*: New marine species of *Anthostomella*, with a list of marine fungi known on *Spartina*. Mycol Res. 2002;106:365–374.

Kohlmeyer J, Volkmann-Kohlmeyer B, Eriksson OE. Fungi on *Juncus roemerianus*. 2. New Dictyosporous Ascomycetes. Bot Mar. 1995a;38:165–174.

Kohlmeyer J, Volkmann-Kohlmeyer B, Eriksson OE. Fungi on *Juncus roemerianus*. 3. New Ascomycetes. Bot Mar. 1995b;38:175–186.

Kohlmeyer J, Volkmann-Kohlmeyer B, Eriksson OE. Fungi on *Juncus roemerianus*. 4. New marine ascomycetes. Mycologia 1995c;87:532–542.

Kohlmeyer J, Volkmann-Kohlmeyer B, Eriksson OE. Fungi on *Juncus roemerianus*. New marine and terrestrial ascomycetes. Mycol Res. 1995d;100:393–404.

Kohlmeyer J, Volkmann-Kohlmeyer B, Eriksson OE. Fungi on *Juncus roemerianus*. 8. New bitunicate ascomycetes. Can J Bot. 1996;74:1830–1840.

Kohlmeyer J, Volkmann-Kohlmeyer B, Eriksson OE. Fungi on *Juncus roemerianus*. 9. New obligate and facultative marine Ascomycotina. Bot Mar. 1997;40:291–300.

Kohlmeyer J, Baral H-O, Volkmann-Kohlmeyer B. Fungi on *Juncus roemerianus*. 10. A new *Orbilia* with ingoldian anamorph. Mycologia 1998a;90:303–309.

Kohlmeyer J, Volkmann-Kohlmeyer B, Eriksson OE. Fungi on *Juncus roemerianus*. 11. More new ascomycetes. Can J Bot. 1998b;76:467–477.

Kohlmeyer J, Volkmann-Kohlmeyer B, Eriksson OE. Fungi on *Juncus roemerianus*. 12. Two new species of *Mycosphaerella* and *Paraphaeosphaeria* (Ascomycotina). Bot Mar. 1999;42:505–511.

Kohlmeyer J, Volkmann-Kohlmeyer B, Tsui CKM. Fungi on *Juncus roemerianus* 17. New ascomycetes and the hyphomycete genus *Kolletes* gen. nov. Bot Mar. 2005;48:306–317.

Lyons JI, Newell SY, Brown RP, Moran MA. Screening for bacterial–fungal associations in a south-eastern US salt marsh using pre-established fungal monocultures. FEMS Microbiol Ecol. 2005;54:179–187.

Lyons JI, Alber M, Hollibaugh JT. Ascomycete fungal communities associated with early decaying leaves of *Spartina* spp. from central California estuaries. Oecologia 2010;162:435–442.

Maccubbin AE, Hodson RE. Mineralization of detrital lignocelluloses by salt marsh sediment microflora. Appl Environ Microbiol. 1980;40:735–740.

McLusky DS, Elliott M. The Estuarine Ecosystem – Ecology, threats, and management. 3rd ed. Oxford: Oxford University Press, 2004.

Menéndez M, Sanmartí N. Geratology and decomposition of *Spartina versicolor* in a brackish Mediterranean marsh. Estuar Coast Shelf S. 2007;74:320–330.

Newell SY. Minimizing ergosterol loss during preanalytical handling and shipping of samples of plant litter. Appl Environ Microbiol. 1995;61:2794–2797.

Newell SY. Established and potential impacts of eukaryotic mycelial decomposers in marine/terrestrial ecotones. J Exp Mar Biol Ecol. 1996;200:187–206.

Newell SY. Spore-expulsion rates and extents of blade occupation by ascomycetes of the smooth-cordgrass standing-decay system. Bot Mar. 2001a;44:277–285.

Newell SY. Multiyear patterns of fungal biomass dynamics and productivity within naturally decaying smooth-cordgrass shoots. Limnol Oceanogr. 2001b;46:573–583.

Newell SY, Fallon RD. Litterbags, leaf tags, and decay of nonabscised intertidal leaves. Can J Bot. 1989;67:2324–2327.

Newell SY, Porter D. Microbial secondary production from saltmarsh-grass shoots, and its known and potential fates. In: Weinstein MP, Kreeger DA, eds. Concepts and controversies in tidal marsh ecology. Amsterdam: Kluwer, 2002:159–185.

Newell SY, Wall VD. Response of saltmarsh fungi to presence of mercury and polychlorinated biphenyls at a Superfund Site. Mycologia 1998;90:777–784.

Newell SY, Wall VD, Maruya KA. Fungal biomass in saltmarsh-grass blades at two contaminated sites. Arch Environ Con Tox. 2000a;38:268–273.

Newell SY, Wasowsky J. Sexual productivity and spring intramarsh distribution of a key salt-marsh microbial secondary producer. Estuaries 1995;18:241–249.

Newell SY, Zakel KL. Measuring summer patterns of ascospore release by saltmarsh fungi. Mycoscience 2000;41:211–215.

Newell SY, Fallon RD, Miller JD. Decomposition and microbial dynamics for standing, naturally positioned leaves of the salt-marsh grass *Spartina alterniflora*. Mar Biol. 1989;101:471–481.

Newell SY, Arsuffi TL, Palm LA. Misting and nitrogen fertilization of shoots of a saltmarsh grass: effects upon fungal decay of leaf blades. Oecologia 1996a;108:495–502.

Newell SY, Porter D, Lingle WL. Lignocellulolysis by ascomycetes (Fungi) of a saltmarsh grass (smooth cordgrass). Microsc Res Techniq. 1996b;33:32–46.

Newell SY, Arsuffi TL, Palm LA. Seasonal and vertical demography of dead portions of shoots of smooth cordgrass in a south-temperate saltmarsh. Aquat Bot. 1998;60:325–335.

Newell SY, Blum LK, Crawford RE, Dai T, Dionne M. Autumnal biomass and productivity of saltmarsh fungi from 29 to 43 degrees north latitude along the United States Atlantic coast. Appl Environ Microbiol. 2000b;66:180–185.

Pointing SB, Hyde KD. Lignocellulose-degrading marine fungi. Biofouling 2000;15:221–229.

Poon MOK, Hyde KD. Biodiversity of intertidal estuarine fungi on *Phragmites* at Mai Po Marshes, Hong Kong. Bot Mar. 1998a;41:141–155.

Poon MOK, Hyde KD. Evidence for the vertical distribution of saprophytic fungi on senescent *Phragmites australis* culms at Mai Po marshes. Bot Mar. 1998b;41:285–292.

Raghukumar S. The role of fungi in marine detrital processes. In: Ramaiah N, ed. Marine microbiology: Facets and opportunities. Goa: Nat Inst Ocean, 2004:91–101.

Samiaji J, Barlocher F. Geratology and decomposition of *Spartina alterniflora* Loisel in a New Brunswick saltmarsh. J Exp Mar Biol Ecol. 1996;201:233–252.

Simas T, Nunes JP, Ferreira JG. Effects of global climate change on coastal salt marshes. Ecol Model. 2001;139:1–15.

Teal JM. Energy flow in the salt marsh ecosystem of Georgia. Ecology 1962;43:614–624.

Torzilli AP. Polysaccharidase production and cell wall degradation by several salt marsh fungi. Mycologia 1982;74:297–302.

Torzilli AP, Andrykovitch G. Degradation of *Spartina* lignocellulose by individual and mixed cultures of salt-marsh fungi. Can J Bot. 1986;64:2211–2215.

Torzilli AP, Sikaroodi M, Chalkley D, Gillevet PM. A comparison of fungal communities from four salt marsh plants using automated ribosomal intergenic spacer analysis (ARISA). Mycologia 2006;98:690–698.

Van Ryckegem G, Verbeken A. Fungal diversity and community structure on *Phragmites australis* (Poaceae) along a salinity gradient in the Scheldt estuary (Belgium). Nova Hedw. 2005a;80:173–197.

Van Ryckegem G, Verbeken A. Fungal ecology and succession on *Phragmites australis* in a brackish tidal marsh. I. Leaf sheaths. Fungal Divers. 2005b;19:157–187.

Van Ryckegem G, Verbeken A. Fungal ecology and succession on *Phragmites australis* in a brackish tidal marsh. II. Stems. Fungal Divers. 2005c;20:209–233.

Van Ryckegem G, Van Driessche G, Van Beeumen JJ, Verbeken A. The estimated impact of fungi on nutrient dynamics during decomposition of *Phragmites australis* leaf sheaths and stems. Microb Ecol. 2006;52:564–574.

Van Ryckegem G, Gessner MO, Verbeken A. Fungi on leaf blades of *Phragmites australis* in a brackish tidal marsh: diversity, succession, and leaf decomposition. Microb Ecol. 2007;53:600–611.

Volkmann-Kohlmeyer B, Kohlmeyer J. A new *Aniptodera* from saltmarsh *Juncus*. Bot Mar. 1994;37:109–113.

Walker AK, Campbell J. Marine fungal diversity: a comparison of natural and created salt marshes of the north-central Gulf of Mexico. Mycologia 2010;102:513–521.

Wall VD, Alberts JJ, Moore DJ, Newell SY, Pattanayek M, Pennings SC. The effect of mercury and PCBs on organisms from lower trophic levels of a Georgia salt marsh. Arch Environ Con Tox. 2001;40:10–17.

Wong MKM, Hyde KD. Diversity of fungi on six species of Gramineae and one species of Cyperaceae in Hong Kong. Mycol Res. 2001;105:1485–1491.

Wong MKM, Hyde KD. Fungal saprobes on standing grasses and sedges in a subtropical aquatic habitat. In: Hyde KD, ed. Fungi in marine environments. Hong Kong: Fungal Divers. 2002:195–212.

Wong MKM, Poon MOK, Hyde KD. *Phragmitensis marina* gen. *et* sp. nov., an intertidal saprotroph from *Phragmites australis* in Hong Kong. Bot Mar. 1998;41:379–382.

19 Diversity and ecology of marine-derived fungi

Purnima Singh, Xin Wang, Keming Leng and Guangyi Wang

Fungi are a key component of the biosphere and fulfill a wide range of important biogeochemical and ecological functions, including those related to nutrient cycling processes in both aquatic and terrestrial environments (Christensen 1989; Pang and Mitchell 2005). Fungi in marine environments have been referred to as "marine-derived" since the marine ecosystem comprises obligate and facultative marine species (Bugni and Ireland 2004; Jones 2011a, 2011b). Obligate marine fungi can grow and sporulate in a marine environment; while facultative ones are those from terrestrial or freshwater habitats that may also grow in the marine environment (Kohlmeyer and Volkmann-Kohlmeyer 2003; Bugni and Ireland 2004). Although marine fungi are known to be distinct in their physiology and morphology compared to those from freshwater habitats (Meyers 1996), they are still considered an ecologically, not taxonomically, defined microbial group (Kohlmeyer and Volkmann-Kohlmeyer 2003). Compared with other microorganisms in the ocean, fungi are still one of the most under-studied microbial groups. Recent investigations on fungi in seawater (Gao et al. 2008; Gutierrez et al. 2010, 2011), deep-sea environments (Le Calvez et al. 2009; Burgaud et al. 2010; Raghukumar et al. 2010), sponges (Gao et al. 2008) and other habitats (selected chapters in this book), have identified novel diversity and implicated important biogeochemical functions. This chapter summarizes the recent knowledge on the diversity and ecology of fungi derived from marine sponges and deep-sea environments.

Diversity and ecology of fungi derived from marine sponges

Marine sponges as a microbial habitat

Sponges (phylum *Porifera*) include an estimated 15,000 species in three taxonomic classes: *Calcarea* (calcareous sponges), *Hexactinellida* (glass sponges), and *Demospongiae* (demosponges) (Hooper et al. 2005). They constitute an important component of benthic communities in the world's oceans because of both their biomass and biogeochemical functions in benthic or pelagic processes (Maldonado et al. 2005a; Wang 2006; Kochkina et al. 2007; Taylor et al. 2007). As sedentary filter-feeding organisms, sponges have a remarkable ability to obtain food from the surrounding water and a 1-kg sponge can pump up to 24,000 liters of seawater per day (Hentschel et al. 2006; Wang 2006; Taylor et al. 2007). A typical milliliter of seawater contains over 10^3 fungal cells, 10^6 bacteria, and 10^7 viruses (Gao et al. 2008). Any planktonic microbe can become a resident in sponges, provided that it can survive their digestion and immune response, and is capable of growing in the microenvironment inside the mesohyl (Zhu et al. 2008). Marine sponges have recently become an interesting target of intensive investigations for three major reasons.

(i) First, sponges harbor a wide variety of microorganisms. In high-microbial-abundance sponges, the bacterial population constitutes as much as 40% of the sponge tissue volume, and microbial densities can reach over 10^9 cells per ml of sponge tissue (Hentschel et al. 2006), a magnitude several orders higher than the number typically found in seawater. Cultivation-dependent and molecular approaches have identified phylogenetically-diverse microbial groups, including all three domains of life, i.e. *Bacteria, Archaea,* and *Eukarya,* some of which are sponge specific (for reviews, see Hentschel et al. 2006; Wang 2006; Taylor et al. 2007; Webster and Blackall 2009; Webster et al. 2010).

(ii) Second, sponges are a rich source of bioactive secondary metabolites and other biotechnological products. For example, sponge-derived fungi account for the largest number (33%) of total bioactive compounds in the literature and have the overall highest number of novel metabolites (Bugni and Ireland 2004).

(iii) Finally, sponges play a significant role in the biogeochemical processes of marine ecosystems, particularly in coral reef ecosystem (Maldonado et al. 2005a; Hoffmann et al. 2009).

Previous studies of sponge-associated microbes have been largely focused on prokaryotes, i.e. *Bacteria* and *Archaea* (Taylor et al. 2007). Eukaryotic microbes, such as dinoflagellates and diatoms, have also been reported to be present in marine sponges (Gao et al. 2008). In the next section, we focus on fungi associated with marine sponges.

Culturable fungi derived from marine sponges

Sponges display a ubiquitous association with fungi (Wang 2006). Studies on sponge-derived fungi have mostly centered on natural product chemistry, with relatively few investigations dealing with their biology (Höller et al. 2000; Bugni and Ireland 2004). In an effort to explore the natural compounds from fungi associated with marine sponges, a total of 681 fungal strains were obtained from 16 species of sponges collected at six different locations, including temperate, subtropical, and tropical regions (Höller et al. 2000). The isolated fungi belonged to 13 genera of Ascomycota, two of Zygomycota, and 37 of asexual fungi.

Marine sponges from different geographical locations have been reported to harbor diverse forms of fungi. Members of the genera *Acremonium, Arthrium, Coniothyrium, Fusarium, Mucor, Penicillium, Phoma, Trichoderma,* and *Verticillium* were identified from almost all locations, although different fungal genera were dominant at different locations. The fungal diversity associated with a similar species of marine sponge from different locations varied greatly.

Certain fungal genera were dominant in some sponges, and these varied among different sponge species collected at the same location. A study of filamentous fungi associated with marine sponges in the coral reefs of Palau and Bunaken Island (Indonesia) indicated that the fungal diversity associated with sponges varied from one site to another (Namikoshi et al. 2002). The lack of proper identification of fungal isolates and sponges render it difficult to evaluate the diversity of fungi associated with these marine sponges. In a survey of filamentous fungi from Australian coral reefs, 208 fungal isolates were cultured from 70 sponge samples (Morrison-Gardiner 2002), where *Alternaria, Aspergillus, Cladosporium, Fusarium*, and *Penicillium* were found to be dominant fungal genera. However, fungal diversity associated with the different sponge species was not apparent because sponge samples collected at various locations were combined without identification of individual sponge species. Overall, the majority of fungal isolates identified from marine sponges using morphology-based approaches belong to the genera commonly found in terrestrial habitats (e.g. *Acremonium, Alternaria, Aspergillus, Cladosporium, Fusarium, Trichoderma,* and *Penicillium*). The diversity of culturable fungal assemblages varied greatly, ranging from zero to 21 fungal genera per sponge species (Höller et al. 2000; Morrison-Gardiner 2002; Namikoshi et al. 2002; Wang 2006).

Fungi isolated from natural substrates/hosts, including sponges, in many cases appear as vegetative mycelia, thus posing an identification challenge (Gao et al. 2008). Recent advances in identifying species at the molecular level have substantially improved our understanding of the species concept of fungi. The phylogenetic diversity of cultured fungi from the Hawaiian marine sponges *Suberites zeteki* and *Gelliodes fibrosa* has been reported to be greater than that of other marine sponge species mentioned earlier (Höller et al. 2000; Namikoshi et al. 2002; Wang et al. 2008). In an another study (Li and Wang 2009), the overall phylogenetic composition of fungal genera in the Hawaiian sponges *G. fibrosa, Haliclona caerulea*, and *Mycale armata* was quite different from that in the study by Höller et al. (2000). This difference in the diversity of culturable fungi may be ascribed to the different protocols used for their isolation and identification (Li and Wang 2009). A real difference in fungal communities between geographical locations is also possible. Furthermore, the phylogenetic analysis of fungal isolates from these three Hawaiian sponges revealed that ~17% of these isolates were closely affiliated with marine fungi from other habitats, indicating their obligate marine nature. Thus, molecular studies have further improved our understanding of the diversity of culturable fungal communities in marine sponges.

Molecular analysis of fungal communities associated with sponges

Many primers, designed to specifically target a fungal 18S rRNA gene or ITS regions can also amplify the rRNA gene sequences of *Porifera* (Pang and Mitchell 2005). Thus, direct molecular analysis of the fungal community in marine sponges using polymerase chain reaction (PCR)-based approaches has been challenging. A molecular diversity investigation of the fungal communities associated with the Hawaiian sponges *S. zeteki* and *M. armata* was initiated for the first time by Gao et al. (2008). In that study, most of the universal fungal primers designed to preferentially amplify fungal rRNA genes from environmental samples yielded largely sponge rRNA gene sequences. The nested PCR strategy using the primer pairs ITS1F/ITS4 and ITS3/ITS4-GC was reported to give a satisfactory result in discriminating against sponge rRNA genes while maintaining a broad range of fungal compatibility. Denaturing gradient gel electrophoresis, in combination with clone library construction, revealed novel fungal diversity in *S. zeteki* and *M. armata* and resulted in the identification of 11 taxonomic orders belonging to two phyla: Basidiomycota and Ascomycota (Figure 19.1 and Table 19.1). Five groups (Malasseziales, Corticiales,

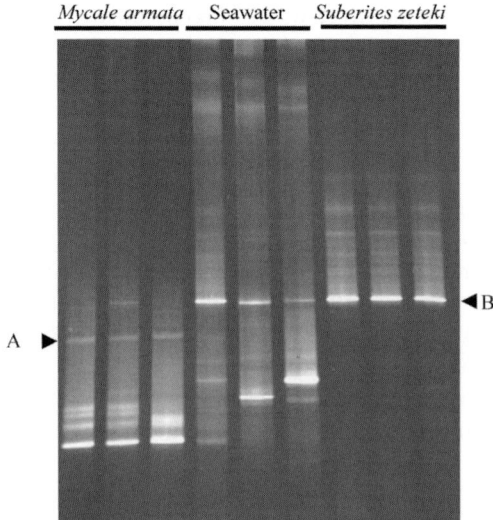

Figure 19.1 Denaturing gradient gel electrophoresis analysis of uncultured fungi derived from marine sponges (*Mycale armata* and *Suberites zeteki*) and seawater (Gao et al. 2008).

Polyporales, Agaricales, and Dothideomycetes and Chaetothyriomycetes *incertae sedis*) were identified for the first time in marine sponges. The study by Gao et al. (2008) reported the greatest diversity of *Malassezia* spp. from a single host so far (Figure 19.2).

Overall, diverse and unculturable fungi have been identified in the marine sponges *S. zeteki* and *M. armata* using molecular approaches. Although a small fraction of these fungi belong to the known genera of culturable fungi in sponges, the majority are fungal taxa identified in marine sponges for the first time. The identification of several "marine" fungal clades in these two marine sponges supports the existence of "marine fungal phylotypes" in sponges. Clearly, marine sponges represent an ecologically-important hot spot for fungal diversity.

Symbiosis and ecological function of sponge-derived fungi

In contrast to symbiotic or otherwise-associated bacteria whose presence in sponges is well documented, our knowledge of the symbiotic relationship between fungi and sponges is rather limited (König et al. 2006; Taylor et al. 2007). Fungal isolates are easily cultured from the inner tissues of sponges, but the ecological relationship between the host and the fungi remains largely unknown. Recently, sponge-inhabiting fungi were classified into two categories: "resident fungi" and "transient fungi" (Li and Wang 2009). "Resident fungi" are sponge-specific, while "transient fungi" are those resulting from "wash-in spores" trapped in sponges during the filter-feeding process. As of now, hardly any direct evidence has been reported on the specificity of sponge-associated fungi. Nevertheless, evidence appears to support a true symbiotic association between sponges and fungi:

Table 19.1 Taxonomic distribution of ITS clone sequences derived from seawater and the marine sponges *Suberites zeteki* and *Mycale armata* (Gao et al. 2008).

Species ID[a]	Taxon		Closest relative (accession no.)	% Identity	No. of clones in libraries
	Phylum	Order			
SZ10.2.1	Ascomycota	Capnodiales	*Cladosporium oxysporum* (DQ912837)	99	1
SZ10.2.11		Capnodiales	*Cladosporium* sp. strain 7306 (EF120415)	99	1
MA10.2.I2		Capnodiales	*Cladosporium* sp. strain 7306 (EF120415)	99	1
SW10.4.09		Capnodiales	*Cladosporium* sp. strain 7306 (EF120415)	99	4
MA10.2.09		Capnodiales	*Cladosporium oxysporum* (DQ875018)	98	2
MA10.3.08		Capnodiales	*Cladosporium oxysporum* (DQ875018)	99	2
SW10.2.12		Capnodiales	*Cladosporium oxysporum* (DQ875018)	98	1
SW10.1.09		Capnodiales	*Cladosporium oxysporum* (DQ875018)	98	1
SW10.5.07		Capnodiales	*Cladosporium* sp. HKA30 (DQ092512)	99	1
MA10.3.06		Dothideales	*Hortaea werneckii* ATCC 36317 (DQ168665)	100	2
SW10.3.4		Dothideales	*Hortaea werneckii* ATCC 36317 (DQ168665)	99	1
SZ10.3.5		incertae sedis	*Aureobasidium pullulans* CBS110374 (AY139394)	98	2

Table continued on next page.

Table 19.1 (Continued)

Species ID[a]	Taxon	Closest relative (accession no.)	% Identity	No. of clones in libraries
SW10.5.4	*incertae sedis*	*Aureobasidium pullulans* CBS110374 (AY139394)	98	1
SZ10.1.2	Eurotiales	*Penicillium brevicompactum* (DQ123641)	99	1
SZ10.1.3	Eurotiales	*Penicillium restrictum* NRRL 25744 (AF033459)	99	1
SZ10.1.4	Eurotiales	*Penicillium restrictum* FRR 332 (AY373928)	100	1
SZ10.2.12	Eurotiales	*Aspergillus restrictus* ATCC 16912 (AY373864)	99	1
SZ10.4.11	Eurotiales	*Penicillium oxalicum* (DQ681323)	99	1
SZ10.5.8	Eurotiales	*Penicillium steckii* NRRL 35367 (DQ123665)	100	3
MA10.1.5	Eurotiales	*Aspergillus flavus* FG38 (EU030347)	99	1
MA10.2.13	Eurotiales	*Aspergillus flavus* FG38 (EU030347)	98	2
SW10.3.06	Eurotiales	*Aspergillus flavus* FG38 (EU030347)	100	1
SZ10.2.6	Hypocreales	*Hypocreales* sp. strain LM9 (EF060402)	99	1
SZ10.5.13	Hypocreales	*Hypocreales* sp. strain LM9 (EF060402)	98	1

Table continued on next page.

Table 19.1 (Continued)

Species ID[a]	Taxon		Closest relative (accession no.)	% Identity	No. of clones in libraries
SW10.5.1		Hypocreales	*Gibberella moniliformis* NRRL 43697 (EF453174)	99	1
SW10.4.12		Saccharomycetales	*Candida tropicalis* isolate16 (EF216862)	100	1
MA10.3.10		Pleosporales	Uncultured fungus clone 24b (EU003082)	99	1
SW10.4.4		Pleosporales	Uncultured fungus clone 24b (EU003082)	99	1
SW10.45		Pleosporales	Uncultured fungus clone 24b (EU003082)	99	1
MA10.2.15		Pleosporales	*Ascomycota* sp. strain LM107 (EF060476)	99	1
SW10.4.08		Pleosporales	*Phoma putaminum* isolate P-14 (AM691009)	98	1
SZ10.7.4	Basidiomycota	Agaricales	*Schizophyllum commune* HNO34 (AF280758)	99	1
SZ10.2.2		Corticiales	*Phlebia* sp. Strain olrim964 (AY787680)	86	1
SZ10.2.9		Malasseziales	*Malassezia restricta* CBS 7877 (AY743636)	98	3
SZ10.2.13		Malasseziales	*Malassezia restricta* CBS 7877 (AY743636)	98	2
SZ10.5.3		Malasseziales	*Malassezia restricta* CBS 7877 (AY743636)	98	1

Table continued on next page.

Table 19.1 (Continued)

Species ID[a]	Taxon	Closest relative (accession no.)	% Identity	No. of clones in libraries
SZ10.5.5	Malasseziales	*Malassezia restricta* CBS 7877 (AY743636)	99	2
MA10.3.17	Malasseziales	*Malassezia restricta* CBS 7877 (AY743636)	99	2
SZ10.3.12	Malasseziales	*Malassezia* clone gbCTA7_004 (DQ900961)	99	1
SZ10.4.4	Malasseziales	*Malassezia* clone ghCTA7_004 (DQ900961)	98	2
SZ10.4.5	Malasseziales	*Malassezia* clone gbCTA7_004 (DQ900961)	99	1
SW10.1.1	Malasseziales	*Malassezia* clone gbCTA7_004 (DQ900961)	98	7
SW10.1.5	Malasseziales	*Malassezia* clone gbCTA7_004 (DQ900961)	98	1
MA10.2.08	Malasseziales	*Malassezia* clone gbCTA7_004 (DQ900961)	99	1
SZ10.5.6	Malasseziales	Fungus clone (AM260792)	99	1
MA10.2.5	Malasseziales	Fungus clone (AM260792)	99	3
SZ10.1.5	Malasseziales	*Malassezia restricta* FRR 332 (AJ437695)	99	13
MA10.3.4	Malasseziales	*Malassezia restricta* FRR 332 (AJ437695)	99	2
MA10.3.5	Malasseziales	*Malassezia restricta* FRR 332 (AJ437695)	99	1
MA10.3.11	Malasseziales	*Malassezia restricta* FRR 332 (AJ437695)	100	1

Table continued on next page.

Table 19.1 (Continued)

Species ID[a]	Taxon		Closest relative (accession no.)	% Identity	No. of clones in libraries
MA10.3.16		Malasseziales	*Malassezia restricta* FRR 332 (AJ437695)	99	4
SW10.3.5		Malasseziales	*Malassezia restricta* FRR 332 (AJ437695)	99	1
SW10.2.9		Malasseziales	*Malassezia restricta* FRR 332 (AJ437695)	100	1
SW10.4.07		Malasseziales	*Malassezia restricta* FRR 332 (AJ437695)	99	2
MA10.3.1		Malasseziales	*Malassezia globosa* CBC1 (AY387136)	99	1
MA10.3.3		Malasseziales	*Malassezia sympodialis* CBS 8740 (EF140668)	87	1
MA10.2.06		Malasseziales	*Malassezia sympodialis* WF42 (AY387181)	85	1
SW10.5.3		Malasseziales	*Malassezia sympodialis* WF42 (AY387181)	85	3
MA10.1.11		Malasseziales	*Malassezia sympodialis* MA 477 (AY743639)	85	1
MA10.3.12		Malasseziales	*Malassezia sympodialis* MA 477 (AY743639)	85	1
SW10.2.5		Malasseziales	*Malassezia sympodialis* MA 477 (AY743639)	85	2
SW10.2.08		Malasseziales	*Malassezia sympodialis* MA 477 (AY743639)	85	1
SW10.5.09		Malasseziales	*Malassezia sympodialis* MA 477 (AY743639)	85	1

Table continued on next page.

Table 19.1 (Continued)

Species ID[a]	Taxon	Closest relative (accession no.)	% Identity	No. of clones in libraries
MA10.1.12	Malasseziales	Uncultured Malasseziales clone BD19 (DQ317381)	99	1
SW10.3.13	Malasseziales	Uncultured AMF fungus (AY267224)	99	1
SW10.5.2	Malasseziales	Malassezia globosa CBS 7966 (AY743630)	99	1
SW10.5.06	Malasseziales	Malassezia globosa CBS 7966 (AY743630)	97	1
SW10.5.08	Malasseziales	Malassezia globosa CBS 7966 (AY743630)	98	1
SW10.5.11	Malasseziales	Malassezia globosa CBS 7966 (AY743630)	98	1
SW10.2.10	Malasseziales	Malassezia sympodialis MA 424 (AY743654)	83	1
SZ10.7.2	Polyporales	Basidiomycete sp. Strain LC5 (AY605709)	99	4

[a]The species ID number indicates the sample source (SZ, *Suberites zeteki*; MA, *Mycale armata*; SW, seawater) and clone number.

(i) First, yeasts associated with sponges (e.g. *Chondrilla* spp.) have been observed in both adult sponge tissue and reproductive structures, indicating the vertical transmission of the yeast symbionts (Maldonado et al. 2005b). The results of this study provide compelling evidence for the true symbiotic relationship between yeasts and sponges.

(ii) Second, the sponge *Suberites domuncula* has been reported to possess a molecular mechanism for recognizing fungi via the D-glucan carbohydrates on their surfaces (Perovic-Ottstadt et al. 2004).

(iii) Third, phylogenetic analysis of the *COI* protein sequence and the intron open reading frame indicates that the intron of the sponge *COI* gene may have been

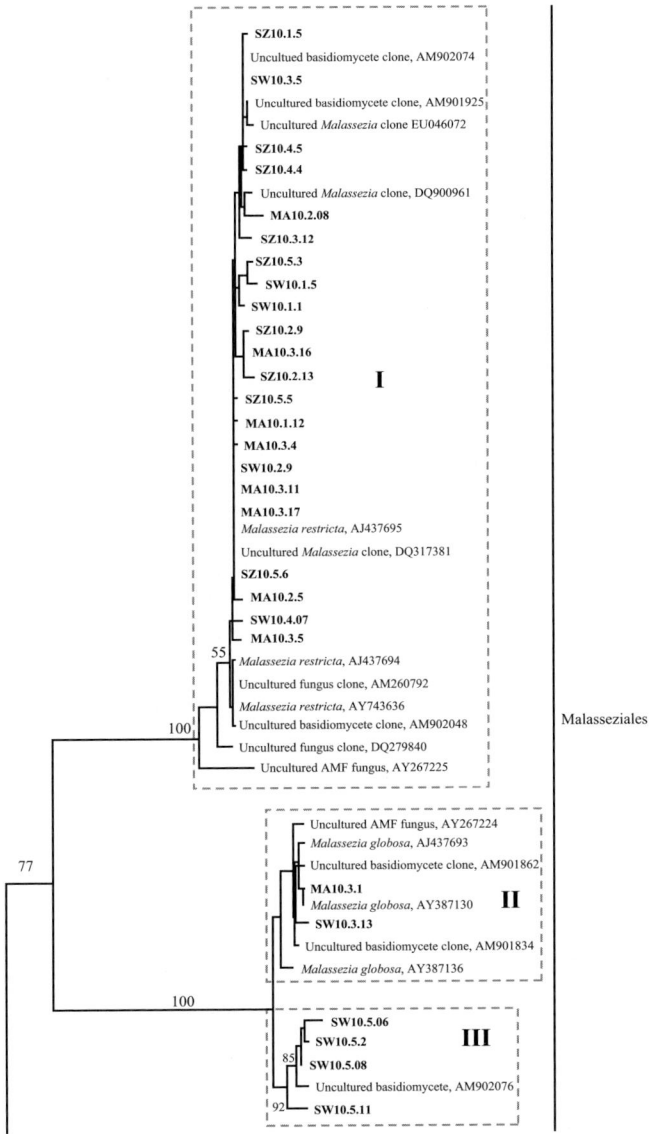

Figure 19.2 (Continued)

Figure 19.2 (Continued)

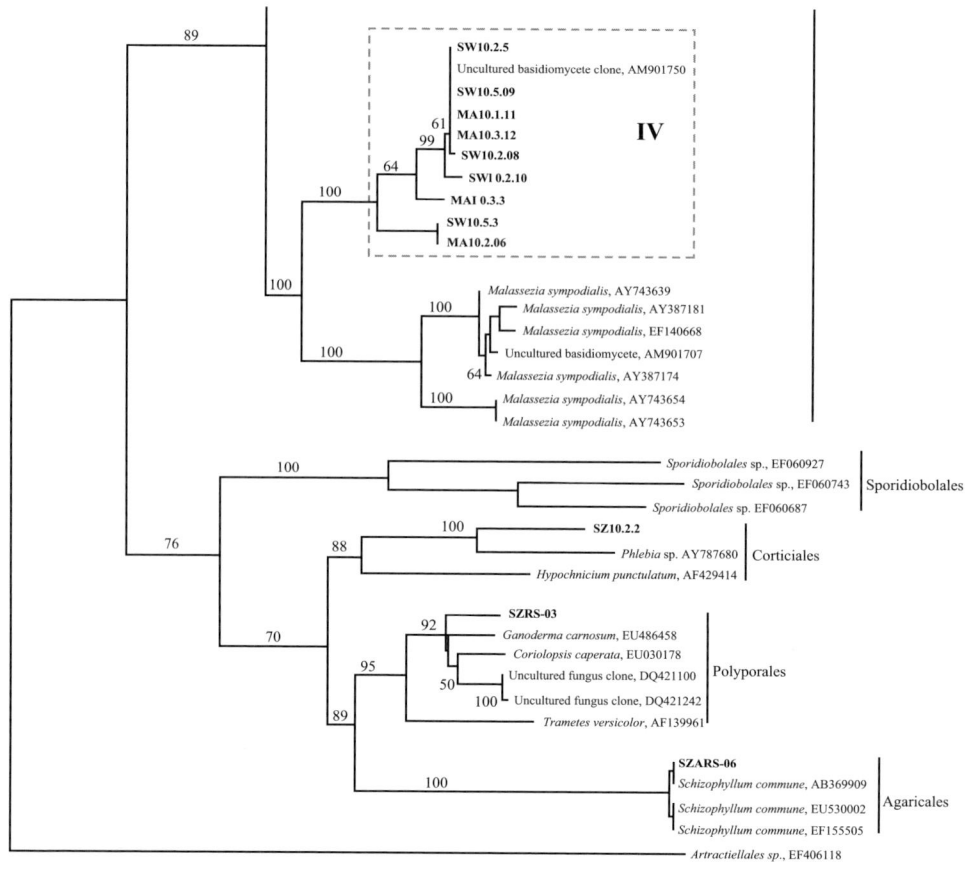

Figure 19.2 Neighbor-joining phylogenetic tree of the *Basidiomycota* based on rRNA-ITS sequences (represented in bold) derived from marine sponges and seawater (Gao et al. 2008).

transmitted horizontally from a fungus donor (Rot et al. 2006). Horizontal gene transfer of a mitochondrial intron suggests a "true symbiotic" relationship between fungi and sponges.

(iv) Finally, marine ascomycetes of the genus *Koralionastes* have been reported to have a unique association with crustaceous sponges, with the capacity to develop their ascomata on or within these sponges (Kohlmeyer and Volkmann-Kohlmeyer 1990). It is worth mentioning that *Koralionastes* spp. were also isolated from coral rock (Kohlmeyer and Volkmann-Kohlmeyer 1987).

Furthermore, certain fungal phylotypes have been identified to exclusively affiliate with a sponge species even though no solid phylogenetic information supports the existence of "sponge-specific fungi" (Gao et al. 2008). Large-scale phylogenetic analysis of sponge-derived fungal sequences may shed additional light on the relationship between sponges and fungi (Taylor et al. 2000).

Studies on the ecological function of fungi in sponges are relatively rare. Bacteria have been thought to be involved in the production of natural active compounds that contribute to the chemical defense and ecological functions of sponges (Proksch et al. 2003; König et al. 2006; Wang 2006). Fungi have long been considered causal agents of sponge diseases. The diseased tissues of commercial sponge species have been ascribed to unidentified fungi (Galstoff 1942). Isolation of fungi from diseased sponge tissues implies that they may contribute to the disease development, or are opportunistic and/or secondary colonizers of diseased tissues predisposed by other infections or stresses (Sparks 1985; Vacelet et al. 1994). The increased occurrence of marine Thraustochytriaceae has been observed in the dying marine sponge *Halichondria panicea* (Richter 1985). These studies, however, were based on histopathology and no specific pathogens were identified from the diseased tissues (Webster 2007). Recently, the marine sponge *Spongia obscura* has been reported to be a reservoir of the fungal pathogen *Aspergillus sydowii* (Ein-Gil et al. 2009), which is the causative agent of epidemics that affect gorgonian corals (sea fans; Geiser et al. 1998), see Chapter 16. This sponge can also take up and retain viable conidia of this pathogen without notably affecting the health of the sponge. Thus, fungi are likely to be secondary invaders of diseased sponge tissues instead of the primary cause of disease.

Characteristics of deep-sea habitats

The deep-sea environment is characterized by extreme conditions of temperature, hydrostatic pressure and nutrient availability. It covers ~65% area of the Earth's total surface (Svendrup et al. 1942). The hydrostatic pressure increases at a rate of 0.1 MPa (Mega Pascal) for every 10 m in depth, so it ranges from 2–100 MPa for different deep-sea environments. In contrast, temperature decreases with increasing depth and remains at a constant value (isothermal) of ~2–3°C in deep-sea habitats, except for the transition zones between deep-sea hot and cold waters. These zones include thermocline areas between surface and deep-sea waters and the transition between the cold and hot waters from the deep-sea floor and hydrothermal vents, respectively. This environment is also characterized by low nutrients and therefore is supposed to be physically stable (Sanders 1968). The salinity is generally 35‰ (completely marine) in most deep-sea habitats; however, in some hypersaline deep-sea habitats the salinity value reaches over 39‰ (Mediterranean and Red Sea) and even 300‰ in Orca Basin in the Gulf of Mexico (Shokes et al. 1976). Deep-sea habitats are devoid of any photosynthetic activities by organisms thriving there as the penetration of light does not occur below 250 m. The main source of oxygen in these habitats is the surface water that carries dissolved oxygen generated by phytoplankton in the euphotic zone, as well as exchange with the surrounding waters. Deep-sea currents generated due to the descending movement of surface water are the main force bringing oxygenated water to the lower depths. Usually the dissolved oxygen concentration in the bottom water is ~5–6 ml L^{-1}; whereas in surface waters it is ~5–8 ml L^{-1} (Gewin 2010). However, it varies with depth in different deep-sea sediments (Reimers 1987). The deep-sea harbors various kinds of microorganisms, comprising bacteria, archaea and fungi, which are known to play an important role in the recycling of nutrients (Snelgrove et al. 1997). Among these, bacteria and archaea have been studied in detail (Li et al. 1999; Takai and Horikoshi 1999; Delong and Pace 2001; Sogin et al. 2006; Xu et al. 2008). Fungi are comparatively less well studied in deep-sea environments.

A total of 97,330 species of fungi have been described (Kirk et al. 2008), representing only about 5% of the estimated 1.5 million species worldwide (Hawksworth 1991). The estimated number of marine fungal species is 10,000, in comparison with 200,000 marine animals and 20,000 marine plants (Jones 2011a). The documentation of marine fungi is still at the inventory stage, however, and many new taxa have yet to be discovered (Jones 2011a, 2011b). Fungi in the marine ecosystem occupy an important position as they are ubiquitous and involved in the decomposition and mineralization of organic matter (Kohlmeyer and Kohlmeyer 1979; Hyde 1989; Newell 2001).

Fungi in the deep-sea environment

The occurrence of fungi in the deep sea was first reported in shells collected from waters at a depth of 4,610 m (Höhnk 1961, 1969). Isolation of fungi has been reported from water samples collected from the subtropical Atlantic Ocean, from the surface to a depth of 4,500 m using sterile van Dorn bags or Niskin samplers (Roth et al. 1964). Lignicolous fungi from the deep sea were recovered from submerged wooden panels at 1,615–5,315 m depth (Kohlmeyer 1977); while more recently Dupont et al. (2009) recovered two fungi (*Alisea longicolla* and *Oceanitis scuticella*) from a depth of 1,000 m in the Pacific Ocean off the Vanuatu Islands. Fungi inside molluscan shells have also been reported (Poulicek et al. 1986; 1988). Several marine yeasts, such as *Debaryomyces, Rhodotorula* and *Rhodosporidium* spp., were isolated from a range of temperature and hydrostatic pressure conditions by Lorenz and Molitoris (1997). Fungi have also been reported from the Mariana trench at a depth of ~11,000 m in the Pacific Ocean (Takami et al. 1997).

Fungi have been reported from other deep-sea environments, including hydrocasts near hydrothermal plumes from the Mid-Atlantic Ridge near the Azores (Gadanho and Sampaio 2005) and in Pacific sea-floor sediments (Nagahama et al. 2001b). Fungi from both habitats were dominated by unicellular forms, commonly designated as yeasts. Many of the fungi isolated from the ocean beds are undescribed species. For example, Nagahama and coworkers have described a number of novel yeast species from deep-sea sediments (Nagahama et al. 2001a, 2003, 2006). Various filamentous fungi were isolated from calcareous sediments and their spores were shown to germinate at elevated hydrostatic pressure (Raghukumar and Raghukumar 1998). Fungal filaments were directly detected in calcareous fragments using the fluorescent brightener calcofluor under an epifluorescence microscope (Raghukumar and Raghukumar 1998). Isolation and direct detection of fungal mycelium was reported from the Chagos trench in the Indian Ocean from deep-sea sediments (Raghukumar et al. 2004). Over 200 fungi were isolated from deep-sea sediments of the Central Indian Basin using different techniques including dilution plating, particle plating and pressure-enrichment techniques (Damare et al. 2006). The presence of fungi in sediments was also demonstrated by direct detection and immunofluorescence techniques (Damare et al. 2006). The recovery of fungi from similar locations by culture-dependent as well as culture-independent approaches has also been reported (Singh et al. 2010, 2011a, 2011b).

Diversity of fungi from deep-sea environments

Fungi reported from deep-sea environments mostly belong to the phyla Ascomycota and Basidiomycota (Table 19.2). A total of 181 fungi were isolated in culture using

Table 19.2 Fungal phyla reported from various deep-sea habitats by culture-dependent as well as culture-independent approaches.

Habitat	Target gene for molecular identification	No. of culturable isolates	No. of non-culturable phylotypes	Phyla	References
Deep-sea environment	Taxonomic identification	181	–	Ascomycota Basidiomycota	Damare et al. (2006)
	18S rRNA and ITS	15 13	–	Ascomycota Basidiomycota	Singh et al. (2010)
	ITS	–	23 11 3 2	Ascomycota Basidiomycota Chytridiomycota Glomeromycota	Nagano et al. (2010)
	18S rRNA	–	5	Basidiomycota	Edgcomb et al. (2010)
	18S and ITS rRNA	–	31 8	Ascomycota Basidiomycota	Singh et al. (2011a)
	18S and ITS rRNA	–	11 16	Ascomycota Basidiomycota	Singh et al. (2011b)
Hydrothermal vent environment	Eukaryotic primer set for 18S rRNA	–	1 4	Ascomycota Basidiomycota	Edgcomb et al. (2002)
	Eukaryotic 18S rRNA	–	1 1	Ascomycota Basidiomycota	López-García et al. (2003)
	D1/D2 region of 28S rRNA	12 7	–	Ascomycota Basidiomycota	Gadanho and Sampaio (2005)
	Eukaryotic 18S rRNA	–	1 4	Ascomycota Basidiomycota	López-García et al. (2007)
	18S rRNA	–	6 11 1	Ascomycota Basidiomycota Chytridiomycota	Bass et al. (2007)
	ITS and D1/D2 of 28S rRNA	3 5	–	Ascomycota Basidiomycota	Connell et al. (2009)
	18S rRNA	–	7 2	Basidiomycota Chytridiomycota	Le Calvez et al. (2009)
	18S, ITS or 28S rRNA	60 1	–	Ascomycota Basidiomycota	Burgaud et al. (2009)
	D1/D2 region of 26 rRNA	11 21	–	Ascomycota Basidiomycota	Burgaud et al. (2010)

Table continued on next page.

Table 19.2 (Continued)

Habitat	Target gene for molecular identification	No. of culturable isolates	No. of non-culturable phylotypes	Phyla	References
Anoxic sites	18S rRNA	–	1	Ascomycota	Stoeck et al. (2003)
	Eukaryotic 18S rRNA	–	4	Ascomycota and Chytridiomycota	Stoeck and Epstein (2003)
	Eukaryotic 18S rRNA	–	5	Ascomycota	Stoeck et al. (2006)
			3	Chytridiomycota	
	18S rRNA	21	27	Ascomycota	Jebaraj et al. (2010)
		5	20	Basidiomycota	
		–	1	Zygomycota	
Methane hydrate cold seeps	Eukaryotic 18S rRNA and ITS	–	1	Basidiomycota	Takishita et al. (2006)
	ITS	–	20	Ascomycota	Lai et al. (2007)
			4	Basidiomycota	
	Eukaryotic 18S rRNA	–	3	Ascomycota	Takishita et al. (2007)
			2	Basidiomycota	
			14	Chytridiomycota and Zygomycota	
	SSU rRNA	–	17	Ascomycota	Nagahama et al. (2011)
			6	Basidiomycota	
			2	Rozella and LKM11 clade	

various techniques from deep-sea sediments of the Central Indian Basin (Damare et al. 2006). These fungi were only identified morphologically, however, and *Aspergillus* spp. were found to be dominant. A total of 28 fungal isolates were isolated and identified by sequence analysis of 18S and ITS regions of the rDNA from the same site of Central Indian Basin (Singh et al. 2010). These isolates included filamentous fungi and yeasts belonging to different classes of Ascomycota and Basidiomycota (Table 19.2). The recovery of *Sagenomella, Exophiala, Capronia* and *Tilletiopsis* spp. was reported for the first time from deep-sea sediments (Singh et al. 2010). This suggests that there is adaptation of deep-sea isolates under such extreme conditions. The majority of these species isolated from deep-sea environments were reported earlier from terrestrial environments (Smit et al. 1999). This strengthens the hypothesis of the transportation of terrestrial fungal hyphae and spores into the ocean and their gradual adaptation to deep-sea conditions (Raghukumar and Raghukumar 1998; Damare et al. 2006; Bass et al. 2007; Burgaud et al. 2009).

Nagano et al. (2010) used a culture-independent approach to document the fungal diversity from deep-sea sediments near Japanese islands and recovered highly novel fungal phylotypes within the deeply-branched Chytridiomycota and related phyla, in addition to Ascomycota and Basidiomycota (Table 19.2). The phylum Chytridiomycota is known to be related to the possible earliest diverging fungal lineage (see Chapter 7). A detailed study of such organisms may provide deeper insights into the evolutionary history of fungi in deep-sea environments. Another phylotype, belonging to the deep-sea fungal clone group 1 (DSF1) was reported in the study by Nagano and colleagues. This group consists of highly novel fungal sequences that have been exclusively reported from different deep-sea environments (Bass et al. 2007; Takishita et al. 2007). The recovery of this group from other anoxic sites suggests that they have an important role in the nitrogen cycle of the deep-sea ecosystem (Nagano et al. 2010).

Fungi were reported as major eukaryotes existing in the deep-sea marine sub-surface water sediments of the Peru Margin and Peru Trench by analysis of the eukaryotic 18S rRNA as well as cDNA environmental clone libraries (Edgcomb et al. 2010). The dominant fungal sequences belonged to the Basidiomycota, and included members of the Ustilaginomycetes. Recovery of these sequences in both rRNA and cDNA libraries indicated that they are metabolically active components of deep-sea sediments. In addition, the detection of sequences affiliated with pathogenic forms, such as *Cryptococcus, Malassezia* and *Trichosporon*, suggests that they may be opportunistic pathogens of deep-sea mammals. Biological communities in the deep sea have shown dependence on variables such as substrate availability and type, biogeochemistry, nutrient input, productivity and hydrological conditions on a regional scale (Levin et al. 2001). Recovery of a halophilic fungal phylotype, e.g. *Wallemia sebi*, has been reported from deep-sea environments (Singh et al. 2011a, 2011b). This species has previously been found in the hypersaline water of salterns and was reported to adapt successfully to living in extreme saline environments (Kunčič et al. 2010). The isolation of *Sagenomella* sp., using both culture-dependent (Singh et al. 2010) and culture-independent approaches (Singh et al. 2011a, 2011b), suggests that this organism is an active member of deep-sea environment. The occurrence of common ubiquitous fungal forms, such as *Aspergillus, Penicillium, Cladosporium* and *Trichosporon* species, has been reported from many deep-sea habitats. The deep sea may be a significant reservoir of these saprobic forms, playing a major role in biogeochemical cycles under such extreme conditions.

Fungi in hydrothermal vent ecosystems

Background

Hydrothermal vents are areas of the sea floor where the spreading of tectonic plates occurs, which results in cracks and fissures. As the surrounding water goes down, it comes into contact with the superheated molten magma present below the Earth's crust and gets superheated. This hot sea water returns through the fissures carrying minerals leached from the crustal rock below. The superheated seawater then ejects out of the holes from the crust, rising quickly above the colder, denser waters of the deep ocean. As the hot and cold seawater meet, the minerals suspended in the hot water precipitate out at the vent opening. The temperature of this hot water is in the range of 60–600°C. In spite of such high temperatures, the surrounding water is prevented from boiling due to the extreme hydrostatic pressure. The precipitated minerals form geologically-unique structures called chimneys. The atmosphere and hydrosphere are relatively oxidizing and the basaltic rocks that form the oceanic crust are relatively reduced. The high-temperature fluid–rock interaction forms reduced gases (H_2S, H_2, and CH_4) that dissolve in the hydrothermal fluid (Zierenberg et al. 2000).

Fungal diversity

In spite of such extreme conditions, dense animal communities have been reported to cluster around these hot springs, supported by the chemolithoautotrophic activities of prokaryotes (Joergensen and Boetius 2007). However, fungal communities at these sites are understudied. Recently there has been increasing interest in studying the fungal diversity of hydrothermal vents (Burgaud et al. 2009; Le Calvez et al. 2009).

Fungal diversity has been described by a culture-dependent approach from hydrothermal vent environments (Gadanho and Sampaio 2005; Connell et al. 2009; Burgaud et al. 2009, 2010). Culturable yeasts were isolated and identified by sequence analysis of the D1/D2 region of the 26S rRNA gene from the Mid-Atlantic ridge (Gadanho and Sampaio 2005). These included *Candida, Pichia, Rhodosporidium, Rhodotorula, Exophiala* and *Trichosporon* species. Yeasts were also isolated and identified from iron-oxide mats and rock surfaces from the crater of the volcanically active Vailulu'u seamount (Samoan chain; Connell et al. 2009). Many of these yeast isolates were demonstrated to produce siderophores, a class of molecules used to acquire and utilize Fe(III). One of the isolates, *Rhodotorula graminis*, oxidized Mn(II) to Mn(IV), suggesting a functional role of fungi in the sea-floor alteration and cycling of metals present under such extreme conditions.

Marine-culturable yeasts were isolated and identified from hydrothermal vent fauna by Burgaud et al. (2010). These isolates belonged to the genera *Rhodotorula, Rhodosporidium, Candida, Debaryomyces* and *Cryptococcus*. Some of these yeast species were detected inside the vent animals, when labeled with suitable probes using a technique known as FISH (fluorescent *in situ* hybridization). The existence of these yeast species within hydrothermal vent fauna suggests that they have a role in the decomposition of organic matter entrapped in the filaments of mussel byssi (Burgaud et al. 2010).

Earlier studies have reported a very low diversity of fungi by analysis of environmental clone libraries targeting the eukaryotic 18S rRNA gene (Edgcomb et al. 2002; López-García et al. 2003, 2007). However, the fungal sequences recovered represented highly novel phylotypes belonging to Ascomycota and Basidiomycota (Table 19.2). Fungal diversity from different deep-sea hydrothermal vent habitats was reported by a culture-independent approach (Bass et al. 2007). A total of ten environmental clone libraries targeting the 18S rRNA gene were constructed, resulting in only 18 fungal phylotypes (Table 19.2). The sequences recovered in this study belonged to the phyla Ascomycota, Basidiomycota and Chytridiomycota. Among these sequences, basidiomycetous yeasts were dominant in deep-sea environmental libraries. The percentage recovery of Chytridiomycota was, however, comparatively lower than for the other two phyla. The fungal community from hydrothermal vents was reported to be composed mostly of phylotypes belonging to Basidiomycota and Chytridiomycota (Le Calvez et al. 2009). Among the basidiomycetes, sequences affiliating with *Cryptococcus* and *Filobasidium* sp. were the most common. Highly novel sequences belonging to Chytridiomycota were also present. A further detailed analysis of such ecologically-important signatures is required to understand their functional role, along with the evolutionary history of fungi in the deep-sea ecosystem.

Fungal diversity at anoxic sites

Background

Anoxic sites are areas of sea or fresh water that are depleted of dissolved oxygen and they result from the restriction of water exchange in these areas, the presence of silt and sometimes heavier hypersaline waters at lower depths resulting in stratification. Anoxic conditions may also arise if organic matter oxidized by the bacterial communities is at a higher rate than the supply of dissolved oxygen. Some examples of anoxic zones are in the eastern Pacific Ocean and Arabian Sea. In these areas, surface productivity is exceptionally high so that an unusually high supply of organic carbon sediment precipitates down to thermocline depths.

The oxygen minimum zones present in these areas are ascribed to a combination of the high productivity described above and limited mixing or circulation with poorly oxygenated waters (Naqvi et al. 1998). The oxygen minimum zone of the Arabian Sea is largely influenced by the south-west and north-east monsoons, resulting in bottom water upwelling and therefore high productivity (Naqvi 1994; Warren 1994). This has resulted in the formation of one of the world's largest perennially oxygen-depleted environments in the open ocean and a seasonal hypoxic zone along the western coast of India. The coastal hypoxic condition develops along the continental shelf down to 200 m, leading to intense sedimentary denitrification activity (Naqvi et al. 2000). Dissolved oxygen, which is normally 180 mM, drops to a concentration of ~20 mM during anoxic conditions (Naqvi et al. 2006).

Fungal diversity

Fungal diversity studies have been neglected in such oxygen-deficient sites due to the underestimation of their role in the denitrification process. The growth of fungi has been demonstrated under oxygen-deficient conditions, however, suggesting the anaerobic denitrification mechanism acquired by these fungi for survival (Jebaraj and Raghukumar 2009). Fungal phylotypes have been reported from different anoxic sites by studying eukaryotic diversity targeting the 18S rRNA gene from environmental samples (Stoeck and Epstein 2003; Stoeck et al. 2003, 2006). These studies reported fungal phylotypes related to Ascomycota and Chytridiomycota (Table 19.2); however, the actual fungal diversity at these anoxic sites was underestimated due to the domain-targeted PCR techniques. Therefore, a multiple-primer approach was implemented including fungal as well as domain-specific primers, to study the fungal diversity of the oxygen minimum zones of the Arabian Sea (Jebaraj et al. 2010). In the above study, Jebaraj et al. reported the presence of fungal phylotypes belonging to Ascomycota and Basidiomycota by culture-dependent and culture-independent approaches. Only one phylotype affiliating with Zygomycota was detected by a culture-independent approach (Table 19.2). The recovery of phylotypes affiliating with *Saccharomyces* and *Fusarium* spp. was reported. These species are known to be able to cope with anoxic conditions by producing ethanol. They also show denitrification capability. The majority of the ascomycetous and basidiomycetous sequences clustered with those previously reported from anoxic sites (Jebaraj et al. 2010). Using a culture-dependent approach, two of the isolates showed high divergence from their closest relative in the *National Center for Biotechnology Information* database, suggesting that they are novel. A detailed study on such novel isolates by multigene analysis and their enzymes related to the nitrate-reducing pathway may shed light on their ecological role under anoxic conditions (Jebaraj et al. 2010).

Diversity of fungi from methane-hydrate bearing cold seep

Background

These are areas rich in hydrogen sulfide, methane and fluid containing other hydrocarbons. Seepage of this fluid in different areas of the deep sea often results in the formation of brine pools. These are depressions formed on the sea floor where the salinity is very high in comparison with the surrounding water. Most of these habitats are located in the Atlantic, Eastern and Western Pacific Ocean and Mediterranean Sea.

The cold seep habitats are characterized by the occurrence of rocks and reefs composed of carbonates. These rocks are created due to the reactions between methane-rich thermogenic or biogenic fluid and the surrounding sea water. Apart from this, the activities of certain bacteria are responsible for the generation of these rocks.

Fungal diversity

Among eukaryotes, basidiomycetous fungi, e.g. *Cryptococcus curvatus*, were reported to be the dominant species in the sediments of Kuroshima Knoll methane seep by culture-independent approach (Takishita et al. 2006). *C. curvatus* is known as an opportunistic pathogen of animals, including humans (Dromer et al. 1995). The existence of a fungal community was reported in the methane hydrate-bearing deep-sea sediments of the South China Sea and composed mainly of *Phoma glomerata* and yeasts (Lai et al. 2007). Analysis of environmental clone libraries targeting the ITS region of the fungal rDNA confirmed the existence of novel fungal phylotypes having no known closely-related cultured taxa (Lai et al. 2007). The improved method used to obtain these fungi in culture may provide an insight into their ecological role in carbon cycling.

Anaerobic methanotrophy plays a significant role in the successful establishment of microbial communities at the methane-seep by converting methane into more readily accessible carbon and energy substrates (Orphan et al. 2002). Microbial communities associated with methane hydrate could play a significant role in global methane and carbon cycles by the consumption of hydrocarbons.

Other studies have reported fungal phylotypes belonging to Ascomycota, Basidiomycota and Chytridiomycota (Takishita et al. 2007; Nagahama et al. 2011). The recovery of sequences affiliating with psychrophilic *Penicillium, Cladosporium, Phoma* and *Geomyces* spp. has been reported (Nagahama et al. 2011). These species are known to possess enzyme activities for lignocellulose degradation (Rodríguez et al. 1996; Claus and Filip 1998; Rice et al. 2006; Junghanns et al. 2009). These fungi may therefore play an important ecological role in the biodegradation of lignin and its derivatives under such extreme conditions.

Conclusions

Recent advancement in molecular biology has significantly improved our understanding of fungal diversity and their ecological function in many marine habitats. Sponges and deep-sea environments, two of the most poorly-understood marine habitats, are the hot spots of novel fungal diversity. With the further application of advanced molecular and genomic approaches, more novel fungal species are expected to be identified from these habitats.

Considering the role of fungi in marine ecosystem as a potential source of useful natural products, their actual diversity needs to be examined. The novel potential forms of the above fungi present in extreme environments, such as deep-sea sediments, waters, hydrothermal vent sites and anoxic sites, must be explored. Such fungi may serve as a significant reservoir of industrially-important bioactive metabolites as well as enzymes.

The extreme conditions of deep-sea environments, such as low temperature and high pressure, may drive the evolution of organisms existing there, making them suitable for the production of more efficient secondary metabolites. Therefore, attempts to discover the actual diversity of fungi in such extreme environments must be employed along with their isolation in culture.

Several studies support the occurrence of true "marine fungi" in sponges and their symbiotic relationship. Molecular approaches, such as a large-scale phylogenetic analysis of fungal ribosomal RNA gene sequences, will surely provide answers to questions regarding the ecological relationship between the host and the fungi. However, the ecological function of sponge-inhabiting fungi remains a challenging topic for future studies.

With the unique physical and chemical features of these habitats, modern environmental genomics approaches in combination with novel sequencing method will likely give a complete understanding of the diversity and biogeochemical functions of fungi in these underexplored marine habitats.

References

Bass D, Howe A, Brown N, Barton H, Demidova M, Michelle H, Li L, Sander H, Watkinson SC, Willcock S, Richards TA. Yeast forms dominate fungal diversity in the deep oceans. Proc R Soc B. 2007;274:3069–3077.

Bugni TS, Ireland CM. Marine-derived fungi: a chemically and biologically diverse group of microorganisms. Nat Prod Rep. 2004;21:143–163.

Burgaud G, Calvez TL, Arzur D, Vandenkoornhuyse P, Barbier G. Diversity of culturable marine filamentous fungi from deep-sea hydrothermal vents. Environ Microbiol. 2009;11:1588–1600.

Burgaud G, Arzur D, Durand L, Cambon-Bonavita M-A, Barbier G. Marine culturable yeasts in deep-sea hydrothermal vents: species richness and association with fauna. FEMS Microbiol Ecol. 2010;73:121–133.

Christensen M. A view of fungal ecology. Mycologia 1989;81:1–19.

Claus H, Filip Z. Degradation and transformation of aquatic humic substances by laccase-producing fungi *Cladosporium cladosporioides* and *Polyporus versicolor*. Acta Hydrochim Hydrobiol. 1998;26:180–185.

Connell L, Barrett A, Templeton A, Staudigel H. Fungal diversity associated with an active deep sea volcano: Vailulu'u Seamount, Samoa. Geomicrobiology J. 2009;26:597–605.

Damare SR, Raghukumar C, Raghukumar S. Fungi in deep-sea sediments of the Central Indian Basin. Deep-Sea Res. 2006;Pt 1, 53:14–27.

DeLong EF, Pace NR. Environmental diversity of Bacteria and Archaea. Syst Biol. 2001;50: 470–478.

Dromer F, Moulignier A, Dupont B, Gueho E, Baudrimont M, Improvisi L, Provost F, Gonzalez-Canali G. Myeloradiculitis due to *Cryptococcus curvatus* in AIDS. AIDS 1995;9:395–396.

Dupont J, Magnin S, Rousseau F, Zbinden M, Frebourg G, Samadi S, Richer de Forges B, Jones EBG. Molecular and ultrastructural characterization of two ascomycetes found on sunken wood off Vanuatu Islands in the deep Pacific Ocean. Mycol Res. 2009;113:1351–1364.

Edgcomb VP, Kysela DT, Teske A, de Vera Gomez A, Sogin ML. Benthic eukaryotic diversity in the Guaymas Basin hydrothermal vent environment. Proc Nat Acad Sci USA. 2002;99:7658–7662.

Edgcomb VP, Beaudoin D, Gast R, Biddle JF, Teske A. Marine subsurface eukaryotes: the fungal majority. Environ Microbiol. 2010;13:172–183.

Ein-Gil N, Ilan M, Carmeli S, Smith GW, Pawlik JR, Yarden O. Presence of *Aspergillus sydowii*, a pathogen of gorgonian sea fans in the marine sponge *Spongia obscura*. ISME J. 2009;3: 752–755.

Gadanho M, Sampaio JP. Occurrence and diversity of yeasts in the mid-Atlantic ridge hydrothermal fields near the Azores Archipelago. Microb Ecol. 2005;50:408–417.

Galstoff PS. Wasting disease causing mortality of sponges in the West Indies and Gulf of Mexico. Proceedings of the VIII American Science Congress 1942;3:411–421.

Gao Z, Li BL, Zheng CC, Wang GY. Molecular detection of fungal communities in the Hawaiian marine sponges *Suberites zeteki* and *Mycale armata*. Appl Environ Microbiol. 2008;74: 6091–6101.

Geiser DM, Taylor JW, Ritchie KB, Smith GW. Cause of sea fan death in the West Indies. Nature 1998;394:137–138.

Gewin V. Dead in the water. Nature 2010;466:812–814.

Gutierrez MH, Pantoja S, Quinones RA, Gonzalez RR. First record of filamentous fungi in the coastal upwelling ecosystem off central Chile. Gayana 2010;74:66–73.

Gutierrez MH, Pantoja S, Tejos E, Quinones RA. The role of fungi in processing marine organic matter in the upwelling ecosystem off Chile. Mar Biol. 2011;158:205–219.

Hawksworth DL. The fungal dimension of biodiversity: magnitude, significance and conservation. Mycol Res. 1991;95:641–655.

Hentschel U, Usher KM, Taylor MW. Marine sponges as microbial fermenters. FEMS Microb Ecol. 2006;55:167–177.

Hoffmann F, Radax R, Woebken D, Holtappels M, Lavik G, Rapp HT, Schläppy ML, Schleper C, Kuypers MM. Complex nitrogen cycling in the sponge *Geodia barretti*. Environ Microbiol. 2009;11:2228–2243.

Höhnk W. A further contribution to the oceanic mycology. Rapp P-V Reun Cons Int Expolr Mer. 1961;149:202–208.

Höhnk W. Uber den pilzlichen Befall Kalkiger Hartteile von Meerestieren. Ber Dtsch Wiss Komm Meeresforsch. 1969;20:129–140.

Höller U, Wright AD, Matthee GF, König GM, Draeger S, Aust HJ, Schulz B. Fungi from marine sponges: diversity, biological activity and secondary metabolites. Mycol Res. 2000;104: 1354–1365.

Hooper DU, Chapin FS, Ewel JJ, Hector A, Inchausti P, Lavorel S. Effects of biodiversity on ecosystem functioning: A consensus of current knowledge. Ecol Monogr. 2005;75:3–35.

Hyde KD. Ecology of tropical marine fungi. Hydrobiol. 1989;178:199–208.

Jebaraj CS, Raghukumar C. Anaerobic denitrification in fungi from the coastal marine sediments off Goa, India. Mycol Res. 2009;113:100–109.

Jebaraj CS, Raghukumar C, Behnke A, Stoeck T. Fungal diversity in oxygen-depleted regions of the Arabian Sea revealed by targeted environmental sequencing combined with cultivation. FEMS Microb Ecol. 2010;71:399–412.

Joergensen BB, Boetius A. Feast and famine – microbial life in the deep-sea bed. Nat Rev Microbiol. 2007;5:770–781.

Jones EBG. Are there more marine fungi to be described? Bot Mar. 2011a;54:343–354.

Jones EBG. Fifty years of marine mycology. Fungal Divers. 2011b;50:73–112.

Junghanns C, Pecyna M, Böhm D, Jehmlich N, Martin C, von Bergen M, Schauer F, Hofrichter M, Schlosser D. Biochemical and molecular genetic characterisation of a novel laccase produced by the aquatic ascomycete *Phoma* sp. UHH 5-1-03. Appl Microbiol Biotechnol. 2009;84: 1095–1105.

Kirk PM, Cannon PF, Minter DW, Stalpers JA. Dictionary of the Fungi. Wallingford, UK: CABI, 2008.

Kochkina GA, Ivanushkina NE, Akimov VN, Gilichinskii DA, Ozerskaya SM. Halo- and psychrotolerant *Geomyces* fungi from Arctic cryopegs and marine deposits. Microbiology 2007;76: 31–38.

Kohlmeyer J. New genera and species of higher fungi from the deep sea (1615–5315m). Rev Mycol. 1977;41:189–206.

Kohlmeyer J, Kohlmeyer E. Marine mycology: The higher fungi, New York: Academic Press, 1979.

Kohlmeyer J, Volkmann-Kohlmeyer B. Koralionastetaceae fam. nov. (ascomycetes) from coral rock. Mycologia 1987;79:764–778.

Kohlmeyer J, Volkmann-Kohlmeyer B. New species of *Koralionastes* (Ascomycotina) from the Caribbean and Australia. Can J Bot. 1990;68:1554–1559.

Kohlmeyer J, Volkmann-Kohlmeyer B. Mycological research news. Mycol Res. 2003;107: 385–387.

König GM, Kehraus S, Seiber SF, Abdel-Lateff A, Muller D. Natural products from marine organisms and their associated microbes. ChemBioChem. 2006;7:229–238.

Kunčič MK, Kogej T, Drobne D, Gunde-Cimerman N. Morphological response of the halophilic fungal genus *Wallemia* to high salinity. Appl Environ Microbiol. 2010;76:329–337.

Lai X, Cao L, Tan H, Fang S, Huang Y, Zhou S. Fungal communities from methane hydrate-bearing deep-sea marine sediments in South China Sea. ISME J. 2007;1:756–762.

Le Calvez T, Burgaud G, Mahe S, Barbier G, Vandenkoornhuyse P. Fungal diversity in deep-sea hydrothermal ecosystems. Appl Environ Microbiol. 2009;75:6415–6421.

Levin LA, Etter RJ, Rex MA, Gooday AJ, Smith CR, Pineda J, Stuart CT, Hessler RR, Pawson D. Environmental influences on regional deep-sea species diversity. Annu Rev Ecol Syst. 2001;132:51–93.

Li L, Kato C, Horikoshi K. Bacterial diversity in deep-sea sediments from different depths. Biodivers Conserv. 1999;8:659–677.

Li Q, Wang G. Diversity of fungal isolates from three Hawaiian marine sponges. Microbiol Res. 2009;164:233–241.

López-García P, Philippe H, Gail F, Moreira D. Autochthonous eukaryotic diversity in hydrothermal sediment and experimental microcolonizers at the Mid-Atlantic Ridge. Proc Natl Acad Sci USA. 2003;2:697–702.

López-García P, Vereshchaka A, Moreira D. Eukaryotic diversity associated with carbonates and fluid-seawater interface in Lost City hydrothermal field. Environ Microbiol. 2007;9:546–554.

Lorenz R, Molitoris HP. Cultivation of fungi under simulated deep-sea conditions. Mycol Res. 1997;11:1355–1365.

Maldonado M, Carmona C, Velasquez Z, Puig A, Cruzado A, Lopez A, Young CM. Siliceous sponges as a silicon sink: An overlooked aspect of benthopelagic coupling in the marine silicon cycle. Limnol Oceanogr. 2005a;50:799–809.

Maldonado M, Cortadellas N, Trillas MI, Ruetzler K. Endosymbiotic yeast maternally transmitted in a marine sponge. Biol Bull. 2005b;209:94–106.

Meyers SP. Fifty years of marine mycology: highlights of the past, projections for the coming century. SIMS News. 1996;46:119–127.

Morrison-Gardiner S. Dominant fungi from Australian coral reefs. Fungal Divers. 2002;9:105–121.

Nagahama T, Hamamoto M, Nakase T, Horikoshi K. *Rhodotorula lamellibrachii* sp. nov., a new yeast species from a tubeworm collected at the deep-sea floor in Sagami bay and its phylogenetic analysis. Antonie Leeuwenhoek Int J G. 2001a;80:317–323.

Nagahama T, Hamamoto M, Nakase T, Takami H, Horikoshi K. Distribution and identification of red yeasts in deep-sea environments around the northwest Pacific Ocean. Anton Leeuw Int J G. 2001b;80:101–110.

Nagahama T, Hamamoto M, Nakase T, Takaki Y, Horikoshi K. *Cryptococcus surugaensis* sp. nov., a novel yeast species from sediment collected on the deep-sea floor of Suruga Bay. Int J Syst Evol Microbiol. 2003;53:2095–2098.

Nagahama T, Hamamoto M, Horikoshi K. *Rhodotorula pacifica* sp. nov., a novel yeast species from sediment collected on the deep-sea floor of the north-west Pacific Ocean. Int J Syst Evol Microbiol. 2006;56:295–299.

Nagahama T, Takahashi E, Nagano Y, Abdel-Wahab MA, Miyazaki M. Molecular evidence that deep-branching fungi are major fungal components in deep-sea methane cold-seep sediments. Environ Microbiol. 2011;13:2359–2370.

Nagano Y, Nagahama T, Hatada Y, Nunoura T, Takami H, Miyazaki J, Takai K, Horikoshi K. Fungal diversity in deep-sea sediments-the presence of novel fungal groups. Fungal Ecol. 2010;3:316–325.

Namikoshi M, Akano K, Kobayashi H, Koike Y, Kitazawa A, Rondonuwu AB, Pratasik SB. Distribution of marine filamentous fungi associated with marine sponges in coral reefs of Palau and Bunaken Island, Indonesia. J Tokyo Univ Fish. 2002;88:15–20.

Naqvi SWA. Denitrification processes in the Arabian Sea. Proc Ind Acad Sci. 1994;103:279–300.

Naqvi SWA, Yoshinari T, Jayakumar DA, Altabet MA, Narvekar PV, Devol AH, Brandes JA, Codispoti LA. Budgetary and biogeochemical implications of N_2O isotope signatures in the Arabian Sea. Nature 1998;394:462–464.

Naqvi SWA, Jayakumar DA, Narvekar PV, Naik H, Sarma VVSS, DeSouza W, Joseph S, George MD. Increased marine production of N_2O due to intensifying anoxia on the Indian continental shelf. Nature 2000;408:346–349.

Naqvi SWA, Naik H, Pratihary A, DeSouza W, Narvekar PV, Jayakumar DA, Devol AH, Yoshinari T, Saino T. Coastal versus open-ocean denitrification in the Arabian Sea. Biogeosciences 2006;3:621–633.

Newell SY. Multiyear patterns of fungal biomass dynamics and productivity within naturally decaying smooth cordgrass shoots. Limnol Oceanogr. 2001;46:573–583.

Orphan VJ, House CH, Hinrichs KU, Mckeegan KD, Delong EF. Multiple archaeal groups mediate methane oxidation in anoxic cold seep sediments. Proc Natl Acad Sci USA. 2002;99:7663–7668.

Pang KL, Mitchell JI. Molecular approaches for assessing fungal diversity in marine substrata. Bot Mar. 2005;48:332–347.

Perovic-Ottstadt S, Adell T, Proksch P, Wiens M, Korzhev M, Gamulin V, Müller IM, Müller WE. (1 -> 3)-beta-D-glucan recognition protein from the sponge *Suberites domuncula* – Mediated activation of fibrinogen-like protein and epidermal growth factor gene expression. Eur J Biochem. 2004;271:1924–1937.

Poulicek M, Machiroux R, Toussaint C. Chitin diagenesis in deep-water sediments. In: Muzzarelli R, Jeunix C, Gooday GW, eds. Chitin in nature and technology, New York: Plenum Press, 1986: 523–530.

Poulicek M, Goffinet G, Jeuniaux CH, Simon A, Voss-Foucart MF. Early diagenesis of skeletal remains in marine sediments: a 10 year study. Researches Oceanographiques en mer Mediterranee (Biologie, Chimie, Geologie, Physique). 1988;57:107–124.

Proksch P, Ebel R, Edrada RA, Wray V, Steube K. Bioactive natural products from marine invertebrates and associated fungi. In: Müller WEG, ed. Sponges. Heidelberg, Berlin, Germany: Springer Verlag, 2003:117–142.

Raghukumar C, Raghukumar S. Barotolerance of the fungi isolated from the deep-sea sediments. Aquat Microb Ecol. 1998;15:153–163.

Raghukumar C, Raghukumar S, Sheelu G, Gupta SM, Nath BN, Rao BR. Buried in time: Culturable fungi in a deep-sea sediment core from the Chagos Trench, Indian Ocean. Deep-Sea Res. 2004;Pt 1, 51:1759–1768.

Raghukumar C, Damare SR, Singh P. A review on deep-sea fungi: occurrence, diversity and adaptations. Bot Mar. 2010;53:479–492.

Reimers CE. An in-situ microprofiling instrument for measuring interfacial pore water gradients: method and oxygen profiles from the North Pacific Ocean. Deep-Sea Res. 1987;34:2019–2035.

Rice AV, Tsuneda A, Currah RS. In vitro decomposition of *Sphagnum* by some microfungi resembles white rot of wood. FEMS Microb Ecol. 2006;56:372–382.

Richter W. Marine sponges as substrate for *Thraustochytriaceae* marine lower fungi. Veroeff Inst Meeresforschung Bremerh. 1985;20:141–150.

Rodríguez A, Falcón MA, Carnicero A, Perestelo F, La Fuente GD, Trojanowski J. Laccase activities of *Penicillium chrysogenum* in relation to lignin degradation. Appl Microbiol Biotechnol. 1996;45:399–403.

Rot C, Goldfarb I, Ilan M, Huchon D. Putative cross-kingdom horizontal gene transfer in sponge (Porifera) mitochondria. BMC Evol Biol. 2006;6:17.

Roth FJ, Orpurt PA Jr, Ahearn DG. Occurrence and distribution of fungi in a subtropical marine environment. Can J Bot. 1964;42:375–383.

Sanders HL. Marine benthic diversity: a comparative study. Am Nat. 1968;102:243–282.

Shokes RF, Trabant PK, Presley BJ, Reid DF. Anoxic, hypersaline basin in northern Gulf of Mexico. Science 1976;196:1443–1446.

Singh P, Raghukumar C, Verma P, Shouche Y. Phylogenetic diversity of culturable fungi from the deep-sea sediments of the Central Indian Basin and their growth characteristics. Fungal Divers. 2010;40:89–102.

Singh P, Raghukumar C, Verma P, Shouche Y. Fungal community analysis in the deep-sea sediments of the Central Indian Basin by culture-independent approach. Microb Ecol. 2011a;61:507–517.

Singh P, Raghukumar C, Verma P, Shouche Y. Assessment of fungal diversity in deep-sea sediments by multiple primer approach. World J Microbiol Biotechnol. 2011b;28:659–667.

Smit E, Leeflang P, Glandorf B, van Elsas JD, Wernars K. Analysis of fungal diversity in the wheat rhizosphere by sequencing of cloned PCR-amplified genes encoding 18S rRNA and temperature gradient gel electrophoresis. Appl Environ Microbiol. 1999;65:2614–2621.

Snelgrove PVR, Blackburn TH, Hutchings P, Alongi D, Grassle JF, Hummel H, King G, Koike I, Lambshead PJD, Ramsing NB, Solis-Weiss V, Freckman DW. The importance of marine sediment biodiversity in ecosystem processes. Ambio. 1997;26:578–582.

Sogin M, Morrison HG, Huber JA, Welch DM, Huse SM, Neal PR, Arrieta JM, Herndl GJ. Microbial diversity in the deep sea and underexplored "rare biosphere". Proc Natl Acad Sci USA. 2006;103:12115–12120.

Sparks AK. Synopsis of invertebrate pathology: exclusive of insects. New York: Elsevier Science Publisher, 1985.

Stoeck T, Epstein S. Novel eukaryotic lineages inferred from small-subunit rRNA analyses of oxygen-depleted marine environments. Appl Environ Microbiol. 2003;69:2657–2663.

Stoeck T, Taylor GT, Epstein SS. Novel eukaryotes from the permanently anoxic Cariaco Basin (Caribbean Sea). Appl Environ Microbiol. 2003;69:5656–5663.

Stoeck T, Hayward B, Taylor GT, Varela R, Epstein SS. A multiple PCR-primer approach to access the microeukaryotic diversity in environmental samples. Protist 2006;157:31–43.

Svendrup HU, Johnson MW, Fleming RH. The oceans. Englewood Cliffs, New Jersey: Prentice-Hall, 1942.

Takai K, Horikoshi K. Genetic diversity of *Archaea* in deep-sea hydrothermal vent environments. Genetics 1999;152:1285–1297.

Takami H, Inoue A, Fuji F, Horikoshi K. Microbial flora in the deepest sea mud of the Mariana Trench. FEMS Microb Lett. 1997;152:279–285.

Takishita K, Tsuchiya M, Reimer JD, Maruyama T. Molecular evidence demonstrating the basidiomycetous fungus *Cryptococcus curvatus* is the dominant microbial eukaryote in sediment at the Kuroshima Knoll methane seep. Extremophiles 2006;10:165–169.

Takishita K, Yubuki N, Kakizoe N, Inagaki Y, Maruyama T. Diversity of microbial eukaryotes in sediment at a deep-sea methane cold seep: surveys of ribosomal DNA libraries from raw sediment samples and two enrichment cultures. Extremophiles 2007;11:563–576.

Taylor JW, Jacobson DJ, Kroken S, Kasuga T, Geiser DM, Hibbett DS, Fisher MC. Phylogenetic species recognition and species concepts in fungi. Fungal Genet Biol. 2000;31:21–32.

Taylor MW, Radax R, Steger D. Sponge associated microorganisms: evolution, ecology, and biotechnological potential. Microbiol Mol Biol R. 2007;71:295–347.

Vacelet J, Vacelet E, Gaino E, Gallissian MF. Bacterial attach of spongin skeleton during the 1986–1990 Mediterranean sponge disease. In: van Soest RWM, van Kempen TMG, Braekman JC, eds. Sponges in time and space. Rotterdam: AA Balkema, 1994:355–362.

Wang G. Diversity and biotechnological potential of the sponge-associated microbial consortia. J Ind Microbiol Biot. 2006;33:545–551.

Wang P, Xiao X, Zhang H, Wang F. Molecular survey of sulphate-reducing bacteria in the deep-sea sediments of the west Pacific Warm Pool. J Ocean Uni China. 2008;7:269–275.

Warren BA. Context of the suboxic layer in the Arabian Sea. In: Lal D, ed. Biogeochemistry of the Arabian Sea. Bangalore: Indian Academy of Sciences, 1994:203–216.

Webster NS. Sponge disease: a global threat? Environ Microbiol. 2007;9:1363–1375.

Webster NS, Blackall LL. What do we really know about sponge-microbial symbioses? ISME J. 2009;3:1–3.

Webster NS, Taylor MW, Behnam F, Luecker S, Rattei T, Whalan S, Horn M, Wagner M. Deep sequencing reveals exceptional diversity and modes of transmission for bacterial sponge symbionts. Environ Microbiol. 2010;12:2070–2082.

Xu HX, Min W, Xiaogu W, Junyi Y, Chunsheng W. Bacterial diversity in deep-sea sediment from north eastern Pacific Ocean. Acta Ecologica Sinica 2008;28:479–485.

Zhu P, Li Q, Wang G. Unique microbial signatures of the alien Hawaiian marine sponge *Suberites zeteki*. Microb Ecol. 2008;55:406–414.

Zierenberg RA, Adams MWW, Arp AJ. Life in extreme environments: Hydrothermal vents. Proc Natl Acad Sci USA. 2000;97:12961–12962.

Application of marine fungi

20 Natural products from marine-derived fungi

Rainer Ebel

Natural products continue to play a vital role in developing novel drugs. Even in the age of combinatorial chemistry, secondary metabolites are either used unchanged, or act as lead structures, and for certain indications such as anti-invectives or anticancer agents, more than 60% of all drugs admitted worldwide between 1981 and 2006 are based directly or indirectly on structures from nature (Newton and Abraham 1955; Butler 2008).

Traditionally, the focus in the search for new bioactive constituents from nature was on higher terrestrial plants. Some thirty years ago, however, natural product chemists became increasingly aware of the vast and largely unexplored potential of bioactive compounds from marine organisms. In this area, marine invertebrates such as sponges, tunicates or bryozoans rather than algae soon emerged as the most promising chemical resource in the oceans. However, after more than three decades of research into marine natural products, nowadays it is becoming increasingly difficult to identify novel compounds even from this group of organisms. For this reason the search for novel drug lead structures from nature should follow an alternative and innovative approach and thus include "unusual" sources which have not been thoroughly investigated so far.

Of particular interest, especially from an industrial point of view, are microorganisms since at least in principle, interesting natural products could be produced by fermentation in almost unlimited amounts, even though it should not be neglected that establishing of suitable large scale cultivation protocols is a tedious and very time consuming process. Traditionally, bacteria from soil samples, most of them belonging to the phylum actinomycetes, have played an enormous role in the drug discovery process, and still nowadays a vast array of actinomycete-derived secondary metabolites are produced by fermentation to ensure revenues in the order of billions of dollars per year for pharmaceutical companies (Butler 2008). Similarly, fungi have long since been established as producers of commercially important and currently used drugs. Besides the famous penicillins, which still nowadays continue to play a vital role for antibiotic therapy, further relevant examples include statins (used to lower cholesterol levels to treat cardiovascular diseases) or immunosuppressive agents, such as, cyclosporine or mycophenolic acid. However, comparable to the situation described above for terrestrial plants or marine invertebrates, the conventional approach to discover novel lead structures for the drug development pipeline has slowed down since investigation of bacteria or fungi from easily accessible terrestrial habitats tends to yield known metabolites with increasing probability. Thus, also in the sector of microorganisms a shift towards the investigation of new, vastly untapped sources of organisms can clearly be identified, with major emphasis on strains obtained from marine habitats which to date are still largely underexplored.

Historical development

1

2

3

4

Historically, the beginning of research into secondary metabolites from marine-derived fungi goes back to the period after the Second World War, when Italian researchers obtained an isolate of a *Cephalosporium* sp. from seawater close to a sewage outlet off the Sardinian coast, which caught their attention due to its pronounced antibacterial activity (Demain and Elander 1999). Cephalosporin C (**1**) was established as the active principle only in 1955, thus almost ten years later, by Newton and Abraham at Oxford University (Newton and Abraham 1955). Even though cephalosporin C is no longer used as such for clinical applications, its discovery marked the emergence of a new class of modern day antibiotics which still play a considerable role in the treatment of bacterial infections today. But it has to be stated that this important discovery did not immediately lead to a more systematic investigation of the chemical potential of marine-derived fungi, and the following decades are characterized by only occasional reports on this subject. Initially the focus was on primary metabolites such as choline, its sulfate, cholesterol, and fatty acids (Kirk and Catalfomo 1970; Block et al. 1973; Catalfomo et al. 1973; Kirk et al. 1974). What probably was the first report of a true secondary metabolite from a marine fungus, *Dendryphiella salina*, appeared in 1975, reporting 2,6-dimethoxy-1,4-benzoquinone (**2**) (Fukuzumi et al. 1975), a metabolite of widespread occurrence in wood and woody tissues, most likely formed by degradation of lignin. During the next decade, a handful of antibiotics, all of them already known from terrestrials sources, were detected in marine-derived fungi, including siccayne (Kupta et al. 1981), melinacidins III and IV as well as gancidin W (Furuya et al. 1985). The first new natural product from marine fungi was discovered by Jon Clardy's group, leptosphaerin (**3**), an oxidatively modified 2-aminohexose derivative from *Leptosphaeria orae-maris* (Schiehser et al. 1986), while in the following years, Francesco Pietra's group reported a systematic chemical study of the chemistry of the obligate marine asexual fungus *Dendryphiella salina* (Guerriero et al. 1988), which turned out to be a prolific source of unusual terpenoid derivatives, exemplified by the trinoreremophilane dendryphiellin A (**4**), and which are described in more detail below. The beginning of a more systematic chemical investigation of marine-derived fungi dates back to the early 1990's, which very soon resulted in a drastic increase in publications in this subject per year. Much more recently, this field experienced a further boost when Chinese researchers became interested in their rich fungal biodiversity, mostly focusing on the unique mangrove habitats of the tropical island of Hainan. As of the end of 2010, there have been more than 1100 new natural products described from marine-derived fungi, and there is still a considerable increase in the number of publications as well as the number of new structures reported per year (Rateb and Ebel 2011).

From the number of new compounds published each year (Figure 20.1), it is evident that natural products chemistry of marine-derived fungi is a rapidly developing discipline. The total number of new natural products from marine-derived fungi currently (at the end of 2010) exceeds 1100. While until the year 2002, 272 new structures had been reported, in 2009 alone almost two thirds of this figure was reached. However, in 2010 the total number of new published structures per year dropped slightly, and it remains to be seen whether this indicates that a peak has been reached, or if this trend is more due to a temporary dip, and the number will continue to rise again.

Metabolic diversity of marine-derived fungi

Compounds by structural class

Figure 20.2 gives an overview of new chemical structures from marine-derived fungi reported in the literature until the end of 2010, based on their putative biogenetic origin. In most cases, assignment of a given metabolite to a certain category is only based on structural considerations, and only represents the author's personal judgment. Not too surprisingly, polyketides play a dominant role, and if prenylated polyketides and nitrogen-containing polyketides (grouped as alkaloids in this overview) are taken into account, their total share will exceed 50% of all new natural products. Obviously, in some cases it was necessary to compromise in order to avoid too many different categories, as exemplified by the very common group of diketopiperazines, which have been listed as (di-)peptides, rather than alkaloids.

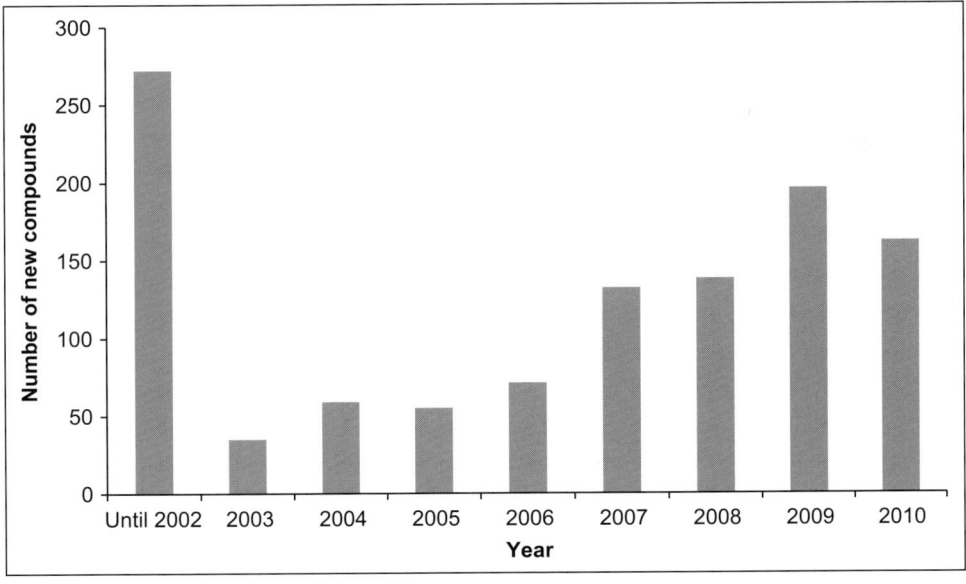

Figure 20.1 New compounds from marine-derived fungi, divided by year of publication (updated from Rateb and Ebel 2011).

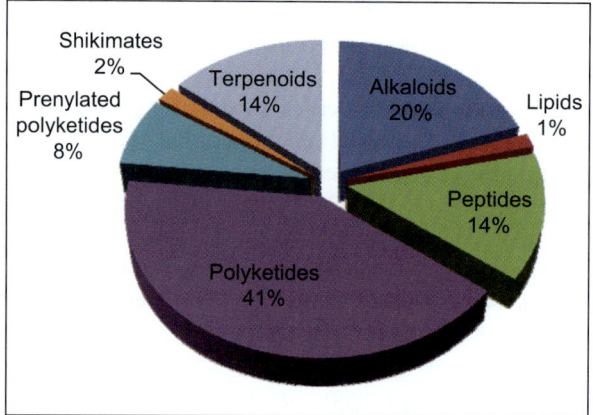

Figure 20.2 New compounds from marine-derived fungi until 2010, sorted by structural/biogenetic types (updated from Rateb and Ebel 2011).

Similar to terrestrial fungi, the chemistry of marine-derived fungi is clearly dominated by polyketides, while alkaloids, peptides (including cyclic dipeptides=diketopiperazines) and terpenoids also contribute considerably to their chemical diversity. Comprehensive review articles covering the chemistry of marine-derived fungi have been published elsewhere (Bugni and Ireland 2004; Saleem et al. 2007; Ebel 2010a, 2010b; Rateb and Ebel 2011). In the following, selected examples for each of these biogenetic groups are presented, with this choice representing this author's personal judgment with regard to chemically or biologically interesting representative examples.

Polyketides

An example of a marine-derived strain rich in polyketides belonging to different biogenetic groups is the fungus *Monodictys putredinis*, isolated from a marine green alga collected in Tenerife, Spain. An initial screening for compounds with potential cancer chemopreventive effects led to the isolation of four new monomeric xanthones, monodictysins A (**5**), B (**6**), C (**7**), monodictyxanthone (**8**), and the benzophenone monodictyphenone (**9**) (Krick et al. 2007). Reinvestigation of the same fungus yielded the two novel dimeric chromanones monodictyochromes A (**10**) and B (**11**) (Pontius et al. 2008), consisting of two uniquely modified xanthone-derived units probably coupled by phenol oxidative coupling in a regioselective manner. It appears that this coupling should also occur under strict stereoselectivity, since **10** was identified as the (*P*)-atropisomer, while **11** represents the (*M*)-stereoisomer. Monodictysin B (**6**), monodictyphenone (**9**) and monodictyochrome B (**11**) inhibited cytochrome P450 1A activity which is involved in the metabolic conversion of procarcinogens into carcinogens. In addition, monodictysins B (**6**) and C (**7**) as well as monodictyochromes A (**10**) and B (**11**) moderately induced the carcinogen-detoxifying enzyme NAD(P)H:quinone reductase, and the latter three compounds also weakly inhibited aromatase activity.

Prenylated polyketides/meroterpenoids

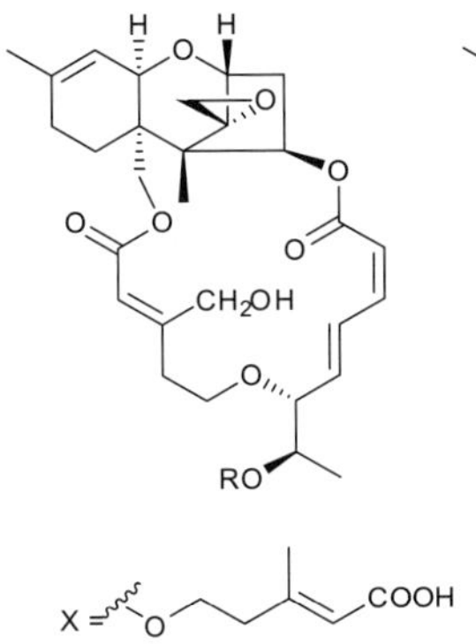

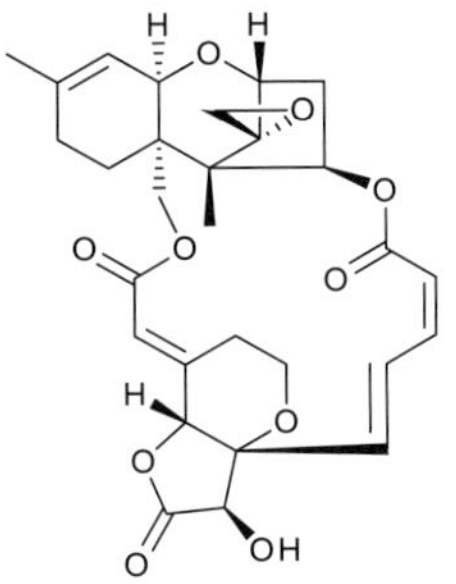

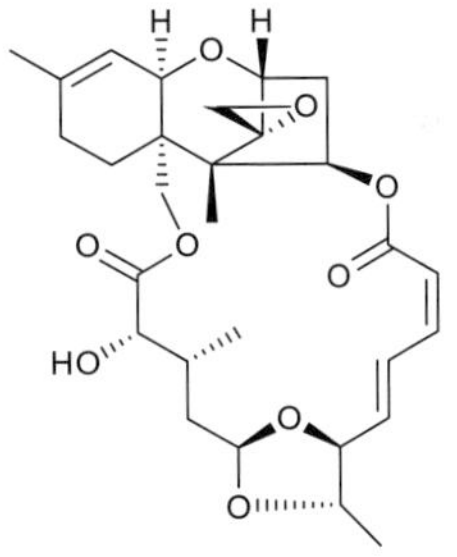

12 (R = OH)
13 (R = X)

14

15

Trichothecenes are a commercially important class of mycotoxins, mainly produced by *Fusarium* spp., but also by other genera including *Cephalosporium, Cylindrocarpon, Myrothecium, Stachybotrys, Trichoderma, Trichothecium,* and *Verticimonosporium* (Bräse et al. 2009). Biogenetically, they comprise a sesquiterpenoid skeleton, to which attached are one or more polyketide-derived chains, which in turn can be connected to one another to form a macrocyclic ring system. Although originally discovered form terrestrial habitats, there are quite a few examples of trichothecenes obtained from marine-derived fungi. Three new macrocyclic trichothecenes, 12′-hydroxyroridin E (**12**), roridin Q (**13**), and 2′,3′-deoxyroritoxin D (**14**), were produced by a strain of *Myrothecium roridum*, isolated from a wood sample collected in Palau, while the new roridin R (**15**) was obtained from the culture filtrate of *Myrothecium* sp., isolated from an unidentified Indonesian sponge (Xu et al. 2006). **13** possesses an unusual ether moiety at position C-**13**′ not present in most other trichothecenes. **12**, **13** and **15** exhibited medium to strong cytotoxic activity towards L1210 murine leukemia cells, while **15** also displayed antifungal activity against *Saccharomyces cerevisiae*.

Terpenoids

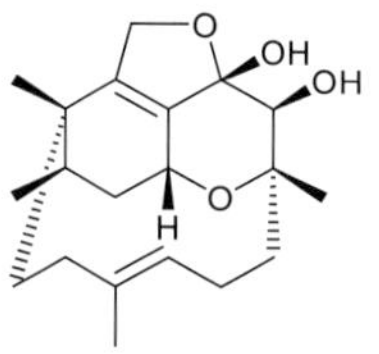

16

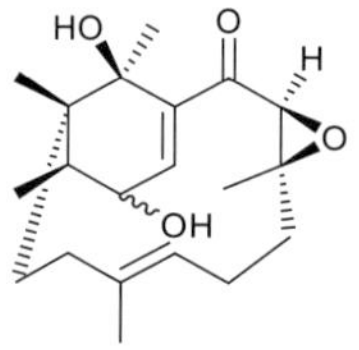

17 (α-OH)
18 (β-OH)

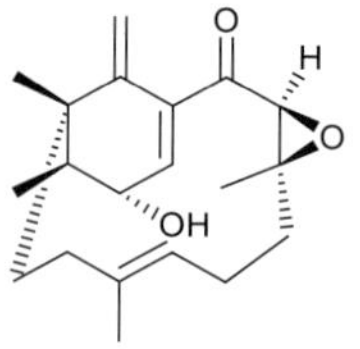

19

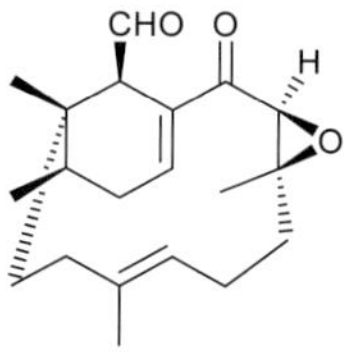

20

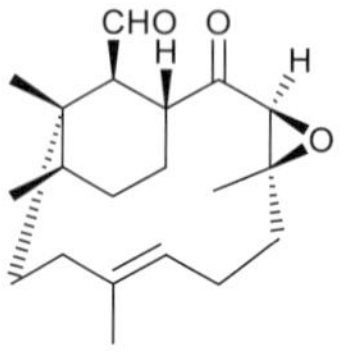

21

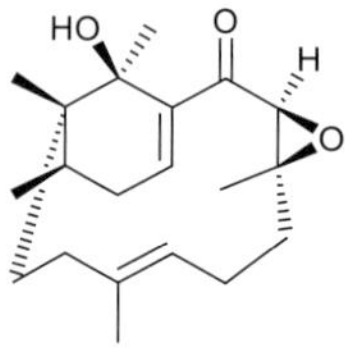

22

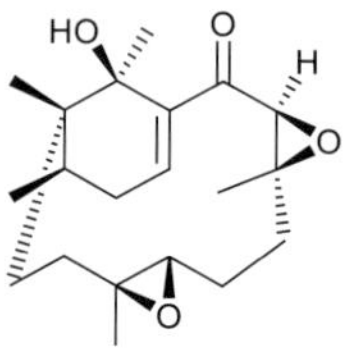

23

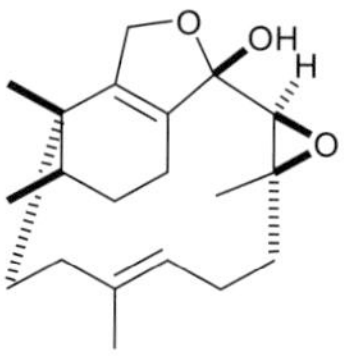

24

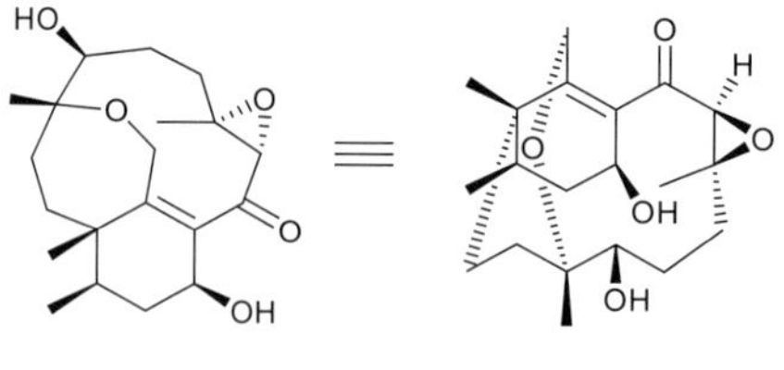

25

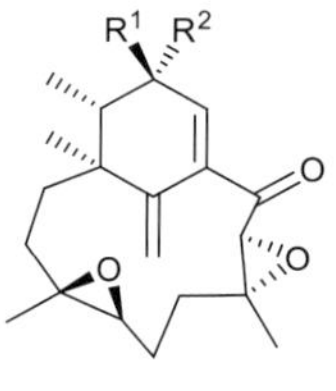

26 (R^1 = OH, R^2 = H)
27 (R^1 = H, R^2 = OH)

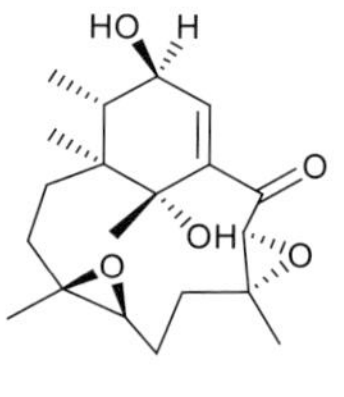

28

Terpenes are frequently encountered as secondary metabolites of marine-derived fungi, and their chemistry has been reviewed in detail (Ebel 2010b). As an example of their structural diversity, phomactins are an intriguing class of fungal diterpenes, mainly due to their pronounced activity as antagonists of platelet-activating factor. Initially, phomactins were discovered in the extract of the fungus *Phoma* sp. obtained from the shell of the crab *Chionoecetes opilio*, including phomactin A (**16**) (Sugano et al. 1991), B (**17**), B1 (**18**), B2 (**19**), C (**20**), D (**21**) (Sugano et al. 1994), and subsequently phomactins E (**22**), F (**23**) and G (**24**) (Sugano et al. 1995). The most active congener, phomactin D (**21**) was found to inhibit PAF receptor binding with an IC$_{50}$ value of 0.12 µM, and platelet aggregation with an IC$_{50}$ value of 0.80 µM, respectively. Parallel investigations of a terrestrial *Phoma* sp. carried out by researchers from Schering Plough in the US revealed further phomactin derivatives Sch 47918 (identical to phomactin C, **20**), Sch 49026, Sch 49027 and Sch 49028 (later found to be identical to phomactin A, **16**) (Chu et al. 1992, 1993). Interestingly, an unidentified marine-derived fungus not closely related to the genus *Phoma*, that was isolated from the surface of the marine brown alga *Ishige okamurae*, was found to produce phomactins H (**25**) (Koyama et al. 2004), I (**26**), 13-*epi*-phomactin I (**27**), and phomactin J (**28**) (Ishino et al. 2010). Numerous efforts into the synthesis of phomactins which have greatly provided to our current understanding of structure-activity relationships have been reviewed elsewhere (Goldring and Pattenden 2006).

Peptides including diketopiperazines

A series of new lipodepsipeptides, acremolides A–D (**29–32**), were obtained together with known chaetoglobosins from the sediment-borne fungal strain *Acremonium* sp. MST-MF558a, which was isolated from a Tasmanian estuarine habitat, and based on its rDNA sequence was found to represent a new species (Ratnayake et al. 2008). The absolute stereochemistry of amino acid residues in **29–32** was determined using a new C3 Marfey's method for amino acid analysis. **29–32** displayed no antibacterial, antifungal, or cytotoxic properties, but were shown to synergize chaetoglobosin cytotoxicity.

Arguably the single most important natural product from marine-derived fungi is the simple diketopiperazine halimide (**33**), discovered by Bill Fenical's group in the 1990's (Fenical et al. 1998), and which due to its tubulin-depolymerizing activity served as a lead structure for the development of the synthetic derivative analogue plinabulin (NPI-2358) (**34**). **34** is a novel vascular disrupting agent, displaying promising activity for the treatment of non-small cell lung cancer (Aren et al. 2010). Phase I (Mita et al. 2010) and phase II clinical trials of NPI-2358, also in combination with docetaxel have recently been completed, and there is evidence that **34** also triggers JNK-mediated apoptosis besides inhibiting angiogenesis (Singh et al. 2011).

Alkaloids and other nitrogen-containing metabolites

35 (R^1 = Ac, R^2 = H)
36 (R^1 = X, R^2 = CH_3)
37 (R^1 = X, R^2 = H)
38 (R^1 = X, R^2 = CHO)

Communesins are a group of structurally complex indole alkaloids, including communesins A (**35**) and B (**36**) originally described from the fungus *Penicillium* sp. which was obtained from the green alga *Enteromorpha intestinalis* (Numata et al. 1993). Later on, communesins C (**37**) and D (**38**) were reported from an undescribed *Penicillium* sp. from the Mediterranean sponge *Axinella verrucosa* (Jadulco et al. 2004). In general, most communesin derivatives have been shown to possess antiproliferative activity against different leukemia cell lines. Communesins have been the target of various synthetic approaches, which at least in part is due to the report of the structure of nomofungin, described as a disruptor of microfilament assembly, the story of which has been summarized recently (Siengalewicz et al. 2008). The originally reported structure of nomofungin (Ratnayake et al. 2001, 2003), a metabolite of an endophytic fungal strain of the fig tree *Ficus microcarpa*, was questioned soon after its publication, and based on biosynthetic arguments in connection with a biomimetic synthesis of a related model compound (May et al. 2003), and an independent synthetic study (Crawley and Funk 2003) it was demonstrated that nomofungin was in fact identical to communesin B (**36**). Nowadays it appears that communesins occur quite frequently in terrestrial fungi, and in fact appear to be typical metabolites of *Penicillium expansum*, including strains obtained from food materials such as fruit juice, potato pulp, or fermented soybean (Larsen et al. 1998; Andersen et al. 2004; Hayashi et al. 2004).

Shikimate-derived metabolites

Shikimate-derived natural products ("phenylpropanoids") have not too frequently been reported form marine-derived fungi. One example are a series of terprenins, prenylterphenyllin (**39**), 4″-deoxyprenylterphenyllin (**40**) and 4″-deoxyisoterprenin

39 (R^1 = OH, R^2 = prenyl)
40 (R^1 = H, R^2 = prenyl)
44 (R^1 = OH, R^2 = H)

41 (R^1 = R^3 = H, R^2 = prenyl)
42 (R^1 = R^2 = H, R^3 = prenyl)
43 (R^1 = OH, R^2 = H, R^3 = prenyl)

prenyl

(**41**), which were described from a Japanese sediment-borne isolate of *Aspergillus candidus* IF1, collected off the Gokasyo Gulf (Wei et al. 2007). **39–41** as well as the also encountered known 4″-deoxyterprenin (**42**) were moderately active against KB3–1 cells. These compounds show structural similarity to terprenin (**43**) and terphenyllin (**44**) which were previously reported from terrestrial strains of *A. candidus* (Marchelli and Vining 1975; Kamigauchi et al. 1998; Liu 2006). For the latter, it has been demonstrated through labeling studies with phenylalanine that the terphenyl ring system is formed by self-condensation of two phenylpropanoid units (Chandra et al. 1966).

Lipids

45 (R = H)
46 (R = β-D-glucopyranosyl)

47 (R = β-D-glucopyranosyl)

48 (R = β-D-glucopyranosyl)

Quite surprisingly, the number of lipids described from marine-derived fungi is rather low, but this might be due to most natural product chemists focusing on more polar metabolites, and also to the fact that it is difficult to make a clear distinction between primary and true secondary metabolites for lipid-like compounds. An isolate of *Aspergillus niger*, obtained from the Chinese brown alga *Colpomenia sinuosa* produced the moderately antifungal sphingolipid, asperamide A (**45**) and its corresponding cerebroside, asperamide B (**46**) (Zhang et al. 2007). **45** and **46** are structurally unique because they contain a 9-methyl-C_{20}-sphingosine moiety, whereas sphingolipids containing a 9-methyl-C_{18}-sphingosine moiety have been described from natural sources on various occasions (Tan and Chen 2003). Another strain of *Aspergillus niger*, isolated from sea water in China, yielded structurally related cerebrosides, asperiamides B (**47**) and C (**48**) (Wu et al. 2008).

Miscellaneous

49

Only very few reported examples exist in the literature, describing the ability of marine natural products to bind to metals (Wright et al. 2008), although this ability might be crucial in understanding their potential ecological role, given the fact many essential trace metals are only available at extremely low concentrations in marine habitats, and therefore marine organisms have been shown to have developed sophisticated strategies of metal binding, sequestering and uptake (Butler 1998). As far as marine-derived fungi are concerned, the first example of a metal-binding natural product is ZZF51(A) (**49**), a copper complex composed of two fusaric acid moieties, which was reported from the fungus *Fusarium* sp., an endophyte of the Chinese mangrove plant *Castanopsis fissa* (Tan et al. 2008). Free, i.e., uncomplexed fusaric acid had already been reported as a mycotoxin in the 1930s in various *Fusarium* strains, and is known to exert a series of effects on the mammalian nervous, cardiovascular and immune system (Wang and Ng 1999). The ZZF51(A)-producing strain was shown to tolerate concentrations of Cu(II) ions up to 300 ppm, and also to be capable of active biosorption of copper. **49** exhibited mild antibiotic activity towards four bacterial strains, and was cytotoxic against three different tumor cell lines.

Compounds by the origin or the source of the producing fungal strains

Figure 20.3 gives an overview of new chemical structures from marine-derived fungi reported in the literature until the end of 2010, sorted by the source of the respective fungal strains as indicated in the relevant publications. Almost 80% of all new compounds reported from marine fungi are based on isolates from living matter, with an approximately even split between plant and animal sources in the broadest term, while the remaining compounds are due to fungi from non-living sources, most notably sediments. Within the individual groups, algae are the predominant source for fungal diversity, closely followed by sponges and mangrove habitats. One of the best studied habitats are mangrove areas on the subtropical island of Hainan, which have come under intense scrutiny by Chinese scientist since approximately one decade ago, but obviously still continue to yield chemically interesting fungal strains, as evident from the large number of annual publications in the chemical literature. An interesting newly emerging source is the deep sea, and it is only very recently that this habitat has attracted more attention (listed in the sediment category for the purpose of this overview).

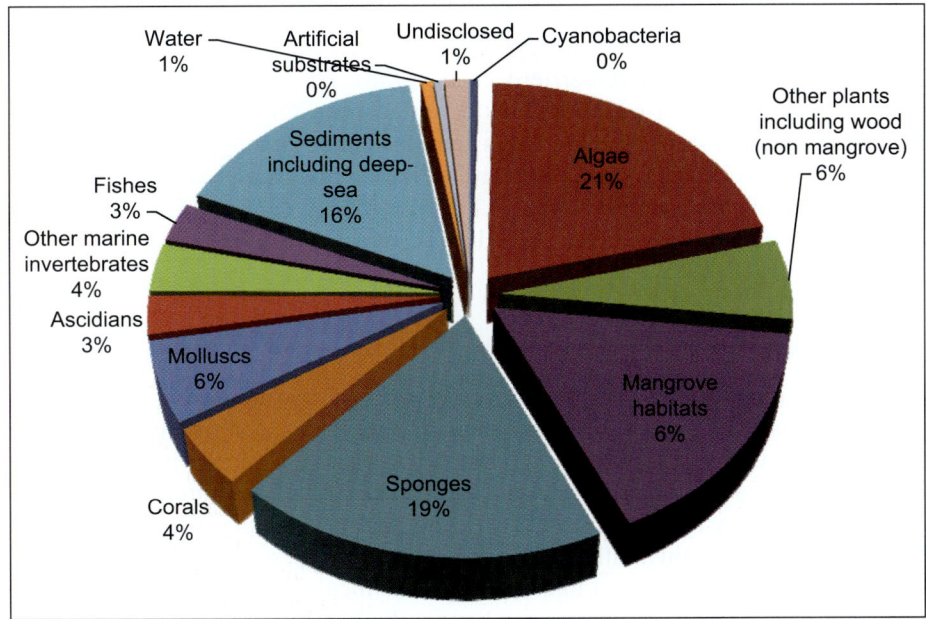

Figure 20.3 New compounds from marine-derived fungi until 2010, divided by sources of the fungal strains (updated from Rateb and Ebel 2011).

Obligate versus facultative marine-derived fungi

By far the majority of studies in the chemical literature focused on facultative marine fungal strains, whereas obligate marine fungi have been investigated for their metabolic potential far less frequently (Jensen and Fenical 2002). A discussion of the distinction between both groups is beyond the focus of this chapter, but is presented elsewhere in this book.

4 (R^1 = CH_2OH, R^2 = H, R^3 = H)
50 (R^1 = COOH, R^2 = H, R^3 = H)
51 (R^1 = CH_3, R^2 = OH, R^3 = H)
52 (R^1 = CH_3, R^2 = H, R^3 = H)
53 (R^1 = CH_3, R^2 = H, R^3 = OH)

54

55

56

57

58

Probably only the second new natural product ever to be reported from obligate marine fungi was the terpene dendryphiellin A (**4**), a metabolite of the asexual fungus *Dendryphiella salina* (Guerriero et al. 1988). At the time of its isolation, **4** represented the first trinor-eremophilane identified from fungi in general, and also its fatty acid-derived ester side chain was unprecedented among fungal secondary metabolites. In the following years, Francesco Pietra's research group disclosed further trinor-eremophilanes, dendryphiellins A1 (**50**), B (**51**), C (**52**), and D (**53**), together with the intact eremophilanes, dendryphiellins E (**54**), E1 (**55**), E2 (**56**), F (**57**), and G (**58**) (Guerriero et al. 1989, 1990).

Geographic origin

Since the beginning of the chemical investigation of marine-derived fungi (see above), virtually all geographic locations, including temperate and tropical areas have been sampled for chemically promising strains. Of course, a certain bias caused by the location of researchers active in this field can be stated, and areas such as the Mediterranean, the coastal areas of the US, Japan, and in particular China, have been investigated quite thoroughly, although, as the latter examples clearly shows, there is still a great deal of underexplored microbial and chemical diversity. From a chemical point of view, only a few geographic areas have so far hardly been studied at all, most notably the polar regions.

59 ($R^1 = R^2 = R^3 = R^4 = R^5 = CH_3, R^6 = H$)
60 ($R^1 = R^6 = H, R^2 = R^3 = R^4 = R^5 = CH_3$)
61 ($R^1 = R^2 = R^3 = R^4 = CH_3, R^5 = R^6 = H$)
62 ($R^1 = R^2 = R^3 = R^5 = CH_3, R^4 = R^6 = H$)
63 ($R^1 = R^2 = R^3 = R^4 = R^5 = CH_3, R^6 = OH$)
64 ($R^1 = R^4 = R^5 = CH_3, R^2 = R^6 = H, R^3 = CH(CH_3)_2$)

The psychrophilic fungus *Trichoderma asperellum*, isolated from a sediment sample collected in the Antarctic, was found to produce six new peptaibols, asperelines A–F (**59–64**), characterised by an uncommon prolinol residue at the *C*-terminus (Ren et al. 2009). In the course of the structure elucidation of **59–64**, the absolute configurations of some of the amino acid residues were determined using a newly developed approach based on ^{1}H NMR spectroscopic comparison of complexes formed with the chiral shift reagent Ru(D$_4$-Por*)CO. The new peptaibols displayed mild antibacterial activity towards *Staphylococcus aureus* and *Escherichia coli*, and also inhibited the growth of the phytopathogenic fungi *Alternaria solani* and *Pyricularia oryzae*.

Non-living sources

Mangrove habitats

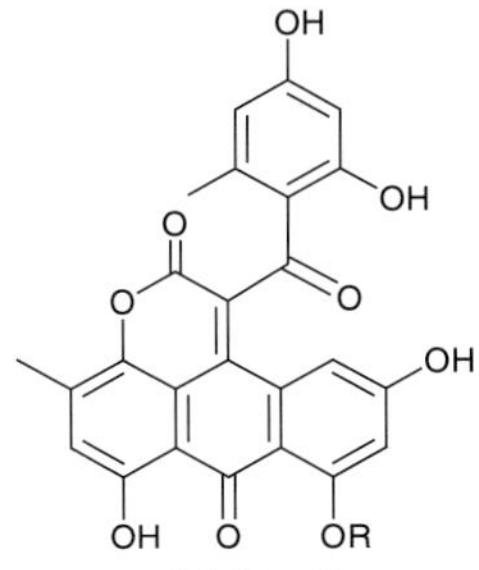

65 (R = H)
66 (R = CH₃)

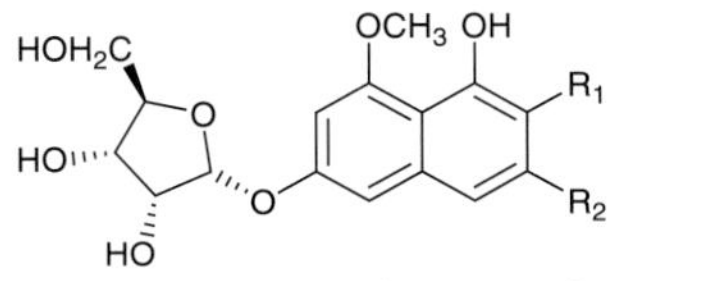

67 (R¹ = acetyl, R² = CH₃)
68 (R¹ = H, R² = CH₃)
69 (R¹ = R² = H)

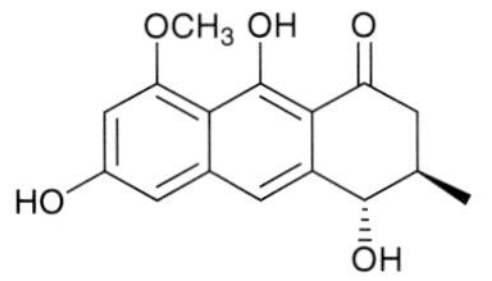

70

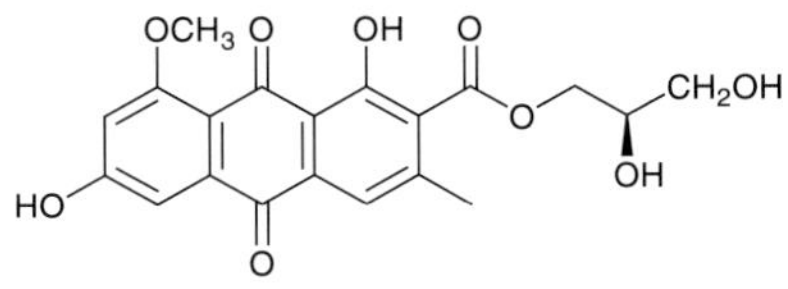

71

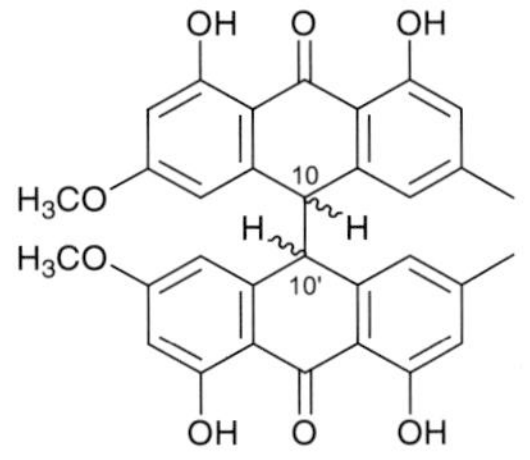

72 (H-10,H-10' *trans*)
73 (H-10,H-10' *cis*)

Mangrove areas probably represent the single most studied habitat in terms of chemical diversity of fungal strains. This includes both fungal strains living as epiphytes on or as endophytes within mangrove plants, and strains derived from sediment samples collected in mangrove habitats. An example for the latter is the fungus *Aspergillus glaucus*, cultured from sediment around mangrove roots in China, which yielded a novel anthraquinone derivative, aspergiolide A (**65**). This compound features an unprecedented naphtho[1,2,3-*de*]chromene-2,7-dione skeleton, and exhibited cytotoxicity at submicromolar concentrations towards A-549 and HL-60 cells (Du et al. 2007). Feeding experiments with labeled acetate units revealed that the biosynthesis of **65** follows the canonical acetate-malonate pathway of aromatic polyketides, and that it proceeds *via* coupling of an octaketide and a pentaketide precursor (Tao et al. 2009). Further chemical investigation of the same fungus resulted in the discovery of another congener, aspergiolide B (**66**), together with three new naphthyl furanosides (**67–69**), the two new anthranoids isoasperflavin (**70**), the enantiomer of the known variecolorquinone A (**71**), and two new physicon-emodin bisanthrones (**72** and **73**) (Du et al. 2008). The cytotoxic activity of aspergiolide B (**66**) was comparable to that of aspergiolide A (**65**), while **72** and **73** were only moderately active.

Sediments

As one example of a sediment-borne strain, the Chinese fungus *Spicaria elegans* deserves mentioning, since it proved an extremely rich source of new cytochalasin derivatives. Cytochalasins Z_7–Z_9 (**74–76**) feature a rare 12-membered macrolactone ring, and displayed mild cytotoxicity towards P388 and A-549 cells in the MTT assay (Liu et al. 2006). Further analysis of the same fungal strain yielded cytochalasins Z_{10}–Z_{15} (**77–82**), the first cytochalasin derivatives from nature in which the normally present 11–14-membered macrocyclic ring is opened to give a C_8-side chain (Liu et al. 2008). **78** and **79** exhibited moderate cytotoxicity, whereas the remaining compounds were inactive. Subsequently, this isolate of *Spicaria elegans* was subjected to a systematic variation of culture conditions following the OSMAC (one strain-many compounds) approach. As a result, the novel spicochalasin A (**83**) with an unusual pentacyclic ring system, and aspochalasins M–Q (**84–88**) were obtained (Lin et al. 2009a). **83** and **84** showed modest cytotoxic activity towards HL-60 cells. Cultivation of the same fungus in the presence of the cytochrome P-450 inhibitor metyrapone led to the discovery of two new compounds, 7-deoxy-cytochalasin Z_7 (**89**) and 7-deoxy-cytochalasin Z_9 (**90**) (Lin et al. 2009b). Thus, while oxidation at C-7 was prevented, the 'Baeyer-Villiger'-type oxidation at C-9 still occurred. **89** was moderately active against A-549 cells, while **90** was inactive. Finally, fermentation of the fungus for a longer time using resulted in the production of further aspochalasin congeners, aspochalasins R – T (**91–93**), all of which were devoid of cytotoxic activity (Lin et al. 2010).

Deep sea fungi

Fungal strains from the deep-sea fungi have only rarely been investigated chemically, and the best-studied example so far is an isolate of *Phialocephala* sp. that was obtained from a sediment sample collected from a depth of 5059 m. Chemical analysis of the culture filtrate resulted in the discovery of the new bisorbicillinoids oxosorbiquinol (**94**) and its 10,11-dihdro derivative (**95**), which exhibited weak cytotoxic activity towards five different cancer cell lines (Li et al. 2007a). The first sorbicillin trimer, trisorbicillinone A (**96**), was reported shortly after, and likewise found to possess moderate cytotoxicity (Li et al. 2007b). Re-examination of the fermentation broth resulted in the discovery of three additional derivatives, trisorbicillinones B – D (**97–99**), all of which showed only mild cytotoxic activity (Li et al. 2010). Recently, the new sorbicillin dimer dihydrotrichodermolide (**100**), the new monomer dihydrodemethylsorbicillin (**101**), and the biogenetically related benzofuranone derivative phialofurone (**102**) were reported from the same fungus (Li et al. 2011). While **100** exhibited moderate cytotoxicity, **101** and **102** were active in the submicromolar range.

Non-natural sources

The fungus *Ascochyta* sp. NGB4 was isolated from a rather "exotic" source, i.e., floating scrap of festering rope in a Japanese fishing port. Chemical analysis revealed the presence of ascochytatin (**103**), a new spirodioxynaphthalene (Kanoh et al. 2008). The compound was found to be active in an assay targeting the bacterial two-component regulatory system (TCS), probably the YycG/YycF-TCS, since it was more active towards the temperature-sensitive mutant of *Bacillus subtilis* CNM2000 than against a wild-type strain. Moreover, **103** displayed strong antimicrobial activity against Gram-positive bacteria and *Candida albicans*, and also exhibited considerable cytotoxicity.

Biological sources: plants and algae

Algae

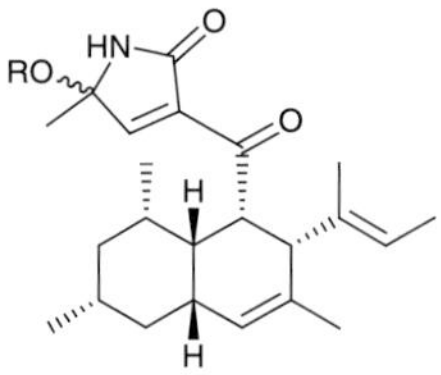

104 (R = *n*-butyl)
105 (R = CH$_3$)

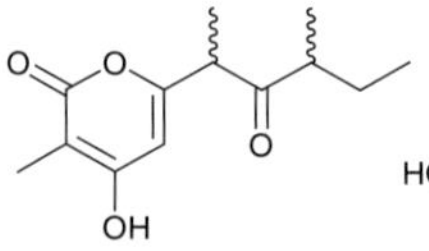

106

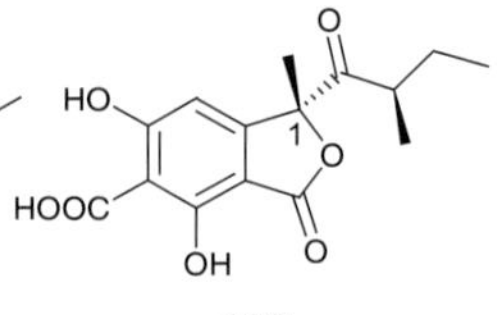

107
108 (1-epimer)

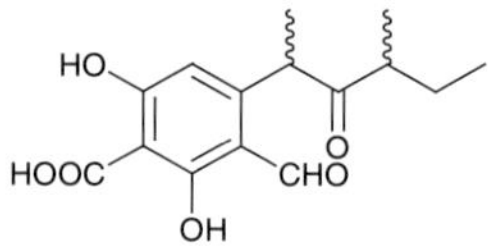

109

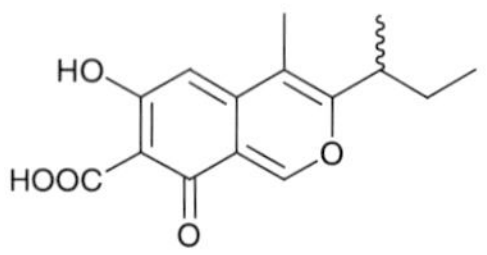

110

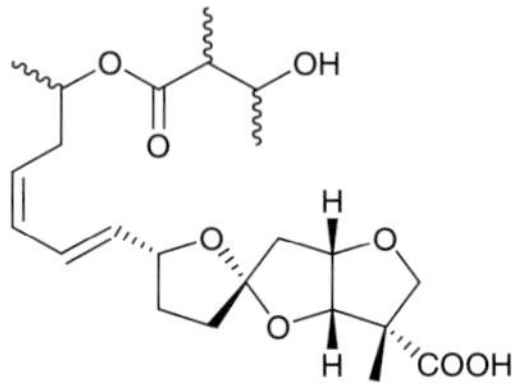

111

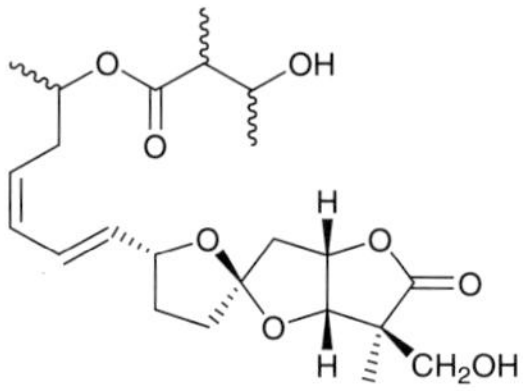

112

To date, algae are the most prominent source of marine-derived fungal strains that have been analyzed for their secondary metabolites. Over a period of a few years, Gabriele König's group extensively characterized the obligate marine endophyte *Ascochyta salicorniae* which was isolated from the green alga *Ulva* sp., collected from the German North Sea. Two new deoxytetramic acids, ascosalipyrrolidinones A (**104**) and B (**105**), and the new α-pyrone ascosalipyrone (**106**) were obtained (Osterhage et al. 2000). While **105** and **106** proved inactive, **104** exhibited a series of biological activities, including antimicrobial activity against *Bacillus megaterium*, *Mycotypha microsporum* and *Microbotryum violaceum*, moderate antiplasmodial activity against two strains of *Plasmodium falciparum*, significant activity against *Trypanosoma cruzi* and *Trypanosoma brucei* subsp. *rhodesiense*, cytotoxic activity against two cell lines, and was also found to inhibit tyrosine kinase p56(lck). When fermented under different culture conditions, the same fungus yielded two new epimeric lactones, ascolactones A (**107**) and B (**108**), in addition to biogenetically related polyketides including the new ascochitin (**109**) and the known ascochital (**110**) (Seibert et al. 2006), the latter having previously been reported from the marine ascomycete *Kirschsteiniothelia maritima* (Kusnick et al. 2002). Through feeding experiments with labeled precursors, it was established that **106** and **110** are biosynthesized via a single pentaketide and hexaketide chain, respectively, and that all branching methyl groups in **106** are derived from *S*-adenosylmethionine (Seibert et al. 2006). Predictions based on an *in silico* screening for protein phosphatase inhibitory properties of **106–110** were experimentally confirmed for **110**, which inhibited MPtpB (mycobacterial protein tyrosine phosphatase B) and PTP1B (protein tyrosine phosphatase 1B) at a micromolar range (Seibert et al. 2006). A later reinvestigation led to the discovery of ascospiroketals A (**111**) and B (**112**), two biogenetically unique cycloethers, which based on their structures should be formed via a geminal biomethylation via *S*-adenosyl-methionine (Seibert et al. 2007).

Seagrasses

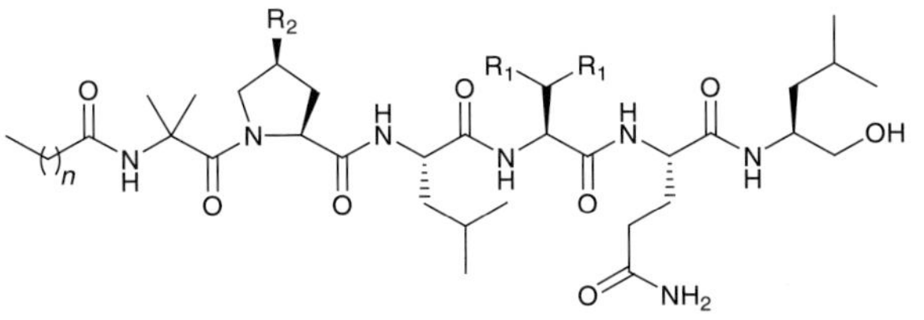

113 (R^1 = CH_3, R^2 = OH, n = 12)
114 (R^1 = H, R^2 = OH, n = 12)
115 (R^1 = CH_3, R^2 = H, n = 12)
116 (R^1 = CH_3, R^2 = OH, n = 10)
117 (R^1 = CH_3, R^2 = H, n = 10)

True higher plants are quite rare in the marine environment, and thus it is not too surprising that they are not a very prominent source of isolated fungal strains for subsequent chemical studies. One example of the latter is the fungus *Scytalidium* sp., obtained from the Caribbean seagrass *Halodule wrightii*, which was found to produce a series of novel lipophilic, linear hexapeptides named halovirs A – E (**113–117**) (Rowley et al. 2003). **113–117** strongly inhibited Herpes simplex viruses 1 and 2 *in vitro*, thought to act through membrane destabilization of the viral lipid envelope. Extensive NMR studies, including measurement of deuterium exchange rates, temperature, and solvent dependence of NH proton chemical shifts, and NOE/ROESY observations, allowed insight into the solution conformation of **113**, which was found to adopt a 3_{10}-helical configuration. Subsequent synthetic efforts by the same group of authors produced a series of analogues, which provided important information with regard to structure-activity relationships of the halovirs (Rowley et al. 2004). For example, acylation of the *N*-terminus as well as the presence of the dipeptide unit of the two unusual amino acids α-aminoisobutyric acid and hydroxyproline proved a requirement for anti HSV-1 activity.

Biological sources: marine invertebrates and other animals

Sponges

Sponge-derived fungi are a very popular source of interesting new natural products. A prominent example is the fungus *Gymnascella dankaliensis*, isolated from *Halichondria japonica* collected off Osaka in Japan, which has been studied over many years and has yielded a group of unusual nitrogen-containing polyketides, gymnastatins as well as structurally unusual steroids. Gymnastatins A – E (**118–122**) were reported first, and were found to exhibit pronounced cytotoxicity (Numata et al. 1997; Amagata et al. 1998).

118 (R = Cl) **119** (R = Cl) **120** (R = Cl) **121** **122**
127 (R = Br) **128** (R = Br) **129** (R = Br)

126

123 (R¹ = OH, R² = H, R³ = OCH₃) **124** **125** **132** **133**
130 (R¹ = H, R² = OH, R³ = OCH₃)
131 (R¹ = OH, R² = H, R³ = H)

134 **135** **136** (R¹ = OH, R² = H) **138** **139**
         137 (R¹, R² = O)

Further chemical investigation led to the discovery of gymnastatins F – H (**123–125**), and gymnamide (**126**), essentially the side chain of the gymnastatin congeners (Amagata et al. 2006). **123** and **124** possess a unique bicyclo[3.3.1]nonane ring system which differs from that of previously reported gymnastatins, and the compounds were found to exhibit pronounced cytotoxic activity against the P388 cancer cell line. Cultivation of *Gymnascella dankaliensis* in a bromine-containing medium led to the production of the brominated analogues gymnastatins I – K (**127–129**), all of them exhibiting potent cytotoxicity, of which **127** and **128** displayed pronounced activity against a panel of 39 human cancer cell lines (Amagata et al. 2010). In a similar manner, a different metabolic profile was obtained when the fungus was cultivated using soluble starch instead of glucose as a carbon source. Under these conditions, the two new gymnastatins Q (**130**) and R (**131**), together with the related dankastatins A (**132**) and B (**133**) were obtained (Amagata et al. 2008). All of these compounds were active against murine P388 cells, while **130** also inhibited the growth of human BSY-1 (breast) and MKN7 (stomach) cancer cell lines. Besides the gymnastatins, *Gymnascella dankaliensis* was also identified as a rich source of structurally unusual steroid-type compounds, the pattern of which varied depending on media composition. In a malt extract medium, gymnasterones A – D (**134–137**) were produced, while replacing glucose by soluble starch resulted in the production of dankasterones A (**138**) and B (**139**) (Amagata et al. 1999, 2007). The latter compounds possess a very unusual 13(14→8) *abeo*-8-ergostane skeleton, while gymnasterone A (**134**) is a unique steroid alkaloid with an additional ring and an amide-linked side chain that is made of gymnamide (**126**), and thus is reminiscent of the gymnastatins. Most of these steroids displayed pronounced cytotoxicity against the murine P388 cancer cell line, while dankasterone A (**138**) was also active towards human cancer cell lines.

Corals

140 **141** **142** **143**

Coral-derived fungi have occasionally been investigated for their chemical potential. For example, the fungus *Nigrospora* sp. which was isolated from the sea fan *Annella* sp. in Thailand yielded three new epoxydon esters, nigrospoxydons A – C (**140–142**), and the new pyrone, nigrosporapyrone (**143**) (Trisuwan et al. 2008). **140** is an ester of the plant hormone abscisic acid and epoxydon, both of which were also detected in free form in the culture medium. The latter is a well-known metabolite originally described from a terrestrial *Phoma* sp. (Closse et al. 1966). Nigrospoxydon A (**140**) exhibited weak antibiotic activity against *Staphylococcus aureus* and methicillin-resistant *S. aureus*.

Bryozoans

144 **145** **146** (R^1 = OH, R^2 = H)
147 (R^1 = H, R^2 = OH)
148 (R^1, R^2 = O)
149

Six new ergosterols, including 3β-hydroxy-(22E,24R)-ergosta-5,8,22-trien-7,15-dione (**144**), 3β-hydroxy-(22E,24R)-ergosta-5,8,14,22-tetraen-7-one (**145**), 3β,15β-dihydroxy-(22E,24R)-ergosta-5,8(14),22-trien-7-one (**146**), 3β,15β-dihydroxy-(22E,24R)-ergosta-5,8(14),22-trien-7-one (**147**), 3β-hydroxyl-(22E,24R)-ergosta-5,8(14),22-trien-7,15-dione (**148**), and 5α,8α-epidioxy-23,24(R)-dimethylcholesta-6,9(11),22-trien-3β-ol (**149**) were detected in the culture broth of *Rhizopus* sp., a fungus that was isolated from the bryozoan *Bugula* sp., collected in Jiaozhou Bay, China (Wang et al. 2008). When tested for cytotoxic activity, all compounds were active to varying degrees towards four different cancer cell lines.

Echinoderms

Even though fungi from echinoderms have only very rarely been studied by natural product chemists, an isolate of *Acremonium striatisporum* that was obtained from superficial mycobiota of the sea cucumber, *Eupentacta fraudatrix* collected in the Sea of Japan, proved to be an exceptionally rich source of new isopimaradiene diterpene glycosides. Repeated analysis over the course of eight years resulted in the discovery of virescenosides M (**150**), N (**151**), O (**152**), P (**153**), Q (**154**), R (**155**), S (**156**), T (**157**), U (**158**), V (**159**), W (**160**), and X (**161**) (Afiyatullov et al. 2000, 2002, 2004, 2006), together with the known virescenosides A (**162**), B (**163**), and C (**164**), metabolites of terrestrial strains of *Oospora virescens*. Most virescenosides are glycosides containing the unusual sugar β-D-altropyranose, and display cytotoxic effects on developing eggs of the sea urchin *Strongylocentrotus intermedius*, which is very often associated with cytotoxic activity against Ehrlich carcinoma cells.

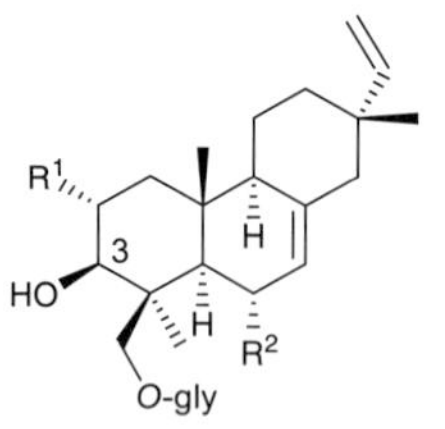

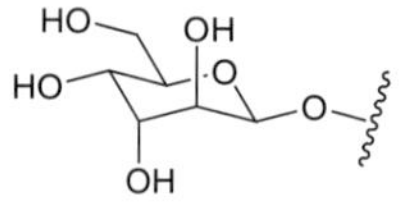

alt = β-D-altropyranosyl-

man = β-D-mannopyranosyl-

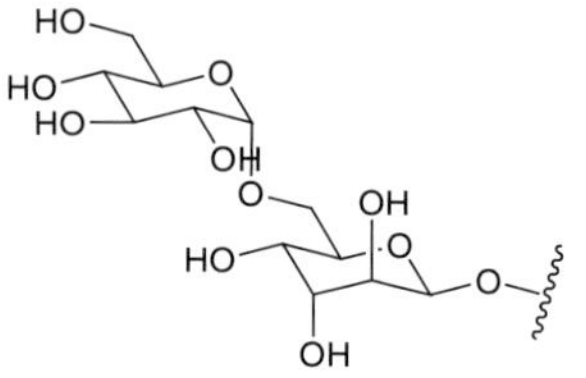

154 (R^1 = H, R^2 = H, gly = man)
155 (R^1 = OH, R^2 = H, gly = alt-glc)
160 (R^1 = OH, R^2 = OH, gly = alt)
162 (R^1 = OH, R^2 = H, gly = alt)
163 (R^1 = H, R^2 = H, gly = alt)
164 (R^1 = H, R^2 = H, gly = alt, 3-ketone)

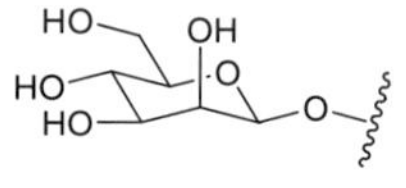

alt-glc = α-D-glucopyranosyl(1→6)-β-D-altropyranosyl-

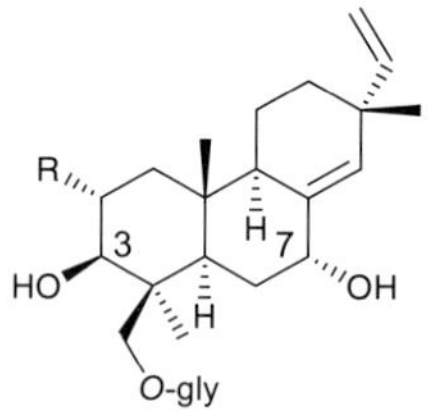

151 (R = OH, gly = alt)
152 (R = H, gly = alt)
156 (R = H, gly = alt, 3-ketone)
158 (R = H, gly = alt, 3,7-diketone)
159 (R = OH, gly = alt, 7-ketone)

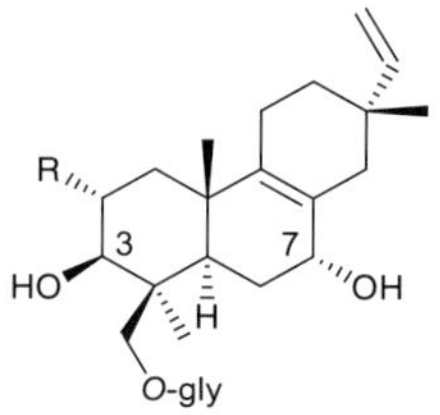

150 (R = OH, gly = alt, 7-ketone)
153 (R = H, gly = alt, 7-ketone)
157 (R = H, gly = alt, 3,7-diketone)
161 (R = OH, gly = alt)

Mollusks

Notoamides form a large class of structurally complex prenylated diketopiperazines that are produced by an *Aspergillus* sp., which was originally isolated from the common mussel, *Mytilus edulis*. The first compounds reported belonging to this group were notoamides A – D (**165–168**) (Kato et al. 2007), of which the first two share an unusual spiro-indolinone skeleton with sclerotiamide and paraherquamide, toxins produced by terrestrial isolates of *Aspergillus sclerotiorum* and *Penicillium paraherquei*, respectively.

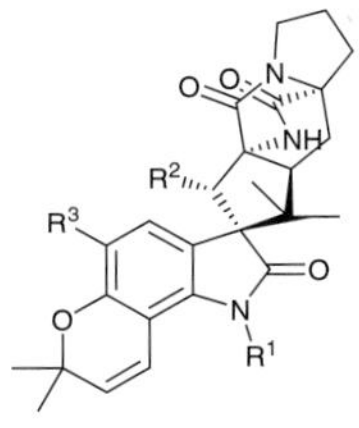

165 (R¹ = OH, R² = R³ = H)
166 (R¹ = R² = R³ = H)
175 (R¹ = R² = OH, R³ = H)
181 (R¹ = R² = H, R³ = Cl)

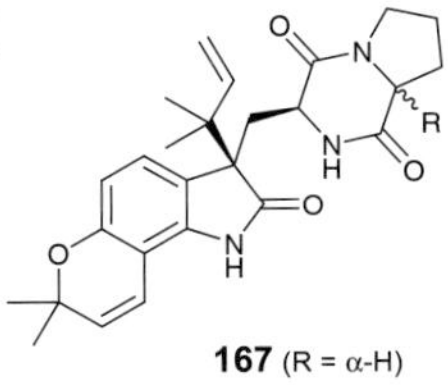

167 (R = α-H)
180 (R = OH)
185 (R = OCH₃)

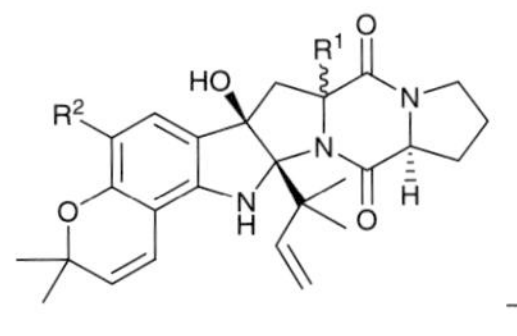

168 (R¹ = α-H, R² = H)
178 (R¹ = β-OH, R² = H)
184 (R¹ = β-OH, R² = Br)

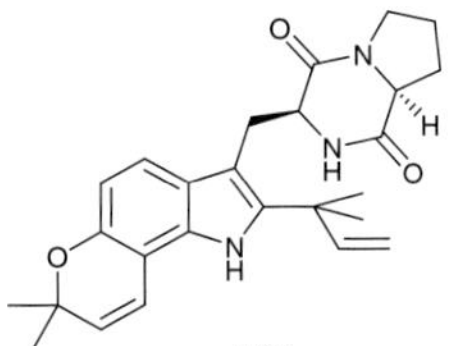

169

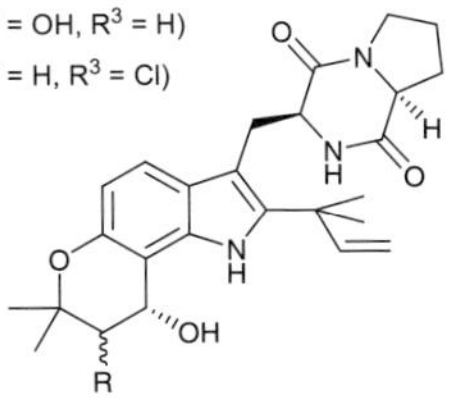

170 (R = β-OH)
171 (R = α-OH)

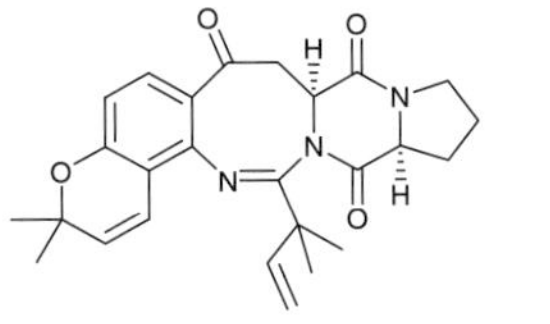

172

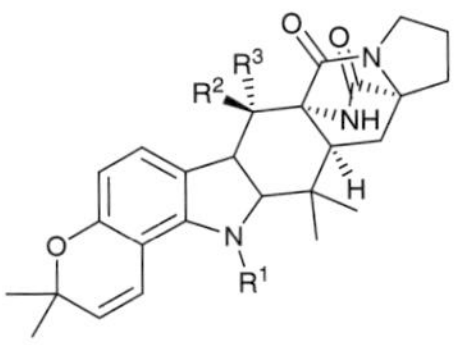

173 (R¹ = R² = H, R³ = OCH₃)
174 (R¹ = OH, R² = H, R³ = OCH₃)
176 (R¹ = H, R², R³ = O)
186 (R¹ = R² = H, R³ = OH)

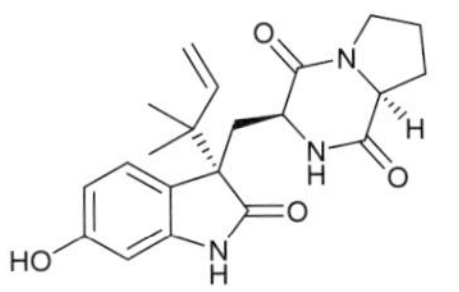

177

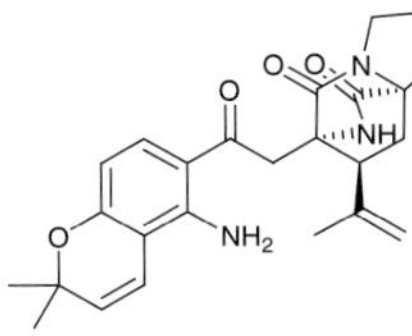

179

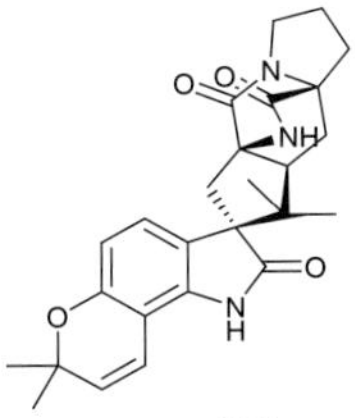

182

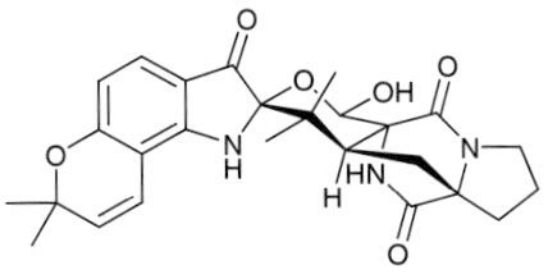

183

Among the subsequently reported notoamides F – K (**173–178**) (Tsukamoto et al. 2008) there was another representative of this group, notoamide H (**175**), while notoamides F (**173**), G (**174**) and I (**176**) are closely related to stephacidin A, a metabolite of a terrestrial isolate of *Aspergillus ochraceus*. Notoamides A – C (**165–167**) and I (**176**) displayed moderate cytotoxicity towards HeLa cells, while the remaining compounds were inactive, and notoamide C (**167**) also induced G2/M-cell cycle arrest. Shortly afterwards, reinvestigation of the original fungal strain resulted in the discovery of the short-lived biogenetic intermediate notoamide E (**169**), together with notoamides E2 (**170**), E3 (**171**) and E4 (**172**) which were obtained upon feeding synthetic doubly ^{13}C-labeled **169** (Tsukamoto et al. 2009a). In a subsequent study, the structures of notoamides L – N (**179–181**) and (-)-versicolamide B (**182**) were reported (Tsukamoto et al. 2009b), the latter quite remarkably representing the enantiomer of a natural product formed by a terrestrial strain of *A. versicolor*. The latest addition to the series to date are notoamides O – Q (**183–185**) (Tsukamoto et al. 2010), while notoamide R reported in the same communication is identical to 21-hydroxystephacidin A (**186**), discovered shortly before from a marine isolate of *A. ostianus*, obtained from an unidentified Japanese sponge (Kito et al. 2009). Notoamide O (**183**) features a hemiacetal/hemiaminal ether function which is unprecedented in the notoamide class of compounds, whereas notoamide P (**184**) is the first brominated congener in this series.

Fish

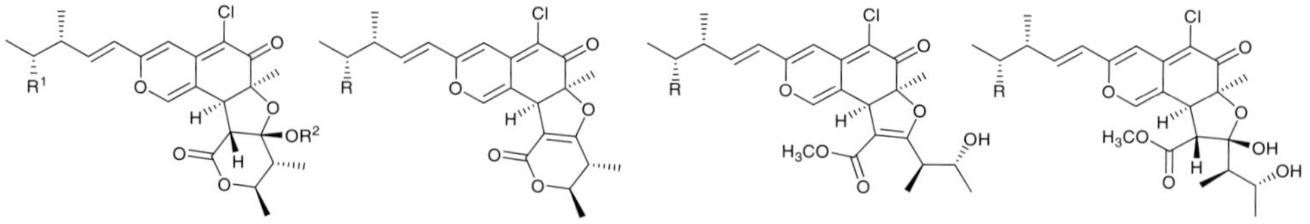

187 (R¹ = OH, R² = H)
188 (R¹ = OH, R² = CH₃)
190 (R¹ = H, R² = H)
191 (R¹ = H, R² = CH₃)

189 (R = OH)
192 (R = H)

193 (R = OH)
194 (R = H)

195 (R = OH)
196 (R = H)

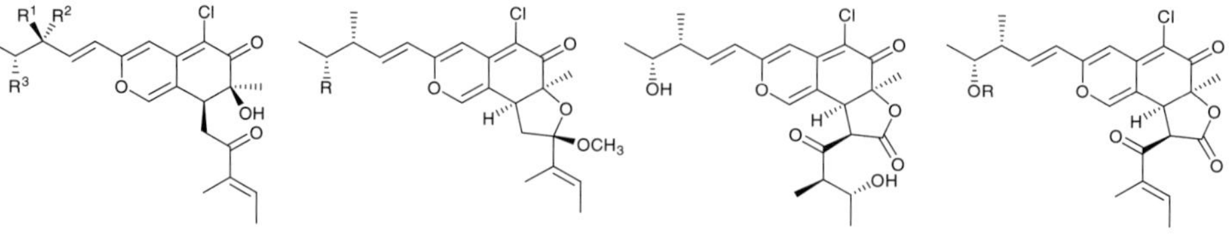

197 (R¹ = H, R² = CH₃, R³ = OH)
198 (R¹ = H, R² = CH₃, R³ = H)
209 (R¹ = CH₃, R² = H, R³ = OH)

199 (R = H)
200 (R = OH)

201

202 (R = OH)
203 (R = H)

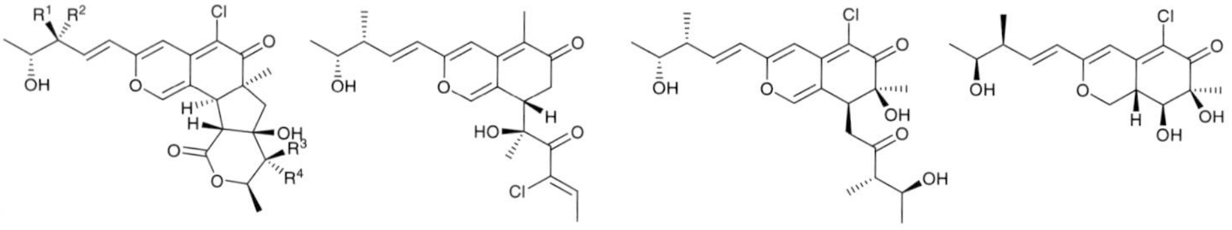

204 (R¹ = CH₃, R² = H, R³ = H, R⁴ = CH₃)
205 (R¹ = H, R² = CH₃, R³ = CH₃, R⁴ = H)

206

207

208

The vast majority of the fungal strains derived from animals which have been investigated for their natural products so far were isolated from marine invertebrates, whereas vertebrate animals have hardly been studied. A notable exception is the fungus *Chaetomium globosum* which was obtained from the Japanese fish *Mugil cephalus*. It was found to produce a series of complex azaphilones named chaetomugilins, most of them exhibiting significant cytotoxic activity. The initial report disclosed the structures of chaetomugilins A – C (**187–189**) (Yamada et al. 2008), soon followed by chaetomugilins D – F (**190–192**) (Yasuhide et al. 2008). Against a panel of 39 human cancer cell lines, chaetomugilin F (**192**) displayed pronounced growth inhibitory properties in combination with selective cytotoxic activity, and analysis of the activity profiles obtained using the COMPARE program indicated that chaetomugilins A (**187**), C (**189**) and F (**192**) might act through an unprecedented mode of action. Further chemical analyses of the same fungus resulted in the discovery of chaetomugilins G (**193**), H (**194**) (Yamada et al. 2009b), and subsequently *seco*-chaetomugilins A (**195**) and D (**196**) (Yamada et al. 2009a), of which **193** and **196** were the most active. In the following year, the structures of chaetomugilins I – O (**197–203**) were reported, all of them with the exception of chaetomugilin L (**200**) displaying significant cytotoxic activity against different cancer cell lines (Muroga et al. 2009). The activity profile of chaetomugilin I (**197**) suggested that it possessed a novel mechanism of action. The two stereoisomers of chaetomugilin A (**187**), 11-*epi*-chaetomugilin A (**204**)

Natural products from marine-derived fungi | 433

and 4'-*epi*-chaetomugilin A (**205**) proved to be less potent (Muroga et al. 2010), while the most recent congeners chaetomugilins P – R (**206–208**) and 11-*epi*-chaetomugilin I (**209**) displayed similar levels of activity in comparison to the other congeners of the chaetomugilin series (Yamada et al. 2011). In contrast to all other chaetomugilins, the positions of one chlorine atom and one methyl group are exchanged in chaetomugilin P (**206**), which might be due to an unusual biogenetic pathway involving skeletal rearrangement *via* both a Diels-Alder and a retro-Diels-Alder reaction, whereas chaetomugilin R (**208**) lacks one of the two side chains attached to the azaphilone system.

Enhancing metabolic diversity by competing co-culture experiments

As shown in the previous section, the metabolic diversity of marine-derived fungi is enormous. However, there are a few selected examples reported in the chemical literature when fermentation of a fungal strain in the presence of another microorganism led to even further enhanced metabolic profiles. With all likelihood, this phenomenon is due to the activation of "hidden" or silenced biosynthetic pathways at the level of the corresponding gene clusters, and further studies at the molecular level will provide more insight into microbial communication and potential ecological roles for fungal secondary metabolites, which are usually assumed to play a role in terms of a chemical defense mechanism or even of a "chemical warfare" to ward off competitors for space or resources.

In a pioneering study, the induction the biosynthesis of a new chlorinated benzophenone antibiotic, pestalone (**210**), was observed when a marine-proteobacterium (strain CNJ-328) was added to the culture of the fungus *Pestalotia* sp., isolated from the surface of the brown alga *Rosenvingea* sp. collected in the Bahamas Islands (Cueto et al. 2001). When cultured alone, neither the fungal strain nor the bacterium produced this compound, but traces could be detected when the fungus was exposed to small amounts of ethanol as an external trigger. **210** displayed potent antibiotic activity against methicillin-resistant *Staphylococcus aureus* and vancomycin-resistant *Enterococcus faecium*. Interestingly, it was the same bacterial strain that induced the formation of four new pimarane diterpenoids, libertellenones A–D (**211–214**) when added to an established 3-day-old culture of the marine-derived fungus *Libertella* sp., isolated from an ascidian that was collected in the Bahamas (Oh et al. 2005). Based on structural considerations, libertellenones A–D (**211–214**) are most likely produced by the fungus, since pimarane diterpenes have never been reported from bacteria. However, compounds **211–214** did neither display antibiotic activity towards the inducing marine bacterium, nor were active against multidrug-resistant human pathogenic bacterial strains, but libertellenone D (**214**) showed pronounced cytotoxic activity towards HCT-116 human adenocarcinoma cells. Emericellamides A (**215**) and B (**216**) are cyclic lipodepsipeptides that were detected upon co-cultivating the fungus *Emericella* sp., isolated from the Papua New Guinean green alga *Halimeda* sp., with the marine actinomycete *Salinispora arenicola* (Oh et al. 2007). **215** and **216** were produced in low yields by *Emericella* sp. alone, but upon co-culture their levels were enhanced by 100-fold. Both compounds showed modest antibacterial activities against methicillin-resistant *Staphylococcus aureus*. More recently, the biogenetic gene cluster for the production of **215** comprising a PKS-NRPS hybrid synthase has been identified through genomic data mining in the fully sequenced genome of *Aspergillus nidulans* (Chiang et al. 2008). In what appears to be the first example of a competing co-culture experiment involving two marine-derived fungi, it was demonstrated that marinamide (**217**) and its methyl ester (**218**) were only produced when two unidentified fungal strains, obtained from an undisclosed plant growing in an estuarine mangrove in Hong Kong, were fermented together, but not when either of the two fungi was cultured by itself (Zhu and Lin 2006).

Conclusion

Marine-derived fungi are an intriguing source for the discovery of pharmacologically active novel secondary metabolites. The most prominent example is plinabulin (NPI-2358), a compound closely related to the marine fungal-derived natural product halimide, and currently in advanced stages of clinical trials, although it is difficult to predict the outcome of its clinical evaluation. Among the most prolific producers of new chemistry are facultative marine species, especially ubiquitous genera such as *Aspergillus* and *Penicillium*. There are many reported cases in which alterations of culture conditions led to the discovery of even more new natural products, for example the production of halogenated metabolites by addition of halide salts to the culture medium, the systematic evaluation of the OSMAC (one strain, many compounds) approach, or the activation of previously silenced biosynthetic gene clusters through competing co-culture with other microorganisms. The number of published compounds in the chemical literature steadily rose until 2009, but in 2010 dropped slightly, albeit it at a very high level. There is still an enormous discrepancy between the number of cultivated and chemically characterized fungal strains on the one hand, and the estimated biodiversity of fungi in marine habitats on the other hand (Jones 2011), and thus it can be expected that marine-derived fungi will continue to provide new natural products with novel structures and biologically relevant activities in the future.

References

Afiyatullov SS, Kuznetsova TA, Isakov VV, Pivkin MV, Prokof'eva NG, Elyakov GB. New diterpenic altrosides of the fungus *Acremonium striatisporum* isolated from a sea cucumber. J Nat Prod. 2000;63:848–850.

Afiyatullov SS, Kalinovsky AI, Kuznetsova TA, Isakov VV, Pivkin MV, Dmitrenok PS, Elyakov GB. New diterpene glycosides of the fungus *Acremonium striatisporum* isolated from a sea cucumber. J Nat Prod. 2002;65:641–644.

Afiyatullov SS, Kalinovsky AI, Kuznetsova TA, Pivkin MV, Prokof'eva NG, Dmitrenok PS, Elyakov GB. New glycosides of the fungus *Acremonium striatisporum* isolated from a sea cucumber. J Nat Prod. 2004;67:1047–1051.

Afiyatullov SS, Kalinovsky AI, Pivkin MV, Dmitrenok PS, Kuznetsova TA. New diterpene glycosides of the fungus *Acremonium striatisporum* isolated from a sea cucumber. Nat Prod Res. 2006;20:902–908.

Amagata T, Doi M, Ohta T, Minoura K, Numata A. Absolute stereostructures of novel cytotoxic metabolites, gymnastatins A-E, from a *Gymnascella* species separated from a *Halichondria* sponge. J Chem Soc, Perkin Trans. 1998;1:3585–3599.

Amagata T, Doi M, Tohgo M, Minoura K, Numata A. Dankasterone, a new class of cytotoxic steroid produced by a *Gymnascella* species from a marine sponge. Chem Commun. 1999;14:1321–1322.

Amagata T, Minoura K, Numata A. Gymnastatins F-H, cytostatic metabolites from the sponge-derived fungus *Gymnascella dankaliensis*. J Nat Prod. 2006;69:1384–1388.

Amagata T, Tanaka M, Yamada T, Doi M, Minoura K, Ohishi H, Yamori T, Numata A. Variation in cytostatic constituents of a sponge-derived *Gymnascella dankaliensis* by manipulating the carbon source. J Nat Prod. 2007;70:1731–1740.

Amagata T, Tanaka M, Yamada T, Minoura K, Numata A. Gymnastatins and dankastatins, growth inhibitory metabolites of a *Gymnascella* species from a *Halichondria* sponge. J Nat Prod. 2008;71:340–345.

Amagata T, Takigawa K, Minoura K, Numata A. Gymnastatins I-K, cancer cell growth inhibitors from a sponge-derived *Gymnascella dankaliensis*. Heterocycles 2010;81:897–907.

Andersen B, Smedsgaard J, Frisvad JC. *Penicillium expansum*: consistent production of patulin, chaetoglobosins, and other secondary metabolites in culture and their natural occurrence in fruit products. J Agric Food Chem. 2004;52:2421–2428.

Aren O, Matamala L, Reyes M, Santini A, McArthur K, Lloyd GK, Spear MA. Plinabulin (NPI-2358), a novel vascular disrupting agent (VDA), in non-small cell lung cancer (NSCLC). J Thorac Oncol. 2010;5:S91–S92.

Block JH, Catalfomo P, Constantine GH, Kirk PW. Triglyceride fatty acids of selected higher marine fungi. Mycologia 1973;65:488–491.

Bräse S, Encinas A, Keck J, Nising CF. Chemistry and biology of mycotoxins and related fungal metabolites. Chem Rev. 2009;109:3903–3990.

Bugni TS, Ireland CM. Marine-derived fungi: a chemically and biologically diverse group of microorganisms. Nat Prod Rep. 2004;21:143–163.

Butler A. Acquisition and utilization of transition metal ions by marine organisms. Science 1998;281:207–210.

Butler MS. Natural products to drugs: natural product-derived compounds in clinical trials. Nat Prod Rep. 2008;25:475–516.

Catalfomo P, Block JH, Constantine GH, Kirk PW. Choline sulfate (ester) in marine higher fungi. Mar Chem. 1973;1:157–162.

Chandra P, Read G, Vining LC. Studies on the biosynthesis of volucrisporin. II. Metabolism of some phenylpropanoid compounds by *Volucrispora aurantiaca* Haskins. Can J Biochem Cell Biol. 1966;44:403–413.

Chiang YM, Szewczyk E, Nayak T, Davidson AD, Sanchez JF, Lo HC, Ho WY, Simityan H, Kuo E, Praseuth A, Watanabe K, Oakley BR, Wang CCC. Molecular genetic mining of the *Aspergillus* secondary metabolome: discovery of the emericellamide biosynthetic pathway. Chem Biol. 2008;15:527–532.

Chu M, Patel MG, Gullo VP, Truumees I, Puar MS, McPhail AT. Sch-47918: a novel PAF antagonist from the fungus *Phoma* sp. J Org Chem. 1992;57:5817–5818.

Chu M, Truumees I, Gunnarsson I, Bishop WR, Kreutner W, Horan AC, Patel MG, Gullo VP, Puar MS. A novel class of platelet activating factor antagonists from *Phoma* sp. J Antibiot. 1993;46:554–563.

Closse A, Mauli R, Sigg HP. Die Konstitution von Epoxydon. Helv Chim Acta. 1966;49:204–213.

Crawley SL, Funk RL. A synthetic approach to nomofungin/communesin B. Org Lett. 2003;5: 3169–3171.

Cueto M, Jensen PR, Kauffman C, Fenical W, Lobkovsky E, Clardy J. Pestalone, a new antibiotic produced by a marine fungus in response to bacterial challenge. J Nat Prod. 2001;64: 1444–1446.

Demain AL, Elander RP. The β-lactam antibiotics: past, present, and future. Antonie Leeuwenhoek 1999;75:5–19.

Du L, Zhu TJ, Fang YC, Liu HB, Gu QQ, Zhu WM. Aspergiolide A, a novel anthraquinone derivative with naphtho[1,2,3-de]chromene-2,7-dione skeleton isolated from a marine-derived fungus *Aspergillus glaucus*. Tetrahedron 2007;63:1085–1088.

Du L, Zhu TJ, Liu HB, Fang YC, Zhu WM, Gu QQ. Cytotoxic polyketides from a marine-derived fungus *Aspergillus glaucus*. J Nat Prod. 2008;71:1837–1842.

Ebel R. Natural product diversity from marine fungi. In: Mander L, Liu HW, eds. Comprehensive natural products II: chemistry and biology. Oxford: Elsevier, 2010a:223–262.

Ebel R. Terpenes from marine-derived fungi. Mar Drugs 2010b;8:2340–2368.

Fenical W, Jensen PR, Cheng XC. Halimide, a cytotoxic marine natural product, and derivatives thereof. US Pat 6069146, 1998.

Fukuzumi T, Miyauchi Y, Tubaki K, Minami K. Metabolic products in the medium of sulfite pulp waste liquor by a marine fungus. *Dendryphiella salina*. Mokuzai Gakkaishi 1975;21:558–562.

Furuya K, Okudaira M, Shindo T, Sato A. *Corollospora pulchella*, a marine fungus producing antibiotics, melinacidins III, IV and gancidin W. Sankyo Kenkyusho Nenpo. 1985;37:140–142.

Goldring WPD, Pattenden G. The phomactins. A novel group of terpenoid platelet activating factor antagonists related biogenetically to the taxanes. Acc Chem Res. 2006;39:354–361.

Guerriero A, Dambrosio M, Cuomo V, Vanzanella F, Pietra F. Dendryphiellin A, the first fungal trinor-eremophilane. Isolation from the marine deuteromycete *Dendryphiella salina* (Sutherland) Pugh et Nicot. Helv Chim Acta. 1988;71:57–61.

Guerriero A, Dambrosio M, Cuomo V, Vanzanella F, Pietra F. Novel trinor-eremophilanes (dendryphiellin B, C, and D), eremophilanes (dendryphiellins E, F, and G), and branched C9-carboxylic acids (dendryphiellic acid A and B) from the marine deuteromycete *Dendryphiella salina* (Sutherland) Pugh et Nicot. Helv Chim Acta. 1989;72:438–446.

Guerriero A, Cuomo V, Vanzanella F, Pietra F. A novel glyceryl ester (glyceryl dendryphiellate A), a trinor-eremophilane (dendryphiellin A1), and eremophilanes (dendryphiellins E1 and E2) from the marine deuteromycete *Dendryphiella salina* (Sutherland) Pugh et Nicot. Helv Chim Acta. 1990;73:2090–2096.

Hayashi H, Matsumoto H, Akiyama K. New insecticidal compounds, communesins C, D and E, from *Penicillium expansum* link MK-57. Biosci Biotechnol Biochem. 2004;68:753–756.

Ishino M, Kiyomichi N, Takatori K, Sugita T, Shiro M, Kinoshita K, Takahashi K, Koyama K. Phomactin I, 13-epi-phomactin I, and phomactin J, three novel diterpenes from a marine-derived fungus. Tetrahedron 2010;66:2594–2597.

Jadulco R, Edrada RA, Ebel R, Berg A, Schaumann K, Wray V, Steube K, Proksch P. New communesin derivatives from the fungus *Penicillium* sp. derived from the Mediterranean sponge *Axinella verrucosa*. J Nat Prod. 2004;67:78–81.

Jensen PR, Fenical W. Secondary metabolites from marine fungi. In: Hyde KD, ed. Fungi in marine environments. Hong Kong: Fungal Diversity Press, 2002:293–315.

Jones EBG. Are there more marine fungi to be described? Bot Mar. 2011;54:343–354.

Kamigauchi T, Sakazaki R, Nagashima K, Kawamura Y, Yasuda Y, Matsushima K, Tani H, Takahashi Y, Ishii K, Suzuki R, Koizumi K, Nakai H, Ikenishi Y, Terui Y. Terprenins, novel immunosuppressants produced by *Aspergillus candidus*. J Antibiot. 1998;51:445–450.

Kanoh K, Okada A, Adachi K, Imagawa H, Nishizawa M, Matsuda S, Shizuri Y, Utsumi R. Ascochytatin, a novel bioactive spirodioxynaphthalene metabolite produced by the marine-derived fungus, *Ascochyta* sp. NGB4. J Antibiot. 2008;61:142–148.

Kato H, Yoshida T, Tokue T, Nojiri Y, Hirota H, Ohta T, Williams RM, Tsukamoto S. Notoamides A-D: prenylated indole alkaloids isolated from a marine-derived fungus, *Aspergillus* sp. Angew Chem Int Ed Engl. 2007;46:2254–2256.

Kirk PW, Catalfomo P. Marine fungi: occurrence of ergosterol and choline. Phytochem. 1970;9: 595–597.

Kirk PW, Catalfomo P, Block JH, Constantine GH. Metabolites of higher marine fungi and their possible ecological significance. Veroeff Inst Meeresforsch Bremerhaven. 1974;Suppl 5:509–518.

Kito K, Ookura R, Kusumi T, Namikoshi M, Ooi T. X-ray structures of two stephacidins, heptacyclic alkaloids from the marine-derived fungus *Aspergillus ostianus*. Heterocycles 2009;78: 2101–2106.

Koyama K, Ishino M, Takatori K, Sugita T, Kinoshita K, Takahashi K. Phomactin H, a novel diterpene from an unidentified marine-derived fungus. Tetrahedron Lett. 2004;45:6947–6948.

Krick A, Kehraus S, Gerhäuser C, Klimo K, Nieger M, Maier A, Fiebig HH, Atodiresei I, Raabe G, Fleischhauer J, König GM. Potential cancer chemopreventive in vitro activities of monomeric xanthone derivatives from the marine algicolous fungus *Monodictys putredinis*. J Nat Prod. 2007;70:353–360.

Kupta J, Anke T, Steglich W, Zechlin L. Antibiotics from basidiomycetes. XI. The biological activity of siccayne, isolated from the marine fungus *Halocyphina villosa* J. & E. Kohlmeyer. J Antibiot. 1981;34:298–304.

Kusnick C, Jansen R, Liberra K, Lindequist U. Ascochital, a new metabolite from the marine ascomycete *Kirschsteiniothelia maritima*. Pharmazie 2002;57:510–512.

Larsen TO, Frisvad JC, Ravn G, Skaaning T. Mycotoxin production by *Penicillium expansum* on black currant and cherry juice. Food Addit Contam. 1998;15:671–675.

Li DH, Wang FP, Cai SX, Zeng X, Xiao X, Gu QQ, Zhu WM. Two new bisorbicillinoids isolated from a deep-sea fungus, *Phialocephala* sp. FL30r. J Antibiot. 2007a;60:317–320.

Li DH, Wang FP, Xiao X, Fang YC, Zhu TJ, Gu QQ, Zhu WM. Trisorbicillinone A, a novel sorbicillin trimer, from a deep ocean sediment derived fungus, *Phialocephala* sp. FL30r. Tetrahedron Lett. 2007b;48:5235–5238.

Li DH, Cai SX, Zhu TJ, Wang FP, Xiao XA, Gu QQ. Three new sorbicillin trimers, trisorbicillinones B, C, and D, from a deep ocean sediment derived fungus, *Phialocephala* sp. FL30r. Tetrahedron 2010;66:5101–5106.

Li DH, Cai SX, Zhu TJ, Wang FP, Xiao X, Gu QQ. New cytotoxic metabolites from a deep-sea-derived fungus, *Phialocephala* sp., strain FL30r. Chem Biodiv. 2011;8:895–901.

Lin ZJ, Zhu TJ, Wei HJ, Zhang GJ, Wang H, Gu QQ. Spicochalasin A and new aspochalasins from the marine-derived fungus *Spicaria elegans*. Eur J Org Chem. 2009a:3045–3051.

Lin ZJ, Zhu TJ, Zhang GJ, Wei HJ, Gu QQ. Deoxy-cytochalasins from a marine-derived fungus *Spicaria elegans*. Can J Chem. 2009b;87:486–489.

Lin ZJ, Zhu TJ, Chen L, Gu QQ. Three new aspochalasin derivatives from the marine-derived fungus *Spicaria elegans*. Chin Chem Lett. 2010;21:824–826.

Liu JK. Natural terphenyls: developments since 1877. Chem Rev. 2006;106:2209–2223.

Liu R, Gu QQ, Zhu WM, Cui CB, Fan GT, Fang YC, Zhu TJ, Liu HB. 10-phenyl-[12]-cytochalasins Z7, Z8, and Z9 from the marine-derived fungus *Spicaria elegans*. J Nat Prod. 2006;69: 871–875.

Liu R, Lin ZJ, Zhu TJ, Fang YC, Gu QQ, Zhu WM. Novel open-chain cytochalasins from the marine-derived fungus *Spicaria elegans*. J Nat Prod. 2008;71:1127–1132.

Marchelli R, Vining LC. Terphenyllin, a novel p-terphenyl metabolite from *Aspergillus candidus*. J Antibiot. 1975;28:328–331.

May JA, Zeidan RK, Stoltz BM. Biomimetic approach to communesin B (a.k.a. nomofungin). Tetrahedron Lett. 2003;44:1203–1205.

Mita MM, Spear MA, Yee LK, Mita AC, Heath EI, Papadopoulos KP, Federico KC, Reich SD, Romero O, Malburg L, Pilat M, Lloyd GK, Neuteboom STC, Cropp G, Ashton E, LoRusso PM. Phase 1 first-in-human trial of the vascular disrupting agent plinabulin (NPI-2358) in patients with solid tumors or lymphomas. Clin Cancer Res. 2010;16:5892–5899.

Muroga Y, Yamada T, Numata A, Tanaka R. Chaetomugilins I-O, new potent cytotoxic metabolites from a marine-fish-derived *Chaetomium* species. Stereochemistry and biological activities. Tetrahedron 2009;65:7580–7586.

Muroga Y, Yamada T, Numata A, Tanaka R. 11-and 4′-epimers of chaetomugilin A, novel cytostatic metabolites from marine fish-derived fungus *Chaetomium globosum*. Helv Chim Acta. 2010;93:542–549.

Newton GGF, Abraham EP. Cephalosporin C, a new antibiotic containing sulfur and D-α-aminoadipic acid. Nature 1955;175:548.

Numata A, Takahashi C, Ito Y, Takada T, Kawai K, Usami Y, Matsumura E, Imachi M, Ito T, Hasegawa T. Communesins, cytotoxic metabolites of a fungus isolated from a marine alga. Tetrahedron Lett. 1993;34:2355–2358.

Numata A, Amagata T, Minoura K, Ito T. Gymnastatins, novel cytotoxic metabolites produced by a fungal strain from a sponge. Tetrahedron Lett. 1997;38:5675–5678.

Oh DC, Jensen PR, Kauffman CA, Fenical W. Libertellenones A-D: induction of cytotoxic diterpenoid biosynthesis by marine microbial competition. Bioorg Med Chem. 2005;13: 5267–5273.

Oh DC, Kauffman CA, Jensen PR, Fenical W. Induced production of emericellamides A and B from the marine-derived fungus *Emericella* sp. in competing co-culture. J Nat Prod. 2007;70: 515–520.

Osterhage C, Kaminsky R, König GM, Wright AD. Ascosalipyrrolidinone A, an antimicrobial alkaloid, from the obligate marine fungus *Ascochyta salicorniae*. J Org Chem. 2000;65: 6412–6417.

Pontius A, Krick A, Mesry R, Kehraus S, Foegen SE, Müller M, Klimo K, Gerhäuser C, König GM. Monodictyochromes A and B, dimeric xanthone derivatives from the marine algicolous fungus *Monodictys putredinis*. J Nat Prod. 2008;71:1793–1799.

Rateb ME, Ebel R. Secondary metabolites of fungi from marine habitats. Nat Prod Rep. 2011;28:290–344.

Ratnayake AS, Yoshida WY, Mooberry SL, Hemscheidt TK. Nomofungin: a new microfilament disrupting agent. J Org Chem. 2001;66:8717–8721.

Ratnayake AS, Yoshida WY, Mooberry SL, Hemscheidt TK. Nomofungin: a new microfilament disrupting agent. [Erratum for Vol. 66, 2001]. J Org Chem. 2003;68:1640.

Ratnayake R, Fremlin LJ, Lacey E, Gill JH, Capon RJ. Acremolides A-D, lipodepsipeptides from an Australian marine-derived fungus, *Acremonium* sp. J Nat Prod. 2008;71:403–408.

Ren JW, Xue CM, Tian L, Xu MJ, Chen J, Deng ZW, Proksch P, Lin WH. Asperelines A-F, peptaibols from the marine-derived fungus *Trichoderma asperellum*. J Nat Prod. 2009;72:1036–1044.

Rowley DC, Kelly S, Kauffman CA, Jensen PR, Fenical W. Halovirs A-E, new antiviral agents from a marine-derived fungus of the genus *Scytalidium*. Bioorg Med Chem. 2003;11:4263–4274.

Rowley DC, Kelly S, Jensen P, Fenical W. Synthesis and structure-activity relationships of the halovirs, antiviral natural products from a marine-derived fungus. Bioorg Med Chem. 2004;12: 4929–4936.

Saleem M, Ali MS, Hussain S, Jabbar A, Ashraf M, Lee YS. Marine natural products of fungal origin. Nat Prod Rep. 2007;24:1142–1152.

Schiehser GA, White JD, Matsumoto G, Pezzanite JO, Clardy J. The structure of leptosphaerin. Tetrahedron Lett. 1986;27:5587–5590.

Seibert SF, Eguereva E, Krick A, Kehraus S, Voloshina E, Raabe G, Fleischhauer J, Leistner E, Wiese M, Prinz H, Alexandrov K, Janning P, Waldmann H, König GM. Polyketides from the marine-derived fungus *Ascochyta salicorniae* and their potential to inhibit protein phosphatases. Org Biomol Chem. 2006;4:2233–2240.

Seibert SF, Krick A, Eguereva E, Kehraus S, König GM. Ascospiroketals A and B, unprecedented cycloethers from the marine-derived fungus *Ascochyta salicorniae*. Org Lett. 2007;9:239–242.

Siengalewicz P, Gaich T, Mulzer J. It all began with an error: the nomofungin/communesin story. Angew Chem Int Ed. 2008;47:8170–8176.

Singh AV, Bandi M, Raje N, Richardson P, Palladino MA, Chauhan D, Anderson KC. A novel vascular disrupting agent plinabulin triggers JNK-mediated apoptosis and inhibits angiogenesis in multiple myeloma cells. Blood 2011;117:5692–5700.

Sugano M, Sato A, Iijima Y, Oshima T, Furuya K, Kuwano H, Hata T, Hanzawa H. Phomactin A: a novel PAF antagonist from a marine fungus *Phoma* sp. J Am Chem Soc. 1991;113:5463–5464.

Sugano M, Sato A, Iijima Y, Furuya K, Haruyama H, Yoda K, Hata T. Phomactins, novel PAF antagonists from marine fungus *Phoma* sp. J Org Chem. 1994;59:564–569.

Sugano M, Sato A, Iijima Y, Furuya K, Kuwano H, Hata T. Phomactin E, F, and G: new phomactin-group PAF antagonists from a marine fungus *Phoma* sp. J Antibiot. 1995;48:1188–1190.

Tan N, Pan JH, Peng GT, Mou CB, Tao YW, She ZG, Yang ZL, Zhou SN, Lin YC. A copper coordination compound produced by a marine fungus *Fusarium* sp. ZZF51 with biosorption of Cu(II) ions. Chin J Chem. 2008;26:516–521.

Tan RX, Chen JH. The cerebrosides. Nat Prod Rep. 2003;20:509–534.

Tao K, Du L, Sun X, Cai M, Zhu T, Zhou X, Gu Q, Zhang Y. Biosynthesis of aspergiolide A, a novel antitumor compound by a marine-derived fungus *Aspergillus glaucus* via the polyketide pathway. Tetrahedron Lett. 2009;50:1082–1085.

Trisuwan K, Rukachaisirikul V, Sukpondma Y, Preedanon S, Phongpaichit S, Rungjindamai N, Sakayaroj J. Epoxydons and a pyrone from the marine-derived fungus *Nigrospora* sp. PSU-F5. J Nat Prod. 2008;71:1323–1326.

Tsukamoto S, Kato H, Samizo M, Nojiri Y, Onuki H, Hirota H, Ohta T. Notoamides F-K, prenylated indole alkaloids isolated from a marine-derived *Aspergillus* sp. J Nat Prod. 2008;71: 2064–2067.

Tsukamoto S, Kato H, Greshock TJ, Hirota H, Ohta T, Williams RM. Isolation of notoamide E, a key precursor in the biosynthesis of prenylated indole alkaloids in a marine-derived fungus, *Aspergillus* sp. J Am Chem Soc. 2009a;131:3834–3835.

Tsukamoto S, Kawabata T, Kato H, Greshock TJ, Hirota H, Ohta T, Williams RM. Isolation of antipodal (-)-versicolamide B and notoamides L-N from a marine-derived *Aspergillus* sp. Org Lett. 2009b;11:1297–1300.

Tsukamoto S, Umaoka H, Yoshikawa K, Ikeda T, Hirota H. Notoamide O, a structurally unprecedented prenylated indole alkaloid, and notoamides P-R from a marine-derived fungus, *Aspergillus* sp. J Nat Prod. 2010;73:1438–1440.

Wang F, Fang Y, Zhang M, Lin A, Zhu A, Gu Q, Zhu W. Six new ergosterols from the marine-derived fungus *Rhizopus* sp. Steroids 2008;73:19–26.

Wang HX, Ng TB. Pharmacological activities of fusaric acid (5-butylpicolinic acid). Life Sci. 1999;65:849–856.

Wei H, Inada H, Hayashi A, Higashimoto K, Pruksakorn P, Kamada S, Arai M, Ishida S, Kobayashi M. Prenylterphenyllin and its dehydroxyl analogs, new cytotoxic substances from a marine-derived fungus *Aspergillus candidus* IF10. J Antibiot. 2007;60:586–590.

Wright SH, Raab A, Tabudravu JN, Feldmann J, Long PF, Battershill CN, Dunlap WC, Milne BF, Jaspars M. Marine metabolites and metal ion chelation: intact recovery and identification of an iron(II) complex in the extract of the ascidian *Eudistoma gilboviride*. Angew Chem Int Ed. 2008;47:8090–8092.

Wu ZJ, Ouyang MA, Su RK, Kuo YH. Two new cerebrosides and anthraquinone derivatives from the marine fungus *Aspergillus niger*. Chin J Chem. 2008;26:759–764.

Xu JZ, Takasaki A, Kobayashi H, Oda T, Yamada J, Mangindaan REP, Ukai K, Nagai H, Namikoshi M. Four new macrocyclic trichothecenes from two strains of marine-derived fungi of the genus *Myrothecium*. J Antibiot. 2006;59:451–455.

Yamada T, Doi M, Shigeta H, Muroga Y, Hosoe S, Numata A, Tanaka R. Absolute stereostructures of cytotoxic metabolites, chaetomugilins A-C, produced by a *Chaetomium* species separated from a marine fish. Tetrahedron Lett. 2008;49:4192–4195.

Yamada T, Muroga Y, Tanaka R. New azaphilones, seco-chaetomugilins A and D, produced by a marine-fish-derived *Chaetomium globosum*. Mar Drugs 2009a;7:249–257.

Yamada T, Yasuhide M, Shigeta H, Numata A, Tanaka R. Absolute stereostructures of chaetomugilins G and H produced by a marine-fish-derived Chaetomium species. J Antibiot. 2009b;62:353–357.

Yamada T, Muroga Y, Jinno M, Kajimoto T, Usami Y, Numata A, Tanaka R. New class azaphilone produced by a marine fish-derived *Chaetomium globosum*. The stereochemistry and biological activities. Bioorg Med Chem. 2011;19:4106–4113.

Yasuhide M, Yamada T, Numata A, Tanaka R. Chaetomugilins, new selectively cytotoxic metabolites, produced by a marine fish-derived *Chaetomium* species. J Antibiot. 2008;61:615–622.

Zhang Y, Wang S, Li X-M, Cui C-M, Feng C, Wang B-G. New sphingolipids with a previously unreported 9-methyl-C20-sphingosine moiety from a marine algous endophytic fungus *Aspergillus niger* EN-13. Lipids 2007;42:759–764.

Zhu F, Lin YC. Marinamide, a novel alkaloid and its methyl ester produced by the application of mixed fermentation technique to two mangrove endophytic fungi from the South China Sea. Chin Sci Bull. 2006;51:1426–1430.

21 Enzymes from marine fungi: current research and future prospects

Natarajan Velmurugan and Yang Soo Lee

Marine fungi are well-known to produce a diverse range of novel biologically-important enzymes. They also possess novel physiological characteristics, like high salt tolerance, hyper thermostability, barophilicity, cold adaptivity, and are capable of high-level synthesis of interesting enzymes (Raghukumar 2008). Researchers have isolated lignolytic, chitinolytic, cellulolytic, xylan-hydrolyzing, antibacterial, anti-cancer, and several industrially-significant enzymes and bioactive compounds from marine fungi (Pointing et al. 1999; Sabu et al. 2000; Hou et al. 2006; Beena et al. 2010; Burtseva et al. 2010). Further, extreme locations have been accessed for the search of marine fungi to produce novel enzymes (Raghukumar and Raghukumar 1998). Fungi isolated from materials in marine environments such as mangroves, sediments, sea sand, decayed wood and algae have been found to produce enzymes of commercial interest, and they play an important role as the primary decomposers in marine ecosystems (Rohrmann and Molitoris 1992; Li et al. 2002; Bucher et al. 2004). Fungi decompose colonized organic substrata by synthesizing and secreting enzymes at the hyphal surface and the ability to produce enzymes related to the chemical composition of the substrate is undoubtedly adaptive and of ecological significance (Fenice et al. 1997). In addition, hazardous dyes can be degraded by the enzymes, such as manganese peroxidases (MnP), lignin peroxidases (LiP) and laccases, which have been isolated from fungi growing on marine substrates (Bonugli-Santos et al. 2010). The extracellular enzyme profiles of these fungi have been widely studied by researchers from China, India, Thailand, Hong Kong, Brazil, UK, Japan, Russia, Malaysia, and South Korea (Burtseva et al. 2010). However, only little information is available on a wide range of enzymes produced by marine fungi in comparison with terrestrial fungi.

This chapter deals with the research and development of marine fungal enzymes in relation to their occurrence, distribution, molecular biology, production, activity, scale-up and downstream processing. The usage of marine fungal enzymes in biocatabolism, bioremediation, and biomedical applications is also described.

Ligno-cellulolytic enzymes

A wide range of marine fungi can be found on lignocellulosic materials in marine environments, such as mangrove wood (Bugni and Ireland 2004; Jones et al. 2009). Marine fungi obtain their nutrients by the decomposition of lignocellulosic materials, such as prop roots, pneumatophores, branches, leaves, and drift wood in the intertidal region of mangrove stands (Li et al. 2002). The close association of marine fungi with these substrates suggests that they may have an important role in the degradation of organic materials in marine and hypersaline ecosystems via the production of lignocellulolytic enzymes (Pointing et al. 1999). Marine lignicolous fungi have been shown to possess a wide range of enzymes (cellulases, hemi-cellulases, laccase, tyrosinase and laminarase) capable of utilizing lignocellulose components (Pointing et al. 1999). However, studies on the production of high lignin-degrading potential enzymes by marine fungi in hypersaline environments are sparse compared with those reported for their terrestrial counterparts. The enzymes mediating degradation of cellulose materials by terrestrial fungi are well studied (Coughlan 1991; Bucher et al. 2004) and the degradation pathways are expected to be similar in marine fungi. Degradation is initiated by cleavage of the β-1-4 linkages within the accessible amorphous regions of the cellulose chains by endoglucanases, followed by the removal of oligosaccharides from the reducing ends of the partially cleaved cellulose chains by exoglucanases and cello-biohydrolases. The degradation of cellulose is complete when the cellobiose and cellotriose are converted to glucose by β-glucosidase (Glick and Pasternak 2003).

Hemicellulose hydrolysis is attained by the action of the hydrolytic enzymes, xylanases, mannanases, and other hydrolases with broad substrate specificity (Bucher et al. 2004). The plant cell wall mainly contains hemi-cellulose (xylan), which comprises a backbone of β-1,4-linked xylopyranosyl residues that are highly branched with acetyl groups and various sugars (Hou et al. 2006). Xylanase hydrolyzes the xylan backbone, with the action of several other enzymes (Hou et al. 2006). In addition, lignin compounds are mineralized by peroxidases (lignin peroxidases and manganese dependent peroxidases) and laccases.

Pointing et al. (1999) provided the first quantitative data relating to cellulolytic enzyme production by mangrove fungi. Extracellular endoglucanase, cellobiohydrolase, and β-glucosidase were obtained from five lignicolous mangrove fungi (*Halorosellinia oceanica, Julella avicenniae, Lignincola laevis, Savoryella lignincola*, and *Trematosphaeria mangrovei*) in a defined liquid growth medium with a supplement of salts (Table 21.1). Bucher et al. (2004) tested 45 mangrove marine fungi for their potential to produce cellulase and xylanase. Enzyme production assays in solid medium revealed that 89% of the marine fungi tested were capable of producing cellulolytic enzymes and 84% were positive for the production of xylanolytic enzymes. This study significantly increased our knowledge on the *in vitro* production of enzymes by marine fungi and enzyme activity varied considerably among taxonomic groups (Table 21.1).

In an earlier study, Burtseva et al. (2003) recorded the O-glycosyl hydrolases' occurrence in 90 marine fungal strains belonging to 42 species. Among the 90 marine fungal strains, 87 strains were found to produce amylases; 70 strains produced laminarinases (1,3-β-D-glucanases); 40 strains pustunalases (1,6-β-D-glucanases); nine strains fucoidan hydrolases; six strains cellulases; two strains alginases; 80 strains β-D-glucosidase; 45 strains N-acetyl-β-D-glucosaminidases; 26 strains β-D-galactosidases; and 11 strains β-D-mannosidases (Table 21.1). Recently, Burtseva et al. (2010) reported the occurrence of O-glycosyl hydrolases in 78 marine fungi strains isolated from the Sea of Okhotsk and 14 strains from the Sea of Japan (Table 21.1). A higher-level occurrence of O-glycosyl hydrolases and enzyme activity was found in marine fungi surviving at low temperatures (in the Sea of Okhotsk and the Sea of Japan) than those living in warmer regions (off the Kurills; Burtseva et al. 2010).

The extracellular ligno-cellulose degrading enzymes of marine fungi isolated from different substrates exhibit different activity. For example, laminarinase and N-β-D-glucosaminidase activities were higher in marine fungal strains collected from sea ground than in strains associated with sea sponges, while the reverse was reported for pustulanase and β-D-glucosidase activity (Burtseva et al. 2010).

Table 21.1 List of enzymes isolated from marine or marine-derived fungi.

Enzyme	Marine or marine-derived fungus	Source of isolation of fungus	Sea	Optimum pH for enzyme activity	Optimum temperature for enzyme activity (°C)	Reference
Lignin peroxidase (LiP; EC 1.11.1.14) Manganese-dependent peroxidase (MnP; EC 1.11.1.13) Laccase (Lac; EC 1.10.3.2)	*Flavodon flavus* (Klotzsch) Ryvarden (312)	Seagrass from a coral lagoon	West Coast of India	ND	ND	Raghukumar (2000)
	Phlebia sp. (MG-60)	Mangrove stands	Okinawa, Japan	ND	ND	Li et al. (2003)
	Aspergillus sclerotiorum G.A. Huber (CBMAI 849)	*Palythoa variabili*	Town of Sao Sebastiao, Northern Coast of the State of Sao Paulo, Brazil	ND	ND	Bonugli-Santos et al. (2010)
	Cladosporium cladosporioides Fresen. (CBMAI 857)	*Palythoa caribaeorum*				
	Mucor racemosus Fresen. (CBMAI 847)	*Mussismilia hispida*				
	Sordaria fimicola (Roberge) Cesati et De Notaris (#298)	Decaying mangroves and sea grass	West Coast of India	ND	ND	Raghukumar et al. (1996)
	Saagaromyces ratnagiriensis (Patil et Borse) KL Pang et EBG Jones					
	Unidentified basidiomycetous fungus (#312)					

Table continued on next page.

Table 21.1 (Continued)

Enzyme	Marine or marine-derived fungus	Source of isolation of fungus	Sea	Optimum pH for enzyme activity	Optimum temperature for enzyme activity (°C)	Reference
Cellulase (EC 3.2.1.4)	*Cytopaliosphaeria phragmaticola* Poon et K.D. Hyde	Mangroves	The University of Hong Kong Culture Collection or City University of Hong Kong Culture Collection	ND	ND	Bucher et al. (2004)
	Dactylaria sp. (HKUCC 6728)					
	Dendryphiella salina (G.K. Sutherl.) Pugh et Nicot (CY 2723)					
	Periconia prolifica Anastasiou (HKUCC 6724)					
	Phoma sp. ((HKUCC 6725)					
	Phomopsis sp. (HKUCC 6155)					
	Zalerion varium Anastasiou (HKUCC 5485)					
	Acrocordiopsis patilii Borse et K.D. Hyde (HKUCC 9145)					
	Aniptodera salsuginosa Nakagiri et Tad. Ito (HKUCC 6729)					
	Ascocratera manglicola Kohlm. (HKUCC 9174)					

Astrosphaeriella striatispora
(K.D. Hyde) K.D. Hyde
(HKUCC 5200)

A. striatispora
(HKUCC 7651)

Bathyascus grandisporus
K.D. Hyde (HKUCC 6868)

Botryosphaeria sp.
(HKUCC 8019)

Corollospora maritima
Werderm. (CY 1520)

Cryptovalsa halosarceicola
K.D. Hyde (HKUCC 9142)

Dactylospora mangrovei
E.B.G. Jones et al.
(HKUCC 9141)

Eutypa sp. (CY GJ 94)

Oceanitis cincinnatula
(Shearer et J.L. Crane)
J. Dupont et E.B.G. Jones

Helicascus nypae K.D.
Hyde (HKUCC 5788)

Kallichroma tethys Kohlm.
et E. Kohlm. (HKUCC 6084)

Hypocrea sp.
(HKUCC 9144)

Leptosphaeria sp.
(HKUCC 6004)

Lignicola laevis Höhnk
(OM) (HKUCC 6066)

Table continued on next page.

Table 21.1 (Continued)

Enzyme	Marine or marine-derived fungus	Source of isolation of fungus	Sea	Optimum pH for enzyme activity	Optimum temperature for enzyme activity (°C)	Reference
	L. laevis Höhnk (OM) (HKUCC 6867)					
	L laevis Höhnk (OM) (HKUCC 6737)					
	Linocarpon bipolaris K.D. Hyde (HKUCC 5790)					
	Lulworthia grandispora Meyers (CY 1303)					
	Lulworthia sp. (HKUCC 8054)					
	Lulworthia sp. (HKUCC 8055)					
	Marinosphaera mangrovei K.D. Hyde (HKUCC 6914)					
	Massarina achostrichi K.D. Hyde (HKUCC 6727)					
	Halomassarina thalassiae (Kohlm. et Volkm.-Kohlm.) Suetrong et al.					
	Morosphaeria velatispora (K.D. Hyde et Borse) Suetrong et al. (HKUCC 5793)					
	Neptunella longirostris (Cribb et J.W. Cribb) K.L. Pang et E.B.G. Jones					

		Marine habitats		
	Phragmitensis marina K.M. Wong, Poon et K.D. Hyde (HKUCC 6730)			
	Quintaria sp. (HKUCC 6726)			
	Rhizophila marina K.D. Hyde et E.B.G. Jones (HKUCC 9143)			
	Salsuginea ramicola K.D. Hyde (HKUCC 6915)			
	Savoryella lignicola E.B.G. Jones et R.A. Eaton (HKUCC 9176)			
	Verruculina enalia (Kohlm.) Kohlm. et Volkm.-Kohlm. (HKUCC 6869)			
	Penicillium chrysogenum Thom.	The Kurill Islands and Pociet Bay of the Sea of Japan	ND	ND	Burtseva et al. (2003)
	P. verrucosum var. *verrucosum*		–		
Endoglucanase (EC 3.2.1.4)	*Hypoxylon oceanicum* S. Schatz (CY 325)	The City University of Hong Kong Culture Collection and University of Portsmouth Culture Collection	6.2	50	Pointing et al. (1999)
	Julella avicenniae (Borse) K.D. Hyde (CY 109)		6.6	40	

Table continued on next page.

Table 21.1 (Continued)

Enzyme	Marine or marine-derived fungus	Source of isolation of fungus	Sea	Optimum pH for enzyme activity	Optimum temperature for enzyme activity (°C)	Reference
	Lignincola laevis Höhnk (UP 5745)			6.2	60	
	Savoryella lignicola E.B.G. Jones et R.A. Eaton (UP 6076)			6.2	50	
	Trematosphaeria mangrovei Kohlm. (UP 4745)			6.2	50	
	Chaetomium sp. (NIOCC #36)	Mangrove wood litters	Chorao and Zuari, West Coast Goa, India	5.0, 7.0 and 12.0	50	Ravindran et al. (2010)
Carboxymethyl cellulase (EC 3.2.1.4)	*Corollospora maritima* Werderm.	Information not revealed	Information not revealed	ND	ND	Cuomo et al. (1987)
Filter paper cellulase (EC 3.2.1.4)	*Corollospora maritima* Werderm.	Information not revealed	Information not revealed	ND	ND	Cuomo et al. (1987)
Hemi cellulase (EC 3.2.1.4)	*Corollospora maritima* Werderm.	Information not revealed	Information not revealed	ND	ND	Cuomo et al. (1987)
Cellobiohydrolase or exoglucanase (EC 3.2.1.91)	*Hypoxylon oceanicum* S. Schatz (CY 325)	The City University of Hong Kong Culture Collection and University of Portsmouth Culture Collection	—	6.2	50	Pointing et al. (1999)

Enzyme	Fungi	Habitats			Reference
	Julella avicenniae (Borse) K.D. Hyde (CY 109)		6.4	40	
	Lignincola laevis Höhnk (UP 5745)		6.2	60	
	Savoryella lignicola E.B.G. Jones et R.A. Eaton (UP 6076)		6.2	50	
	Trematosphaeria mangrovei Kohlm. (UP 4745)		6.2	50	
β-1,3-glucanase (EC 3.2.1.39)	Aspergillus versicolor (Vuill.) Tirab.	Marine habitats	ND	ND	Burtseva et al. (2010)
	Geomyces pannorum (Link) Sigler et J.W. Carmich.				
	G. pannorum (Link) Sigler et J.W. Carmich.*				
	Penicillium chrysogenum Thom				
	P. chrysogenum Thom*				
	P. commune Thom				
	P. crustosum Thom				
	P. cyaneum (Bainier et Sartory) Biourge				
	P. implicatum Biourge				
	P. islandicum Sopp				
	Aspergillus flavipes (Bainier et Sartory) Thom et Church	The Kuril Islands and Pociet Bay of the Sea of Japan	ND	ND	Burtseva et al. (2003)
	A. flavus Link				
	A. fumigatus Fresen.				

Table continued on next page.

Table 21.1 (Continued)

Enzyme	Marine or marine-derived fungus	Source of isolation of fungus	Sea	Optimum pH for enzyme activity	Optimum temperature for enzyme activity (°C)	Reference
	A. sydowii (Bainier et Sartory) Thom et Church					
	A. wentii Wehmer					
	Lautisporopsis cicumvestita (Kohlm.) E.B.G. Jones, Yusoff et S.T. Moss					
	Chaetomium indicum Corda					
	C. olivaceum Cooke et Ellis					
	Cladosporium sphaerospermum Penzig					
	Echinobotrium atrum Corda					
	Fusarium oxysporum var. *orthoceras* (Appel et Wollenw.) Bilai					
	Fusarium sp.					
	Geomyces pannorum (Link) Sigler et J.W. Carmich.					
	Myceliophthora lutea Costantin					
	Oidiodendron truncatum Barron					
	Penicillium chrysogenum Thom					

	P. cordubense C. Ramírez et A.T. Martínez				
	Penicillium verrucosum var. *cyclopium* (Westling) Samson, Stolk et Hadlok				
	Penicillium verrucosum var. *verrucosum*				
	Trichoderma viridae Pers.				
	Wardomyces inflatus (Marchal) Hennebert				
β-glucosidase (EC 3.2.1.21)	*Chaetomium* sp. (NIOCC 36)	Mangrove wood litters	Chorao and Zuari, West Coast, Goa, India	5.0, 7.0 and 12.0	50	Ravindran et al. (2010)
	Halorosellinia oceanica (S. Schatz) Whalley, E.B.G. Jones, K.D. Hyde & Læssøe (CY 325)	The City University of Hong Kong Culture Collection and University of Portsmouth Culture Collection	–	6.0	50	Pointing et al. (1999)
	Julella avicenniae (Borse) K.D. Hyde (CY 109)			6.4	50	
	Lignincola laevis Höhnk (UP 5745)			6.0	60	
	Savoryella lignicola E.B.G. Jones et R.A. Eaton (UP 6076)			6.0	50	
	Trematosphaeria mangrovei Kohlm. (UP 4745)			6.0	50	

Table continued on next page.

Table 21.1 (Continued)

Enzyme	Marine or marine-derived fungus	Source of isolation of fungus	Sea	Optimum pH for enzyme activity	Optimum temperature for enzyme activity (°C)	Reference
	Aspergillus flavipes (Bainier et Sartory) Thom et Church	Marine habitats	The Kuril Islands and Pociet Bay of the Sea of Japan	ND	ND	Burtseva et al. (2003)
	A. flavus Link					
	A. fumigatus Fresen.					
	A. hallophilicus					
	A. niger P. Micheli ex Haller					
	A. sydowii (Bainier et Sartory) Thom et Church					
	A. varians Wehmer					
	A. versicolor (Vuill.) Tirab.					
	A. wentii Wehmer					
	Lautisporopsis cicumvestita (Kohlm.) E.B.G. Jones, Yusoff et S.T. Moss					
	Chaetomium indicum Corda					
	Cladosporium algarum Cooke et Massee					
	C. atrospetum Pidopl.					
	C. brevicompactum Pidopl. et Deniak					
	C. sphaerospermum Penzig					
	Echinobotrium atrum Corda					

Species	Marine habitats			Reference
Fusarium oxysporum var. *orthoceras* (Appel et Wollenw.) Bilai				
Fusarium sp.				
Myceliophthora lutea Costantin				
Oidiodendron truncatum Barron				
Penicillium chrysogenum Thom				
P. citrinum Thom				
P. cordubense C. Ramírez et A.T. Martínez				
Penicillium verrucosum var. *verrucosum*				
Phoma sp. (*Epicoccum* sp.)				
Trichoderma viridae Pers.				
Wardomyces inflatus (Marchal) Hennebert				
Aspergillus chevalieri (Mangin) Thom. et Church	The Sea of Japan and The Sea of Okhotsk	ND	ND	Burtseva et al. (2010)
A. versicolor (Vuill.) Tirab.				
Geomyces pannorum (Link) Sigler et J.W. Carmich.				
G. pannorum (Link) Sigler et J.W. Carmich.				
Penicillium chrysogenum Thom				
P. chrysogenum Thom*				

Table continued on next page.

Table 21.1 (Continued)

Enzyme	Marine or marine-derived fungus	Source of isolation of fungus	Sea	Optimum pH for enzyme activity	Optimum temperature for enzyme activity (°C)	Reference
	P. chrysogenum Thom*					
	P. chrysogenum Thom*					
	P. commune Thom					
	P. crustosum Thom					
	P. cyaneum (Bainier et Sartory) Biourge					
	P. implicatum Biourge					
	P. isolandicum Sopp					
	P. isolandicum Sopp*					
	Aspergillus (strain SA 58)	Marine sediments	West Coast of India	5.0	35	Elyas et al. (2010)
	Corollospora maritima Werderm.	Information not revealed	Information not revealed	ND	ND	Cuomo et al. (1987)
N-acetyl-β-glucosaminidase (EC 3.2.1.30)	Phoma glomerata (Corda) Wollenw. et Hochafel	Collection of marine microorganism (Pacific Institute of Bioorganic Chemistry)	–	6.0–8.0	55	Zhuravleva et al. (2004)
	Aspergillus chevalieri (Mangin) Thom. et Church	Marine habitats	The Sea of Japan and the Sea of Okhotsk	ND	ND	Burtseva et al. (2010)
	A. versicolor (Vuill.) Tirab.			ND	ND	
	Penicillium canescens Sopp.			4.5	45	
	P. chrysogenum Thom*			ND	ND	

Species	Marine habitats				Reference
P. chrysogenum Thom*		ND	ND	ND	
P. chrysogenum Thom*		ND	ND	ND	
Aspergillus flavipes (Bainier et Sartory) Thom et Church	The Kuril Islands and Pociet Bay of the Sea of Japan	ND	ND	ND	Burtseva et al. (2003)
A. flavus Link					
A. fumigatus Fresen.					
A. hallophilicus					
A. niger P. Micheli ex Haller					
A. sydowii (Bainier et Sartory) Thom et Church					
A. varians Wehmer					
A. versicolor (Vuill.) Tirab.					
Ceriosporiopsis cicumvestita (Kohlm.) Kohlm.					
Chaetomium indicum Corda					
Cladosporium algarum Cooke et Massee					
C. brevi-compactum Pidopl. et Deniak					
C. sphaerospermum Penzig					
Oidiodendron truncatum Barron					
Penicillium chrysogenum Thom					
P. verrucosum var verrucosum					

Table continued on next page.

Table 21.1 (Continued)

Enzyme	Marine or marine-derived fungus	Source of isolation of fungus	Sea	Optimum pH for enzyme activity	Optimum temperature for enzyme activity (°C)	Reference
	Phoma sp. (*Epicoccum* sp.) *Trichoderma viride* Pers.					
β-galactosidase (EC 3.2.1.23)	*Aspergillus flavipes* (Bainier et Sartory) Thom et Church	Marine habitats	The Kuril Islands and Pociet Bay of the Sea of Japan	ND	ND	Burtseva et al. (2003)
	A. flavus Link *A. niger* P. Micheli ex Haller *Chaetomium globosum* Kunze ex Fr. *C. indicum* Corda					
	Penicillium chrysogenum Thom* *P. chrysogenum* Thom* *P. commune* Thom *P. implicatum* Biourge *P. islandicum* Sopp	Marine habitats	The Sea of Japan and the Sea of Okhotsk	ND	ND	Burtseva et al. (2010)
Chitinase (EC 3.2.1.14)	*Plectosphaerella* sp. (MF-1)	Sea Shells	Yellow Sea of Korea	3.0–4.0	10 and 37	Velmurugan et al. (2011)
	Beauveria bassiana (Bals.) Vuill.	Marine sediments	South Coast, Kochi, India	9.20	27	Suresh and Chandrasekaran (1999)
	Verticillium lecanii (Zimm.) Viegas (A3)	Moss samples	Continental Antarctica	4.0	5 and 25	Fenice et al. (1998a)

Enzyme	Species	Source			Reference
	Trichoderma harzianum Rifai (P1)	Istituto di Patologia Vegetale, University of Napoli	4.0	5 and 25	
	Trichoderma harzianum Rifai (T22)	Istituto di Patologia Vegetale, University of Napoli			
α-mannosidase (EC 3.2.1.24)	Penicillium chrysogenum Thom	Marine habitats	ND	ND	Burtseva et al. (2010)
	Aspergillus flavus Link	Marine habitats	ND	ND	Burtseva et al. (2003)
	A. niger P. Micheli ex Haller				
	A. versicolor (Vuill.) Tirab.				
	Myceliophthora lutea Costantin				
Pustulanase or β-1,6-endoglucanase (EC 3.2.1.75)	Aspergillus versicolor (Vuill.) Tirab.	Marine habitats	ND	ND	Burtseva et al. (2010)
	Geomyces pannorum (Link) Sigler et J.W. Carmich.				
	Penicillium chrysogenum Thom				
	P. commune Thom				
	P. implicatum Biourge				
	P. islandicum Sopp				
	Aspergillus fumigatus Fresen.	Marine habitats	ND	ND	Burtseva et al. (2003)

Table continued on next page.

Table 21.1 (Continued)

Enzyme	Marine or marine-derived fungus	Source of isolation of fungus	Sea	Optimum pH for enzyme activity	Optimum temperature for enzyme activity (°C)	Reference
Fucoidan hydrolase (EC 3.2.1.44)	*Cladosporium brevi-compactum* Pidopl. et Deniak *Geomyces pannorum* (Link) Sigler et J.W. Carmich. *Penicillium chrysogenum* Thom *P. commune* Thom *P. crustosum* Thom *P. fuscum* (Sopp) Raper et Thom *P. implicatum* Biourge *Pencillium* sp.	Marine habitats	The Sea of Japan and the Sea of Okhotsk	ND	ND	Burtseva et al. (2010)
Xylanase (EC 3.2.1.8)	*Cytoplacosphaeria phragmaticola* Poon et K.D. Hyde (HKUCC 6722) *Cytospora rhizophorae* Kohlm. et E. Kohlm. (HKUCC 6012) *Dactylaria* sp. (HKUCC6728) *Dendryphiella salina* (G.K. Sutherl.) Pugh et Nicot (CY 2723)	Mangroves	The University of Hong Kong Culture Collection or City University of Hong Kong Culture Collection	ND	ND	Bucher et al. (2004)

Periconia prolifica Anastasiou (HKUCC 6724)
Phoma sp. (HKUCC 6725)
Phomopsis sp. (HKUCC 6155)
Zalerion varium Anastasiou (HKUCC 5485)
Aigialus grandis Kohlm. et S. Schatz (HKUCC 5796)
Aniptodera salsuginosa Nakagiri et Tad. Ito (HKUCC 6729)
Ascocratera manglicola Kohlm. (HKUCC 9174)
Astrosphaeriella striatispora K.D. Hyde (HKUCC 5700)
A. striatispora K.D. Hyde (HKUCC 7651)
Bathyascus grandisporus K.D. Hyde (HKUCC 6868)
Botryosphaeria sp. (HKUCC 8019)
Corollospora maritima Werderm. (CY 1520)
Cryptovalsa halosarceicola (HKUCC 9142)
Dactylospora mangrovei E.B.G. Jones et al. (HKUCC 9141)

Table continued on next page.

Table 21.1 (Continued)

Enzyme	Marine or marine-derived fungus	Source of isolation of fungus	Sea	Optimum pH for enzyme activity	Optimum temperature for enzyme activity (°C)	Reference
	Eutypa sp. (CYGJ 94)					
	Oceanitis cincinnatula (Shearer et J.L. Crane) J. Dupont et E.B.G. Jones 2009 (HKUCC 6731)					
	Helicascus nypae K.D. Hyde (HKUCC 5788)					
	Kallichroma tethys Kohlm. et E. Kohlm. (HKUCC 6084)					
	Leptosphaeria sp. (HKUCC 6004)					
	Lignicola laevis Höhnk (OM) (HKUCC 6066)					
	L. laevis Höhnk (OM) (HKUCC 6867)					
	L. laevis Höhnk (OM) (HKUCC 6737)					
	Linocarpon bipolaris K.D. Hyde (HKUCC 5790)					
	Lulworthia grandispora Meyers (CY 1303)					
	Lulworthia sp. (HKUCC 8054)					
	Marinosphaera mangrovei K.D. Hyde (HKUCC 8089)					

M. mangrovei K.D. Hyde (HKUCC 6914)

Massarina achostrichi K.D. Hyde (HKUCC 6727)

Halomassarina thalassiae (Kohlm. et Volkm.-Kohlm.) Suetrong et al. (HKUCC 9140)

Morosphaeria velatispora (K.D. Hyde et Borse) Suetrong et al. (HKUCC 5793)

Neptunella longirostris (Cribb et J.W. Cribb) K.L. Pang et E.B.G. Jones

Phragmitensis marina K.M. Wong, Poon et K.D. Hyde (HKUCC 6730)

Quintaria sp. (HKUCC 6726)

Rhizophila marina K.D. Hyde et E.B.G. Jones (HKUCC 9143)

Salsuginea ramicola K.D. Hyde (HKUCC 6915)

Savoryella lignicola E.B.G. Jones et R.A. Eaton (HKUCC 9176)

Verruculina enalia (Kohlm.) Kohlm. et Volkm.-Kohlm. (HKUCC 6869)

Table continued on next page.

Table 21.1 (Continued)

Enzyme	Marine or marine-derived fungus	Source of isolation of fungus	Sea	Optimum pH for enzyme activity	Optimum temperature for enzyme activity (°C)	Reference
	Penicillium sp. (FS010)	Marine sediments	Yellow Sea	5.5 (for recombinant xylanase)	25 (for recombinant xylanase)	Hou et al. (2006)
Amylase (EC 3.2.1.8)	*Corollospora maritima* Werderm.	Information not revealed	Information not revealed	ND	ND	Cuomo et al. (1987)
	Aspergillus versicolor (Vuill.) Tirab.	Marine habitats	The Sea of Japan and the Sea of Okhotsk	ND	ND	Burtseva et al. (2010)
	Geomyces pannorum (Link) Sigler et J.W. Carmich.*					
	G. pannorum (Link) Sigler et J.W. Carmich.*					
	P. canescens Sopp.					
	P. chrysogenum Thom*					
	P. chrysogenum Thom*					
	P. chrysogenum Thom*					
	P. commune Thom					
	P. crustosum Thom					
	P. cyaneum (Bainier et Sartory) Biourge					
	P. implicatum Biourge					
	P. islandicum Sopp*					

Species	Marine habitats			Reference
P. islandicum Sopp*				
Penicillium sp.				
Aspergillus flavus Link	The Kuril Islands and Pociet Bay of the Sea of Japan	ND	ND	Burtseva et al. (2003)
A. fumigatus Fresen.				
A. hallophilicus				
A. sydowii (Bainier et Sartory) Thom et Church				
A. versicolor (Vuill.) Tirab.				
A. wentii Wehmer				
Beauveria albae (Limber) Saccas				
Lautisporopsis cicumvestita (Kohlm.) E.B.G. Jones, Yusoff et S.T. Moss				
Chaetomium indicum Corda				
C. olivaceum Cooke et Ellis				
Cladosporium algarum Cooke et Massee				
C. brevi-compactum Pidopl. et Deniak				
C. sphaerospermum Penzig				
Echinobotrium atrum Corda				
Fusarium sp.				
Geomyces pannorum (Link) Sigler et J.W. Carmich.				
Myceliophthora lutea Costantin				

Table continued on next page.

Table 21.1 (Continued)

Enzyme	Marine or marine-derived fungus	Source of isolation of fungus	Sea	Optimum pH for enzyme activity	Optimum temperature for enzyme activity (°C)	Reference
	Oidiodendron truncatum Barron					
	Penicillium chrysogenum Thom					
	P. citrinum Thom					
	P. verrucosum var. *cyclopium* (Westling) Samson, Stolk et Hadlok					
	verrucosum var. *verrucosum*					
	Trichoderma viride Pers.					
	Wardomyces inflatus (Marchal) Hennebert					
Pectinase (EC 3.2.1.15)	*Corollospora maritima* Werderm.	Information not revealed	Information not revealed	ND	ND	Cuomo et al. (1987)
Alginase (EC 3.2.1.16)	*Asteromyces cruciatus* Moreau et Moreau ex Hennebert	The Culture Collection of Marine Fungi in Bremerhaven (KMPB)	—	6.0	25	Schaumann and Weide (1990)

Enzyme	Organism	Habitat	pH	Temperature (°C)	Reference
	Corollospora intermedia Schmidt		ND	ND	
	Dendryphiella salina (G.K. Sutherl.) Pugh et Nicot		6.0	25	
	Aspergillus versicolor (Vuill.) Tirab.	Marine habitats	ND	ND	Burtseva et al. (2010)
	Penicillium cyaneum (Bainier et Sartory) Biourge				
Tannase (EC 3.1.1.20)	Aspergillus awamori (Nakaz.) (BTMFW032)	Sea water	2.0 and 8.0	30	Beena et al. (2010)
L-glutaminase (EC 3.5.1.2)	Beauveria sp. (DTMFS10)	Marine sediment	ND	ND	Sabu et al. (2000)
Protease (EC 3.4.21–24) (no group-specific identification)	Aspergillus ustus (Bain.) Thom et Church	Deep-sea sediments	ND	10 (under 100 bar pressure)	Raghukumar and Raghukumar (1998)
	Graphium sp.	Deep-sea sediments	ND	10 (under 100 bar pressure)	

*Isolated from different habitats

ND: not determined

Recombinant expression of a marine fungal enzyme (xylanase)

Hou et al. (2006) first reported a thermo-labile xylanase extracted from the psychrotrophic marine fungus *Penicillium chrysogenum* (FS010) isolated from the Yellow Sea. They cloned and expressed the cold-active xylanase with fusion partner glutathione-S-transferase in *Escherichia coli* (BL21). The expressed recombinant marine fungal xylanase was highly active at 25°C and pH 5.5. The recombinant xylanase showed its maximum activity (80%) at 4°C (Table 21.1).

Chitinolytic enzymes

Marine fungi attract enormous interest as producers of O-glycosyl hydrolases, catalyzing hydrolysis of O-glycoside bonds in various carbohydrate-containing compounds (Rohrmann and Molitoris 1992; Burtseva et al. 2003; Raghukumar et al. 2004). Such compounds contain chitin, a widely-distributed polysaccharide consisting of N-acetyl-D-glucosamine residuals bound with β-1, 4-bonds (Burtseva et al. 2010). Chitin is an essential component of fungal cell walls, and the exoskeletons of insects and crustaceans. Gooday (1990) reported that non-degradable chitin was produced yearly in the order of 10^{10}–10^{11} tons.

Chitinase and N-acetyl-β-D-glucosaminidases are two groups of enzymes responsible for the complete degradation of chitin.

Chitin degradation is first initiated by chitinases to produce soluble oligosaccharides, with further hydrolyzation to monosaccharides by N-acetyl-β-D-glucosaminidases. Chitinase is a widely-occurring enzyme, mainly produced by fungi, bacteria and plants (Bidochka et al. 1993; Kang et al. 1999; Suresh and Chandrasekaran 1999; Park et al. 2000; Wang et al. 2008a, 2008b). The effects of environmental factors and process parameters on chitinase production have been investigated by Suresh and Chandrasekaran (1999) and Gkargkas et al. (2004). Of various marine sources, microbes obtained from sea shells are the main producers of extracellular chitinases (Wang et al. 2008a, 2008b).

Several chitinases have been purified and characterized from fungi. Bidochka et al. (1993) and Harman et al. (1993) presented the earliest reports on the purification and characterization of fungal chitinase from *Beauveria bassiana* (GK2016) and *Trichoderma harzianum* (P1). Similarly, N-acetyl-D-glucosaminidase, endo-chitinase and exo-chitinase were isolated from *Metarhizium anisopliae* (ATCC 20500) (St. Leger et al. 1996; Pinto et al. 1997; Kang et al. 1999). Further, chitinobiosidases, endo-chitinases, and N-acetyl-D-glucosaminidase were isolated from *Beauveria bassiana* (RD 101) and *Fusarium oxysporum* (F3) (Suresh and Chandrasekar 1999; Gkargkas et al. 2004). Suresh and Chandrasekar (1999) first reported chitinase production from the marine derived fungal strain *B. bassiana* by solid state fermentation (Table 21.1).

Distribution of chitinases in marine fungi of the Sea of Japan and Sea of Okhotsk is recorded by Burtseva et al. (2010), see Table 21.1. The authors reported β-D-glucosidases from 80 strains, N-acetyl- β-D-glucosaminidases from 45 strains, β-D-glucosamindases from 45 strains, β-D-galactosidases from 26 strains and β-D-mannosidases from 11 strains (Table 21.1).

Other reports on the production of chitinase from marine fungi are scarce (Fenice et al. 1998a; Suresh and Chandrasekaran 1999), although Fenice et al. (1998a) reported chitinase activity at lower temperatures from an Antarctic strain of *Verticillium lecanii* (A3), see Table 21.1. Fenice et al. (1998a) recorded the production of a 45 kDa chitinase from *V. lecanii* from Antarctic waters. The highest yields of *V. lecanii* (A3) were 0.22 U ml^{-1} at 25°C and 0.20 U ml^{-1} at 5°C.

Velmurugan et al. (2011) reported the production of a novel chitinase (67 kDa) from the marine fungus *Plectosphaerella* sp. MF-1 (Table 21.1) in a liquid medium supplemented with 0.5% colloidal chitin to induce chitinase production. Chitinase activity was identified on solid medium at a low temperature (10°C) and extracellular chitinase was extracted from culture filtrates and characterized. Active chitinases at low temperature are essential for biological applications in various fields, including the biocontrol of phytopathogens in cold environments and the treatment of chitin-containing materials at low temperatures.

Most filamentous fungi have only been studied on solid substrates and/or in submerged stage fermentation conditions (Sabu et al. 2000; Beena et al. 2010). Liquid-stage fermentation conditions must be developed to overcome the limitations of solid substrate and submerged-stage fermentation, such as heat restriction and mass transfer.

Other enzymes

L-glutaminase (L-glutamine amidohydrolase) was isolated from the marine-derived fungus *Beauveria* sp. under solid-state fermentation (Sabu et al. 2000). L-glutaminase is a unique enzyme because of its potential as an anticancer agent and also for its use as a flavor-enhancing agent in the food industry (Sabu et al. 2000). A maximum yield of 49.89 U ml^{-1} was obtained after 96 h of incubation in seawater-based medium supplemented with 0.25% (w/v) L-glutamine and 0.5% D-glucose (0.5% w/v) at pH 9.0 at 27°C (Sabu et al. 2000), see Table 21.1.

The marine-derived fungus *Aspergillus awamori* (BTMFW032) isolated from seawater was found to produce tannase as an extracellular enzyme under submerged culture conditions (Beena et al. 2010). Tannase (tannin acyl hydrolase) is an industrially-relevant enzyme because of its usage in the manufacture of instant tea and coffee-flavored soft drinks, improvement of flavor in grape wine, and clarification of beer and fruit juices (Beena et al. 2010). Beena et al. (2010) produced this enzyme with a specific activity of 2769.98 IU mg^{-1} under submerged culture conditions. The optimum temperature and pH for maximum enzyme activity were 30°C and, pH 2 and 8, respectively (Table 21.1). The gene encoding tannase was characterized from the marine fungus and revealed that the open reading frame consisted of one stretch of 1122 bp (374 amino acids) in the −1 strand.

Schaumann and Weide (1990) screened 72 strains belonging to 19 species of marine fungi for the production of alginase (alginate hydrolase and alginate lyase) in order to decompose sodium alginate, calcium alginate and/or freshly prepared calcium alginate gel. A total of 18 strains (25% of the total tested) were found to produce alginase and these strains belonged to three different species: *Asteromyces cruciatus*, *Corollospora intermedia*, and *Dendryphiella salina* (Table 21.1). Raghukumar and Raghukumar (1998) isolated a barotolerant protease from the deep-sea fungi *Aspergillus ustus* and *Graphium* sp. The protease was produced when the fungi were grown at 100 bar pressure at 10 and 30°C, respectively (Table 21.1).

The role of marine fungal enzymes in bioremediation and biomedical applications

Various xenobiotic compounds cause environmental pollution and become a global challenge and a serious threat to human health. Oil, oil products, chlorinated compounds, synthetic dyes, and polycyclic aromatic hydrocarbons (PAHs) are the most hazardous pollutants found in the oceans of the world. Heavy industrial emissions and textile industries contribute to the significant amount of these hazardous compounds released, which cause destructive effects in ecosystems (Raghukumar 2000). Marine ecosystems represent a largely unexplored niche for unidentified fungi that could potentially be used in biotechnological processes to degrade these pollutants (Jones 2011).

Bioremediation of these pollutants can be achieved by using salt-tolerant fungi and their salt-tolerant enzymes (Passarini et al. 2011). Many researchers have studied the potential of marine fungi to degrade pollutants and evaluated their enzymes (Raghukumar et al. 1999; Passarini et al. 2011). Most of the pollutant degrading enzymes of marine fungi are produced as extracellular enzymes. Most research has focused on the involvement of extracellular lignin-degrading enzymes produced by marine fungi during bioremediation.

Raghukumar et al. (1999) first reported the involvement of extracellular enzymes of a marine fungus *Flavodon flavus* in bioremediation. The enzymes included MnP and laccases (Raghukumar et al. 1999). *Flavodon flavus* (strain 312) was shown to efficiently degrade the dyes Poly-B, Poly-R, Congo Red, Remazol Brilliant Blue R and Azure (Raghukumar et al. 1999). The same marine fungus was reported to produce LiP in a low-nitrogen medium for Azo dye degradation (Raghukumar 2000). Raghukumar et al. (2006) reported the involvement of lignin-degrading enzymes and exopolysaccharides produced by the marine fungus *F. flavus* (NIOCC #312) for the removal of PAHs, removing 60–70% of the three-ring PAH phenanthrene at a concentration of 12 mg l^{-1} (12 ppm). This isolate also produced all three lignin degrading enzymes: LiP, MnP and laccase. It was shown to decolorize bleach plant effluent at pH 4.5 as well as at pH 8.5 (Raghukumar 2008). The same research group isolated two fungi – *Sordaria fimicola* (NIOCC #298) from mangrove sediment and *Saagaromyces ratnagiriensis* (NIOCC #321) from mangrove wood – that were shown to decolorize bleach plant effluent from the paper and pulp industries via the production of laccase as the main lignin-degrading enzyme and MnP to a lesser extent (Table 21.1).

Passarini et al. (2011) isolated eight potential marine-derived fungi having the ability to degrade Remazol Brilliant Blue R dye and PAHs. These fungal strains were evaluated in a medium containing large benzene ring PAHs. *Aspergillus sclerotiorum* (CBMAI 849) was reported to be responsible for 99.7% degradation of the four-ring pyrene and 76.6% degradation of the five-ring benzo[a]pyrene after eight and 16 days of exposure, respectively. In addition, significant amounts of benzo[a]pyrene depletion (more than 50.0%) was also achieved by another marine derived fungus, *Mucor racemosus* (CBMAI 847). The mechanism behind the degradation of PAHs was found to be mediated by the enzyme cytochrome P450 monooxygenase followed by conjugation with sulfate ions.

Research on marine fungal intracellular enzymes is still in its infancy. In natural marine habitats, free-living microbes are faced with a low nutrient and high-salt environment, which is characterized by moderate changes in temperature, pH, pressure and redox gradients (Schweder et al. 2008). In order to meet these challenges, marine microbes have adapted their physiology to the adverse conditions (Schweder et al. 2008). Furthermore, microbes require a distinct mechanism in order to survive or resist potentially toxic compounds such as PAHs, industrial effluents and hazardous metals. The cellular response to these environmental changes involves drastic changes in fungal gene expression. Despite this, the biological effects of these toxic environmental compounds and the mechanism of its toxicity are still not fully understood. Hence, the study of protein expression upon exposure to toxic compounds is essential for understanding the mechanisms of microbial resistance. The application of two-dimensional gel electrophoresis (2-DGE), in combination with mass spectrometry (MS), has proven to be an excellent tool for this purpose. Intracellular proteome analysis of the toxic environmental response in terrestrial microbes was carried out to address the adaptive response mechanism of microbes (Vido et al. 2001).

Using 2-DGE, we can identify a core set of proteins that appear to be essential for general stress, starvation, and toxic resistance. This is a promising way to discover biomarker proteins of exposure that could act as "detectors" of pollution (Figure 21.1). Salt-tolerant marine fungi could be potential microorganisms for the proteome-based development of biomarkers for bioremediation and biomedical applications (Figure 21.1). The two-dimensional separation of proteins in sodium dodecyl sulfate gels significantly enhances the identification rates of proteins. The up-regulated and/or down-regulated (presence and/or absence) of individual proteins spots on the two-dimensional gels allow a better understanding of microbial responses to environmental stimuli. In recent years, the computer-based evaluation of two-dimensional page gel images has significantly increased. Specific software tools facilitate a precise quantitative assessment of proteome patterns, offering the additional opportunity to numeralize and relativize changes in protein spot volumes. This has opened a new door for precise comparative proteome analysis. Very recently our research group has focused on the evaluation of PAHs stress response in marine-derived fungi. Proteomic analysis in response to different molecular weight PAHs in the marine-derived fungus *Paecilomyces* sp. (SF-8) using 2-DGE was carried out. The best degradation of low molecular weight PAHs, phenanthrene, anthracene, and high molecular weight PAHs, benzo[a]anthracene, benzo[b]fluoranthene was after seven and 14 days, respectively. 2-DGE approach was used to analyze the stress response mechanisms of strain SF-8 after exposure to phenanthrene, anthracene, benzo[a]anthracene, and benzo[b]fluoranthene. Differential expression patterns of intracellular proteins were found under PAHs stress conditions. A total of 304 individual protein spots were found to be differentially expressed (unpublished data). The differentially-expressed proteins could be precisely identified by using liquid chromatography-MS, and matrix-assisted laser desorption/ionization-time of flight MS or Q-time-of-flight-MS. Discovery of specific enzymes involved in the activation and/or regulation of stress response could be the right choice to develop a biomarker (Kakker and Jaffery 2005). Thus, our study potentially leads to the discovery of biomarkers of exposure and helps to gain insights into the underlying mechanisms of toxicity. With biomedical applications in particular, the development of antibodies to some specific regulated proteins (under specific PAH, metal or other toxic stress response) can provide potential biomarkers that are more relevant

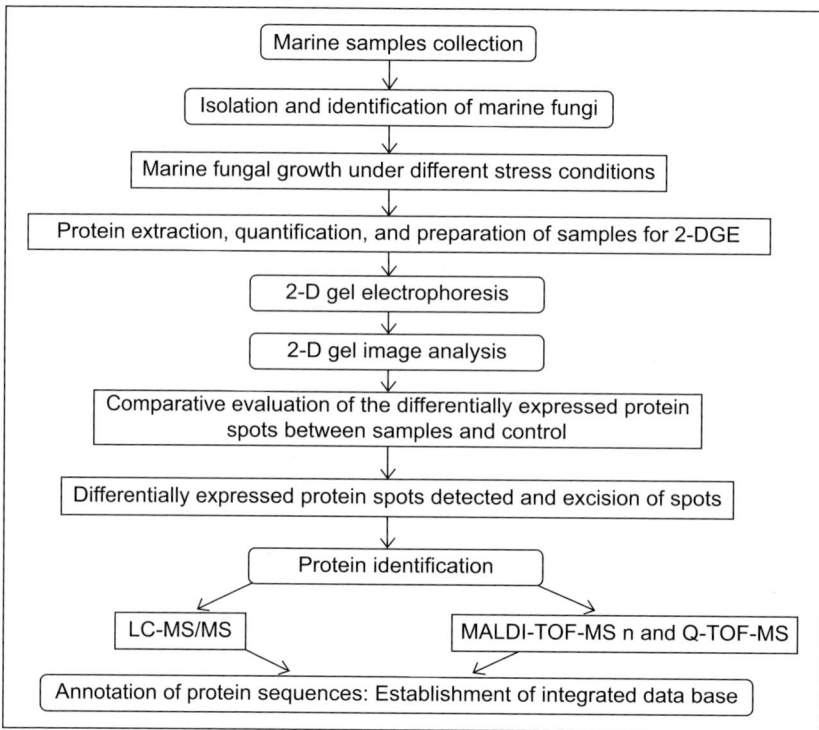

Figure 21.1 Flow chart on two-dimensional gel electrophoresis-based marine fungal proteome techniques for various applications.

to specific diseases. The selection of specific high-affinity antibodies can be achieved by screening synthetic or immune antibodies libraries (Yildirim et al. 2011).

2-DGE can also be used to study the isoforms of extracellular proteins produced during bioremediation processes (Figure 21.1). For example, Raghukumar et al. (1999) reported that the marine fungus *F. flavus*, grown in low-nitrogen medium, produced four isoforms of LiP, and it has been suggested that these isozymes contributed to the versatility of the lignin-degrading fungi in degrading a wide variety of lignocellulose substrates and xenobiotics. This can clarify the specific role of these isozymes in biodegradation.

High-level production of marine fungal enzymes

Chitinase (Suresh and Chandrasekaran 1999), L-glutaminase (Sabu et al. 2000) and β-glucosidases (Elyas et al. 2010) are some of the marine fungal enzymes significantly produced at high-levels of 246.6 U g^{-1}, 49.89 U ml^{-1} and 6200 U g^{-1}, respectively, in solid-state fermentation (SSF). The conventional SSF needs an inert support material for the production of enzymes, and may either be a nutritional element or packaging material (e.g., polystyrene; Sabu et al. 2000). Utilization of SSF for the production of marine fungal enzymes has limitations because of inert material. Other major limitations include process parameters, scale-up strategies and various bioprocess engineering steps. Earlier, we reported the production of marine fungal chitinase in liquid (M9 salt) medium supplemented with colloidal chitin. However, most of the marine fungal enzymes were produced in flask cultivation. No reports studied the production of marine fungal enzymes using a bioreactor. Production of marine fungal enzymes in bioreactors is essential to meet the demand for the industrialization of enzyme products. The production of chitinase of *Penicillium janthinellum* (P9) in a bench-top bioreactor has been reported (Fenice et al. 1998b). Two liters of standard medium production medium containing colloidal chitin 15 g l^{-1} and corn steep liquor (CSL) 15 g l^{-1} were used in a stirrer tank bioreactor for the production of fungal chitinase. During the fermentation, temperature, pH, dissolved oxygen content and stirred speed (rpm) were controlled by microcomputer. This resulted in the highest level of enzyme activity achieved (686 U l^{-1} as compared to 415 U l^{-1} obtained under the initial culture conditions). Both agitation and aeration significantly influenced enzyme production (Fenice et al. 1998b). Production level, purification strategy and efficient recovery of enzymes in an active form are crucial issues for commercialization.

Conclusion

To date, various industrially-important extracellular marine fungal enzymes have been isolated and characterized. These include cellulolytic, lignin degrading and pharmaceutically-important enzymes. Significantly, the discovery of barotolerant protease from deep-sea fungi has initiated a platform for the utilization of novel enzymes of marine fungi for more creative applications.

Further expansion of enzyme discovery from fungi unique to marine environments can facilitate the development of new industrial processes. However, detailed identification of marine fungal enzymes at the molecular level is necessary to find novel and extremophilic enzymes with distinctive properties. Further, the establishment and development of large-scale production systems (innovative bioreactor technologies) for marine fungal enzymes is now highly desired.

Acknowledgements

Authors are very grateful to Drs Gareth Jones and Ka Lai Pang for the invitation to write this chapter.

References

Beena PS, Soore MB, Elyas KK, Bhat Sarita G, Chandrasekaran M. Acidophilic tannase from marine *Aspergillus awamori* BTMFW032. J Microbiol Biotechnol. 2010;20:1403–1414.
Bidochka MN, Tong KI, Khachatourians GG. Partial purification and characterization of two extracellular N-acetyl-D-glucosamindiases produced by the entomopathogenic fungus *Beauveria bassiana*. Can J Microbiol. 1993;39:41–45.
Bonugli-Santos RC, Durrant LR, deSilva M, Sette LD. Production of laccase, manganese peroxidase and lignin peroxidase by Brazilian marine-derived fungi. Enzy Microb Technol. 2010;46:32–37.
Bucher VVC, Hyde KD, Pointing SB, Reddy CA. Production of wood decay enzymes, mass loss and lignin solubilization in wood by marine ascomycetes and their anamorphs. Fungal Divers. 2004;15:1–14.

Bugni TS, Ireland CM. Marine-derived fungi: a chemically and biologically diverse group of microorganisms. Nat Prod Rep. 2004;21:143–163.

Burtseva YV, Verigina NS, Sova VV, Pivkin MV, Zvyagintseva TN. Filamentous marine fungi as producers O-glycosylhydrolases: β-1,3-glucanase from *Chaetomium indicum.* Mar Biotechnol. 2003;5:349–359.

Burtseva YV, Sova VV, Pivkin MV, Anastyuk SD, Gorbach VI, Zvyagintseva TN. Distribution of O-glycosylhydrolases in marine fungi of the Sea of Japan and the Sea of Okhotsk: characterization of exocellular N-acetyl-β-D-glucosaminidase of the marine fungus *Pencillium canescens.* Appl Biochem Microbiol. 2010;46:648–656.

Coughlan MP. Mechanisms of cellulose degradation by fungi and bacteria. Animal Feed Sci Techn. 1991;32:77–100.

Cuomo V, Pagano S, Pecorella MA, Parascandola P. Evidence of the active role of ligno-cellulosic enzymes of marine fungi in degradation of *Posidonia oceanica* leaves. Biochem Systemat Ecol. 1987;15:635–637.

Elyas KK, Mathew A, Sukumaran RK, Manzur APP, Sapna K, Kumar SR, Mol KFR. Production optimization and properties of beta glucosidases from a marine fungus *Aspergillus*-SA 58. New Biotechn. 2010;27:347–351.

Fenice M, Selbmann L, Zucconi L, Onofri S. Production of extracellular enzymes by Antarctic fungal strains. Polar Biol. 1997;17:275–280.

Fenice M, Selbmann L, Di Giambattista R, Federici F. Chitinolytic activity at low temperature of an Antarctic strain (A3) of *Verticillium lecanii.* Res Microbio. 1998a;149:289–300.

Fenice M, Leuba JL, Federici F. Chitinolytic enzyme activity of *Pencillium janthinellum* P9 in bench-top bioreactor. J Fermen Bioengineer. 1998b;86:620–623.

Gkargkas K, Mamma D, Nedev G, Topakas E, Christakopoulos P, Kekos D, Macris BJ. Studies on a N-acetyl-β-D-glucosaminidase produced by *Fusarium oxysporum* F3 grown in solid-state fermentation. Process Biochem. 2004;39:1599–1605.

Glick BR, Pasternak JJ. Molecular biotechnology: Principles and applications of recombinant DNA. 3rd ed. Washington DC: ASM Press, 2003.

Gooday GW. Physiology of microbial degradation of chitin and chitosan. Biodegradation 1990;1:177–190.

Harman GE, Hayes CK, Lorito M, Broadway RM, Di Pietro A, Peterbauer C, Tronsmo A. Chitinolytic enzymes of *Trichoderma harzianum*: purification of chitobiosidase and endochitinase. Phytopathol. 1993;83:313–318.

Hou YH, Wang TH, Long H, Zhu HY. Novel cold-adaptive *Penicillium* strain FS010 secreting thermolabile xylanase isolated from Yellow Sea. Acta Biochimi Biophys Sinica. 2006;38:142–149.

Jones EBG. Fifty years of marine mycology. Fungal Divers. 2011;50:73–112.

Jones EBG, Sakayaroj J, Suetrong S, Somrithipol S, Pang KL. Classification of marine Ascomycota, anamorphic taxa and Basidiomycota. Fungal Divers. 2009;35:1–187.

Kakker P, Jaffery FN. Biological markers for metal toxicity. Environ Toxicol Pharmacol. 2005;19:335–349.

Kang SC, Park S, Lee DG. Purification and characterization of a novel chitinase from the entomopathogenic fungus *Metarhizium anisopliae.* J Invertebr Pathol. 1999;73:276–281.

Li X, Kondo R, Sakai K. Studies on hypersaline-tolerant white-rot fungi I: screening of lignin-degrading fungi in hypersaline conditions. J Wood Sci. 2002;48:147–152.

Li X, Kondo R, Sakai K. Studies on hypersaline-tolerant white-rot fungi III: biobleaching of unbleached kraft pulp hypersaline-tolerant manganese peroxidase from a marine white rot isolate, *Phlebia* sp. MG-60. J Wood Sci. 2003;49:42–46.

Park SH, Lee JH, Lee HK. Purification and characterization of chitinase from a marine bacterium, *Vibrio* sp. 98CJ11027. J Microb. 2000;38:224–229.

Passarini MRZ, Rodrigues MVN, deSilva M, Sette LD. Marine-derived filamentous fungi and their potential application for polycyclic aromatic hydrocarbon bioremediation. Mar Pollut Bull. 2011;62:364–370.

Pinto DS, Barreto CC, Schrank A, Ulhoa CJ, Vainstein MH. Purification and characterization of an extracellular chitinase from the entomopathogen *Metarhizium anisopliae*. Can J Microbiol. 1997;43:322–327.

Pointing SB, Buswell JA, Jones EBG, Vrijmoed LLP. Extracellular celluloytic enzyme profiles of five lignicolous mangrove fungi. Mycol Res. 1999;103:696–700.

Raghukumar C. Fungi from marine habitats: an application in bioremediation. Mycol Res. 2000;104:1222–1226.

Raghukumar C. Marine fungal biotechnology: an ecological perspective. Fungal Divers. 2008;31:19–35.

Raghukumar C, Raghukumar S. Barotolerance of fungi isolated from deep-sea sediments of the Indian Ocean. Aquat Microb Ecol. 1998;15:153–163.

Raghukumar C, Chandrmohan D, Michel Jr FC, Reddy CA. Degradation of lignin and decolorization of paper mill bleach plant effluent (BPE) by marine fungi. Biotechn Lett. 1996;18:105–106.

Raghukumar C, D'Souza TM, Thorn RG, Reddy CA. Lignin-modifying enzymes of *Flavodon flavus*, a basidiomycete isolated from a coastal marine environment. Appl Environ Microbiol. 1999;65:2103–2111.

Raghukumar C, Muraleedhara U, Gaud VR, Miishra R. Xylanases of marine fungi of potential use for biobleaching of paper pulp. J Ind Microbiol Biotechnol. 2004;31:433–441.

Raghukumar C, Shailaja MS, Parameswaran PS, Singh SK. Removal of polycyclic aromatic hydrocarbons from aqueous media by the marine fungus NIOCC # 312: involvement of lignin-degrading enzymes and exopolysaccharides. Ind J Mar Sci. 2006;35:373–379.

Ravindran C, Naveenan T, Varatharajan G. Optimization of alkaline cellulose production from marine derived fungi *Chaetomium* sp., using agricultural and industrial wastes as substrates. Bot Mar. 2010;53:275–282.

Rohrmann S, Molitoris HP. Screening of wood-degrading enzymes in marine fungi. Can J Bot. 1992;70:2116–2123.

Sabu A, Keerthi TR, Rajeev Kumar S, Chandrasekar M. L-Glutaminase production by marine *Beauvreia* sp. under solid state fermentation. Process Biochem. 2000;35:705–710.

Schaumann K, Weide G. Enzymatic degradation of alginate by marine fungi. Hydrobiol. 1990;204–205:589–596.

Schweder T, Market S, Hecker M. Proteomics of marine bacteria. Electrophoresis 2008;29:2603–2616.

St. Leger RJ, Joshi L, Bidochka MJ, Rizzo NW, Roberts DW. Characterization and ultrastructural localization of chitinase from *Metarhizium anisopliae, M. flavoviride* and *Beauveria bassiana* during fungal invasion of host (*Manduca sexta*) cuticle. Appl Environ Microbiol. 1996;62:907–912.

Suresh PV, Chandrasekaran M. Impact of process parameters on chitinase production by an alkalophilic marine *Beauveria bassiana* in solid state fermentation. Process Biochem. 1999;34:257–267.

Velmurugan N, Kalpana D, Han JH, Cha HJ, Lee YS. A novel low temperature chitinase from the marine fungus *Plectosphaerella* sp. strain MF-1. Bot Mar. 2011;54:75–81.

Vido K, Spector D, Lagniel G, Lopez S, Toledano MB, Labarre J. A proteome analysis of the cadmium response in *Saccharomyces cerevisiae*. J Biol Chem. 2001;276:8469–8474.

Wang G, Li Q, Zhu P. Phylogenetic diversity of culturable fungi associated with the Hawaiian Sponges *Suberites zeteki* and *Gelliodes fibrosa*. Antonie Leeuwehoek 2008b;93:163–174.

Wang SL, Chen SJ, Wang CL. Purification and characterization of chitinase and chitosanases from a new species strain *Pseudomonas* sp. TKU015 using shrimp shells as a substrate. Carbohydr Res. 2008a;343:1171–1179.

Yildirim V, Ozcan S, Becher D, Buttner K, Hecker M, Ozcangiz G. Characterization of proteome alterations in *Phanerochaete chrysosporium* in response to lead exposure. Proteo Sci. 2011;9: 12–27.

Zhuravleva NV, Luk'yanov PA, Pivkin MV. N-Acetyl-β-D-hexosaminidase secreted by the marine fungus *Phoma glomerata*. Appl Biochem Microbiol. 2004;40:448–453.

22 Decomposition of materials in the sea

Kandikere R. Sridhar

The export of organic matter to the offshore environment is one of the important functions of mangrove ecosystems (Logo and Snedaker 1974). Sustainability and economic benefits in mangroves depend on the import as well as the pattern of decomposition of plant litter. Lignocellulosic materials in mangrove habitats serve as important substrates for mineralization and energy flow to higher trophic levels. The importance of the mangrove ecosystem in the export of plant detritus and elevation of faunal biomass are well documented by Lee (1995). The major organic matter that serves as a recyclable source of energy in mangroves includes woody and leaf litter, sedge detritus and animal remains.

Sustainable ecological services and economic benefits in mangroves are dependent on the rate of input and pattern of woody litter decomposition. A variety of autochthonous and allochthonous woody debris reaches the mangrove floor and serves as a persistent substrate for fungal and faunal colonization. The web-like root structure of mangrove vegetation traps and retains woody debris for a considerable duration at different depths of the intertidal zone. Besides woody litter, senescent and dead wood in attached (standing dead) state are also available for fungal colonization in terrestrial, semi-aquatic and aquatic zones. Naturally-deposited woody litter in mangroves provides a great opportunity to evaluate the pattern of fungal colonization and decomposition under different tidal ranges. Standing dead wood and immersed woody litter may also be evaluated to understand the fungal colonization and pattern of decomposition in mangrove habitats. Despite a slow turnover, woody debris constitutes valuable long-term source for the food webs and nutrient cycles (N, P and K) of mangroves.

Similar to woody litter, mangrove leaf litter serves as a rich source of carbon; the availability of nitrogen is generally poor (C/N ratio, >100) in leaf litter and many inhibitory compounds (e.g., polyphenolics and tannins) are abundant (Neilson and Richards 1989), making leaf litter a less attractive organic source to mangrove fauna. Several microbes (bacteria, protoctista and fungi) are responsible for the transformation of polymeric compounds into dissolved or particulate organic matter utilizable by the consumers of the food web (Newell and Fell 1997). However, little information is available on the role of fungi in recycling of leaf litter nutrients in a mangrove ecosystem. Measurement of mycelial biomass and productivity, especially in saltmarsh ecosystems, has enabled a better understanding of the role of eukaryotic fungal decomposers in marine ecosystems (Newell 1996). Grasses (Poaceae) and sedges (Cyperaceae) serve as a significant part of mangroves and coastal habitats in the tropics and subtropics. Although information on the fungal interactions with mangroves and mangrove associates are available, studies on grasses and sedges are scanty. Interestingly, 48 new species belonging to 14 new genera with a new family of fungi have been reported from *Juncus roemerianus* (Juncaceae) along the Southeast coast of US (Kohlmeyer and Volkmann-Kohlmeyer 2001). *Cyperus malaccensis* (Cyperaceae) is one of the dominant perennial sedges in the mangroves of the Southwest coast of India (Karamchand et al. 2009). Being rhizomatous, these swards occur in a thick mat and contribute a substantial amount of detritus in the form of standing dead culms.

The major objective of the present review is to evaluate the pattern of wood, leaf and sedge litter degradation by higher fungi in mangrove ecosystems.

Decomposition of woody litter

Mass loss

Studies on woody litter decomposition in mangroves are meager. Mackey and Smail (1996) have established a linear relationship between mangrove leaf decomposition (time required for 50% mass loss or half-life or t_{50}) with latitude; and faster decomposition occurs at mid-latitudes than at higher latitudes, which is also applicable to mangrove woody litter due to high temperature and continuous fungal and faunal interactions at mid-latitudes. Wood decomposition studies have adopted methods and models that are employed for leaves and other easily degradable materials (Bärlocher 1997). To describe the mass loss against time, a simple exponential decay model has been used to provide decay coefficient (k) as a single number characterizing the rate of mass loss. Petersen and Cummins (1974) classified leaf litter decay as fast ($k > 0.01$), medium ($k = 0.005$–0.01) and slow ($k < 0.005$). As each chemical class of organic matter decays at a specific rate, more elaborate models have been employed for precise evaluation (Wieder and Lang 1982).

A comparison of the pattern of mass loss of eight woody litters in five mangroves (in Australia, Belize, India and South Africa) is given in Table 22.1. The decay coefficient of *Avicennia officinalis* (0.0011) in the first 11 months (October–August: post-monsoon, summer and monsoon) in the Southwest Indian mangrove is similar to that of *Avicennia marina* (0.0018) of a Northern Australian mangrove during summer at high tide level (Mackey and Smail 1996). However, the half-life of *A. officinalis* decomposition was higher than *A. marina* (627 vs. 421 days) at a South African mangrove (Steinke et al. 1983). Decomposition of twigs of *A. marina* in Australian mangroves showed higher mass loss in summer than in winter in high- and low-tide habitats (t_{50}: 383 and 179 days vs. 1,327 and 1,207 days; Mackey and Smail 1996). Leightley (1980) reported the decay of mangrove timbers (*A. marina* and *Excoecaria agallocha*) exposed to the sea in Queensland by the colonization of fungi and marine borers with an initial loss of 11.7 and 16.0% mass, respectively. In mangroves in Belize, decomposition of the wood stakes of four mangrove plant species (prop roots of *Rhizophora mangle*, branches of *Avicennia, Conocarpus* and *Laguncularia*) did not differ significantly (Kohlmeyer et al. 1995). Bucher et al. (2004) have demonstrated *in vitro* wood decay of *Betula* sp. by 45 species of marine fungi in darkness for 24 weeks under static conditions. The overall mass loss in exposed conditions was significantly higher than in submerged conditions and indicates high decomposition rates in intertidal habitats. However, asexual fungi showed a higher mass loss in submerged than in exposed conditions and reveals the habitat-dependent role of anamorphic and meiosporic fungi in wood decomposition.

To understand mass loss of woody litter over time, a simple exponential decay model has been used. For example, Romero et al. (2005) followed a simple exponential model to explain wood decay in the air, while a two-component model was followed for surface and buried wood. The rate of mass loss of woody litter of *A. officinalis* and *R. mucronata* was low during the first phase and then accelerated; neither simple nor double exponential decay models describe the pattern of decay (Maria et al. 2006). It is presumed that such

Table 22.1 Decay coefficient/percentage mass loss and half-life of wood and root litter of representative mangrove plant species at different geographical locations.

Plant species	Decay coefficient (k/day)	Mass loss/day (%)	Half-life (t_{50}, days)	Location	Reference
Woody litter					
Avicennia germinans	–	0.111	–	Belize	Middleton and Mckee (2001)
A. marina (winter season)	0.0006–0.0005	–	1,207–1,327	Australia	Mackey and Smail (1996)
A. marina (summer season)	0.0039–0.0018	–	179–383	Australia	Mackey and Smail (1996)
A. marina	–	0.100–0.167	–	Australia	Mackey and Smail (1996)
A. marina	–	–	421	South Africa	Steinke et al. (1983)
A. officinalis	0.0055–0.0011	–	125–627	India	Maria et al. (2006)
Laguncularia racemosa	–	0.040	–	Belize	Middleton and Mckee (2001)
Rhizophora apiculata	–	0.074	–	Australia	Robertson and Daniel (1989)
R. mangle	–	0.123	–	Belize	Middleton and Mckee (2001)
R. mucronata	0.0110–0.0006	–	63–1,254	India	Maria et al. (2006)
R. stylosa	–	–	916	Australia	Robertson and Daniel (1989)
Root litter					
Avicennia germinans	–	0.104–0.108	–	Belize	Middleton and Mckee (2001)
A. marina (fibrous roots)	–	0.133	–	Australia	van Der Valk and Attiwill (1984)
Rhizophora mangle	–	0.092–0.108	–	Belize	Middleton and Mckee (2001)

variation in the decay of woody litter was due to conditioning by fungal succession and the physicochemical features of the mangrove habitat.

Decomposition

The decomposition process proceeds in three distinct phases (Gessner et al. 1999):

(i) leaching (rapid loss of soluble compounds: phenolics, sugars, amino acids);
(ii) colonization of microbes (dominated by fungi resulting in higher levels of nitrogen); and
(iii) fragmentation (partly physical process and also contributed by detritivores or shredders).

The pattern of decay differs between wood exposed to tidal inundation and that fully or partially buried in anaerobic mangrove mud. The significant difference of the decay coefficient between *Avicennia* and *Rhizophora* wood indicates a differential pattern of decomposition although they were exposed at the same mangrove in Southwest India (Maria et al. 2006). Drastic decline in the mass of both wood types was seen (12–18 months), which coincided with peaks of fungal richness and diversity (after 10 months; Maria and Sridhar 2004). This shows the necessity of sufficient fungal conditioning of wood for substantial mass loss to occur. The latter period coincided with an increased population of juveniles of *Perna* (a mollusk) on decomposing wood. The initial decline in wood mass may be attributed to leaching of water-soluble compounds like sugars, phenolics and phosphate. In the latter period, there would be a significant loss of structural components due to conditioning of the wood by fungi and the activities of invertebrates. On the southwest coast of India, mangrove wood (twigs and branches) dries up during the summer season (February–May). Due to heavy rain and storm during southwest monsoon (June–September), they detach and reach the mangrove floor (Maria et al. 2006). Maria et al. (2006) followed the pattern of wood decomposition up to 18 months starting from October (post-monsoon season) and found slow mass loss during the initial period in *R. mucronata* and *A. officinalis* (10–12 months) followed by rapid mass loss. The overall mass loss of wood was more rapid in *R. mucronata* than *A. officinalis*.

Fungal colonization and zonation

Sixteen fungi isolated from mangrove wood were tested for their ability to cause soft rot under laboratory conditions (Mouzouras 1986; Mouzouras et al. 1986). All fungi penetrated the wood cell walls, but only eight fungi successfully caused soft rot cavities. A similar study by Mouzouras (1989) using scanning electron microscopy revealed differential activity of fungi (*Halocyphina villosa* and *Trichocladium achrasporum*) on mangrove woods (*A. officinalis* and *Xylocarpus granatum*). Significant decay of *A. officinalis* was seen after 24 weeks of incubation with *H. villosa*, but there was no activity against *X. granatum*. Significant losses of both mangrove woods were caused by the activity of *T. achrasporum*. Another laboratory study on the decomposition of wood (Feeney et al. 1992) demonstrated that the incorporation of potassium nitrate (1 g/l in basal medium) significantly elevated the mass loss of *Fagus sylvatica* by *Chaetomium globosum* (3.68 vs. 28.3%) as well as *Monodictys pelagica* (2.14 vs. 28.2%).

Besides substrate specificity, vertical zonation of fungi on wood at different tidal levels in mangroves has been reported by several researchers (Hyde 1988, 1990a, 1990b; Hyde and Lee 1995; Kohlmeyer et al. 1995). The diversity of fungi was greatest at above mean tide, where wood is usually inundated daily but dries up superficially during low tides. Many fungi sporulated on wood stakes exposed in a Belize mangrove below the water line (0.6–1.2 m above sediment); three fungi sporulated close to the sediment (*Antennospora quadricornuta*, *Lulworthia* sp. and *Phoma* sp.); and *H. villosa*, *Trichocladium alopallonellum* and *Lulworthia* sp. occurred above the high tide line (Kohlmeyer et al. 1995).

Litter decay rate in mangroves depends on the position of the substrate on shore: decay is faster in the lower shore than higher up the shore (Mfilinge et al. 2002). Middleton and McKee (2001) predicted that the standing dead and fallen wood would be processed at different rates and by different pathways, depending on the species composition of mangrove and wood-feeding organisms. Dead twigs of *Rhizophora mangle* persist in the canopy at Belize for up to 18 months before it reaches the floor (Feller and Mathis 1997). Dead twigs of different mangrove tree species might have different periods of exposure to dryness and colonization of fungi in the canopy before they fall onto the floor, which influence the pattern of decomposition in the submerged state.

Bark and wood

In Brunei mangroves, immersed poles of *Xylocarpus granatum* lost their bark within four months (Hyde 1991). However, *Rhizophora apiculata* wood retained bark for longer than 12 months of immersion (Hyde 1991), which is similar to *R. mucronata* wood in a Southwest mangrove (10 months; Maria et al. 2006). Such a delay in the loss of bark can be attributed to the antimicrobial components in bark (e.g., tannin and suberin). The pattern of decomposition of bark differs between timber types in the southwest mangrove of India (Maria et al. 2006). The bark of immersed twigs of *A. officinalis* decayed within four months, started to peel off within six months and fully disappeared by 12 months. In *R. mucronata* twigs, separation of bark took 10 months and became very brittle after 16 months. The damage to the central pith was seen within six months, and a smooth and hollow core was left after eight months. When twigs are submerged, usually fungi decay the phloem and parenchyma cells beneath the bark before the bark is utilized, and very often the bark forms a lose sheath around the wood.

Bark-preferring fungi colonized in the first year of decomposition, followed by colonization by wood decay fungi (Hyde 1991; Kohlmeyer et al. 1995). Similarly, bark-inhabiting fungi colonized first on prop roots (Hyde 1991) and wood decay fungi dominated after bark decomposition or bark detachment (Leong et al. 1991). *Etheirophora blepharospora, Mycosphaerella pneumatophorae, Lulworthia grandispora, Morosphaeria velataspora* and *Rhabdospora avicenniae* were found exclusively on the surface of bark (Kohlmeyer and Kohlmeyer 1979; Hyde and Jones 1988), while *Aniptodera mangrovei, Caryosporella rhizophorae, H. villosa, Halosarpheia marina* and *Leptosphaeria australiensis* were most common on the exposed wood (Hyde and Jones 1988). During wood decomposition, changes in wood texture and color are dependent on the colonized fungal species and the zone of its occupation on the wood (Hyde et al. 1998). For instance, wood colonized by *Halomassarina thalassiae* was soft beneath and bleached white, while it became soft and gray when colonized by *L. australiensis,* and attained a black color on colonization by *Halorosellinia oceanica*. At each stage of fungal decomposition, the replacement of fungi was seen in succession and such fungi transform the conditions of wood less suitable for themselves or more suitable for others (Hudson 1962).

Decomposition of leaf litter

Unlike woody litter, several studies are available on mangrove leaf litter breakdown in different tropical and subtropical regions of the world (Arabian Gulf, Australia, Bahamas, Bangladesh, Belize, Brazil, China, Ecuador, Florida, India, Indonesia, Kenya,

Malaysia, Mozambique, New Zealand, Sri Lanka, South Africa, Tanzania and Thailand), see Table 22.2). Unlike woody litter, the employment of litter bags to study their breakdown is inevitable. According to Boulton and Boon (1991), leaf litter in bags faces different microclimatic conditions compared to naturally decomposing leaf litter. Thus, litter-bag studies predict the possible breakdown rates under field conditions (Imgraben and Dittmann 2008). Fine and coarse mesh litter bag studies are, however, helpful in distinguishing the role of microbes, and microbes along with invertebrates (Graça et al. 2005).

Sequential decay

Among the studies on more than 15 mangrove leaf litters, major studies have been conducted on the decomposition of leaf litter of *Rhizophora* spp. (Table 22.2). Three important phases were recognized by Raghukumar et al. (1995) while studying leaf litter decomposition of *R. apiculata*:

(i) rapid mass loss along with loss of proteins, carbohydrates, reducing sugars, phenolics and cellulose during the first week of exposure;
(ii) increase in fungal biomass and rapid decline of the rest of the organic constituents in about three weeks leading to a decrease in the C/N ratio; and
(iii) decline in fungal and bacterial biomass after three and five weeks, respectively, followed by a decrease in cellulose and lignin.

Comparison of decay with latitude

The mangrove leaf litter decay coefficient (k) per day ranged from 0.0630 (*A. marina*, Queensland, Australia) to 0.0014 (*A. marina*, Newcastle, Australia), while the half-life (t_{50}) was three days (*S. alba*, Gazi Bay, Kenya) to 164 days (*R. mucronata*, Gazi Bay, Kenya), see Table 22.2. The mass loss of leaf litter per day ranged between 0.15% (*A. marina* in Australia) and 4.5% (*Rhizophora* sp. in Malaysia; Ong et al. 1980; Dick and Streever 2001).

Mackey and Smail (1996) showed a linear relationship between half-life vs. latitude and faster decomposition of leaf litter from *A. marina* in mid-latitudes than at higher latitudes due to elevated temperature and the interactions of fungi and fauna. In Bulwer Island (Australia), mass loss of *A. marina* leaf litter was significantly higher at low tide during summer than at high tide during winter (Mackey and Smail 1996). Based on literature, the relationship between leaf litter decay coefficient (k/day) in 24 mangroves ($r^2 = 0.025$, $p < 0.01$), mass loss (%/day) in 27 mangroves ($r^2 = 0.2566$, $p < 0.0001$) and the time required for 50% mass loss (t_{50}) in 34 mangroves ($r^2 = 0.0053$, $p < 0.001$) against latitude depicts a significantly faster rate of decomposition in tropical than in subtropical or temperate latitudes. This supports the observations of Mackey and Smail (1996) that temperature plays an important role in recycling the leaf litter (Figure 22.1). Temperature influences the population level of microbes as well as shredders, which enhances the rates of leaf litter processing and decomposition in the tropics. Such effects may only be expected to occur during the summer season in subtropical and temperate latitudes.

Table 22.2 Decay coefficient, percentage mass loss and half-life of leaf, sedge and seagrass litter in mangroves at different geographical locations.

Plant species	Decay coefficient (k/day)	Mass loss/day (%)	Half-life (t_{50}, days)	Location	Reference
Leaf litter					
Aegiceras corniculatum	0.0065	0.97	107	China	Tam et al. (1990)
A. corniculatum	0.0146	–	48	China	Tam et al. (1998)
Avicennia germinans	–	0.25–0.44	–	Belize	Middleton and Mckee (2001)
A. germinans	–	0.57	–	Florida	Twilley et al. (1986)
A. marina	–	0.31	–	Arabian Gulf	Hegazy (1998)
A. marina	0.0124	–	56	Australia	Goulter and Allaway (1979)
A. marina	0.0087	0.32	80	Australia	van Der Valk and Attiwill (1984)
A. marina	0.0630	–	11	Australia	Robertson (1988)
A. marina	–	0.46–0.54	–	Australia	Mackey and Smail (1996)
A. marina	0.0086–0.0014	0.15–0.26	–	Australia	Dick and Osunkoya (2000)
A. marina	0.0030–0.0023	–	–	Australia	Dick and Streever (2001)
A. marina	–	–	14	Australia	Imgraben and Dittmann (2008)
A. marina	0.1155–0.0277	–	6–25	China	Lu and Lin (1990)
A. marina	0.0126	–	55	China	Tam et al. (1990)
A. marina	0.0188	–	37	New Zealand	Albright (1976)
A. marina	–	0.40	–	New Zealand	Woodroffe (1982)
A. marina	–	–	70	New Zealand	Woodroffe (1985)
A. marina	0.0123–0.0120	–	32–58	South Africa	Steinke and Ward (1987)
A. marina	–	–	21–56	Sri Lanka	Pinto and Swarnamali (1998)
A. marina	0.0045	–	154	Tanzania	Chale (1993)
A. marina	0.0347	–	20	Thailand	Boonruang (1978)
A. officinalis	–	1.79	–	India	Wafar et al. (1997)
A. schaueriana	–	0.27	–	Brazil	Sessegolo and Lana (1991)

Table continued on next page.

Table 22.2 (Continued)

Plant species	Decay coefficient (k/day)	Mass loss/day (%)	Half-life (t_{50}, days)	Location	Reference
Bruguiera gymnorrhiza	0.0133–0.0077	–	56–90	South Africa	Steinke and Ward (1987)
B. parviflora	–	–	70–122	Malaysia	Ashton et al. (1999)
B. parviflora	–	0.67–3.15	–	Malaysia	Hossain and Othman (2005)
Bruguiera spp.	–	4.00	–	Malaysia	Ong et al. (1980)
Ceriops tagal	0.0257	–	27	Australia	Robertson (1988)
C. tagal	–	0.54–1.38	–	Kenya	Woitchik et al. (1997)
Kandelia obovata	0.0114	0.53–0.54	61	China	Lee (1989)
K. obovata	0.0385–0.0124	–	18–56	China	Lu and Lin (1990)
K. obovata	0.0164	–	42	China	Tam et al. (1990)
K. obovata	0.0516	1.29	13	China	Tam et al. (1998)
Laguncularia racemosa	–	0.17–0.43	–	Belize	Middleton and Mckee (2001)
Rhizophora apiculata	–	0.44–0.54	–	China	Tam et al. (1998)
R. apiculata	–	0.95	–	India	Wafar et al. (1997)
R. apiculata	–	1.07	43	Malaysia	Ashton et al. (1999)
R. apiculata	0.0173	–	40	Thailand	Boonruang (1978)
R. apiculata	–	–	66–83	Thailand	Nielsen and Andersen (2003)
R. mangle	–	0.28	–	Brazil	Sessegolo and Lana (1991)
R. mangle	–	0.18–0.43	–	Belize	Middleton and Mckee (2001)
R. mangle	–	0.30	–	Florida	Twilley et al. (1986)
R. mangle	0.0128	–	54	New Zealand	Fell et al. (1975)
R. mucronata	–	0.95	–	India	Wafar et al. (1997)
R. mucronata	0.021–0.018	–	49	India	Ananda et al. (2008)
R. mucronata	–	1.53	–	Kenya	Woitchik et al. (1997)

Table continued on next page.

Table 22.2 (Continued)

Plant species	Decay coefficient (k/day)	Mass loss/day (%)	Half-life (t_{50}, days)	Location	Reference
R. mucronata (bare region)	–	–	164	Kenya	Bosire et al. (2005)
R. mucronata (reforested region)	–	–	26	Kenya	Bosire et al. (2005)
R. mucronata (natural region)	–	–	12	Kenya	Bosire et al. (2005)
R. mucronata	–	0.96–1.96	–	Kenya	Woitchik et al. (1997)
R. mucronata	–	1.25	–	Malaysia	Ashton et al. (1999)
R. stylosa	0.0178	–	39	Australia	Robertson (1988)
R. stylosa	0.011–0.010	–	63–67	Indonesia	Dewiyanti (2010)
Rhizophora spp.	–	0.32	–	Ecuador	Twilley et al. (1997)
Rhizophora spp.	–	3.75–4.50	–	Malaysia	Ong et al. (1980)
Sonneratia alba	–	0.95	–	India	Wafar et al. (1997)
S. alba (bare region)	–	–	31	Kenya	Bosire et al. (2005)
S. alba (reforested region)	–	–	6	Kenya	Bosire et al. (2005)
S. alba (natural region)	–	–	3	Kenya	Bosire et al. (2005)
S. alba	–	–	15–22	Malaysia	Ashton et al. (1999)
Four mixed mangrove leaf litter	–	1–1.73	–	Bangladesh	Hoq et al. (2002)
Four mixed mangrove leaf litter (forest region)	0.031–0.0057	–	–	Malaysia	Ashton et al. (1999)
Four mixed mangrove leaf litter (in cleared region)	0.0204–0.0099	–	–	Malaysia	Ashton et al. (1999)
Mangrove leaf litter	0.03–0.01	–	22–69	Mozambique	Boer (2000)
Sedge litter					
Cyperus malaccensis (basal stem)	0.028	–	10.91	India	Sridhar et al. (2010)
Cyperus malaccensis (top stem)	0.013	–	22.85	India	Sridhar et al. (2010)

Table continued on next page.

Table 22.2 (Continued)

Plant species	Decay coefficient (k/day)	Mass loss/day (%)	Half-life (t_{50}, days)	Location	Reference
Cyperus malaccensis (bract)	0.068	–	4.45	India	Sridhar et al. (2010)
Seagrass litter					
Cymodocea serrulata	0.02	–	30–41	Mozambique	Boer (2000)

Decay in cleared and virgin mangroves

A clear-cut difference in the rate of leaf litter decomposition was noted between cleared vs. virgin mangrove regions in Malaysia (Ashton et al. 1999), see Table 22.2. The t_{50} of leaf litters in cleared regions ranged between 22 (*S. alba*) and 122 (*Bruguiera parviflora* and *R. mucronata*) days against 15 (*S. alba*) and 70 (*B. parviflora*) days in virgin forest. The rate of decomposition was faster on mixing four leaf litters in a virgin forest than in a cleared region (32 days vs. 51 days). The *k* value of leaf litter per day in a cleared region (0.031–0.0057) was higher than forest region (0.0204–0.0099). Similar to this study, the rate of decomposition of *S. alba* and *R. mucronata* leaf litter was followed in bare, reforested and natural regions of Gazi Bay in Kenya (Bosire et al. 2005). Bare locations showed lower decay rates of leaf litter than at other locations during the wet season (t_{50} for *R. mucronata*, with values for bare, reforested and natural habitats of 164, 26 and 12 days; corresponding values for *S. alba* were 31, six and three days). This study clearly demonstrates the impoverishment of bare mangrove locations in terms of rate of leaf litter decomposition compared to replanted and natural mangrove habitats.

Diversity of fungi

Several studies have been undertaken to determine the association and role of various groups of microbes responsible for decomposition of leaf litter in mangrove ecosystems (e.g., Lee and Baker 1972; Newell 1976; Ulken 1984; Vrijmoed and Tam 1990; Singh and Steinke 1992; Bremer 1995; Nakagiri et al. 1996; Ito and Nakagiri 1997; Ananda et al. 2008). Leaf litter possessing low C/N ratios usually support rapid colonization by fungi (Mfilinge et al. 2003) and such fungal conditioning improves the palatability of the litter and also serves as cues to the leaf-shredder communities (Rajendran and Kathiresan 1999; Mfilinge and Tsuchiya 2008). The role of fungal endophytes in mangrove litter degradation is not clearly known (Wilson 2000; Suryanarayanan et al. 2001). Their role as plant litter decomposers has been postulated, as they occur in live and senescent tissues where they initiate tissue decomposition prior to the colonization of saprophytes in terrestrial or aquatic habitats (Petrini et al. 1992; Wilson 2000). Endophytic fungi are known to utilize most substrates present in plant cell walls (Carroll and Petrini 1983; Sieber-Canavesi et al. 1991; White et al. 1991; Kumaresan and Suryanarayanan 2002). Based on the assessment of the enzymatic capabilities of endophytic fungi (*Glomerella* sp. and *Pestalotiopsis* sp.) of *R. apiculata*, Kumaresan and Suryanarayanan (2002) showed that the foliar endophytic fungal assemblage was

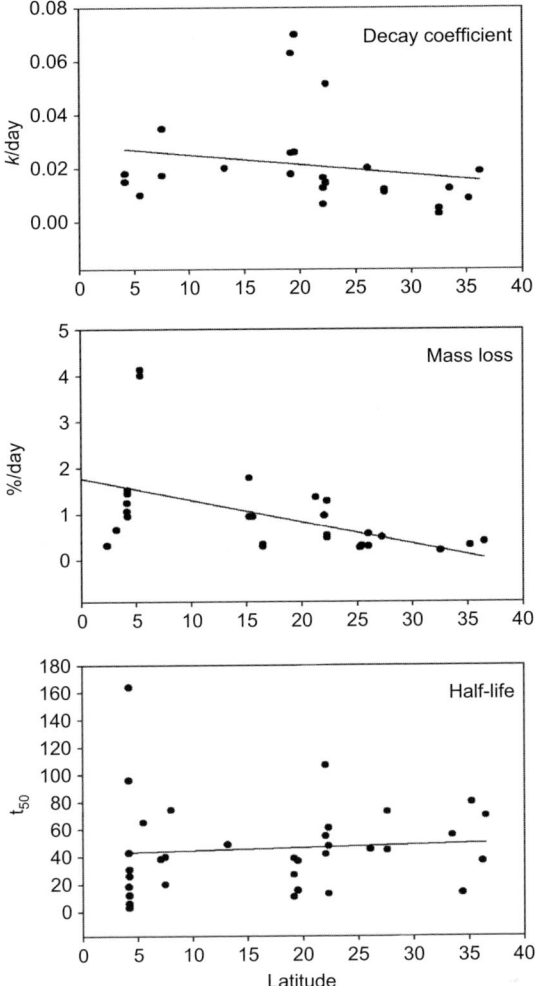

Figure 22.1 The relationship between rate of decomposition of leaf litter based on non-linear regression (as measured by their k/day, % mass loss/day and half-life as t_{50}) against latitude (k, $r^2 = 0.025$, $p = 0.0021$; mass loss, $r^2 = 0.2566$, $p < 0.0001$; half-life, $r^2 = 0.0053$, $p = 0.0003$).

equipped with a complete enzyme array to degrade leaf litter. The number of species and isolates of endophytic fungi that could be recovered from the leaves increased with leaf age and they were elevated or decreased depending on the environment where the leaves settled (e.g., soil or sea water) and influenced the rate of litter decomposition.

Decomposition of sedge litter

Mass loss

Decomposition in the standing dead state predominates in grasses and sedges, especially in wetlands. Among the two substrates (bracts, and basal and top stems) of the sedge

Cyperus malaccensis studied for decomposition, the rate of mass loss was most rapid in bracts, followed by basal stem and top stem (Sridhar et al. 2010). The mass loss rates differed markedly with the substrate and time. In four weeks, 79% mass loss was seen in bracts followed by complete mass loss in five weeks; while in basal and top stems, it took eight weeks for up to 88% and 51% mass loss to occur, respectively. The *k* value ranged between 0.068 (bract) and 0.013 (top stem) and the estimated time of t_{50} ranged between 4.5 days (bract) and 22.9 days (top stem), see Table 22.2. Based on the classification by Petersen and Cummins (1974), the *k* values of this sedge (0.068–0.013) fall into the fast decay category ($k > 0.01$). As a perennial source of organic matter, information on the contribution of *C. malaccensis* to the mangrove energy budgets and food webs is meager. Surprisingly, unlike sedge litter and other vascular macrophytes, decomposition of seagrass was slow (t_{50} of sedge *Cyperus malaccensis* vs. seagrass *Cymodocea serrulata*: 4.5–22.9 vs. 30–41%; Boer 2000; Sridhar et al. 2010). According to Harrison (1989), the rates of decomposition of seagrass litter usually account for <1% dry mass per day.

Decomposition

The initial organic carbon was higher in the bract and top stem, which decreased faster than in the basal stem. The organic carbon loss in *C. malaccensis* was slow in top and basal stems compared to bracts (Sridhar et al. 2010). The C/N ratio was highest in bract, which decreased more steeply than basal stem and top stems. The difference in the chemical constituents of different parts of decomposing *C. malaccensis* indicates that the pattern of fungal colonization and persistence is possibly dependent on the fragile or persistent nature of the substrates and thus showed variation in the bract compared to the stem.

The rate of detritus decomposition in mangroves is dramatically enhanced by the activity of crabs and amphipods (Middleton and McKee 2001). The giant snail *Telescopium telescopium* (Mollusk) is a major detritivore of *C. malaccensis* swards in the southwest mangroves of India (Sridhar et al. 2010). Similarly, *C. malaccensis* serves as a potential food source of detritivorous fish (*Liza macrolepis*) in Taiwan mangroves based on stable isotopic studies (Lin et al. 2007). Further specific studies on the importance of fungi colonizing detritus of *C. malaccensis*, and their role in the nutrition of snails and fishes in mangroves, would be rewarding (see Bärlocher 1981; Sridhar and Sudheep 2011).

Lignocellulose

Fungi

Marine fungi are known to colonize lignocellulosic substrates and thus have great significance in the breakdown of lignocelluloses in marine and mangrove habitats (Hyde and Lee, 1995; Hyde et al. 1998; Bucher et al. 2004). Lignocellulose as heteropolymeric substrate consists of three major components: cellulose, hemicellulose and lignin (Fengel and Wegener 1989; Eaton and Hale 1993).

Luo et al. (2005) categorized 29 isolates of marine/mangrove fungi into three groups based on lignocelluloses degradation:

(i) fungi able to produce three enzymes, such as endoglucanase, xylanase and laccase;
(ii) fungi lacking ability to utilize xylan; and
(iii) fungi without laccase activity.

Soft rot and white rot decay patterns have been reported on wood samples colonized by marine fungi, but brown rot causing-fungi have not been reported in marine habitats as they are unable to colonize wood in waterlogged conditions (Mouzouras 1989).

Production of lignocellulose-degrading enzymes by marine fungi belonging to diverse taxonomic groups has been reported by Rohrmann and Molitoris (1992), Raghukumar et al. (1994), Pointing et al. (1998) and Bucher et al. (2004). Total mineralization of lignin into carbon dioxide has been demonstrated in *Monodictys pelagica* and *Nia vibrissa* by Sutherland et al. (1982). Lignocellulose-degrading fungi in marine habitats have a potential commercial value, especially in bioleaching, biopulping and bioremediation (Raghukumar 2002). The lignin-degrading ability of marine fungi has been confirmed by decolorization of polymeric dyes, e.g., Poly R-478 and Azure B (Raghukumar et al. 1994; Pointing et al. 1998; Bucher et al. 2004) and oxidation of syringaldazine (Bucher et al. 2004). Similarly, using glucose or cellulose as a carbon source, *Julella avicenniae*, *Lignincola laevis*, *Nia vibrissa* and *Stagonospora* sp. totally decolorized polymeric dyes (Pointing et al. 1998). Lignin-modifying enzymes and the decolorization of colored pollutants by obligate and facultative marine fungi have been well documented by Raghukumar (2002).

Enzymes

Many marine fungi have the ability to rapidly utilize cellulose under a wide range of salinities (0–34‰, e.g., *Corollospora maritima*, *Julella avicenniae*, *Lignincola laevis*, *Monodictys pelagica*, *Nia vibrissa* and *Stagonospora* sp.; Rohrmann and Molitoris 1992; Pointing et al. 1998). Interestingly, another set of mangrove fungi, e.g., *Aigialus mangrovei*, *Oceantitis cincinnatula*, *Hydea pygmea*, *Lulworthia grandispora*, *Matsusporium tropicale*, *Passeriniella mangrovei*, *Verruculina enalia* and *Zalerion maritima*, were dominant on immersed woody litter (*A. officinalis* and *R. mucronata*) during the monsoon season. This indicates their role in wood decomposition at low salinities (Sridhar and Maria 2006). Out of 45 marine fungi, 89% showed cellulolytic and 84% showed xylanolytic activities under *in vitro* conditions (Bucher et al. 2004). It is worth noting that *Acrocordiopsis patilii* and *Periconia prolifica* were exclusively cellulolytic, while *Aigialus grandis*, *Marinosphaera mangrovei* and *Salsuginea ramicola* were exclusively xylanolytic.

In vitro cellulolytic activity by some of the non-marine fungi in decomposing leaf litter has been demonstrated by Singh and Steinke (1992). In a southwest Indian mangrove, the occurrence of marine fungi increased after two weeks of submersion of fresh and dry leaf litter of *R. mucronata*, which also resulted in an elevation of proteins, nitrogen and ergosterol (Ananda 2000). Raghukumar et al. (1995) found a peak fungal biomass in *R. apiculata* leaf litter within three to five weeks' immersion (in the second phase).

Similarly, cellulolytic activity increased within six weeks in the decomposition of *Kandelia obovata* leaf litter in Hong Kong (Hodgkiss and Leung 1986). After eight weeks of immersion, fungal sporulation decreased along with protein, nitrogen and ergosterol in the fresh and dry leaf litter of *R. mucronata* (Ananda 2000). Cellulase activity was detected in culture filtrates of the thraustochytrid *Schizochytrium aggregatum* by Bremer (1995). Raghukumar et al. (1994) showed that *Halophytophthora vesicula* degrades phenolics and cellulose by rapidly capturing the labile carbohydrates in water.

During the fourth week of decomposition of *C. malaccensis*, cellulase and xylanase were highest in the basal stem, whereas pectinase peaked in the bracts (Sridhar et al. 2010). Higher enzymatic activity in the first phase of decomposition of the sedge may be attributed to domination by terrestrial fungi already present on the leaves. In the second phase, even though terrestrial fungi were replaced by mangrove fungi, the enzyme activities decreased – perhaps due to low salinity by the inflow of freshwater. The xylanase activity of 11 marine fungi has been studied by Raghukumar et al. (1994). High activity of xylanase by *Aigialus mangrovei*, *Astrosphaeriella mangrovei*, *Gongronella* sp. and *Halorosellinia oceanica* was found. Many fungal isolates of marine origin (e.g., *Aspergillus niger* and *Flavodon flavus*) produced thermostable and wide pH-tolerant cellulase-free xylanases, which are highly useful in the bleaching of paper pulp and textiles (Raghukumar et al. 1996, 1999). Further observations on marine fungal enzymes are detailed in Chapter 21.

Nitrogen, phosphorus and soluble compounds

Nitrogen

Nitrogen enrichment is a common phenomenon during litter decomposition in mangroves. Such enrichment during the decomposition of mangrove wood and leaf litter contributes significantly to the total nitrogen budget of mangroves. The phenomenon of nitrogen enrichment in decomposing litter is not fully understood, although various hypotheses have been postulated to explain immobilization. The nitrogen content of wood reached a peak (four months) prior to the peak of fungal richness and diversity (10 months; Maria and Sridhar 2004; Maria et al. 2006). Similarly, a five-fold increase in the nitrogen of dead trunks of *Rhizophora* spp. was seen in the first two months of decomposition in Australia (Robertson and Daniel 1989). Benner et al. (1990) predicted that when wood breakdown rates are low, absolute amounts of nitrogen exceed that of the initial values due to nitrogen uptake from external sources. A study on the southwest coast of India supports this hypothesis, as the nitrogen peaked when the mass loss of wood was low (Maria et al. 2006). Nitrogen attained a peak around 120 days after the immersion of wood and declined thereafter in both *A. officinalis* and *R. mucronata* woody litter.

Decomposing leaf litter of *R. mucronata* in the laboratory, as well as in mangroves, showed a decrease in organic carbon and ash, while nitrogen, protein and calorific values were elevated (Sumitra et al. 1980). Within seven weeks, the C/N ratio decreased sharply in *R. mucronata* leaf litter (35–42%) in the natural and reforested mangroves of Kenya (Bosire et al. 2005). Compared to leaf litter of *R. mucronata*, leaf litter of *S. alba* possess a lower C/N ratio, which resulted in faster degradation (Rao et al. 1994; Bosire et al. 2005). Similar to woody litter, nitrogen enrichment takes place in leaf litter during decomposition in mangroves. In a southwest Indian mangrove, the nitrogen content of decaying dried leaves of *R. mucronata* doubled within four weeks and subsequently remained constant. Although nitrogen increased up to two weeks in fresh leaf litter, unlike in dried litter, a gradual decrease in nitrogen was seen thereafter (Ananda et al. 2008). In fresh and dry leaf litter of *R. mucronata*, ergosterol and nitrogen content peaked between four and eight weeks. Subsequently, ergosterol decreased but nitrogen was stable in dry litter and decreased in fresh litter, indicating the elevated biomass of higher fungi on dry litter. The nitrogen content of *R. mucronata* leaf litter reached a peak in six weeks during the wet season in a Kenyan mangrove, which was higher than bare and reforested mangrove locations (Bosire et al. 2005). The nitrogen content peaked in five weeks in leaf litter of *S. alba* in a regenerated forest compared to natural and bare locations. Nitrogen content peaked during the 3rd and 4th weeks in natural and bare locations, respectively, however nitrogen content was lower than the reforested location, where it peaked at the 5th week. The total nitrogen content exceeds the initial values during the advanced stages of leaf litter decay, indicating nitrogen immobilization from external sources in mangroves (Benner et al. 1990).

The initial nitrogen content was highest in basal stems of *C. malaccensis* and showed a steady increase in bracts, while it gradually increased in top stem, indicating the differential colonization pattern of fungi and enrichment of nitrogen in this substrate (Sridhar et al. 2010). Enrichment of nitrogen at the advanced stages of decay of sedge litter demonstrates its immobilization by the associated microbes from external sources (Benner et al. 1990).

Phosphorus

Rapid loss of phosphorus was seen within the first two months, followed by gradual elevation during root litter decomposition in Australia (van der Valk and Attiwill 1984). Phosphorus content decreased swiftly at the initial stage of decomposition (two months) and later peaked (13–15 months), which coincided with peaks in fungal richness and diversity (10 months; Maria and Sridhar 2004; Maria et al. 2006). The rapid initial phosphorus loss might be due to leaching effects or due to utilization by fungi for the degradation of wood. The subsequent peak might be the impact of immobilization (or uptake) of phosphorus from colonized fungi from an external source.

Phosphorus content in dry and fresh leaf litter of *R. mucronata* initially decreased and increased thereafter (Ananda et al. 2008). However, phosphorus content in fresh litter was significantly higher than in dry litter during eight weeks of immersion. Within one month, 90% of nitrogen, phosphorus and potassium were released from the fresh leaf litter of *A. marina* in a Sri Lankan estuary, while seven weeks were required to attain 50% loss of dry leaf litter in the mangrove forest (Pinto and Swarnamali 1998). A study carried out in a Thailand mangrove showed an elevation of phosphorus in leaf litter of *R. apiculata* along with the time of decomposition (220% of the initial amount; Nielsen and Andersen 2003). Such elevation of phosphorus in leaf litter was ascribed to the role of humic acids and metal bridging (caused particularly by iron) and its accumulation was five to ten times higher than the initial concentration.

The phosphorus content was higher in bracts than in stems of *C. malaccensis*, which decreased steadily up to two to four weeks followed by an increase and attained the original level within four and seven weeks (Sridhar et al. 2010). Davis et al. (2003) stated that phosphorus content in leaf litter usually decreases during the leaching phase and elevates subsequently due to microbial activities. Steep increases of phosphorus content in bracts of *C. malaccensis* might be due to colonization by fungi, as well as other microbes, due to its fragile nature. In the macrophyte *Juncus effusus* in a freshwater marsh in the

Southeast of the US, the nitrogen and phosphorus contents decreased initially, but increased thereafter (Kuehn et al. 2000, 2008). The seasonal fluctuation of nitrogen and phosphorus was seen in *C. malaccensis* of the Minjiang estuary in China and their concentrations in the above-ground parts were in the order flower > leaf > stem (Zeng et al. 2009), suggesting the differential colonization and biomass of fungi on fragile (flower and leaf) and tough (stem) substrates. Their concentration in spring and winter was higher in live culms than in standing dead culms in autumn and summer. The accumulation of nitrogen and phosphorus in above-ground parts was higher than that of below-ground parts of sedge and reached a peak in autumn and summer, respectively. The N/P ratio reveals that nitrogen serves as a key element that limits the primary productivity of *C. malaccensis* (Sridhar et al. 2010).

Soluble compound

Many of the leachates of mangrove litter (e.g., sugars, phenolics, aliphatic acids and amino acids) serve as nutrients or inhibit fungal colonization or deter detritus feeders.

A steep decline of total phenolics was seen within two months of wood immersion in southwest Indian mangrove (Ananda 2000; Maria et al. 2006). This phenomenon (the loss of phenolics) was also seen in decomposing leaf litter of *A. marina* and *R. mucronata* in mangroves in South Africa (Steinke et al. 1983).

About 30–50% of organic matter of mangrove leaves consists of water-soluble compounds, e.g., tannins and sugars (Cundell et al. 1979). Up to 18% of leachates of *R. mangle* leaf litter (as major fractions of dissolved organic matter) in the Bahamas consists of tannins and other inhibitory phenolics (Benner et al. 1986). In mangroves, up to 42% of such leaf litter leachates incorporated into the food web within two 12 h, which serves as an important source of carbon, nitrogen and energy to the higher trophic levels (Benner et al. 1986). It is known that about 30% of the leachate of mangrove leaf litter was mineralized rapidly into microbial biomass (Benner and Hodson 1985). Decomposition of mangrove leaf litter is reported to be fast due to high leaching rates in the first few weeks of immersion in mangroves (van der Valk and Attiwill 1984; Mfilinge et al. 2005). Imgraben and Dittmann (2008) reported the highest weight loss of leaf litter in the first two weeks; further exposure did not result in considerable weight loss.

The initial quantity of phenolics was higher in dead bracts than in dead basal and top stems of *C. malaccensis* (Sridhar et al. 2010). The phenolics dropped more rapidly in basal stems and bracts than in top stems. The nutrients of *C. malaccensis* (e.g., phenolics, nitrogen and phosphorus) in the first phase of decomposition (one to four weeks) might be a valuable source of nutrients for aquatic fauna. The fungi will benefit by the leachates (e.g., sugars) of the decomposing sedge, and in turn increase their biomass by accumulating nitrogen and phosphorus from the habitat. The dynamics of these constituents differs substantially, however, between fresh and dried leaf litter (Ananda et al. 2008).

Animal remains and fungi

Crab populations play a significant role in the mangrove ecosystem in the shredding, consumption and retention of leaf litter (Alongi 2009). Using markers for fatty acids, it is evident that grapsid crabs supplement their diet by the consumption of fungi and bacteria in addition to mangrove litter (Meziane et al. 2006). Brachyuran crabs are found in high densities (~10.5 individuals/m^2) in Brazilian mangroves (Fratini et al. 2004) and their remains are a substantial source for fungal colonization and degradation. The availability of crab exoskeletons in mangroves serves as a potential source of nutrients for chitin-degrading fungi (e.g., Grant et al. 1996). Higher fungi are also an important component of a wide variety of animal substrata, such as corals, shells (e.g., balanids, bivalves, foraminifers and snails), exoskeletons (e.g., bryozoans, crabs, hydrozoans, shipworms and tunicates), cuttlefish endoskeletons, feathers, snake skin, beetle wings, horse hair and teredinid tunnels (Rees and Jones 1985; Kohlmeyer and Volkmann-Kohlmeyer 1990, 1992; Rosello et al. 1993; Ananda et al. 1998; Ananda and Sridhar 2001).

Dead calcareous, chitin and keratin substrates also constitute potential substrates for colonization and degradation by fungi in mangrove habitats. Calcareous shells, cuttlefish endoskeleton, crab exoskeletons and feathers sampled from three mangroves on the southwest coast of India revealed the occurrence of five ascomycetes and 16 asexual fungi (Ananda and Sridhar 2001). Up to 88% of the feather samples were colonized by 14 fungi. *Corollospora intermedia* was the most frequent fungus (62.5%), followed by *Aspergillus* sp. 1 (34.8%), *Epicoccum nigrum* (33.5%) and *Scolecobasidium* sp. (30.3%). Up to 77% of the cuttlefish endoskeletons on the beaches of the west coast of India harbored marine fungi (Ananda et al. 1998). Arenicolous fungi were abundant on bivalve shells and cuttlefish endoskeletons and the most frequent arenicolous fungus was *C. intermedia*. The animal remains collected from beaches require a prolonged period of incubation when compared with samples from mangroves when assessing fungal colonization (Ananda et al. 1998; Ananda and Sridhar 2001).

Borer activity of shipworms (*Teredo bratschi*) on mangrove stakes in Belize was high due to the abundance of nutrients (ammonia, nitrate and phosphate from guano deposits; Kohlmeyer et al. 1995). In addition, fungal colonization on such substrates likely enhances the nutritive value of woody debris. However, little is known about the role of fungi in the decomposition of animal remains in the sea. Animal cellulose (tunicin) occurs in tunicates and is known to be degraded by marine fungi (Kohlmeyer and Kohlmeyer 1979). Similarly, marine fungi may also invade calcareous substances, such as molluscan shells, the test of barnacles and linings of shipworm burrows (Porter and Lingle 1992; Hyde et al. 1998). It is predicted that thraustochytrids and other oomycetes play an important role in the recycling of elements from animal remains (Hyde et al. 1998).

Conclusions and outlook

Decomposition in mangroves results in the transformation of physical state as well as the composition of coarse particulate organic matter (CPOM). The probable processes in organic matter transformation in a mangrove ecosystem have been conceptualized in Figure 22.2. Mangroves receive CPOM from autochthonous and allochthonous sources. The tree species in mangroves serve as a major source of autochthonous CPOM, while transport from upstream or the ocean serve as an allochthonous source. The major input of CPOM constitutes leaf and woody litter. Production, quality and input of CPOM are influenced by several environmental variables. These also influence the processors of CPOM, especially by microorganisms and shredders.

The community structure, colonization, biomass, reproduction and succession are dependent on the quality of CPOM and also the conditions of the mangroves. Processing of CPOM in mangrove ecosystem is the result of abiotic (environmental) and biotic (microbes

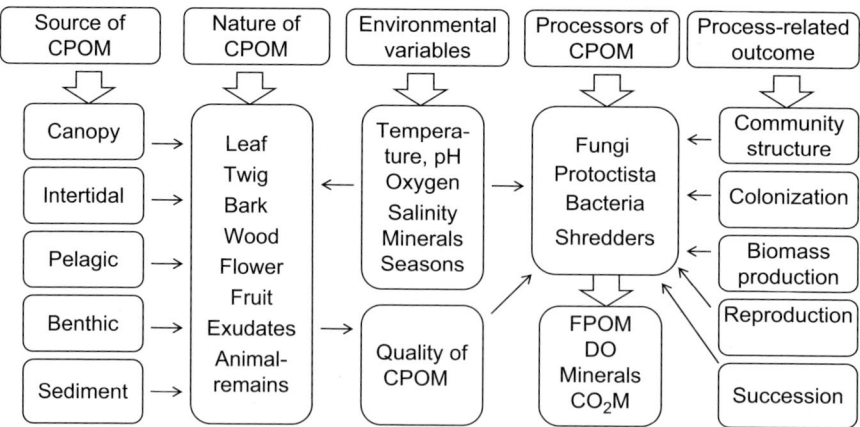

Figure 22.2 Schematic representation of the detritus and detritus processors dominating the decomposition of coarse particulate organic matter in mangroves.

and fauna) variables, which yield biomass (bacteria, fungi and animal), fine particulate organic matter, dissolved organic matter, minerals and carbon dioxide to the energy budget of mangroves.

The proposed model needs fine tuning by mathematical validation and calibration to precisely identify and quantify the organic matter transformation and management of mangroves. The rates of detritus turnover in mangroves are the product of the integration of various sub-processes that are influenced by changes in environment and human interference, and thus the study of detritus decomposition has became a focal issue.

The concept of ecosystem engineering was introduced by Jones et al. (1994, 1997) by defining organisms as ecosystem engineers that directly or indirectly modulate the availability of resources to other species by causing change a in the physical state of biotic materials, and thus modifying/maintaining/creating habitats for other organisms. This concept is highly applicable to mangrove fungi as they convert the quality of mangrove litter into a palatable food resource for mangrove fauna, or fungi themselves serve as a nutritional source supporting the trophic interactions and energy flow. More precisely, mangrove fungi serve as allogenic engineers (meaning that they transform the state of live or dead material into another state) by virtue of mechanical, e.g., softening, or other actions, e.g., enzymatic (Kristensen 2008). Thus, fungal involvement in the degradation of organic materials in a mangrove/marine ecosystem has to be studied in view of structural and functional attributes. Up to 80 species serve as a core group of fungi in different mangroves of the world (Sarma and Hyde 2001; Maria 2003; K.R. Sridhar unpub. obs.). *Dactylospora haliotrepha, H. villosa, Halosarpheia marina, L. laevis* and *Lulworthia grandispora* have been recovered as a core group of fungi in five or more mangrove locations (Sarma and Hyde 2001; Alias and Jones 2009). Thus, it is likely that these fungi serve as keystone species (Hyde et al. 1998), which are indispensable in organic matter turnover as they mediate the tropic interactions in detritus food web.

The C/N ratio is an important yardstick in determining the nutritional quality of mangrove lignocellulosic materials. There is a wide difference in the C/N ratio of organic matter available in mangrove habitats. For example, algae possess a lower C/N ratio (7–10) than animal remains (~10), sedge detritus (15–30), sediments (20–30) and leaf litter (75–100) (Kristensen et al. 1995; Bosire et al. 2005; Kristensen 2008; Lee 2008; Sridhar et al. 2010). Woody litter possesses a much higher C/N ratio than leaf litter, which will be preferred by the fungi (to attain sexual state) as it is persistent in mangroves and also due to their enzymatic capabilities. Enrichment of detritus by fungi and conversion of the detritus status ultimately serves as a valuable resource of protein, nitrogen, lipid and likely growth factors, facilitating energy flow to the higher tropic levels.

Measurement of fungal biomass as an indicator of fungal activity in decaying materials in mangroves has been attempted by many investigators. Studies on fungi on woody litter in the mangrove ecosystem can be improved by adapting various methods of assessment of biomass (Newell 2000), extent of wood decay (Pointing 2000) and quantity of lignocellulolytic enzymes (Pointing 2000). Besides these, direct microscopic observations (West 1988; Raghukumar and Schaumann 1993; Blum et al. 1998) and ergosterol estimation (Newell and Fell 1992; Ananda et al. 2008) in leaves/leaf litter have also been attempted. The decay coefficient (k) and time required for 50% mass loss (half-life, t_{50}) of any organic matter in a mangrove ecosystem are important yardsticks to ascribe the extent and length of residence time in mangrove habitats. Besides applying decay coefficients for the mass loss of organic matter, the dynamics of important constituents of organic matter (loss or gain in C, N and P; ratios of C/N, N/P and C/N/P) help us to understand the process of decomposition more precisely, and in turn appreciate the vulnerability of specific mangroves to environmental change. For example, due to the enrichment of mangroves by guano in Belize (Man-of-War-Cay and Twin Cays), wood processing by microbes and wood borers differed considerably from mangroves elsewhere (Kohlmeyer et al. 1995). Such a natural phenomenon reflects positive or negative impacts on organic matter processing and serves as an early warning of environmental change. These measures also help us to understand the extent of impoverishment (or recovery by rehabilitation) of mangrove locations due to human activities. However, there is no universal method to apply to understand the structure and function of fungi in organic substrates in mangrove habitats. To some extent the dominant fungi (core group) show their major involvement in organic matter processing and transformation; but it is necessary to standardize the methodology to study fungal colonization and decomposition of plant and animal detritus in mangroves. Various methods employed to assess the freshwater fungi in leaf litter are also appropriate for the assessment of mangrove fungi and their activity on organic matter (Tsui et al. 2003; Graça et al. 2005; Sridhar et al. 2008; Ardón et al. 2009). Molecular methods of the assessment of mangrove fungi have been initiated especially to understand their taxonomy, phylogeny and diversity (Pang and Mitchell 2005; Jones et al. 2009). Moreover, molecular studies are providing evidence on the occurrence and role of non-culturable and non-sporulating fungi in the organic substrates of mangrove ecosystems. Future molecular approaches should focus on the role of fungi in organic matter transformation (or bioconversion) in mangroves that drive the energy into the food web.

Acknowledgements

I am grateful to Mangalore University for the facilities to carry out wood and leaf-litter decomposition studies of the southwest Indian mangroves. I also thank Madhu S. Kandikere for assistance in statistical analysis.

References

Albright LJ. *In situ* degradation of mangrove tissues. J Mar Freshw Res. 1976;10:385–389.

Alias SA, Jones EBG. Marine fungi from mangroves of Malaysia. Inst Ocean Earth Studies. 2009;8:109.

Alongi DM. Paradigm shifts in mangrove ecology. In: Perillo GME, Wolanski E, Cahoon DDR, Brinson MM, eds. Coastal wetlands: An integrated ecosystem approach. Elsevier BV, 2009:615–640.

Ananda K. Ecological and biochemical studies on filamentous fungi of West Coast of India, PhD dissertation, Mangalore University, India, 2000.

Ananda K, Prasannarai K, Sridhar KR. Occurrence of higher marine fungi on animal substrates along the west coast of India. Indian J Mar Sci. 1998;27:233–236.

Ananda K, Sridhar KR. Mycoflora on dead animal materials of mangrove habitats of Karnataka Coast, India. Sri Lanka J Aquatic Sci. 2001;6:85–93.

Ananda K, Sridhar KR, Raviraja NS, Bärlocher F. Breakdown of fresh and dried *Rhizophora mucronata* leaves in a mangrove of Southwest India. Wetlands Ecol Manage. 2008;16:1–9.

Ardón M, Pringle CM, Eggert SL. Does leaf chemistry differently affect breakdown in tropical vs. temperate streams? Importance of standardized analytical techniques to measure leaf chemistry. J N Am Benthol Soc. 2009;28:440–453.

Ashton EC, Hogarth PJ, Ormond R. Breakdown of mangrove leaf litter in a managed mangrove forest in peninsular Malaysia. Hydrobiol. 1999;413:77–88.

Bärlocher F. Fungi in the food and in the faeces of *Gammarus pulex*. Trans Br Mycol Soc. 1981;76:14–19.

Bärlocher F. Pitfalls of traditional techniques when studying decomposition of vascular plant remains in aquatic habitats. Limnetica 1997;13:1–11.

Benner R, Hodson RE. Microbial degradation of the leachable and lignocellulosic components of leaves and wood from *Rhizophora mangle* in a tropical mangrove swamp. Mar Eco Prog Ser. 1985;23:221–230.

Benner R, Peele ER, Hodson RE. Microbial utilization of dissolved organic matter from leaves of the red mangrove, *Rhizophora mangle*, in the French Creek Estuary, Bahamas. Est Coast Shelf Sci. 1986;23:607–619.

Benner R, Hatcher PG, Hedges JI. Early diagenesis of mangrove leaves in a tropical estuary: bulk chemical characterization using solid-state ^{13}C NMR and elemental analysis. Geochim Cosmochim Acta. 1990;54:2003–2013.

Blum LK, Mills AL, Zieman JC, Zieman RT. Abundance of bacteria and fungi in seagrasses and mangrove detritus. Mar Ecol Prog Ser. 1998;42:73–78.

Boer WF. Biomass dynamics of seagrasses and the role of mangrove and seagrass vegetation as different nutrient sources for an intertidal ecosystem. Aquat Bot. 2000;66:225–239.

Boonruang P. The degradation rates of mangrove leaves of *Rhizophora apiculata* (Bl.) and *Avicennia marina* (Forsk.) Vierh. at Phuket Island in Thailand. Phuket Mar Biol Cen Res Bull. 1978;26:1–7.

Bosire JO, Dahdouh-Guebas F, Kairo JG, Kazungu J, Dehairs F, Koedam N. Litter degradation and CN dynamics in reforested mangrove plantations at Gazi Bay, Kenya. Biol Conser. 2005;126:287–295.

Boulton AJ, Boon PI. A review of methodology used to measure leaf litter decomposition in lotic environments: time to turn over an old leaf? Aust J Mar Freshw Res. 1991;42:1–43.

Bremer GB. Lower marine fungi (Labyrinthulomycetes) and decay of mangrove leaf litter. Hydrobiol. 1995;295:89–95.

Bucher VVC, Hyde KD, Pointing SB, Reddy CA. Production of wood decay enzymes, mass loss and lignin solubilization in wood by marine ascomycetes and their anamorphs. Fungal Divers. 2004;15:1–14.

Carroll GC, Petrini O. Patterns of substrate utilization by some fungal endophytes from coniferous foliage. Mycologia 1983;75:53–63.

Chale FMM. Degradation of mangrove leaf litter and aerobic conditions. Hydrobiol. 1993;257:177–183.

Cundell AM, Brown MS, Stafford R, Mitchell R. Microbial degradation of *Rhizophora mangle* leaves immersed in the sea. Est Coastal Mar Sci. 1979;9:281–286.

Davis SE III, Corronado-Molina C, Childers DL, Day Jr JW. Temporally dependent C, N, and P dynamics associated with the decay of *Rhizophora mangle* L. leaf litter in oligotrophic mangrove wetlands of the Southern Everglades. Aquat Bot. 2003;75:199–215.

Dewiyanti I. Litter decomposition of *Rhizophora stylosa* in Sabang-Wehsland, Aceh, Indonesia – evidence for mass loss and nutrients. Biodiversitas 2010;11:139–144.

Dick TM, Osunkoya OO. Influence of tidal restriction floodgates on decomposition of mangrove litter. Aquat Bot. 2000;68:273–280.

Dick TM, Streever WJ. Decomposition of *Avicennia marina* on an iron-smelting slag substrate. Austral Ecol. 2001;26:127–131.

Eaton RA, Hale MDC. Wood decay pests and prevention. London: Chapman and Hall, 1993.

Feeney N, Curran PMT, O'Muircheartaigh IG. Biodeterioration of woods by marine fungi *Chaetomium globosum* in response to an external nitrogen source. Internat Biodet Biodegrad. 1992;29:123–133.

Fell JW, Cefalu RC, Master IM, Tallman AS. Microbial activities in the mangrove (*Rhizophora mangle*) leaf detrital system. In: Walsh GE, Snedaker SC, Teas HJ, eds. Proceedings of the international symposium in biology and management of mangroves. Florida: University of Florida, 1975:551–667.

Feller IC, Mathis WN. Primary herbivory by wood-boring insects along a tree-height architectural gradient of *Rhizophora mangle* L. in Belizean mangrove swamps. Biotropica 1997;29:440–451.

Fengel D, Wegener G. Wood: Chemistry, ultrastructure, reactions. New York: Walter De Gruyter, 1989.

Fratini S, Vigiani V, Vannini M, Cannicci S. *Terebralia palustris* (Gastropoda; Potamididae) in a Kenyan mangal: Size structure, distribution and impact on the consumption of leaf litter. Mar Biol. 2004;144:1173–1182.

Gessner MO, Chauvet E, Dobson M. A perspective on leaf litter breakdown in streams. Oikos 1999;85:377–384.

Goulter PFE, Allaway WG. Litterfall and decomposition in a mangrove stand, Avicennia marina (Forsk.) Vierh., in Middle Harbour, Sydney. Aust J Mar Freshw Res. 1979;30:541–546.

Graça MAS, Bärlocher F, Gessner MO. Methods to study litter decomposition – A practical guide. Netherlands: Springer, 2005.

Grant WD, Atkinson M, Burke B, Molloy C. Chitinolysis by the marine ascomycete *Corollospora maritima* Werdermann: Purification and properties of a chitobiosidase. Bot Mar. 1996;39:177–186.

Harrison PG. Detrital processing in seagrass system: A review of factors affecting decay rates, remineralization and detritivory. Aquat Bot. 1989;35:263–288.

Hegazy AL. Perspectives on survival, phenology, litter fall and decomposition, and caloric content of *Avicennia marina* in the Arabian Gulf region. J Arid Env. 1998;40:417–429.

Hodgkiss IJ, Leung HC. Cellulase associated with mangrove leaf decomposition. Bot Mar. 1986;29:467–469.

Hoq ME, Islam ML, Paul HK, Ahmed SU. Decomposition and seasonal changes in nutrient constituents in mangrove litter of Sundarbans mangrove, Bangladesh. Indian J Mar Sci. 2002;31:130–135.

Hossain M, Othman S. Degradation rate of leaf litter of *Bruguiera parviflora* of mangrove forest of Kuala Selangor, Malaysia. Indian J For. 2005;28:144–149.

Hudson HJ. Succession of micro-fungi on aging leaves of *Saccharum officinarum*. Trans Br Mycol Soc. 1962;45:395–423.

Hyde KD. Observations on the vertical distribution of marine fungi on *Rhizophora* spp. at Kampong Danau mangrove Brunei. Asian Mar Biol. 1988;5:77–81.

Hyde KD. A study of vertical zonation of intertidal fungi on *Rhizophora apiculata* at Kampong Kapok mangrove, Brunei. Aquat Bot. 1990a;36:255–262.

Hyde KD. A comparison of the intertidal mycota of five mangrove tree species. Asian Mar Biol. 1990b;7:93–107.

Hyde KD. Fungal colonization of *Rhizophora apiculata* and *Xylocarpus granatum* poles in Kampong Kapok mangrove, Brunei. Sydowia 1991;43:31–38.

Hyde KD, Jones EBG. Marine mangrove fungi. PSZNI Mar Ecol. 1988;9:15–33.

Hyde KD, Lee SY. Ecology of mangrove fungi and their role in the nutrient cycling: what gaps occur in our knowledge? Hydrobiol. 1995;295:107–118.

Hyde KD, Jones EBG, Leaãno E, Pointing SB, Poonyth D, Vrijmoed LLP. Role of fungi in marine ecosystems. Biodivers Conserv. 1998;7:1147–1161.

Imgraben S, Dittmann S. Leaf litter dynamics and litter consumption in two temperate South Australian mangrove forests. J Sea Res. 2008;59:83–93.

Ito T, Nakagiri A. A mycofloral study on mangrove mud in Okinawa, Japan. Inst Ferment Osaka Res Comm. 1997;18:32–39.

Jones CG, Lawton JH, Shachak M. Organisms as ecosystem engineers. Oikos 1994;69:373–386.

Jones CG, Lawton JH, Shachak M. Positive and negative effects of organisms as physical ecosystem engineers. Ecology 1997;78:1946–1957.

Jones EBG, Sakayaroj J, Suetrong S, Somrithipol S, Pang KL. Classification of marine Ascomycota, anamorphic taxa and Basidiomycota. Fungal Divers. 2009;35:1–203.

Karamchand KS, Sridhar KR, Bhat R. Diversity of fungi associated with estuarine sedge *Cyperus malaccensis* Lam. J Agric Technol. 2009;5:111–227.

Kohlmeyer J, Kohlmeyer E. Marine mycology: the higher fungi. New York: Academic Press, 1979.

Kohlmeyer J, Volkmann-Kohlmeyer B. New species of *Koralionastes* (Ascomycotina) from the Caribbean and Australia. Can J Bot. 1990;68:1554–1559.

Kohlmeyer J, Volkmann-Kohlmeyer B. Two Ascomycotina from coral reef in the Caribbean and Australia. Crypto Bot. 1992;2:367–374.

Kohlmeyer J, Bebout B, Volkmann-Kohlmeyer B. Decomposition of mangrove wood by marine fungi and teredinids in Belize. PSZNI Mar Ecol. 1995;16:27–39.

Kohlmeyer J, Volkmann-Kohlmeyer B. The biodiversity of fungi on *Juncus roemerianus*. Mycol Res. 2001;105:1411–1412.

Kristensen E. Mangrove crabs as ecosystem engineers; with emphasis on sediment processes. J Sea Res. 2008;59:30–43.

Kristensen E, Holmer M, Banta GT, Jensen MH, Hansen K. Carbon, nitrogen and sulfur cycling in sediments of the Ao Nam Bor mangrove forest, Phuket, Thailand: A review. Phuket Mar Biol Cent Res Bull. 1995;60:37–64.

Kuehn KA. The role of fungi in the decomposition of emergent wetland plants. In: Sridhar KR, Bärlocher F, Hyde KD, eds. Novel techniques and ideas in mycology. Kunming, PR China: Fungal Diversity Press, 2008:19–41.

Kuehn KA, Lemke MJ, Suberkropp K, Wetzel RG. Microbial biomass and production associated with decaying leaf litter of the emergent macrophyte *Juncus effusus*. Limnol Oceanogr. 2000;45:862–870.

Kumaresan V, Suryanarayanan TS. Endophyte assemblage in young, mature and senescent leaves of *Rhizophora apiculata*: Evidence for the role of endophytes in mangrove litter degradation. Fungal Divers. 2002;9:81–91.

Lee BKH, Baker GE. An ecological study of the soil microfungi in a Hawaiian mangrove swamp. Pac Sci. 1972;26:1–10.

Lee SY. Litter production and turnover of the mangrove *Kandelia candel* (L.) Druce in a Hong Kong tidal shrimp pond. Est Coast Shelf Sci. 1989;29:75–87.

Lee SY. Mangrove outwelling: a review. Hydrobiol. 1995;295:203–212.

Lee SY. Mangrove macrobenthos: Assemblages, services, and linkages. J Sea Res. 2008;59:16–29.

Leightley LE. Wood decay activities of marine fungi. Bot Mar. 1980;23:387–395.

Leong WF, Tan TK, Jones EBG. Fungal colonization of submerged *Bruguiera cylindrica* and *Rhizophora apiculata* wood. Bot Mar. 1991;34:69–76.

Lin H-J, Kao W-Y, Wang YT. Analyses of stomach contents and stable isotopes reveal food sources of estuarine detritivorous fish in tropical/subtropical Taiwan. Est Coast Shelf Sci. 2007;73:527–537.

Logo AE, Snedaker SC. The ecology of mangroves. Ann Rev Ecol Syst. 1974;5:39–64.

Lu CY, Lin P. Studies of litter fall and decomposition of *Bruguiera sexangula* (Lour.) Poir, community on Hainan Island. Chin Bull Mar Sci. 1990;47:459–463.

Luo W, Vrijmoed LLP, Jones EBG. Screening of marine fungi for lignocellulose-degrading enzyme activities. Bot Mar. 2005;48:379–386.

Mackey AP, Smail G. The decomposition of mangrove litter in a subtropical mangrove forest. Hydrobiol. 1996;332:93–98.

Maria GL. Studies on the mangrove mycoflora of West Coast of India. PhD Dissertation, Mangalore University, India, 2003.

Maria GL, Sridhar KR. Fungal colonization of immersed wood in mangroves of the southwest coast of India. Can J Bot. 2004;82:1409–1418.

Maria GL, Sridhar KR, Bärlocher F. Decomposition of dead twigs of *Avicennia officinalis* and *Rhizophora mucronata* in a mangrove in southwest India. Bot Mar. 2006;49:450–455.

Meziane T, d'Agata F, Lee SY. Fate of mangrove organic matter along a subtropical estuary: small-scale exportation and contribution to the food of crab communities. Mar Ecol Prog Ser. 2006;312:15–27.

Mfilinge PL, Tsuchiya M. Effect of temperature on leaf litter composition by grapsid crabs in a subtropical mangrove (Okinawa, Japan). J Sea Res. 2008;59:94–102.

Mfilinge PL, Atta N, Tsuchiya M. Nutrient dynamics and leaf litter decomposition in a subtropical mangrove forest at Oura Bay, Okinawa, Japan. Trees 2002;16:172–180.

Mfilinge PL, Meziane T, Bachok Z, Tsuchiya M. Fatty acids in decomposing mangrove leaves: microbial activity, decay and nutritional quality. Mar Ecol Prog Ser. 2003;265:97–105.

Mfilinge PL, Meziane T, Bachok Z, Tsuchiya M. Litter dynamics and particulate organic matter outwelling from a subtropical mangrove in Okinawa Island, South Japan. Est Coast Shelf Sci. 2005;63:301–313.

Middleton BA, McKee KL. Degradation of mangrove tissues and implications for peat formation in Belizean island forests. J Ecol. 2001;89:818–828.

Mouzouras R. Patterns of timber decay caused by marine fungi. In: Moss ST, ed. The biology of marine fungi. Cambridge Cambridge: University Press, 1986:341–353.

Mouzouras R. Soft rot decay of wood by marine fungi. J Inst Wood Sci. 1989;11:193–201.

Mouzouras R, Jones EBG, Venkatasamy R, Moss ST. Decay of wood by microorganisms in marine environments. Rec Ann Con BWPA. 1986:27–44.

Nakagiri A, Newell SY, Ito T, Tan TK. Biodiversity and ecology of the oomycetous fungus, *Halophytophthora*. In: Turner IM, Diong CH, Lim SSL, Ng RKL, eds. Biodiversity and the dynamics of ecosystems. Tokyo: Intern Network DIVERSITAS in Western Pacific and Asia, 1996: 273–280.

Neilson MJ, Richards GN. Chemical-composition of degrading mangrove leaf litter and changes produced after consumption by mangrove crab *Neosarmatium smithi* (Crustacea, Decapoda, Sesarmidae). J Chem Ecol. 1989;15:1267–1283.

Newell SY. Mangrove fungi: the succession in the mycoflora of red mangrove (*Rhizophora mangle*) seedlings. In: Jones EBG, ed. Recent advances in aquatic mycology. London: Eleck, 1976:51–91.

Newell SY. Establishment and potential impacts of eukaryotic mycelial decomposers in marine/terrestrial ecotones. J Exp Mar Biol Ecol. 1996;200:187–206.

Newell SY. Methods for determining biomass and productivity of mycelia marine fungi. In: Hyde KD, Pointing SB, eds. Marine mycology – A Practical approach. Hong Kong: Fungal Diversity Press, 2000:69–91.

Newell SY, Fell JW. Ergosterol content of living and submerged decaying leaves and twigs of red mangrove. Can J Bot. 1992;38:979–982.

Newell SY, Fell JW. Competition among mangrove oomycetes and between oomycetes and other microbes. Aquat Microb Ecol. 1997;12:21–28.

Nielsen T, Andersen F. Phosphorus dynamics during decomposition of mangrove (*Rhizophora apiculata*) leaves in sediments. J Exp Mar Biol Ecol. 2003;293:73–88.

Ong EJ, Gong WK, Wong CH. Ecological survey of the Sungai Merbok estuarine mangrove. Penang, Malaysia: Universiti Sains Malayaia, 1980.

Pang KL, Mitchell JI. Molecular approaches for assessing fungal diversity in marine substrata. Bot Mar. 2005;48:332–347.

Petersen RC, Cummins KW. Leaf processing in woodland stream. Freshw Biol. 1974;4:343–368.

Petrini O, Sieber TN, Toti L, Viret O. Ecology, metabolite production and substrate utilization in endophytic fungi. Natural Toxins 1992;1:185–196.

Pinto L, Swarnamali PA. Decomposition and nutrient release by *Avicennia marina* (Forsk.) Vierh. in a mangrove islet and a brushpile in Negombo Estuary. J Natn Sci Coun Sri Lanka. 1998;25: 211–224.

Pointing SB. Lignocellulolytic enzyme assays. In: Hyde KD, Pointing SB, eds. Marine mycology – A practical approach. Hong Kong: Fungal Diversity Press, 2000:137–157.

Pointing SB, Vrijmoed LLP, Jones EBG. A qualitative assessment of lignocelluloses degrading enzyme activity in marine fungi. Bot Mar. 1998;41:293–298.

Porter D, Lingle WL. Endolithic thraustochytrid marine fungi from planted shell fragments. Mycologia 1992;84:289–299.

Raghukumar C. Bioremediation of coloured pollutants by terrestrial versus facultative marine fungi. In: Hyde KD, ed. Fungi in marine environment. Hong Kong: Fungal Diversity Press, 2002:317–344.

Raghukumar C, Raghukumar S, Chinaraj A, Chandramohan D, D'Souza TM, Reddy CA. Laccase and other lignocellulose modifying enzymes of marine fungi isolated form the coast of India. Bot Mar. 1994;35:512–527.

Raghukumar C, Chandramohan D, Michel FC Jr, Reddy CA. Degradation of lignin and decolorization of paper mill bleach plant effluent (BPE) by marine fungi. Biotechnol Lett. 1996;18:105–108.

Raghukumar C, D'Souza TM, Thorn RG, Reddy CA. Lignin-modifying enzymes of *Flavodon flavus*, a basidiomycete isolated from a coastal marine environment. Appl Environ Microbiol. 1999;65:2103–2111.

Raghukumar S, Schaumann K. An epifluorescence microscopy method for direct detection of the fungi-like marine protists, the thraustochytrids. Limnol Oceanogr. 1993;38:182–187.

Raghukumar S, Sathe-Pathak V, Sharma S, Raghukumar C. Thraustochytrid and fungal component of marine detritus III. Field studies on decomposition of leaves of the mangrove *Rhizophora apiculata*. Aquat Microb Ecol. 1995;9:117–125.

Rajendran N, Kathiresan K. Do decomposing leaves of mangroves attract fishes? Curr Sci. 1999;77:972–976.

Rao RG, Woitchik AF, Goeyens L, Van Riet A, Kazungu J, Dehairs F. Carbon, nitrogen contents and stable isotope abundance in mangrove leaves from an east African coastal lagoon (Kenya). Aquat Bot. 1994;47:175–183.

Rees G, Jones EBG. The fungi of coastal sand dune system. Bot Mar. 1985;28:213–220.

Robertson AI. Decomposition of mangrove leaf litter in tropical Australia. J Exp Mar Biol Ecol. 1988;116:235–247.

Robertson AI, Daniel PA. Decomposition and annual flux of detritus from fallen timber in tropical mangrove forests. Limnol Oceanogr. 1989;34:640–646.

Rohrmann S, Molitoris HP. Screening for wood decay enzymes in marine fungi. Can J Bot. 1992;70:2116–2123.

Romero LM, Smith III TJ, Fourqurean JW. Changes in mass and nutrient content of wood during decomposition in a south Florida mangrove forest. Ecology 2005;93:618–631.

Rosello MA, Descals E, Cabrer B. *Nia epidermoidea*, a new marine gasteromycete. Mycol Res. 1993;97:68–70.

Sarma VV, Hyde KD. A review on frequently occurring fungi in mangroves. Fungal Divers. 2001;8:1–34.

Sessegolo GC, Lana PC. Decomposition of *Rhizophora mangle*, *Avicennia schaueriana* and *Laguncularia racemosa* leaves in a mangrove of Paranguá Bay (South-eastern Brazil). Bot Mar. 1991;34:285–289.

Sieber-Canavesi F, Petrini O, Sieber TN. Endophytic *Leptostroma* species on *Picea abies, Abies alba* and *Abies balsamea:* a cultural, biochemical, and numerical study. Mycologia 1991;83:89–96.

Singh N, Steinke TD. Colonization of leaves of *Bruguiera gymnorrhiza* (Rhizophoraceae) by fungi, and *in vitro* cellulolytic activity of the isolates. S Afr J Bot. 1992;58:525–529.

Sridhar KR, Maria GL. Fungal diversity on woody litter of *Rhizophora mucronata* in a southwest Indian mangrove. Indian J Mar Sci. 2006;35:318–325.

Sridhar KR, Sudheep NM. Do the tropical freshwater fishes feed on aquatic fungi? Front Agric China. 2011;5:77–86.

Sridhar KR, Bärlocher F, Hyde KD. Novel techniques and ideas in mycology. Fungal Divers Res. Series # 20, Kunming, PR China: Fungal Diversity Press, 2008.

Sridhar KR, Karamchand KS, Sumathi P. Fungal colonization and breakdown of sedge (*Cyperus malaccensis* Lam.) in a southwest mangrove, India. Bot Mar. 2010;53:525–533.

Steinke TD, Ward CJ. Degradation of mangrove leaf litter in the St. Lucia Estuary as influenced by season and exposure. S Afr J Bot. 1987;53:323–328.

Steinke TD, Naidoo G, Charles LM. Degradation of mangrove leaf and stem tissues *in situ* in Mgeni Estuary, South Africa. In: Teas HJ, ed. Tasks for vegetation science, Volume # 8. The Hague: W Junk Publishers, 1983:141–149.

Sumitra·V, Ramadhas V, Kumari KL, Royan JP. Biochemical changes and energy content of the mangrove, *Rhizophora mucronata* leaves during decomposition. Indian J Mar Sci. 1980;9:120–123.

Suryanarayanan TS, Kumaresan V, Johnson JA. Fungal endophytes: the tropical dimension. In: Misra JK, Horn BW, eds. Trichomycetes and other fungal groups. Enfield, New Hampshire: Science Publishers, 2001:197–207.

Sutherland JB, Crawford DL, Speedie MK. Decomposition of 14C-labelled maple and spruce lignin by marine fungi. Mycologia 1982;74:511–513.

Tam NFY, Vrijmoed L, Wong YS. Nutrient dynamics associated with leaf decomposition in a small subtropical mangrove community in Hong Kong. Bull Mar Sci. 1990;47:68–78.

Tam NFY, Wong YS, Lan CY, Wang LN. Litter production and decomposition in a subtropical mangrove swamp receiving wastewater. J Exp Mar Biol Ecol. 1998;226:1–18.

Tsui CKM, Hyde KD, Hodgkiss IJ. Methods for investigating the biodiversity and distribution of freshwater ascomycetes and anamorphic fungi on submerged wood. In: Tsui CKM, Hyde KD, eds. Freshwater mycology. Hong Kong: Fungal Diversity Press, Freshwater mycology 2003: 195–209.

Twilley RR, Lugo AE, Patterson-Zucca C. Litter production and turnover in basin mangrove forests in southwest Florida. Ecology 1986;67:670–683.

Twilley RR, Pozo M, Garcia VH, Rivera VH, Zambrano MR, Bodero A. Litter dynamics in riverine mangrove forests in the Guayas River Estuary, Ecuador. Oecologia 1997;111:109–122.

Ulken A. The fungi of the mangle ecosystem. In: Por FD, Dor I, eds. Hydrobiology of the mangle. The Hague: W Junk, 1984:27–33.

van der Valk AG, Attiwill DM. Decomposition of leaf and root litter of *Avicennia marina* at Westernport Bay, Victoria, Australia. Aquat Bot. 1984;18:205–221.

Vrijmoed LLP, Tam NFY. Fungi associated with leaves of *Kandelia candel* (L.) Druce in litter bags on the mangrove floor of a small subtropical mangrove community in Hong Kong. Bull Mar Sci. 1990;47:261–267.

Wafar S, Untawale AG, Wafar M. Litterfall and energy flux in a mangrove ecosystem. Est Coast Shelf Sci. 1997;44:111–124.

West AW. Specimen preparation, stain type, and extraction and observation procedures as factors in the estimation of soil mycelia lengths and volumes by light microscopy. Biol Fert Soil. 1988;7:88–94.

White IF Jr, Breen JP, Morgan-Jones G. Substrate utilization in selected *Acremonium, Atkinsonella* and *Balansia* species. Mycologia 1991;83:601–610.

Wieder RK, Lang GE. A critique of the analytical methods used in examining decomposition data obtained from litter bags. Ecology 1982;63:1636–1642.

Wilson D. Ecology of woody plant endophytes. In: Bacon CW, White IF Jr, eds. Microbial endophytes. New York: Marcel Dekker, 2000:389–420.

Woitchik AF, Ohowa B, Kazungu JM, Rao RG, Goeyens L, Dehairs F. Nitrogen enrichment during decomposition of mangrove leaf litter in an east African coastal lagoon (Kenya): Relative importance of biological nitrogen fixation. Biogeochem. 1997;39:15–35.

Woodroffe C. Litter production and decomposition in the New Zealand mangrove, *Avicennia marina* var. *resinifera*. NZ J Mar Freshw Res. 1982;16:179–188.

Woodroffe C. Studies of a mangrove basin, Tuff Crater, New Zealand: I. Mangrove biomass and production of detritus. Est Coast Shelf Sci. 1985;20:265–280.

Zeng CS, Zhang LH, Tong C. Seasonal variation of nitrogen and phosphorus concentration and accumulation of *Cyperus malaccensis* in Minjiang River estuary. Chinese J Ecol. 2009;28: 788–794.

23 Culture collections and maintenance of marine fungi

Akira Nakagiri

Safe and long-term preservation of fungal cultures without genetic change is essential for basic and applied research in mycology. Marine fungi, and fungal-like organisms, are potential genetic resources because they have been less explored than terrestrial fungi. A wide taxonomic range of marine fungi including chytridiomycetes, ascomycetes, basidiomycetes and anamorphic fungi, and fungal-like organisms, such as oomycetes and labyrinthulids, are able to grow on artificial media, except for some biotrophic (parasitic and symbiotic) fungi. The basic and general methods for the culture of a wide range of marine fungi and fungal-like organisms, and also the preservation of cultures by freezing for a long period, are mentioned below. Public culture collections maintaining cultures are also introduced.

Culture collections maintaining and providing cultures

Cultures of marine fungi and fungal-like organisms are available from several culture collections in the world. Many of them are maintained in the four culture collections (Table 23.1) in which cultures are preserved in a good condition, mainly by freezing. The cultures are distributed to researchers on order, while culture deposition from researchers is also accepted. The largest number of cultures of marine fungi is maintained in the BIOTEC Culture Collection in Bangkok, Thailand, and many are mangrove fungi. In the NITE Biological Resource Center (NBRC) culture collection in Japan, all strains are well-characterized with DNA sequence data of the rDNA genes, which can be downloaded from the website's online catalogue. When searching for a marine fungal strain, StrainInfo (http://www.straininfo.net/) is a convenient and useful site through which you can find the strain and the collection maintaining that isolate.

Cultivation of marine fungi and fungal-like organisms

Filamentous fungi

Most saprophytic filamentous marine fungi are able to grow on ordinary fungal media made up with seawater instead of distilled water. For marine ascomycetes and basidiomycetes, Seawater Starch Agar (SWSA: soytone 0.1%, soluble starch 1.0% and agar 1.5% in diluted seawater [2% salinity], pH 8.2) can be used to induce the sexual reproduction of lignicolous species. Many produce ascomata or basidiomata in the medium, or on the surface of the glass vessels in the case of common arenicolous species such as *Corollospora* spp., after incubation for about one month (Nakagiri and Tubaki 1982). For asexual (anamorphic) marine fungi, Cornmeal Yeast Extract Seawater Agar (CMYSWA: cornmeal extract 0.2%, yeast extract 0.1% and agar 1.5% in seawater [2% salinity], pH 7.0–7.5) is

Table 23.1 Major culture collections of marine fungi and fungal-like organisms.

Collection name	Details of collection
BIOTEC Culture Collection (BCC)	BIOTEC Central Research Unit National Center for Genetic Engineering and Biotechnology (BIOTEC) 113 Thailand Science Park, Phaholyothin Road, Klong 1, Klong Luang, Pathumthani 12120 Thailand Tel: +66 2 5646700 ext. 3336 Fax: +66 2 5646707 E-mail: bcc@biotec.or.th http://www.biotec.or.th/bcc/index.asp
NITE Biological Resource Center (NBRC)	Biological Resource Center (NBRC), National Institute of Technology and Evaluation 2–5–8 Kazusakamatari, Kisarazu-shi, Chiba, 292–0818 Japan Tel: +81 438 20 5763 Fax: +81 438 52 2329 E-mail: nbrc@nite.go.jp http://www.nbrc.nite.go.jp/e/index.html
Centraalbureau voor Schimmelcultures (CBS)	CBS-KNAW Fungal Biodiversity Centre, P.O. Box 85167, 3508 AD, Utrecht, The Netherlands Tel: +31 (0)30 21 22 600 Fax: +31 (0)30 25 12 097 E-mail: g.verkleij@cbs.knaw.nl http://www.cbs.knaw.nl/collections/DefaultInfo.aspx?Page=Home
American Type Culture Collection (ATCC)	ATCC.ORG 10801 University Boulevard Manassas, Virginia 20110–2209 USA http://www.atcc.org/CulturesandProducts/Microbiology/tabid/175/Default.aspx

a useful medium on which abundant conidial production will occur on sparse mycelia. Unlike freshwater aquatic hyphomycetes, marine anamorphic fungi produce conidia on the surface of the agar media without submersing hyphae in water. For preparation of the agar medium, both natural seawater and artificial seawater can be used to prepare 2% salinity (salinity of oceanic seawater is 3.5%). Most marine fungi adapt to a wide range of salinities (1–4%) and the salinity of the medium may become more concentrated due to drying out of the medium over a prolonged incubation period. A temperature of 15–30°C (especially 20–25°C for many temperate species) is suitable for their growth, but can be modified according to the climate of the natural habitats of the strain.

Oomycetes

In marine and brackish water environments, some oomycetes including *Halophytophthora* and *Pythium* are saprophytic on plant material or parasites of algae. They produce hyphae resembling true fungi on an agar medium, e.g., Vegetable Juice Seawater Agar (VJSWA: filtered vegetable juice 10 or 20% saturated with 0.3% $CaCO_3$ and agar 1.5% in diluted seawater [2% salinity], pH 7.0; Nakagiri 1993). Zoosporangium formation and zoospore release are induced by submerging the mycelia into seawater or diluted seawater, though oospores are often produced in the agar medium without submersion.

Saprophytic oomycetes grow and decline rapidly on agar medium so that regular subculturing is required within one to two weeks. Preservation by freezing is therefore crucial if cultures are to be maintained for a long period of time. Algal-parasitic oomycetes, e.g., *Olpidiopsis porphyrae*, can be cultured in the presence of living thalli of *Porphyra* spp. (Rhodophyta). Such a biotrophic species can only be grown by two-membered culture with its host; therefore it is practically impossible to keep the strain in an active state for a long time. Inactivated preservation method including freezing is essential for maintaining such strains. Freezing preservation in liquid nitrogen, after cooling in a programmable freezer at the constant cooling rate of –1°C/min has been successfully applied to maintain cultures of saprophytic and parasitic marine oomycetes (Nishii and Nakagiri 1991, A. Nakagiri, unpublished data).

Labyrinthulids

Labyrinthula spp. have been known to parasitize the sea grass *Zostera marina* or other marine microorganisms like yeasts, bacteria and diatoms (Johnson and Sparrow 1961). *Labyrinthula* spp. isolates can also be cultured axenically on Seawater Serum Agar (SSA: 1% horse serum, agar 1.2% in diluted seawater [2% salinity]) with a seawater overlay on the surface of the agar medium. The growth of the spindle-shaped gliding cells in the ectoplasmic network is often retarded during repeated subculturing. Therefore, monoxenic culture with marine yeasts or bacteria, which serve as living nutrients for *Labyrinthula*, can be used to maintain cultures. If a marine bacterium, e.g., *Vibrio alginolyticus,* is co-cultured with *Labyrinthula*, the two-membered culture can be grown on Trypticase Yeast Extract Agar (TYS: trypticase peptone or polypeptone 0.01%, yeast extract 0.10%, agar 1.20% in diluted seawater [2% salinity]), or other poor nutrient seawater agar media that prevent overgrowth of the food bacteria and keep a good balance between the host and the parasite. The monoxenic culture is preserved by liquid nitrogen storage after the programmed freezing.

The saprophytic thraustochytridian members, such as *Schizochytrium* and *Aurantiochytrium*, can easily be cultured axenically on medium-H (Honda et al. 1998: glucose 0.20%, yeast extract 0.02%, monosodium glutamate 0.05% and agar 1.20% in seawater) with a seawater overlay in which zoospores swim to expand the colony. These axenic cultures can be well preserved by freezing.

Preservation methods

For long-term preservation of cultures, drying methods and freezing methods are applicable to a wide range of microorganisms.

Drying preservation methods such as lyophilization (freeze-drying) and L-drying (liquid phase drying) are only applicable to some anamorphic marine fungi that produce a large number of relatively small conidia in culture, e.g., *Dendryphiella arenaria*.

The freezing method should be applicable to most marine fungal cultures. The major preservation method for living cultures of marine fungi is freezing and the methods applicable to each group are described below.

Procedures

Filamentous fungi

Most filamentous marine fungi belonging to the ascomycetes, basidiomycetes and anamorphic fungi tolerate freezing, so they can easily be preserved in an electric deep freezer at –80°C or in a liquid nitrogen tank at ~–170°C. The procedure for freezing preservation is also used for terrestrial fungi, except using seawater medium for pre-culture and reviving cultivation:

(i) *Pre-culture* – strains are grown on the agar plate of SWSA or CMYSWA at the appropriate temperature. From the edge of the colony, agar discs (6~7 mm in diameter) containing mycelium are cut with a plastic straw or cork borer. The agar discs are put into a freezing tube, e.g., a 2 ml Cryotube (Nalge Nunc International, USA) containing 1 ml of cryoprotectant (10% glycerol solution). The tubes are placed in refrigerator (4°C) for two days before freezing.

(ii) *Freezing* – the tubes are packed in a freezing (paper) box, then placed into an electric deep freezer (–80°C).

(iii) *Viability check* – one of the frozen tubes of each strain is thawed by bathing in a 30°C water bath for three minutes and the agar discs are inoculated onto fresh agar medium to determine the survival of the strain, i.e., to check whether hyphae grow out from the discs.

(iv) *Storage* – the frozen tubes are stored in a deep freezer or they can be transferred after freezing into a liquid nitrogen storage tank (~–170°C) for more stable preservation.

Oomycetes

Cultures of most oomycetes are sensitive to freezing, but it is known that they survive the freezing process if they are frozen at a controlled cooling rate and stored in a liquid nitrogen storage tank. The procedures for freezing preservation of marine oomycetes are as follows, and are similar to that for terrestrial oomycetes, except the use of a seawater medium.

(i) *Preculture* – the saprobic oomycete strain is cultured onto a plate of VJSWA or other appropriate media. Agar discs containing mycelium are cut from the colony and transferred into tubes containing a cryoprotectant, as in the case of the filamentous marine fungi (see above). The cryoprotectant is a mixture of 10% glycerol and 5% trehalose, as it has been found to achieve a higher survival rate, especially in *Saprolegnia* and other oomycete cultures (A. Nakagiri unpublished data).

Algal-parasitic oomycetes, like *Olpidiopsis porphyrae,* are cultured with the host alga *Porphyra yezoensis* in seawater aerated with an air pump (Sekimoto et al. 2008). The infected algal thalli are submerged in the cryoprotectant (a mixture of 10% glycerol and 5% trehalose) in a Cryotube.

(ii) *Freezing* – the tubes containing agar discs or infected algal thalli with the cryoprotectant are frozen in the programmable freezer at a cooling rate –1°C/min from 20°C to –40°C, then at –2°C/min to –90°C (Figure 23.1).
(iii) *Storage* – the frozen tubes are transferred to a liquid nitrogen storage tank (~–170°C).
(iv) *Viability check* – one of the frozen tubes in the liquid nitrogen tank is thawed by bathing in a 30°C water bath for five minutes; then agar discs are incubated on a new VJSWA plate to examine their viability.

To examine the viability of the frozen material of the algal-parasitic species, after thawing the algal thalli are cultured with uninfected thalli of the host alga for several days (the duration depends on the cycle of zoospore maturation on the thawed thalli). Viability of the frozen parasite is confirmed by the observation of new infection on the newly added thalli.

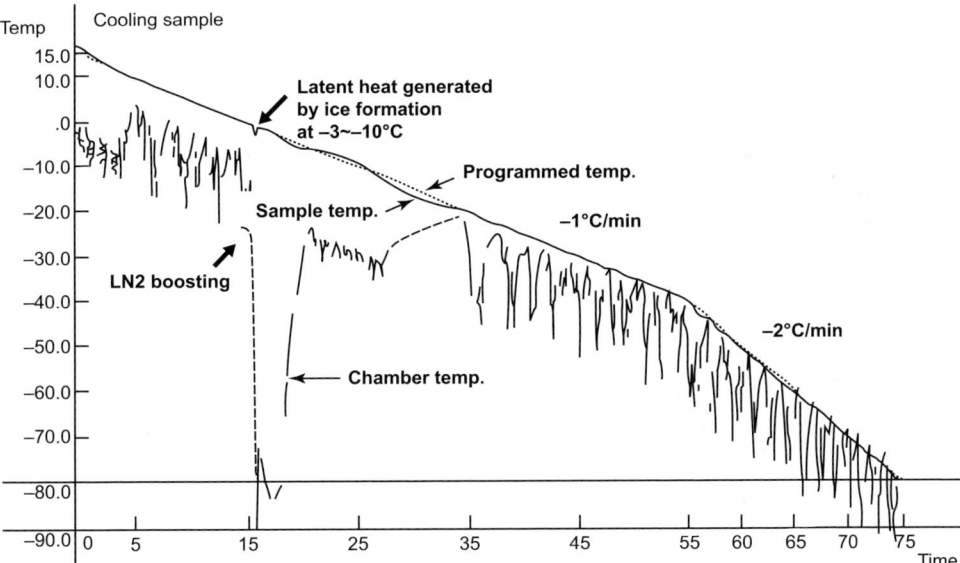

Figure 23.1 Temperature course of the sample and a chamber of programmable freezer in program freezing.

Labyrinthulids

Labyrinthula spp. and thraustochytridian species can be successfully preserved by freezing.

Preculture – *Labyrinthula* strains are co-cultured with marine yeasts or marine bacteria on a TYS plate. Agar discs containing both organisms are transferred to the freezing tubes containing a cryoprotectant (a mixture of 10% glycerol and 5% trehalose).

Thraustochytridian strains are cultured on medium-H, as described above. Smaller sized zoosporangia are more tolerant against freezing damage than well-developed zoosporangia. A synchronized culture about one day after zoospore release, in which the majority are small cyst cells, is therefore appropriate for freezing preservation (A. Nakagiri unpublished data). This is also the case for freezing preservation of chytrid cultures.

(i) *Freezing* – tubes are cooled and frozen in a programmable freezer under the same cooling program as described above for oomycete cultures.
(ii) *Storage* – the frozen tubes can be stored in an electric deep freezer, but more stable preservation in a liquid nitrogen tank is recommended.
(iii) *Viability check* – the frozen tube taken from the liquid nitrogen tank is thawed in 30°C water; then the agar discs are transferred to a new TYS plate. From a successfully survived culture, an ectoplasmic network will appear with sliding spindle cells from the agar disc.

The thraustochytridian strains will need several days after thawing and growing on the new medium before viability can be judged because frozen cells will take some time to grow up as sporangia and release zoospores to establish new colonies.

Overview

The procedures described above are examples of the successful freezing preservation methods currently employed for marine fungi and fungal-like organisms. However, there are still problematical strains that do not survive freezing, such as the abalone-parasitic oomycetous species *Halioticida noduliformans*, and improved cultivation and freezing methods need to be developed.

Programmed freezing is efficient for preservation of freeze-sensitive oomycetes and labyrinthulids. The programmable freezer, however, is not always equipped in a research laboratory. Some handy freezing container to freeze material slowly (at cooling rate around −1°C/min) in an electric deep freezer can be used, e.g., "Mr. Frosty" (Nalgene) may be usable for freezing the less sensitive strains.

Further and continuous improvement of freezing procedures during the preculture, cryoprotectant, cooling rate and reviving culture method is still necessary to properly preserve a wide range of marine fungi and fungal-like organisms.

References

Honda D, Yokochi T, Nakahara T, Erata M, Higashihara T. *Schizochytrium limacinum* sp. nov., a new thraustochytrid from a mangrove area in the west Pacific Ocean. Mycol Res. 1998;102:39–448.
Johnson TW Jr, Sparrow FK Jr. Fungi in oceans and estuaries. Weinheim, Germany: J Cramer, 1961.

Nakagiri A. Growth and reproduction of *Halophytophthora* species. Trans Mycol Soc Jpn. 1993;34:87–99.

Nakagiri A, Tubaki K. A new marine ascomycete and its anamorph from Japan. Trans Mycol Soc Jpn. 1982;23:101–110.

Nishii T, Nakagiri A. Cryopreservation of oomycetous fungi in liquid nitrogen. IFO Res Commun. 1991;15:105–118.

Sekimoto S, Yokoo K, Kawamura Y, Honda D. Taxonomy, molecular phylogeny, and ultrastructural morphology of *Olpidiopsis porphyrae* sp. nov. (Oomycetes, straminipiles), a unicellular obligate endoparasite of *Bangia* and *Porphyra* spp. (Bangiales, Rhodophyta). Mycol Res. 2008;112: 361–374.

24 Epilogue: importance and impact of marine mycology and fungal-like organisms: challenges for the future

Ka-Lai Pang and E.B. Gareth Jones

Marine mycology: current situation

In general, marine fungi are poorly studied in comparison to bacteria, plants and animals. The first marine fungus was described during the period 1840–1849. During the past 150 years, many studies have been published on the diversity, physiology, morphology, ecology, phylogeny and commercial applications of marine fungi. The updated information on these areas is summarized in the various chapters of this volume.

The golden age of marine mycology was during the 1960–1990 period, during which many mycologists were trained by our predecessors. This was the period when most of the more than 500 so-called obligate marine fungi were discovered and described (Jones et al. 2009; Jones 2011). The past decade has seen a decline in the number of research papers published (excluding natural product research, and studies of polyunsaturated fatty acid production by fungal-like organisms, the thraustochytrids) and laboratories studying marine fungi, however, and many well-known marine mycologists have retired.

The current research atmosphere favors applied rather than fundamental research (such as diversity, ecology and physiology) in mycology, and research proposals require an applied 'element' in order to obtain funding by science agencies in most countries. Even so, many projects are simply not funded and consequently the recruitment of students to study diversity/ecology/taxonomy of fungi, especially in marine environments, has suffered. Marine-derived fungi have recently been proven to produce industrial- and pharmaceutical-important metabolites (see Chapters 20 and 21), but currently there is a general lack of expertise in the isolation and cultivation of these organisms. This is in spite of the fact that most of the surface of the earth is covered by marine ecosystems. Some of the laboratories currently studying diversity and taxonomy of marine fungi are listed in Table 24.1. In this chapter, we highlight areas that are poorly studied and warrant further research, and we discuss the importance of marine mycology and the challenges for the future.

Recent research topics

Diversity/ecology

During the 1980–2000 period, over 250 species of marine fungi were described, but there has been a sharp decline in the number of new marine fungi discovered in the past decade (see Figure 1.3, in Chapter 1). Currently, research on the diversity of marine fungi is primarily being conducted in East and Southeast Asia, including Hong Kong, India, Japan, Malaysia, Taiwan and Thailand. Scattered studies are also in progress in Egypt, Mexico and Russia (Abdel-Aziz 2010; Gonzàlez and Hanlin 2010; Zaitsev et al. 2010).

Table 24.1 A list of laboratories currently studying diversity/ecology of marine fungi.

Institution	Major marine mycologist(s)
University of Malaya, Malaysia	Prof E.B. G. Jones
	Dr S. A. Alias
National Center for Genetic Engineering and Biotechnology (BIOTEC), Thailand	Dr J. Sakayaroj
	Dr S. Suetrong
Universidad Nacional Autónoma de México, Mexico	Dr M. González
National Taiwan Ocean University, Taiwan	Dr K. L. Pang
King Saud University, Kingdom of Saudi Arabia	Dr M. A. Abdel-Wahab
Faculdade de Ciências da Universidade de Lisboa, Portugal	Prof M. Barata
Mangalore University, India	Prof K. R. Sridhar
Myko Tech Private Limited, Goa, India	Dr S. Raghukumar
Sir Syed College, Kerala, India	Dr K. Raveendran
University of Calicut, Kerala, India	Dr P. Manimohan
National Institute of Oceanography, Goa, India	Dr C. Raghukumar
Uttamrao Patil College of Arts and Science, Dhule, India	Dr B. D. Borse
University of Santo Tomas, Philippines	Dr T. E. dela Cruz
Qingdao Agricultural University, China	Dr J. Jin
Pondicherry University, India	Dr V. V. Sarma

The most numerous marine fungi have been documented for Taiwan (107 species), Hong Kong (117), Japan (112), the Philippines (103), Malaysia (140) and Thailand (162) (Pang and Jheng 2012). Driftwood and woody substrata used for baiting continue to result in the discovery of new taxa (Abdel-Wahab et al. 2010; Pang et al. 2010; Pang and Jheng 2012), indicating that wider geographical locations should be explored for marine fungi. Detailed studies of marine fungi and their diversity are presently not available for Africa, South America, China (with the exception of Hong Kong), the Caucasus, Russia and Korea. Most biodiversity studies have been restricted to coastal waters, while surveys of offshore habitats, the deep sea and bottom sediments have rarely been attempted. Marine fungi associated with the submerged parts of estuarine palms are diverse and new taxa have constantly been described (Hyde et al. 1999; see Chapter 15). However, many other substrates remain to be sampled for marine fungi such as seaweeds, tropical intertidal grasses and sedges, and coral reefs.

Modern techniques

Traditionally when studying marine fungal diversity, only species with fruiting bodies at the time of collection have been recorded, few have been isolated and unculturable taxa have remained undetected (Jones and Hyde 2002). Molecular techniques have solved the shortcoming of this approach by direct extraction of genomic DNA from the substrates and subsequently using polymerase chain reaction-based methods, such as denaturing-/temperature-gradient gel electrophoresis (Zuccaro et al. 2003; Pang and Mitchell 2005;

Zuccaro et al. 2008). Tag-encoded 454 pyrosequencing of the nuclear ribosomal internal transcribed spacer-1 is a rapidly-evolving technique and has been used recently to study the fungi associated with soil in forests (Buée et al. 2009) and tropical peat swamp forest soil (Kanokratana et al. 2011). These studies have yielded a large number of operational taxonomic units primarily belonging to the Dikarya, often representing new taxa. This technique has been applied in the marine field to the study of exposed and submerged parts of *Avicennia marina* and *Rhizophora stylosa* in New Caledonian mangroves (Arfi et al. 2012). They have confirmed that Ascomycota is the main group associated with mangrove trees, with Dothideomycetes and Sordariomycetes being dominant in the submerged samples.

Many sequences found in environmental surveys are not found in public databases such as GenBank, suggesting either an imbalanced representation of sequences in the GenBank or novel fungal lineages. Likewise, many sequences in the GenBank cannot currently be matched to already-described species and greater taxon sampling is required. Stages in the colonization of wood (or other substrata) in the sea can only be followed by documenting the sporulation of the fungi (Alias and Jones 2000; Jones and Hyde 2002; Panebianco et al. 2002), however environmental sequencing helps to detect unculturable taxa and thus improves our knowledge of fungal ecology. Many unidentified taxa can now be detected in deep water sediments and matched with sequences in the GenBank (see Chapters 7, 9 and 19).

Ecological function

Linking fungal diversities to their functions (niches) in the environment is a daunting task, but recently Buée et al. (2011) were able to characterize functional ecological groups (parasitic, saprotrophic and mutualistic) of fungi. Transcriptome analysis, based on the expression of all RNA molecules (mRNA, tRNA and rRNA), is a popular technique that can be used to study the role/function of organisms in the environment, in conjunction with pyrosequencing.

Taxonomy and phylogeny

Ribosomal RNA gene clusters of the major marine taxa have been sequenced over the past two decades, and results so far suggest that there are at least 11 transitions from the terrestrial environment to the marine milieu in the Ascomycota and the Basidiomycota (see Chapter 1). The taxonomic position of a number of problematic taxa has been verified and many new lineages have been discovered, e.g., TBM clade (Sakayaroj et al. 2005; Schoch et al. 2007), Manglicolaceae (Suetrong et al. 2009, 2011) and Savoryellales (Suetrong et al. 2011). The phylogeny based on one to two genes has been proven to be insufficient to provide significant statistical support for many lineages, especially the terminal nodes in the Halosphaeriaceae (Pang et al. 2008). Sakayaroj et al. (2011) used three genes (*18S, 28S* and *rpb2*) to construct a tree with all available sequences of the Halosphaeriaceae and provided better resolutions of the terminal groups, while Suetrong et al. (2009) used four genes to resolve the taxonomy of the marine Dothideomycetes (*18S, 28S, TEF1* and *RPB2*). Other genes should also be sequenced to give better support, such as mitochondrial genes.

Phylogenomics is a rapidly-evolving area of research and it involves using the whole or a large portion of genome to infer evolutionary relationships and predict gene functions. Spatafora (2011) recently used next-generation sequencing techniques to mine for 71 orthologous protein sequences that are single or low copy for the phylogenetic reconstruction of Ascomycota. The use of this high number of genes has implicated higher phylogenetic informativeness and it will become a trend for the phylogenetic studies of fungi. Many whole-genomic sequencing projects, especially on the Ascomycota, are in progress (Voigt and Kirk 2011). In addition, genome data can also reveal the distribution of functional genes in various taxonomic groups, i.e., predicting their role (Castoe et al. 2007). Currently, only two complete genomes of marine Ascomycota have been sequenced: *Corollospora maritima* and *Lindra thalassiae* (C. Schoch pers. comm.).

Toxicology

Marine mangrove fungi (predominantly filamentous Ascomycota) and fungal-like organisms (*Halophytophthora* and thraustochytrids) are actively involved in element recycling through the degradation of complex polymers, such as cellulose and pectic substances, into simpler molecules. Rapid industrialization, together with rapid population growth, increased urbanization, exploration and exploitation of natural resources, the extension of irrigation and other modern agricultural practices, as well as the lack of environmental regulations has led to increased disposal of wastewater into the environment, especially mangrove ecosystems (ENS 2011). There is a common view that pollutants are infinitely diluted in the oceans of the world, but experience with antifouling coatings containing tributyl tin showed the fallacy of this argument when imposex was reported in mollusks (Gibbs et al. 1987; Evans et al. 2000; Jones 2009). Pollutants in wastewater, including organic compounds and heavy metals, can affect the survival/growth of marine fungi and subsequently the well-being of a mangrove community. Filamentous marine/estuarine fungi, including *Pleospora vagens, Dendryphiella salina, Buergenerula spartinae, Asteromyces cruciatus* and *Drechslera halodes*, were able to tolerate high concentration of Ni^{2+} of up to 500 ppm (Babich and Stotzky 1983). *Halophytophthora* and thraustochytrids, zoosporic fungal-like organisms on decaying mangrove leaves, can tolerate low levels (<10 ppm) of Cu^{2+} and Zn^{2+}, but growth and reproduction were inhibited at high levels (Leaño and Pang 2010; Lin et al. 2010). Many mangrove sediment fungi were also able to degrade organic pollutants, such as phthalate esters (Luo et al. 2009, 2011). Raghukumar et al. (1996) have shown the ability of the marine fungi *Sordaria fimicola* and *Saagaromyces ratnagiriensis* in decolorizing paper mill bleach plant effluent. Little is known about the effect of pollution on marine fungi and much work is required in this area. One factor reported as accounting for the occurrence of fungi in the sea is pH, with a requirement for alkaline conditions (Jones 2000; dela Cruz et al. 2006). The world's oceans are reportedly becoming more acidic, along with an increase in the concentration of carbon dioxide in the atmosphere due to the combustion of fossil fuels and the increase in mean water temperatures. These changes could have an important effect on the composition of marine fungal communities.

Bioactive compounds

For the past decade, marine and marine-related resources have been important sources of novel chemical structures for new therapeutic drugs, antibiotics, pesticides and anti-fouling substances (Sjögren et al. 2004; Jones et al. 2008), see Chapter 20. These marine resources include sponges, corals, algae, thermophilic microbes and endophytes of marine or marine-related plants (Bugni and Ireland 2004; Jones et al. 2008). Rateb and Ebel (2011) noted that some 1100 new chemical structures have been reported from marine and marine-derived fungi; although only a few of these have become commercially successful. However, there is a continued demand for the production of new antibiotics and pesticides, especially when resistance has been demonstrated.

The importance of marine mycology

Fungi and fungal-like organisms in marine environments are important ecologically and industrially. Marine fungi represent a diverse range of saprobes, pathogens and symbionts that form an integral part of coastal systems (Hyde et al. 1998). As saprobes, marine fungi transform complex organic substances into simpler molecules for their own nutrition, and they are food for organisms at higher trophic levels. Unassimilated nutrients are released to support the growth of other organisms in coastal habitats and open oceans (Jones and Pang 2010).

Fungi, and fungal-like organisms, are common inhabitants of marine algae and animals and many cause diseases to their hosts. For example:

(i) *Ostracoblabe implexa* was reported to cause shell disease in the European oyster *Ostrea edulis* and giant oyster *Crassostrea gigas* (Alderman and Jones 1971; Pirkova and Demenko 2008);
(ii) *Thraustochytrium caudivorum* parasitizes *Macrostomum lignano*, a marine free-living flatworm (Scharer et al. 2007);
(iii) marine chytrids infect diverse prokaryotic and eukaryotic algae including diatoms and filamentous species (Gachon et al. 2010; Rasconi et al. 2012); and
(iv) *Pythium porphyrae* can cause red rot disease of the *Porphyra*, a commercially important alga in the production of nori (Park et al. 2006), and has resulted in significant economic losses (Gachon et al. 2010).

Fisher et al. (2012) have highlighted the threat of emerging fungal diseases of animals and plants. This applies to marine ecosystems with sea-fan aspergillosis caused by *Aspergillus sydowii* and the failure of loggerhead turtle eggs to hatch due to infection by *Fusarium solani*. Marine fungi form mutualistic relationships with marine organisms. A lichen is an association between a marine fungus and an alga, for example, the ascomycete *Verrucaria tavaresiae* and the brown alga *Petroderma maculiforme* (Sanders et al. 2004). *Littoraria irrorata*, a snail occurring on leaves of *Spartina alterniflora*, forms a facultative mutualistic relationship with intertidal fungi by grazing live grass for fungal growth and subsequently consuming the invasive fungi (Silliman and Newell 2003). Understanding the interactions of marine fungi and fungal-like organisms with other organisms and the environment is essential in the conservation of these ecologically-important habitats and the management of commercially-important organisms.

Challenges for the future

Our knowledge of marine fungi and fungal-like organisms has increased dramatically over the past 50 years (Jones 2011), with the publication of many edited volumes and substantial reviews. However, much needs to be learned about the biology of these microorganisms as the oceans, mangroves, salt marshes, coral reefs and estuaries come under pressure from pollution, land reclamation, and the tourist industry for leisure activities. These ecosystems are vital for Man's survival and need to be protected and conserved. On one hand, fungi and fungal-like organisms are important in the recycling of nutrients, especially the breakdown of recalcitrant compounds such as lignocellulose. Particulate and dissolved organic matter play an important role in the oceans, providing nutrients for juvenile animals in mangroves and salt marshes, and such organisms are involved in the export of dissolved organic matter to the open oceans. Another vital feature of coastal ecosystems (recently called blue carbon ecosystems, Tatalovic 2012) is the rate of carbon dioxide sequestration and its contribution to a reduction in the effects of climate change. The capacity of mangroves, seagrasses and salt marshes to sequester carbon dioxide is now well documented, and these communities form much of the Earth's blue carbon sinks; yet these areas are under constant pressure, with 35% of mangroves, 30% of seagrass meadows and 20% salt marshes lost due to Man's encroachment (ENS 2011; Herr et al. 2011). Mangroves also play a role in the reduction of coastal erosion, dampening storm surges such as tsunamis, reducing the flooding of coastal planes and lining banks of rivers to prevent erosion (Saenger et al. 1983). Alias and Jones (2009) discuss the importance of replanting of mangrove areas and the restoration of discarded fish and prawn ponds.

Aspects of marine mycology that deserve greater attention include: the diseases of marine animals and algae, deep sea research, the characterization of marine-derived fungi and the establishment of a database of fungal sequences for ease of identification, greater effort in the study of their physiology and biochemistry, and their potential as a source of new fine chemicals, enzymes and bioactive compounds. Phylogenetic studies have established the occurrence of numerous lineages of marine fungi, resulting in the erection of a number of new families and orders, not only of marine but also terrestrial taxa. Ecological studies need to move away from such topics as the documentation of sporulating fungi or isolation of fungi, and focus on the extraction of DNA directly from substrates under study so that unculturable fungi can be detected. Pyrosequencing will enable the establishment of large public data sets for determining fungal communities in sediments and selected substrates.

In many ecologically-important environments, marine fungi play a key role in the decomposition of plant-derived lignocellulose; while in the laboratory they produce a wide range of bioactive secondary metabolites that may have pharmaceutical uses. Pollution and human disturbance are the major threats to marine fungal diversity in coastal habitats and no efforts have been made to conserve this ecological group of fungi. The conservation of fungi (or microbes) has been hindered by the norm 'everything is everywhere' (i.e., the cosmopolitan theory), but many studies have suggested that this is not the case and there are many endemic taxa (Griffith 2011). Increased awareness and education of the public on the importance of marine fungi (or fungi in general) are urgently needed.

Acknowledgements

We thank Dr Frank Gleason for reading a draft of this chapter and for his valuable comments.

References

Abdel-Aziz FA. Marine fungi from two sandy Mediterranean beaches on the Egyptian north coast. Bot Mar. 2010;53:283–289.

Abdel-Wahab MA, Pang KL, Nagahama T, Abdel-Aziz FA, Jones EBG. Phylogenetic evaluation of anamorphic species of *Cirrenalia* and *Cumulospora* with the description of eight new genera and four new species. Mycol Prog. 2010;9:537–558.

Alderman DJ, Jones EBG. Shell disease of oysters. Fishery investigations, London, Ser II. 1971;8:1–19.

Alias SA, Jones EBG. Colonization of mangrove wood by marine fungi at Kuala Selangor mangrove stand, Malaysia. Fungal Divers. 2000;5:9–21.

Alias SA, Jones EBG. Marine fungi from mangroves of Malaysia. Institute of Ocean and Earth Studies, No. 8: Kuala Lumpur, 2009.

Arfi Y, Buée M, Marchand C, Levasseur A, Record E. Multiple markers pyrosequencing reveals highly diverse and host-specific fungal communities on the mangrove trees *Avicennia marina* and *Rhizophora stylosa*. FEMS Microbiol Ecol. 2012;79:433–444.

Babich H, Stotzky G. Nickel toxicity to estuarine/marine fungi and its amelioration by magnesium in sea water. Water Air Soil Pollut. 1983;19:193–202.

Buée M, Reich M, Murat C, Morin E, Nilsson RH, Uroz S, Martin F. 454 Pyrosequencing analyses of forest soils reveal an unexpectedly high fungal diversity. New Phytol. 2009;184:449–456.

Bugni TS, Ireland CM. Marine-derived fungi: a chemically and biologically diverse group of microorganisms. Nat Prod Rep. 2004;21:143–163.

Castoe TA, Stephens T, Noonan BP, Calestani C. A novel group of type I polyketide synthases (PKS) in animals and the complex phylogenomics of PKSs. Gene 2007;392:47–58.

dela Cruz TE, Wagner S, Schulz B. Physiological responses of marine *Dendryphiella* species from different geographical locations. Mycol Prog. 2006;5:108–119.

ENS (Environment News Service), Paris, France. Global blue carbon market prepared by five UN Agencies, November 1, 2011. http://www.ens-newswire.com/ens/nov2011/2011-11-01-01.html

Evans SM, Birchenough AC, Bancari MS. The TBT ban: out of the frying pan into the fire? Mar Pollut Bull. 2000;40:204–211.

Fisher MC, Henk DA, Briggs CJ, Brownstein JS, Madoff LC, McCraw SL, Gurr SJ. Emerging fungal threats to animal, plant and ecosystem health. Nature 2012;484:186–194.

Gachon MM, Sime-Ngando T, Strittmatter M, Chambouvet A, Kim GH. Algal diseases: spotlight on a black box. Trends Plant Sci. 2010;15:633–640.

Gibbs PE, Bryan GW, Pascoe PL, Burt GR. The use of the dog-whelk, *Nucella lapillus*, as an indicator of tributyltin (TBT) contamination. J Mar Biol Assoc UK. 1987;67:507–523.

Gonzàlez MC, Hanlin RT. Potential use of marine arenicolous ascomycetes as bioindicators of ecosystem disturbance on sandy Cancun beaches: *Corollospora maritima* as a candidate species. Bot Mar. 2010;53:577–580.

Griffith GW. Do we need a global strategy for microbial conservation? Trends in Ecol Evol. 2011;27:1–2.

Herr D, Pidgeon E, Laffoley D. (eds). Blue carbon policy framework: based on the first workshop of the International Blue Carbon Policy Working Group. Gland, Switzerland: IUCN and Arlington, USA, 2011.

Index

18S rRNA 385
2-DGE
– LiP 470
– PAHs, Phenanthrene, Anthracene, Benzo[a]anthracene, Benzo[b]fluoranthene, LC-MS/MS 468

abiotic 143
Acrocordiopsis patilii 487
active spore release 259
adaptation 117, 398
aero-aquatic fungi 258
affiliate 394
Agaricales 55, 60, 63
Agaricomycetes 3, 55
Agaricomycetidae 55
Agaricomycotina 54, 91
Aigialaceae 19
Aigialus grandis 487
Aigialus mangrovei 487, 488
Albuginales 192
algae 41, 95, 253, 254, 411, 420, 426, 493, 513, 514
algal host 106, 107, 111
algicolous 329, 330, 337, 338, 339, 340, 341, 342
algicolous fungi 329, 330, 337, 338, 339, 341
alginase, barotolerant protease 467
Aliquandostipitaceae 26, 32
allochthonous 227
Alveolata 2, 167, 183, 184, 185, 193, 205, 207
Amniculicolaceae 19
amoebae 153, 160, 161, 221
Amoebidiales 158
amphibian gut parasites 157
amplification 401
amylase 234, 462
anaerobic 401
anamorphic 91, 99, 100, 273, 274, 277, 284, 289, 330, 337, 339, 340, 341, 384
anamorphic fungi 4, 7, 10, 17, 32, 35, 118, 122, 255, 258, 264, 274, 284, 476, 491, 500, 501, 502, 504
anamorphic marine fungi 65, 84, 501
anamorphs 19, 26

animal exoskeletons 253
Animal remains 490
Animalia 183
Aniptodera mangrovei 479
Anisolpidiaceae 192, 193
anoxic 399
anoxic sites 60
Antarctic 217
Antarctic waters 58
Antennospora quadricornuta 478
anthraquinone 423, 436, 440
antibiotics 513
antifungal 416, 417, 419
antimicrobial activity 305, 314
antimicrobial compounds 330
antioxidants 337
apical chambers 120
Apicomplexa 193, 201, 203, 208, 212
apicomplexans 167, 184, 194, 207
aplanochytrids 215, 216, 217, 218, 221, 223, 224, 227, 230, 234, 237
aplanosporic 186, 204
apothecial 117
application 9, 10
appressoria 257
aquaculture 231, 232, 238
Arabian Sea 401
arachidonic acid 233
Archaea 384
Arctic 217, 226
arenicolous fungi 258, 491
Arthoniomycetes 17
arthroconidia 55
arthropod 157, 158
artificial seawater 502
asci 35, 36, 37, 39, 40, 41, 42, 46
ascogonium initiation 123
ascoma 123
ascomata 35, 36, 37, 39, 40, 41, 42, 117, 118, 121, 123, 124, 394
ascomycetes 19, 26, 29, 31, 32, 33, 65, 117, 118, 121, 122, 123, 124, 130, 131, 132, 273, 274, 284, 290, 329, 330, 338, 342, 343, 344, 501, 504, 517
ascomycetous 49, 91, 93, 95, 96, 100

Ascomycota 1, 3, 4, 7, 12, 13, 17, 31, 32, 33, 34, 35, 42, 43, 44, 45, 46, 49, 59, 62, 63, 84, 89, 117, 123, 131, 132, 133, 277, 289, 384, 511, 512, 516, 517
ascospore 36, 37, 40, 41, 42, 43, 45, 46
ascospore appendage 36
ascospore delimitation 123, 124, 128
ascospore delimiting membranes 124
ascosporogenesis 123, 124, 130
ascus plasmalemma 124, 125
ascus vesicle 124, 125, 126, 128
Aspergillus niger 488
astaxanthin 233, 236, 240
Astrosphaeriella mangrovei 488
Atheliales 56
Australia 476, 477, 479, 480, 481, 482, 483, 488, 489, 496, 499, 500
autochthonous 227
availability of oxygen 256
Avicennia officinalis 476, 497
axenically 503
azaphilones 432, 440

Bacteria 384
bacterioplankton 196
Bahamas 479, 490, 494
ballistoconidia 55, 59
ballistospory 55
bark 254, 257, 262, 264, 479
barnacles 491
basidia 54, 56, 57, 58
basidioma 56
basidiomata 117
basidiomes 49, 55, 56, 121
basidiomycete 58, 95, 97, 330, 338, 343
basidiomycetes 49, 50, 53, 55, 56, 59, 60, 62, 63, 65, 117, 121, 132, 133, 274, 284, 501, 504
basidiomycetous 49, 53, 57, 59, 60, 61, 62, 63, 91, 99, 100, 101
Basidiomycota 1, 3, 4, 7, 10, 12, 35, 45, 49, 56, 59, 62, 63, 84, 89, 91, 101, 117, 121, 132, 277, 289, 385, 511, 516
basidiospory 117
Belize 476, 477, 478, 479, 481, 482, 491, 493, 495, 496, 497
benthic 383
benzophenone antibiotic 433
bicoeceans 219
bioactive 384

bioactive substances 98
biochemistry 103, 104
biodegradation 403
biodiversity 168, 206, 207, 273, 288, 289, 412, 434
biodiversity studies 137, 338
biodiversity survey 273
biogeochemical 384
biogeography 168
bioremediation 487
BIOTEC Culture Collection 501
biotechnological 384
biotechnological application 1
biotic 143
biotrophic 141
bisorbicillinoids 425, 437
bitunicate 121, 122
Blastocladiomycota 103, 104, 105, 109, 114, 138, 145, 148, 154, 155, 163
bleaching 488
blue carbon 514, 515, 517
blue carbon sink 265
bothrosome 215
Botryosphaeriales 65
bottom sediments 510
brackish water palm 56
brown tides 198
Bruguiera parviflora 484, 496
Brunei 479, 496
bryozoans 411
Bryozoans 429
budding 91

C/N ratio 475, 480, 484, 486, 488, 493
calcareous 491
calcareous material 254
calcofluor 396
cancer cell lines 425, 428, 429, 432
Capnodiales 17, 18, 30, 65, 331, 338
carbonaceous ascomata 259
Carboxy methylcellulase, Filter paper cellulase 448
carotenoids 231, 233, 235
Caryosporella rhizophorae 479
cellobiohydrolase 448
cellulase 264, 444, 488
cellulose 480, 486, 487, 488, 491, 497, 498
Central Indian Basin 398
Cephalosporin 412, 438
Cercozoa 139, 148

Chaetomium globosum 478, 495
Chaetothyriomycetidae 28
challenges 509
chemical defense 395
chemolithoautotrophic 400
chemotactic 105, 146
chemotaxis 146, 149, 155, 156, 257
chemotaxonomical 219
chemotypes 142
China 479, 481, 482, 490, 498, 499
chitin 153, 163, 491
chitinase, L-glutaminase, β-glucosidases 466, 470
chitinolytic enzymes
chlorophyll fluorescence kinetics 111
choanoflagellates 153, 156, 161, 162, 163, 164
Choanozoa 157
Chordariales 111
Chromalveolata 2, 10, 139, 167, 168, 185, 188, 215, 216, 234
Chromista 167, 183, 185, 188, 202, 215
chytrid parasite 195
Chytridiomycetes 108, 115, 501
Chytridiomycota 1, 7, 12, 103, 104, 108, 113, 114, 115, 118, 138, 139, 140, 143, 145, 148, 153, 155, 162, 163
chytrids 4, 10, 103, 113, 114, 156, 163
ciliates 167, 184, 202, 206
cladoceran 199
clamp connections 57
classification 1, 9, 10, 11
cleistothecial 117, 119
clinical trials 418, 434, 435
cluster 400
coastal and oceanic environments 94
coastal waters 91, 94, 96, 99, 510
colonization 256, 257, 258, 260, 261, 267, 268, 269, 270, 475, 476, 477, 478, 479, 484, 486, 489, 490, 491, 493, 496, 497, 499
colonization of substrates 122
colonization sequence 260
commensals 184, 215, 227, 230, 231, 235
commercial 219, 231, 232, 234
commercial applications 509
commercial production 219
communesins 418, 436
conditioning 477, 478, 484
conidia 395
conidial development 122

conidiogenous cell 122
conservation 265, 270, 514
copper complex 420
coral mucus 228, 241
coral reefs 253, 384, 510, 514
coral-derived fungi 429
corals 41, 95, 491, 513
core group 492, 493
cornmeal yeast extract seawater 501
Corollospora intermedia 491
Corollospora maritima 487, 495
Coryneliales 29
cosmopolitan 8, 147, 223
CPOM 491, 492
crab 490
crustacean parasite 190
crustaceans 168, 186, 188, 190, 196, 197, 198, 199, 201, 202, 204
cryoprotectant 504, 505, 506
cryptic 7
cryptic species 37, 84
cryptomonads 167
Cryptomycota 2, 10, 103, 138, 139, 140, 153, 154, 155, 161, 162
Cryptophyceae 167
cultivation-dependent 384
culturable 385
Culture Collections 501
culture dependent 396
culture-independent 396
cuttlefish 491
cyclic dipeptides 414
Cymodocea serrulata 484, 486
Cyperus malaccensis 475, 483, 484, 486, 496, 499, 500
cyphellaceous 117
Cystobasidiales 57
Cystobasidiomycetes 57
Cystofilobasidiales 54
cytochalasin derivatives 423
cytotoxic activity 416, 423, 424, 425, 426, 428, 429, 430, 432, 434
cytotoxicity 417, 423, 424, 425, 426, 427, 431
Cyttariales 29

Dactylospora haliotrepha 492
Debaryomycetaceae 92, 93
decay coefficient 476, 478, 480, 493
decay process 345, 346
decaying algae 84

decaying seaweeds   217
decolorization   487, 498
decomposers   168, 190, 191, 194, 195, 196
decomposition   141, 143, 150, 227, 241, 400, 475, 476, 477, 478, 479, 480, 484, 485, 486, 487, 488, 489, 490, 491, 492, 493, 494, 495, 496, 497, 498, 499, 500
deep sea   65, 84, 89, 91, 93, 96, 97, 226, 510, 514
Deep sea fungi   424
deep-sea environments   57, 383
deep-sea sediments   55, 59, 60, 62, 63
degradative enzymes   225, 227, 264
degrading enzymes   231, 233, 234
delimiting membrane   121, 123, 124, 126, 128, 129, 130
dendryphiellin   412, 422, 436
deoxytetramic acids   426
depleted   401
depressions   402
depth   395
derivatives   403
Dermocystida   157, 158, 159, 160, 161
descending   395
Desmarestiales   111
deterioration   340
detritus   254, 264, 265, 270
developing asci   124
Devonian   2, 13
DHA   230, 231, 232, 233, 234, 236, 241, 242
Diaporthales   286, 289
diatoms   188, 195, 206
Dictyosiphonales   111
Didymellaceae   19, 31
diketopiperazines   413, 414, 417
dimorphic life cycles   54
dinoflagellates   145, 150, 167, 168, 182, 184, 193, 194, 199, 202, 203, 207, 209, 210
dipteran larvae   105, 109
direct detection   396
disease   65
diseases   513, 514, 515
diseases of algae   197
dispersal   117, 122, 132
dissolved nutrients   265
distribution   1, 7, 8, 9, 11, 13
diterpenes   416, 434, 436
divergence   402
diverging   399
diverse   384

diversity   256, 257, 258, 259, 260, 261, 262, 265, 266, 267, 268, 269, 270, 484, 496, 497, 499, 509, 510, 514, 515, 516, 517
docosahexaenoic acid   230, 235, 236, 238, 239, 240, 241, 242, 243
docosapentaenoic acid   232, 233
docosatetraenoic acid   233
dolipore septa   121, 131, 338
DOM   492
domain-targeted   401
dominant   400
Dothideales   17, 18, 31
Dothideomycetes   3, 4, 10, 13, 17, 24, 26, 29, 31, 32, 33, 35, 65, 84, 117, 134, 286, 342, 511, 517
Dothideomycetes *incertae sedis*   17
Dothideomycetidae   17, 18, 30
driftwood   20, 31, 84
drug discovery   411
dyes   487

Eccrinales   158, 163
Echinoderms   429
ecological   386
ecology   9, 91, 94, 96, 98, 101, 218, 223, 227, 242, 509, 510, 511, 516
economic losses   95, 199, 329, 513
ecosystems   265, 267, 270, 271, 517
Ectocarpales   111, 114, 116
ectoplasmic filaments   223
ectoplasmic net   215, 216, 217, 218, 219, 220, 222, 223, 224
Ectrogellaceae   185
eicosapentaenoic acid   232, 233
electron microscopy   156
elevated hydrostatic pressure   224
endemic   91, 94, 95, 146, 147
endemic marine species   55
endemic taxa   514
endobiotic   107
endobiotic parasite   140
endoglucanase   447
endoglucanase, cello biohydrolase   442
endophytes   49, 55, 254, 262, 266, 268, 269, 271, 291, 303, 304, 305, 310, 311, 312, 313, 314, 322, 323, 324, 325, 326, 327, 329, 330, 337, 339, 340, 341, 420, 423, 426, 513
endophytic   4, 10, 262, 264, 266, 268, 269, 270, 418, 440, 484, 499
endoskeletons   491

endosymbiosis 167, 185
endosymbiotic 218
energy flow 475, 492, 493
entomopathogenic 264
entrapment 117, 122, 257
Entylomatales 59
environmental clone libraries 400
environmental sequences 4
enzyme 10, 11, 485, 487, 488, 493, 495, 497, 498, 499
epibiotic 106, 107
epifluorescence 223, 241, 396
epiphytes 28, 423
epiplasmic lamellae 126, 127, 130
episporium 120, 121, 126, 128, 129
epoxydon esters 429
ergosterol 439, 487, 489, 493
erosion 265
Erysiphales 29
Erythrobasidiales 57
Euascomycetes 129
eugaric clade 56
eukaryotes 1, 2, 215, 218, 234, 239, 242
eukaryotic 400
euphotic 395
Eurotiales 65
Eurotiomycetes 3, 17, 28, 65, 84
Eurotiomycetidae 28
Eurychamataceae 185
Eurychasmomales 187
evolution 1, 12, 13
evolutionary 399
Excoecaria agallocha 476
Exobasidiomycetes 58
exoskeletons 491
exosporic sheath 118
exosporium 120, 126, 128
extracellular adhesive 257
extracellular enzymes 95, 98, 99
extreme 399

facultative 94, 141, 383, 421, 434
facultative marine fungi 376
faecal pellets 227, 241
Fagus sylvatica 478
Family *incertae sedis* 25
fatty acids 219, 231, 232, 236, 238, 240, 241, 242, 243
feathers 49, 491
feeding rates 98

fermentation
– bioreactor, defined medium 470
filamentous 396
filamentous fungi 329, 338, 339
Filobasidiales 54
filter feeding 386
filter feeding invertebrates 145
filter-feeding 383
fish 432, 486, 491, 497, 499
FISH 400
fish farming 199, 200
fish parasite 156, 157, 159
fission 91
Flavodon flavus 488, 498
Fluorescence *in situ* hybridization 154
food chain 97
fossil 2
fossil records 103
fragmentation 118, 120, 124
freeze-drying 504
freeze-sensitive oomycetes 506
frequency of occurrence 274, 277
freshwater 488, 489, 493, 499, 500
freshwater ascomycete 20
fruits 253
fucoidan-hydrolase 458
functional 400
fungal community 260
fungal competition 257
fungal cultures 501, 504
fungal diversity 420
fungal zonation 260
fungal-like organisms 167, 169, 183, 194, 195, 198, 199, 200, 333, 340, 501, 502, 504, 506
fungi 197, 205, 206, 207, 501, 502, 503, 504, 505, 506, 507
fungus-like microorganisms 137, 138, 139, 140, 141, 142, 144, 147
fungus-like organisms 137, 139, 140, 142, 143, 145, 146

β-galactosidase 456
gasteroid 117, 121, 133
genera 384
genome 512
genomics 403
geographic distribution 196
geographic locations 254, 256
geographical 385
Georgefischeriales 58

global carbon cycle   254
β-1,3-glucanase   449
β-glucosidase   442, 451
glycerol   504, 505, 506
gorgonian   314, 321, 325
gradual   398
grasses   475, 510, 514

hair   491, 494, 499, 500
half-life   476, 477, 480, 481, 485, 493
Haliphthorales   189
Halocyphina villosa   478, 479, 492
Halomassarina thalassiae   479
Halophytophthora vesicula   488
Halorosellinia oceanica   479, 488
Halosarpheia marina   479, 492
Halosphaeriaceae   4, 13, 35, 36, 37, 39, 40, 42, 43, 46, 81, 121, 124, 131, 132, 133, 331, 337, 338, 341, 511, 517
Halosphaeriales   2, 12, 13
hamate appendages   119, 120
Haptoglossales   185
Haptophyta   167, 183
haptophytes   167, 185
Harpellales   158
Hawaiian   385
heavy metals   512
Helotiales   29
hemi cellulase   448
herbaceous plant   258
heterokont zoospores   216, 221
heterotrophic   167, 184, 194, 196, 200, 208, 212
heterotrophic flagellates   194, 208, 212
heterotrophic unicellular eukaryotes   215
higher trophic levels   195, 196
high-level production   470
histochemical analysis   130
Hong Kong   488, 497, 499, 500
horizontal distribution   286
host species   265
host specific   278, 286
host/substrate specificity   347
host-dependent   159, 163
hot springs   400
human pathogens   55, 60
Hydea pygmea   487
hydrocarbons   402
hydrogen sulfide   402

hydrostatic pressure   256, 395
hydrothermal   93, 97, 99, 100, 101
hydrothermal vents   49, 54, 55, 60, 61, 226, 399
hypersaline   395
hyper-saline lakes   226
hypertrophy   58
Hyphochytriaceae   192
Hyphochytriomycota   2, 138, 139, 145, 167, 168, 169, 192, 193, 199, 212
Hypocreales   35, 36, 37, 45, 81, 331, 337, 338
Hypocreomycetidae   39, 42, 43, 44, 45, 46
hypoxic   401
Hysteriales   17

ice pack   97
Ichthyophonida   157, 158, 159, 160, 161
Ichthyosporea   153, 157, 164
immunofluorescence   230, 240
incubation   501
India   475, 476, 477, 478, 479, 481, 482, 483, 484, 486, 487, 488, 490, 491, 494, 496, 497, 498, 499
Indian Ocean   396
indole alkaloids   418, 437, 439
interactions   256, 271
intertidal palm fungi   286
intertidal wood   84
intron   392
invagination   124, 128
invertebrates   105, 109, 110, 113
isolation   385, 509, 514
isopimaradiene diterpene glycosides   429
ITS   385

Jahnulales   17, 24, 26, 31, 33
Julella avicenniae   487
Juncus effusus   489
Juncus roemerianus   345, 347, 373, 374, 375, 376, 475, 496

Kandalia obovata   482, 488
Keissleriella blepharophora   479
Kenya   479, 480, 482, 483, 484, 488, 494, 495, 499, 500
keratinaceous substrates   253
keystone species   492
Kingdom Fungi   1, 137, 139, 155, 215, 216
Kirschsteiniotheliaceae   26, 31
Koralionastetales   35, 41, 44, 65

Laboulbeniomycetes 17
Labyrinthomorpha 193
labyrinthulids 215, 218, 228, 240, 243, 501, 503, 506
Labyrinthulomycetes 141, 142, 143, 147, 150, 215, 216, 217, 218, 219, 223, 224, 225, 226, 227, 228, 229, 230, 231, 233, 234, 240, 241, 242, 243
Labyrinthulomycota 138, 139, 143, 144, 145, 167, 215, 219, 238, 239, 240, 264
Lagenismatales 185
Laminariales 111
large-scale 403
latitude 476, 480, 485
L-drying 504
leachates 490
leaf decomposition 264
leaf litter 475, 476, 479, 480, 483, 484, 485, 487, 488, 489, 490, 493, 494, 495, 496, 497, 498, 499, 500
leaves 253, 262, 263, 264, 266, 268, 269, 270
Lecanorales 29
Lecanoromycetes 3, 17, 29
Lecanoromycetidae 29
Lentitheciaceae 19
Leotiaceae 29
Leotiales 29, 80
Leotiomycetes 17, 29, 31, 80, 84
Leptolegniales 190
Leptomitales 186, 190
Leptosphaeria australiensis 479
Leptosphaeriaceae 19
Leucosporidiales 58
leukemia cell lines 418
L-glutaminase, Tannase 467
lichen 333, 513, 517
lichen associations 337
lichen parasitic 29
lichenized ascomycete 28
Lichinomycetes 17
life cycle 160, 245, 246, 248, 249
light 256, 260
Lignincola laevis 487, 492
Ligno celluloytic enzymes
– Cellulases, Xylanases 441
lignocellulose 402, 486, 487
lineage 28, 33, 154, 164, 184, 185, 186, 189, 190, 192, 194, 206, 399
lineages 1, 2, 3, 4, 17, 28, 29, 30, 156, 161, 215, 218, 219, 238

linear hexapeptides 427
Lip, MnP, Laccase 443
lipase 234
lipodepsipeptides 417, 434, 438
lipophilic 59
liquid phase drying 504
litter fall 254, 266
Liza macrolepis 486
locations 384
lomasome-like structures 126, 127, 130
long-term preservation 11
Lophiostomataceae 19, 20, 25
lower fungi 137, 139, 256
lower littoral zone 259
Lulworthia grandispora 479, 487, 492
Lulworthiaceae 81, 331, 339
Lulworthiales 4, 12, 35, 41, 42, 44, 45, 65, 81, 84, 89, 331, 337, 338
lyophilization 504

macrophytes 486
Magnaporthales 35, 39, 81
Malasseziales 59
Malaysia 480, 482, 483, 484, 494, 496, 498
Manglicolaceae 27, 33, 511, 517
mangrove 1, 4, 8, 10, 11, 12, 13, 95, 100, 101, 117, 118, 131, 132, 133, 141, 143, 146, 147, 148, 149, 150, 278, 281, 286, 288, 289, 475, 476, 477, 478, 479, 480, 481, 483, 484, 486, 487, 488, 489, 490, 491, 492, 493, 494, 495, 496, 497, 498, 499, 500, 515
mangrove associates 253
mangrove fern 258, 267
mangrove flora 253
mangrove forests 59, 273
mangrove habitats 26, 49, 55, 57, 63, 225, 412, 420, 423
mangrove leaves 141, 143, 149, 195, 196, 207, 209, 210, 217, 225, 226, 227, 229, 231, 233, 234, 238, 239
mangrove sedge 258
mangrove swamps 108
mangrove tree 25, 53, 56
mangrove wood 20, 26, 29, 49
mangrove-inhabiting fungi 195
mangroves 49, 54, 56, 62, 63, 84, 93, 253, 254, 256, 257, 258, 259, 260, 261, 262, 265, 266, 267, 268, 269, 270, 271, 338, 475, 476, 480, 486, 488, 490, 491, 492, 493, 495

mangroves, sediments, sea sand, decayed wood, algae 441
α-mannosidase 457
mariculture 98, 99, 197, 198
marine 383
marine aggregates 224, 239, 242
marine algae 65, 106, 107, 112, 115, 215, 218, 223, 237, 239, 240, 241, 329, 339, 342
marine animals 65, 84, 92
marine aquaculture 231
marine derived 330
marine derived fungi 49
marine derived species 65
marine ecosystems 106, 108, 109, 112, 156, 162
marine ecotypes 106
marine environments 19, 28, 54, 55, 56, 91, 94, 95
marine food webs 94
marine fungal-like organisms 167, 168, 183, 198, 200
marine fungi 1, 2, 3, 4, 7, 8, 9, 10, 11, 12, 13
marine iguanas 110, 114
marine inhabitant 95
marine invertebrates 95, 98, 411, 427, 432
marine lichens 29, 330, 338
marine lineages 18, 185
marine microbial communities 215
marine nematodes 147, 150, 186, 198
marine origin 329
marine plankton 195, 196
marine sediments 65, 225
marine slime or net-slime molds 218
marine sponges 383
marine zooplankton 217
marine-derived 411, 412, 413, 414, 415, 416, 417, 418, 419, 420, 421, 422, 426, 433, 434, 436, 437, 438, 439, 440
marine-derived fungi 411, 412, 413, 414, 415, 416, 417, 418, 419, 420, 421, 422, 433, 434, 435, 436, 440
Marinosphaera mangrovei 487
mass loss 476, 477, 478, 480, 481, 485, 486, 488, 493, 495
Massarinaceae 19, 20, 26
Mastodiaceae 29
Matsusporium tropicale 487
mean tide level 259
Mediterranean Sea 402

medium-H 503, 506
Mega Pascal 395
meiosporic 256
meiosporic fungi 476
membrane complex 117, 121, 124, 126, 129, 130
Mesomyceteozoea 139
Mesomycetozoa 138, 140
Mesomycetozoea 153, 156, 157, 158, 159, 160, 161, 162, 164
mesosporium 120, 121
Mesozoic 3
mesozooplankton 230
metabolites 509, 514, 517
metagenomic 140
metal-binding 420
metaphase 124, 125
metazooplankton 196
methane seep 55, 63, 402
methanotrophy 402
Metschnikowiaceae 93
Microascales 4, 35, 36, 37, 40, 43, 81
microbial community 345, 346
microbial food web 143
Microbotryomycetes 57, 58
microenvironment 383
microheterotrophs 217
microorganisms 383
microscopic characteristics 247
Microstromatales 59, 62
Mid-Atlantic 400
mid-littoral zone 259
mineralization 396
minor vegetation 253
mitochondrial 511
mixotrophic 142
MnP, LiP, Laccases 441
moisture 253
molecular 384
molecular phylogeny 17
Mollusc 478, 486
Molluscs 430
molluskan 396
Monoblepharidales 103
Monodictys pelagica 478, 487
monomeric xanthones 415
monsoon 254, 258, 260, 265
Montagnulaceae 19, 20, 21
Morosphaeria velataspora 479

Morosphaeriaceae   19, 20, 21, 26
morphology   35, 36, 37, 40, 41, 43, 117, 119, 509, 516
morphology-based   385
morphotypes   106, 147
Mortierellales   140
motile zoospores   215
multigene   402
mutualists   94, 141, 215
Mycale armata   385
mycelia   385
mycobionts   333
Mycocaliciomycetidae   29
mycologists   509
mycoparasites   57, 60
mycophycobioses   330, 335
mycophycobiosis   18
mycorrhiza   55
Mycosphaerella pneumatophorae   479
mycotoxins   415, 435
Mytilinidiales   25
Myxomycetes   139
Myzocytiopsidaceae   189, 203
N/P ratio   490
N-acetyl-β-glucosaminidase   454
N-acetyl-β-D-glucosaminidases   466

nano   137
nanoplankton   144, 151, 196, 200, 213
natural products   411, 413, 418, 420, 427, 432, 434, 438, 439
near shore waters   95
nearshore   91, 95, 98
nearshore   94
nearshore environments   91
necrotrophic   141
nematode hosts   190
nematode trapping fungi   84
nematode trapping fungus   80
nematodes   96, 101, 105, 109, 110, 113, 114, 168, 182, 186, 189, 197, 198, 201, 203, 204, 205, 210
Neocallimastigales   104, 110
Neocallimastigomycota   103, 138, 145
nested PCR   385
net primary production   254, 267
new lineages   340
new taxa   287
next-generation sequencing   512

Nia vibrissa   487
Niskin   396
nitrate-reducing   402
nitrogen   346, 376, 488, 489, 500
nori   329, 513
notoamides   430, 437, 439
novel compounds   231, 233, 234, 235
novel diversity   383
novel drugs   411
novel taxa   337, 338, 340
nutrient   395
Nypa   253, 268
Nypa fruticans   273, 274, 277, 284, 286, 287, 288, 289, 290

obligate   383
obligate endoparasites   199
obligate marine fungi   17, 376, 421, 422
occurrence   1, 8, 9, 10
ocean currents   98
Oceanitis cincinnatula   487
ochrophytes   219
offshore   95, 98
offshore environments   96
offshore habitats   510
O-glycosylhydrolases   442
Olpidiopsidales   189
olpidiopsidiomycetes   189
omega-3 fatty acid composition   232
oomycete   110, 113, 147, 185, 186, 187, 189, 190, 197, 201, 202, 205, 211, 212
oomycetes   219, 329, 491, 498, 501, 503, 504, 505
Oomycetes   1, 2, 139, 140, 141, 142, 143, 145, 146, 167, 183, 184, 185, 186, 187, 190, 194, 195, 196, 197, 198, 201, 202, 205, 206, 208, 209, 211, 212, 503, 504, 507
Oomycota   103, 115, 138, 144, 145, 155, 164, 167, 168, 169, 185, 192, 201, 202, 206, 207, 209, 210, 264
open ocean   96, 98, 226
Ophiostomatales   35, 39, 40
Opisthokonta   2, 10, 104, 139, 153, 154, 161, 164
opportunistic pathogens   264
Orbiliomycetes   80, 84
organic pollutants   512
organisms   399
origin   2, 3, 4, 9, 13

osmoheterotrophic 215, 225, 234
osmotic effects 256
osmotrophic 142, 145, 159
osmotrophs 184
outer delimiting membrane 126, 128, 130
oxidized 401
oxygen minimum zones 401

Pacific Ocean 396
PAF receptor 416
PAHs
– LiP, MnP, Laccase 468
– MnP, laccases 468
pairwise distance 37, 39, 43
palatability 484
Paleozoic 3
palms 284, 286, 288, 289, 290, 510
paraphyses 39, 40, 41, 42
parasite 141, 144, 145, 148
parasites 84, 94, 99, 105, 106, 107, 108, 109, 110, 112, 113, 114, 115, 140, 141, 144, 145, 146, 147, 148, 149, 153, 156, 158, 159, 161, 162, 163, 164, 168, 184, 186, 187, 188, 189, 190, 192, 193, 194, 195, 197, 198, 199, 201, 202, 203, 204, 207, 208, 209, 210, 211, 215, 218, 227, 228, 234, 235, 240, 241, 254, 329, 330, 333, 503
parasites of seaweeds 186, 189
parasitic 155, 159, 160, 163, 167, 168, 183, 186, 188, 192, 194, 195, 197, 199, 210, 211, 329, 338, 343, 501, 503, 505, 506
parenthosomes 338
particulate 254, 265, 269
particulate matter 254
Passeriniella mangrovei 487
Patellariales 17, 26
pathogen 95, 199, 200, 204, 208, 210, 212, 395
pathogenic 108, 114, 264, 271, 339, 340
pathogenicity 95
pathogens 55, 59, 94, 217, 218, 227, 231, 287, 513
pathogens of marine animals 217, 218
pathology 106
patterns of succession of fungi 375
patterns of vertical distribution 376
PCR-based 385
peat swamp 285, 286, 287, 289
pectinase 488
pectinase, alginase, tannase, L-glutaminase, protease 464

pectinases 263
pelagic 383
penetration of woody 122
Periconia prolifica 487
periphyses 40, 41
perithecoid 117
Perkinsozoa 2, 10, 167, 168, 193, 194, 203, 208
Perna 478
Peronosporales 185, 190, 191, 192, 211
Peronosporomycetes 185, 203, 206, 210
Peronosporomycetidae 190, 203
pesticides 513
Pezizomycotina *incerate sedis* 29
pH 256
Phaeophyta 107
Phaeosphaeriaceae 19, 20, 21
Phagomyxida 246
phagotrophic 142, 145
Phallomycetidae 56
phenolics 475, 477, 478, 480, 488, 490
phoenicoxanthin 233
phomactins 416, 436
phosphorus 489, 498
photobiont 29, 330, 333, 336, 342
Phragmites australis 345, 373, 374, 375, 377
phycobiont 28
phycomycetes 137, 150
Phyllachoraceae 332, 339
Phyllachorales 35, 42, 81, 332, 339
phylloplane 254, 262
phylogenetic 18, 19, 20, 21, 26, 29, 31, 32, 33, 34, 54, 60, 62, 103, 104, 110, 112, 114, 116, 137, 139, 140, 150, 153, 154, 155, 156, 157, 158, 163, 164, 185, 186, 187, 188, 189, 190, 191, 192, 193, 197, 202, 203, 204, 205, 206, 207, 208, 211, 215, 219, 234, 286, 287, 337, 339, 340, 341, 342, 343, 385, 517
phylogenetic analysis 35, 36
phylogenetics 65, 84
phylogenies 139, 140, 185, 186, 189, 190, 194, 207
Phylogenomics 512
phylogeny 2, 9, 10, 12, 13, 17, 18, 32, 33, 34, 35, 36, 37, 41, 42, 43, 45, 46, 47, 94, 98, 103, 110, 113, 114, 115, 137, 139, 140, 147, 148, 149, 154, 156, 157, 163, 164, 183, 188, 202, 210, 212, 216, 218, 219, 222, 238, 239, 242, 243, 338, 341, 343, 509, 511, 516, 517
phylotypes 49, 54, 55, 386
physiology 509, 514

Phytomyxea 138, 139, 141, 144, 145, 167, 245, 246, 248, 249, 253
phytoplankton 224, 225, 238
picoeukaryotes 137, 149
picoplankton 137
pit-connections 123
planktonic 168, 195, 199, 206, 329, 383
Plannistromellaceae 21
plant pathogenic 29, 185
plasmalemma 124, 126, 127, 128, 130
Plasmodiophorida 246
plasmodiophorids 245, 246, 249
Pleomassariaceae 26, 31
Pleosporaceae 19, 21, 26, 32
Pleosporales 17, 18, 20, 25, 26, 31, 32, 33, 34, 65, 331, 338
Pleosporomycetidae 17, 18, 31
plinabulin 418, 434, 438, 439
pneumatophores 254
pollutants 512
pollution 94, 95, 98, 377, 512, 514
pollution indicator species 95
polyphyletic 26, 28, 33, 41, 45, 46, 65, 81, 104, 120
polyphyly 42, 120
Polyporales 56
polysaccharides 225, 234
polyunsaturated fatty acid 195, 225, 236
population densities 96, 98
population dynamics 287
potential 403
poultry eggs 232
predation 95
prenylated diketopiperazines 430
prenylated polyketides 413
preservation 501, 503, 504, 506
preservation methods 504
primary appendages 120
primary productivity 224
probes 400
productivity 253, 254, 267
programmable freezer 503, 505, 506
prokaryotes 384
prophase 124, 125
proproots 254
protease 234
Protista 103, 215
psychrophilic 58, 97, 402, 422
Pucciniomycotina 57, 60, 91
PUFA 231, 233, 234
Pustulanase 457

pyrene, benzo[a]pyrene 468
Pyrenulales 28
pyrosequencing 7, 60, 511, 515
Pythiales 189, 190, 191

rDNA 37, 41, 44, 45, 46
recombinant xylanase 466
recovered 398
recycling 395, 512, 514
red algal ancestor 185
red rot disease 513, 516
reservoir 403
resident fungi 386
resupinate 117
Rhabdospora avicenniae 479
Rhipidiales 186, 189, 190, 192
Rhizaria 1, 10, 184, 185, 205
Rhizidiomycetaceae 192
Rhizophora apiculata 477, 479, 482, 494, 496, 497, 498, 499
Rhizophora mangle 476, 477, 479, 494, 495, 498, 499
Rhizophydiales 107, 114
Rhodophyta 107
Rhytismatales 29
ribotypes 338
richness 256, 258, 265
roots 262, 265, 266
Rozellida 197
rRNA 36, 37, 40, 42, 43
Russulales 56

Saccharomycetales 92
Saccharomycotina 91
sagenogenetosome 215, 216
salinity 183, 193, 195, 196, 199, 257, 260, 263, 265, 395, 501, 503
Salsuginea ramicola 487
salt marsh 253, 265
Salt marsh ecosystem 345, 346, 347, 374, 375, 376, 377
salt marsh grasses 49
salt marsh plant 18, 84
salt marsh sedge 20
salt marshes 196, 273, 288
salt tolerant 286
salterns 399
saltmarsh 475
salt-tolerant plant 253
sampling strategies 257
sand dunes 20, 31

saprobe   106, 115
saprobes   55, 105, 106, 141, 145, 168, 194, 227, 254, 263, 269, 329, 513
saprobic   29, 141, 142, 143, 192, 274, 284, 399
Saprolegniaceae   187, 191, 200, 201, 203, 206, 212
saprolegnialean lineages   189
Saprolegniales   185, 190, 198, 206
Saprolegniomycetes   185
Saprolegniomycetidae   190, 203, 211
saprolegniosis   199
saprophytic   503
saprophytic fungi   262, 269
saprotrophs   195
Savoryellales   35, 37, 39, 44, 511
scanning   118, 119, 124
scanning and electron microscopy   222
scanning and transmission electron microscopy   338
Schizochytrium aggregatum   488
Scytosiphonales   111
sea fan   65, 314, 320, 321, 322, 324, 325, 326, 327
sea foam   84, 253
sea grass   58, 63
sea grasses   65, 84, 144, 215, 227, 228, 253, 254, 265
sea urchins   110
seagrass   303, 304, 313, 320, 323, 326, 481, 486, 494, 495
seasonal   196
seasonality   376
seawater   253, 262, 383
Seawater Serum Agar   503
Seawater Starch Agar   501
seaweeds   1, 18, 49, 141, 143, 144, 147, 198, 202, 253, 254, 329, 330, 337, 339, 344, 510
secondary   395
secondary appendages   120
secondary metabolites   11, 411, 412, 416, 419, 422, 426, 433, 434, 435
sedge   475, 476, 481, 485, 488, 489, 490, 493, 496, 499, 510
sediment   91, 94, 97, 100, 101, 161, 253, 254, 395, 417, 419, 420, 422, 423, 424, 437, 493, 496, 498
seedlings   254, 265
sequence   400
sequence data   337, 339
sesquiterpenoid   415

shells   491
shipworms   491
shredder   477, 480, 484, 491
shrub   253, 258
siderophores   400
signatures   401
significant   403
Silurian   2
skin   491
soft rot   478
Sordariales   35, 37, 40
Sordariomycetes   3, 4, 10, 17, 35, 36, 39, 42, 43, 44, 46, 47, 81, 84, 117, 131, 286, 289, 511
Sordariomycetidae   286, 289
Spartina spp.   345, 346, 347, 373, 374
species   396
Sphacelariales   111
spirodioxynaphthalene   425, 437
sponge-inhabiting   386
sponges   59, 61, 411, 420, 427, 513
spore appendage   257
spore attachment   122
spore dispersal   259
Sporidiobolales   57
Sporochnales   111
Sporormiaceae   19, 20, 21, 25, 30
spring tides   253
squalene   233, 237, 238, 239
Sri Lanka   480, 481, 489, 494, 498
SSU rDNA   398
standing seaweed crops   330
stipitate   117
stramenopila   215, 219
Straminipila   2, 139
straminipiles   167, 185, 194, 195, 196, 197
straminipilous fungi   185, 203
stratification   401
sub-Antarctic waters   222
Suberites zeteki   385
substrate-capture strategy   256
sub-surface   399
subtropical   8, 93, 95, 99, 183, 205, 217, 231, 273, 396
succession   256, 258, 260, 267
superheated   399
symbionts   513
symbiosis   386
symbiotic   501
symptoms of infection   246

synapomorphies   222
syringaldazine   487

Taphrinomycotina   91
targeting   401
taxa   396
taxonomic affinities   194
taxonomy   218, 246
TBM   35, 42
TBM clade   65, 81, 83
tectonic   399
teleomorph/anamorph connections   81, 84
teleomorphic   91, 99, 100
Telescopium telescopium   486
temperate   8, 217, 422
temperate waters   333, 337
temperature   183, 193, 195, 196, 253, 256, 338, 395, 503, 504
terpene   416, 422, 434, 435, 436, 437
terpenoids   414, 415, 434
terprenins   418
terrestrial fungi   258, 260
terrestrial halotolerant fungi   376
Testudinaceae   19, 21, 30
Thailand   480, 481, 482, 489, 494, 496
thermal systems   97
thermocline   401
thermophilic microbes   513
thigmotropic response   257
Thraustochytriaceae   395
thraustochytrid   488, 491, 498
thraustochytrid pathogen QPX   224
thraustochytridian   503, 506
thraustochytrids   215, 216, 219, 225, 226, 227, 228, 230, 235, 236, 238, 240, 241, 509, 512, 516
tidal amplitude   256
tidal regime   376
Tilopteridales   111
toxigenic   264
transcriptome   140, 511
transient fungi   386
transition zones   395
transmission electron microscopy   118, 119
tree of life   137, 139, 148
Tree of Life   1
trehalose   505, 506
Trematosphaeriaceae   19, 21, 27, 33
Tremellales   54, 63
Tremellomycetes   54, 55

Trichocladium acharsporum   478
Trichocladium alopallonellum   478
Trichomycetes   158, 163, 164
Trichosphaeriales   81
Trichosporonales   55
Trichothecenes   415
trophic upgrading   195
tropical   1, 8, 10, 12, 95, 183, 193, 206, 217, 231, 235, 273, 289, 290, 412, 420, 422
true mangroves   253
Trypticase Yeast Extract Agar   503
tunicates   411
tunicin   491

ubiquitous   399
ultrastructural   139, 147, 148
ultrastructure   104, 117, 118, 121, 122, 124, 131, 134
unculturable   510, 514
unculturable fungi   260, 337, 339
under-explored   403
unicellular   396
unidentified lineages   196, 197
unikonta   139
United States   475, 490
unitunicate ascomycetes   121
universal   385
unknown clades   200
unusual fungi   260
upper littoral zone   259
upwelling   401
Urocystales   58
Ustilaginales   58
Ustilaginomycetes   3, 58
Ustilaginomycotina   57, 58, 60, 62, 91

vents   93, 97, 99
Verrucariaceae   28, 31
Verrucariales   18, 28, 29, 31, 331, 338, 341
Verruculina enalia   487
vertical distribution   281, 288
vertical zonation   259

Wallemiales *incertae sedis*   56
wasting disease   228, 238, 239, 241, 242
water chemistry   256
water column   223, 224, 225, 227, 235, 237, 241
water supply   377
wetland forests   253

DATE DUE